André Michaux
*in North America*

# André Michaux
## *in North America*

### Journals & Letters, 1785–1797

*Translated from the French, Edited, and Annotated by*

Charlie Williams, Eliane M. Norman & Walter Kingsley Taylor

*With a Foreword by* James E. McClellan III

The University of Alabama Press

Tuscaloosa

The University of Alabama Press
Tuscaloosa, Alabama 35487-0380
uapress.ua.edu

Publication made possible in part by a generous contribution from
Belmont Abbey College, André Michaux International Society Fund.

Downloadable content is available under this title at www.uapress.ua.edu.

Inquiries about reproducing material from this work should be addressed to
the University of Alabama Press.

Typeface: Adobe Caslon

Manufactured in Korea

Cover art: *Clockwise from upper left*, yellow-wood, by Henri J. Redouté,
from the *North American Sylva*, 1819, by François André Michaux (photo by
Charlie Williams); Michaux's journal account of the climb of Grandfather
Mountain, North Carolina (courtesy of the American Philosophical
Society); fever-tree, by Pierre J. Redouté, from the *North American Sylva*,
1819, by François André Michaux (photo by Charlie Williams); Michaux's
New York–New Jersey map (provided by James E. McClellan III;
reproduction courtesy of Archives nationales, Paris)

Cover design: Michele Myatt Quinn

Cataloging-in-Publication data is available from the Library of Congress.
ISBN: 978-0-8173-2030-0
E-ISBN: 978-0-8173-9244-4

## Bartram's Boxes

We travel at risk of health and untold loneliness
to uncover what has been seen only by Creek
or crocodile among the brambles.

How we love each fruit or flower for its singularity,
the way we love a wife's touch, a son's quick mind,
a daughter's attentiveness.

Ours is the commerce of curiosity, seeds gathered,
sifted, tenderly nestled in moss, until with sunlight
and breath, each will spark like tinder,

reveal its secret—beauty, fragrance, usefulness—
brilliant as sunset, dark as coffee, a balm—
sent to bloom across oceans like the children
who blossom in our absence.

We are men of science. Men of faith.
This is our praise.

Reprinted with permission. © Beth Feldman Brandt, from *Solace* (Greenleaf Press, 2016)

# Contents

*Visit the web page for this book at www.uapress.ua.edu to download searchable PDFs of the tables.*

# Illustrations

MAPS

## FIGURES

*See the appendix for color photographs of a selection of the North American plants André Michaux is known to have encountered.*

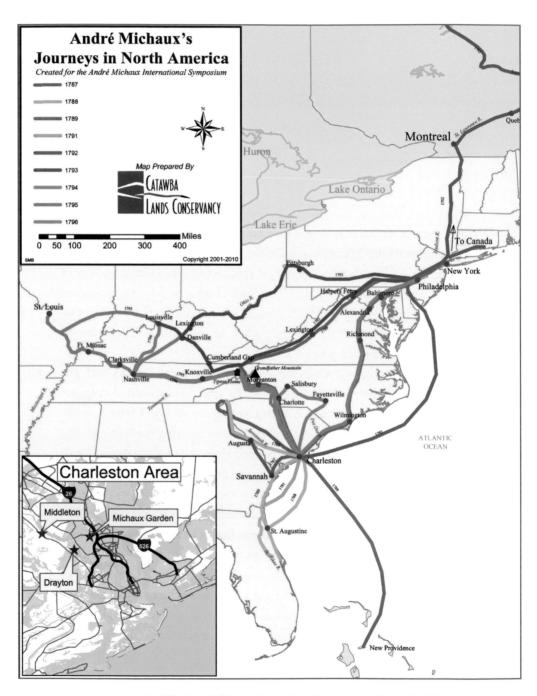

MAP I. André Michaux's North American journeys, color coded by year
(Created by Catawba Lands Conservancy)

# Foreword

WITH A ROYAL COMMISSION BESTOWED on him by King Louis XVI, the French botanist André Michaux sailed to the Americas from Lorient in France on the evening of Tuesday, September 27, 1785. After a turbulent crossing of forty-seven days, Michaux arrived in New York Harbor on Sunday, November 13, 1785, and began his eleven-year botanical odyssey in North America that lasted until his return to France in 1796, when he sailed from Charleston on Saturday, August 13, of that year.

During Michaux's sojourn in the Americas, European contact with the New World passed the three-century mark. By 1792 Spanish and Portuguese incursions in Central and South America had decimated and largely eradicated the pre-Colombian civilizations and cultures thriving there in 1492. By 1792 the heyday of the Iberian conquerors and their decades of rapine of gold and silver had largely passed, and their Central and South American holdings were sleepy backwaters. In the Caribbean, by 1792, the anarchic era of the pirates had passed, too, replaced by thriving French and British plantation colonies, notably in Jamaica and Saint-Domingue (today's Haiti). At great human cost, untold wealth in sugar, coffee, indigo, and cotton was generated from consumption in Europe that fueled the ongoing slave trade with Africa, an inhuman commerce that reached its height just as Michaux reached America. In North America in 1792, the Spanish still held Louisiana and Florida, while the thirteen British colonies along the Atlantic coast had successfully established themselves as an independent, if still nascent, nation. At the same time, along the Saint Lawrence in Canada, the British had ousted the French and nurtured that comparatively minor settlement colony. To the west and north in 1792 still lay a great continent hardly touched by European influence.

Modern world history since 1492 has been shaped by two great historical movements: the expansion of Europe outside its continental boundaries and the Scientific Revolution. The former gave us European colonialism and imperialism, whose effects still rattle the world. Starting with Nicolaus Copernicus (1473–1543), the latter gave us a sun-centered cosmology (heliocentrism), modern science, and the modern scientific worldview. The work that launched the Scientific Revolution, Copernicus's *De revolutionibus orbium coelestium*, appeared on his deathbed in 1543, although his earlier manuscript work the *Commentariolus* dates from 1514, more coincident with Columbus's voyages and the discovery of the New World.

Was there a connection between Columbus and Copernicus? Did the growth of modern science owe something to European overseas expansion? Did the success of European incursions beyond Europe's traditional geographical limits somehow depend on the new movement in the natural sciences?

The answer to these questions is a resounding yes. Over the last generation, historians of science have investigated in detail the ways in which science and European expansion reciprocally benefited each and allowed them to march together to transform the world, not just in the three centuries down to 1792, but well into the nineteenth and twentieth centuries and arguably past decolonization in the era following World War II. Indeed, this work in the historiography of science has proved so potent and productive over the last generation as to have formed the recognized subfield of science and empire studies.

A major finding of science and empire studies has been to reveal the Janus-faced character of how science and empire interacted. It cannot be said, in other words, that European colonial expansion came first and then brought science in its wake; nor can it be said that science somehow separately led the way in colonial development. Rather, the growth of science from the fifteenth century and the growth of empire in the same period went hand in hand. Thus, one facet of this relationship shows empire succeeding because scientific and related expertise was applied in its service. The other facet shows the enterprise of science prospering because empire opened up to it an expanded contact with the world beyond Europe.

The details of this bargain are not hard to identify. European overseas colonies prospered because their Continental masters supported and directed toward the ends of colonization work in astronomy, cartography, hydrography, geography, botany, natural history, agronomy, meteorology, and medicine, to name the most major. Similarly, each of these domains of science and knowledge grew dramatically by dint of enlarged experience overseas in the service of colonial and imperial powers. By 1792, men (and women) of science knew very much more of the world than they did in 1492. Not only did the solution to the problem of longitude, for example, or finding one's position at sea, solved by contemporary science and technology and put into play in the 1760s, allow ships to better find their way to port, but the new techniques allowed for unprecedented, exquisitely accurate cartographical work to proceed. Self-evidently, such precise new maps served science and empire both, simultaneously. Similarly, dealing with new diseases encountered outside Europe not only enriched contemporary medicine but grew the coffers of empire back home in Europe; works by physicians dealing with the health of sailors, colonists, and enslaved people provide ready examples. The case of botany hardly needs to be mentioned in this context; the cultivation of commodity products such as sugar, indigo, tobacco, cotton, and various spices was not directly the fruit of science perhaps, but contemporary botanists and agronomists studied these plants and how best to grow them. The promise of botany and natural history in procuring valuable new cultivars such as the cochineal insect or *Cinchona officinalis* ("Peruvian bark") for the treatment of malaria remained deeply alluring for government patrons of science. Deep in book 2 of the *Principia*, Newton even argued that his new physics would be an aid to ship design!

The concept of expertise perhaps serves better for our understanding of the success of European colonial expansion than does science per se, and in these considerations we should not exclude the world of technology, as in the printing press or clock making. If medicine is to be thought of separately from science, then medicine needs to be included in this mix, and the skills of a physician also played their parts in the success of empire, as did those of a ship's captain navigating the high seas. But expertise in this larger sense required literacy and training that produced the experts and expert knowledge that colonial powers could then tap, and such expertise was always on tap, though not on top.

It should be emphasized that in the centuries surrounding Michaux's mission to America, contemporary support for science and expertise as a tool for colonialism took place largely through state-sponsored institutions. The Portuguese Casa da Índia (1501) is an early example, as are the several such institutions that arose in Spain in the sixteenth century: the Casa de Contratación (1503), the Consejo de Indias (1524), and the Academia de Matemáticas, founded in Madrid by Philip II in 1582. The great trading companies of the English, the Dutch, and the French (the East India Company, the Hudson's Bay Company, the Verenigde Oost-Indische Companie, or the several French *compagnies des Indes*) likewise played their parts in underwriting colonial development and scientific expertise in support of the same. So, too, did various branches of national navies as in naval hospitals, map depositories, or nautical instrument manufacturing. The great royal observatories that arose in France (1671) and in England (1675) similarly provided institutional support for overseas expansion, not to mention the various royal academies and societies of science in Europe, such as the Royal Society of London (1662) or the Paris Académie royale des sciences (1666), all of which had more of an outward orientation toward the world overseas than is usually ascribed to them. Various public and private botanical gardens such as the Jardin du Roi (1635), the botanical garden of Uppsala University (Linnaeus's garden, 1655), the Real Jardín Botánico de Madrid (1755), and the royal gardens at Kew (1759) likewise were institutional centers to and through which passed increasingly vast amounts of botanical specimens and knowledge flowing in from overseas.

Such perspectives are intended to suggest the larger geopolitical, scientific, and institutional contexts in which we need to situate André Michaux, his American experience, and the travel journals that he wrote. When he stepped ashore in New York in 1785, Michaux carried with him the momentum of nearly three hundred years of European scientific and colonial development. He stood at the end of a long line of explorers who had come to the New World underwritten by national patrons in support of science and empire. In addition to earlier adventurers such as Fernández de Oviedo, who wrote *Historia general y natural de las Indias* (1535), and the first specifically scientific expedition to the New World in 1570–77, led by Francisco Hernández and sponsored by the Spanish crown, many others followed in subsequent decades and centuries, including, to pick just three further examples, Hans Sloane to Jamaica in 1687, Joseph Banks to Newfoundland and Labrador in 1766, and the Spanish expedition to Peru in 1777, which was not the last such Spanish scientific undertaking in the Americas. In 1785 Michaux was merely the most recent of explorers sent westward with the mission to advance science and empire.

All that being true, the distinctly French background to Michaux's expedition to America cannot be overemphasized. Michaux was largely solitary in his travels, but his ties to France and the complex scientific establishment there were and are essential to understanding the specifics of his enterprise.

As Charlie Williams, Eliane Norman, and Walter Taylor describe in the introduction of the present work, Michaux was a royal French botanist who had been thoroughly trained in the royal botanical institutions of the Jardin du Roi in Paris and the royal gardens at the Trianon at Versailles. He had already completed a significant botanical mission to the Near East, and he was sent to the United States in 1785 as an agent of the French government with specific instructions of what he was to do there. He had bosses and a formal bureaucratic connection to France and the crown through the major government administrative office of the King's Buildings, the Bâtiments du Roi. Michaux's connections extended further not only to the Jardin du Roi and the professors and botanical staff there, but notably also to the royal transplantation gardens at Rambouillet not thirty miles outside of Paris.

Through the eighteenth century, no country in Europe had invested more in science and in the enlistment of science and expertise in the service of its colonial enterprise than had France, and the result was an unrivaled institutional apparatus, what has been labeled a "colonial machine," that was aligned with and directed toward the advancement of science and the French colonial enterprise. The botanical component of the French colonial machine was unmatched by any other European power, and that infrastructure included not just the aforementioned Jardin du Roi and the gardens at Trianon and Rambouillet but also other royal gardens and nurseries in the provinces and in the colonies (two of which were founded by Michaux) and the international circle of botanical correspondents established by the head of the Jardin du Roi, the Count de Buffon (1707–88). But the botanical formed just one element of the contemporary French colonial machine. We have already mentioned the Académie royale des sciences and related scientific personnel as key institutional elements of this scientifico-colonio-bureaucratic agglomeration, to which we might add a number of specialized royal academies and societies (medicine, surgery, agriculture) in Paris and provincial academies, especially Rouen, Bordeaux, and the extraordinary royal scientific society founded in 1784 in Cap-François in colonial Saint-Domingue in the Caribbean. The French Royal Navy (the Marine royale) was itself an important component of the colonial machine with its own academy of naval sciences in Brest (1752) and astronomical observatories in Marseille (1749) and Paris (1754); in addition, the navy possessed an extensive set of naval medical schools and hospitals in France and in the colonies; its Dépôt des cartes et plans in Paris (1720) and two related map depositories in the port cities of Brest and Lorient employed paid staffs of royal geographers and hydrographers. In addition, the navy operated a manufacturing unit in Brest that produced scientific instruments (compasses, thermometers, barometers, and lightning rods). Largely, but not exclusively through the navy, the king directly licensed and deployed cadres of royal physicians, surgeons, and engineers at court, elsewhere in France, and in the colonies.

Yet, this fantastic contemporary colonial machine was not wholly or strictly a government-supported entity in that the church and trading companies likewise incorporated and tapped scientific experts and scientific expertise in fulfilling their missions. The French Catholic Church and the Jesuit order in particular integrated considerable botanical, medical, and astronomical expertise at home and in the colonies, the best example perhaps being the Dominican priest Jean-Baptiste Labat (1666–1738), who was mostly stationed in Martinique in the years 1694 to 1706 and who wrote virtual monographs on sugar, cotton, cacao, ginger, vanilla, and other commodity products. Similarly, the great French trading company the Compagnie des Indes established and staffed hospitals and botanical gardens in France, in Canada, and in the Indian Ocean colonies of Île de France and Île Bourbon (present-day Mauritius and Réunion).

In a word, André Michaux was a product and an agent of the French colonial machine, and his life and career need to be seen and evaluated first and foremost in the context of the contemporary political and scientific establishment that sent him to America with high hopes for his mission. He was already such an agent when he was dispatched to Persia in 1782, but he was hardly the first traveler sent out by the colonial machine to harvest plants and knowledge from the world outside France. In addition to French physician-botanists (*médecins-botanistes*) and other agents (often correspondents of the Académie des sciences and/or the Jardin du Roi) stationed in the Americas and the Indian Ocean, Michaux followed in the footsteps of such botanical emissaries, among others, as Joseph Pitton de Tournefort (the Levant, 1700–1702), Augustin Lippi (Africa, 1704–5), Jean-André Peyssonnel (Egypt, 1714; Barbary Coast, 1724), Laurent Garcin (East Indies, 1720s), Claude Granger (Egypt, 1730–32, 1733–37), Michel Adanson (Senegal, 1749–53), the abbé Gallois (China, 1764–69), Philibert Commerson (Île de France, 1769–73), Sonnini de Manoncourt (Guiana, 1772–73; Egypt and Asia Minor, 1777–80), Pierre Sonnerat (Indian Ocean and China, 1774–81), and Joseph Dombey (South America, 1776–85). And, contemporaneous with Michaux's own botanizing in North America, other French botanical agents simultaneously pursued their researches elsewhere around the globe: Louis-Claude Richard (Cayenne,

Guadeloupe, and Martinique, 1781–89), Palisot de Beauvois (Africa and the Caribbean, 1786–1803), Du Puget d'Orval (Antilles and Guiana, 1784–86), Jacques-Julien Houtou de La Billardière (the Levant, 1786–88), and Pierre-Rémy Willemet (India, 1788).

So wonderfully presented to us in the present translation, Michaux's journals perfectly capture the Janus-faced character of his mission. On the one hand, as we read, during the period of his stay in America, Michaux shipped tens of thousands of sapling trees and hundreds of cases of seeds back to France for transplantation and propagation at Rambouillet. The idea behind this activity was not so much the advancement of science but an applied-science project to relieve the timber famine besetting France on the eve of the French Revolution. Shipbuilding in particular ate whole forests, and the dearth of naval timber, particularly of oak, became an increasingly serious problem keenly felt by contemporaries. Earlier efforts from the 1720s by the colonial machine to import ships' masts and construction timber from Canada and Louisiana had proven largely ineffectual, and in the 1780s authorities conceived the alternative of importing seeds and seedlings from America for propagation in nurseries in France. Louis XVI purchased Rambouillet in 1784 and sent Michaux to America the following year to fulfill this project. In this way Michaux's mission was plainly intended to benefit empire.

On the other hand, Michaux was a trained botanist who was interested in advancing his science and the cutting edge of knowledge in botany. Here, his travel journals tell the fascinating tale of a man of science on the frontier of nature looking for and finding new treasures. In this capacity Michaux worked to advance science more so than empire. One wonders whether Michaux himself was conscious of the dual character of his mission. If he did feel tensions along these lines, perhaps he sought to resolve them in preparing his *Histoire des chênes de l'Amérique* (1801), which was both a contribution to science and a handbook for imperial seagoing powers whose needs for oak hardly diminished before the coming of ironclad ships later in the nineteenth century.

Williams, Norman, and Taylor deserve praise for producing this translation and edition of André Michaux's travel notebooks. This was self evidently no easy task, considering the historical, scientific, and linguistic expertise necessary to bring this work to fruition. I find part of the beauty of the present work to be that it can be read on many levels. The mere fact that Michaux's complete journal is now available in English is no small matter, and I say "complete" because, by incorporating contemporary letters and other documents from the period 1785–86, Williams, Norman, and Taylor have more than made up for the lost first volume of the ten notebooks that Michaux kept.

The benefits of having this source now available to us in English are several. On the one hand, it will serve as a useful corrective to so much of the Anglocentric literature concerning early America and science in early America, which, beyond hat tips to the Marquis de Lafayette, tends to overlook the role of France and French science in the history of science and the making of America. More in particular, *André Michaux in North America* is a notable contribution to the literature of science and empire studies. It is notable because it draws more attention to Michaux's American adventure and because, as a field journal, it provides a fine-grain, day-to-day focus on the activities of a European scientist working in a colonial context. Michaux is not unknown in the worlds of science and historical scholarship, but with this publication of his journal he will become better known, and historians of science and empire now have available to them a nontrivial, detailed source and case study for evoking ever more refined understandings of how science facilitated colonial development and how the experience of the world outside Europe enriched the natural sciences beyond measure.

Another major strength of this work is that it will be especially appreciated by botanical scientists and students and lovers of botany and natural history. Michaux's notebooks are not just travel diaries but also scientific logbooks. In his journal he records thousands of plant names, many of them new to him and to science, and thus his notebooks are primary source data for the history of botany that professional and amateur botanists and historians of botany cognizant of the science involved will find rewarding to read and consult.

But one hardly needs to be an expert botanical scientist to enjoy reading Michaux's notebooks as presented to us here by Williams, Norman, and Taylor. The present work is fascinating in and of itself, even mesmerizing, as we follow Michaux in his travels. The introduction to each chapter provides an overview of the notebook, orienting us to the journey ahead, and the substantial notes provided by the editors enrich and contextualize Michaux's sometimes scattered details and references. What emerges is an uncanny feel for the reality of Michaux's experience of and life in the new United States in the 1780s. Meeting Benjamin Franklin in 1786. Huddling in the rain in Georgia in 1787. Staying at a Cherokee Indian village in South Carolina in 1788 and learning about hominy from the locals. Visiting the Bartrams' garden in Philadelphia in 1789 and dealing with the expense of stabling his horses in the city. Deciding not to proceed into Kentucky likewise in 1789 because "several travelers were killed by Indians." And so on, as these absorbing examples can be multiplied until they become the journal itself.

Charlie Williams, Eliane Norman, and Walter Taylor are to be congratulated for having combined their many talents to vivify the science and lived experience of one extraordinary French botanist, André Michaux, as he ventured for France and for science into the new American republic.

JAMES E. McCLELLAN III
STEVENS INSTITUTE OF TECHNOLOGY

# Preface

Why did we decide on André Michaux's American diary, *Journal de mon voyage*, as the subject of our decade-long study? It certainly was not the literary style of the author but rather the fascinating details behind the Frenchman's encounters with plants, people, and places. The historical setting could not be more fascinating—the early evolving United States on the one hand, and the French Revolution on the other. In Michaux we found a physically and mentally rugged European-trained botanist let loose in a virtually unknown botanical paradise with a license to explore and document it. His work ethic, coupled with an unquenchable thirst for new plant discoveries, drove him to endure almost unimaginable hardships as he traveled many hundreds of miles on the American frontier. He seems to have been at ease whether he was among the elite of Charleston or Philadelphia or in the company of lonesome settlers on the frontier. It is not without cause that he has been called a "man for the ages." Researching his travels, identifying the people and plants he encountered, and discovering his routes of travel led each of us on rewarding and inspiring journeys of our own. We hope you will come to share our fascination with this extraordinary man.

The heart of our work is an annotated translation of the notebooks that Michaux kept while in America from 1785 to 1796. Many pertinent letters or excerpts of letters are included to reveal the explorer's activities and his thinking. We have also inserted a short biography at the beginning of the work and an epilogue detailing Michaux's activities after he left America.

Although Michaux was one of the most important botanists in early America and author of two landmark books on North American plants, his career here has not been studied extensively. Unlike his son and scientific heir, François André, whose *Voyage à l'ouest des monts Alleghanys* was translated into English shortly after its French publication, the elder Michaux did not live to write a polished memoir of his travels. However, he left a wealth of documents that enable us to piece together the details of his story. Michaux's journal in ten small notebooks is a record of his decade traveling throughout eastern North America. The nine surviving notebooks were transcribed and published in French in 1889. The journal constitutes the botanist's rough field notes in which he typically included directions and distances to places visited, plants encountered, and people met. He wrote these notes in his own blend of eighteenth-century French and botanical Latin peppered with phonetically spelled English personal names and place-names using his own abbreviations. There are, however, chronological and informational gaps; his first notebook, covering the initial seventeen months in America, was lost in a 1796 shipwreck. He kept no journal in 1790 and made no entries between late May 1791 and late March 1792, and there are other omissions.

A less well-known source of information is the extensive correspondence that Michaux maintained. Dozens of his letters have survived in several archives in Europe and America. They show an organized mind focused on plants. Many letters are reports to his principal supervisor in Louis XVI's government concerning collecting, growing, and shipping plants to France, but a few indicate his ambitions and plans for the future. The letters to friends and colleagues after the collapse of the monarchy reveal his state of mind during this crucial period. We have also found some letters to Michaux himself, as well as some concerning him, and also some diaries in which he is mentioned.

Michaux made a herbarium of thousands of plant specimens that were preserved for later study. He often annotated these with names, descriptive phrases, and locations of collection. Today this herbarium is maintained as part of the historic collections at the Muséum national d'histoire naturelle in Paris and is available on microfilm.

There is no published English version of the full contents of the journal; only three of his ten major journeys in America have been translated. Of these, only his three-month voyage to Florida has been carefully studied. The text of his two journeys to Kentucky, the first in 1793 and the second in 1795–96, is available but with few annotations. Much of a 1794 exploration to the North Carolina mountains was also translated but in a heavily edited form without notes. Only a few of Michaux's letters have been published so far.

The American Philosophical Society (APS) Library in Philadelphia made the original notebooks of the journal available and provided us with a microfilm of the notebooks and two microfilms containing a large file of Michaux-related letters and other documents from the Archives nationales (AN) in Paris. We have obtained copies of letters from other archives. Régis Pluchet, a descendant of the Michaux family, has assisted us in locating and transcribing additional valuable documents.

We have translated the nine surviving notebooks of the journal, filling chronological gaps left by the lost journal and omissions with Michaux's own letters and expense account. Other letters and occasional diary entries by people who met the explorer have also been interspersed with the journal. Without letters and other documents from these periods, neither Michaux's proposed expedition to the Pacific Ocean in 1793 nor his subsequent journey to Kentucky for Citizen Genet could have been adequately reported.

The notebooks and letters have been rendered into English as close to the language of the original as possible. Michaux capitalized words inconsistently and used incorrect spellings, as well as many unique abbreviations. We have standardized the capitalization to modern practice and expanded abbreviations and contractions into full words, with the exception of military titles, which remain abbreviated or spelled out according to Michaux's use. He often omitted terminal punctuation and used less punctuation than we would use today. We have silently added punctuation as needed. Michaux footnoted his own journal, and we have preserved these footnotes, presenting our own annotations as endnotes at the back of the book. He often spelled place-names in a way that is either incorrect or no longer in use. We have used modern geographic names but reported the original if it is completely different from the contemporary one. The spelling of personal names in the eighteenth century was often inconsistent, and to complicate matters, Michaux frequently spelled personal names phonetically and sometimes referred to the same individual using a different spelling from one entry to the next. We have retained Michaux's spellings for personal names in the text, but whenever we were able to identify the individuals mentioned, we have provided in our annotations other spellings found in the historical record. We also preserved Michaux's inconsistent use of spelled-out numbers versus numerals. Historical material originally in English has not been changed in any way.

Michaux recorded over two thousand plant names in his journal and letters, and he described many plants that were new to him. He was not always consistent, as he sometimes used different epithets for the same plant on subsequent pages. Only some of these names were validated later in Michaux's *Flora Boreali-Americana*. Sometimes he added a brief description to a genus name.

While these descriptions may look like plant names, they are likely mnemonic devices for describing and naming the plant at a later date.

We have retained Michaux's original spelling of plant names and have translated many short descriptions from both Latin and French. Our annotations often connect a plant name in the journal to a specimen in Michaux's herbarium and to his published books. Reference to individual specimens in the microfilm of his herbarium published by the Inter Documentation Company are placed in parentheses and abbreviated "IDC" followed by the fiche number with a hyphen before the number of the image on the sheet. Brief notations from the sheet may be inserted after the IDC. We have attempted to identify as many of the plants that Michaux referred to in his text and letters as possible, using modern nomenclature, and we have provided in the appendix a table with the old and the new scientific names, as well as one or more common names.

No complete single modern botanical work covers the geographic scope of Michaux's travels. Our principal authority for plant names is Alan Weakley's *Flora of the Southern and Mid-Atlantic States* (working draft of May 21, 2015, posted at www.herbarium.unc.edu). In other cases, we have followed the available sections of the Flora of North America, as indicated in the online databases of the Missouri Botanical Garden (Tropicos), the International Plant Names Index (IPNI), and the US Department of Agriculture (PLANTS), or the advice of specialists.

Michaux's geographic notations were generally minimal, no more than distances and directions from one landmark to the next. In many locales, we have provided references to period maps and added details about his route of travel and the places that he noted. We have provided maps to enable the reader to more easily retrace his journeys.

More than three hundred people are mentioned in Michaux's letters and journal, some famous, most not. He usually only gave their surnames, but he sometimes added a military rank or title, which aided in their identification. He also sometimes used the French abbreviation "M." for "monsieur" before the surnames of his countrymen, but at other times he used nothing at all or the English "Mr." We have searched a variety of sources in order to correctly identify these persons and provide information, which we hope will be of value to local historians and genealogists. Michaux traveled among people who were directly involved in the American Revolution and whose descendants have formed patriotic groups such as the Daughters of the American Revolution and the Sons of the American Revolution. There may be many descendants of these people, who, while interested in history, may be unaware that their ancestor met and often hosted the outstanding botanist.

Our objective throughout this work has been to allow this story to unfold in Michaux's voice while making that story as accessible as possible to modern readers. We have attempted to illuminate the career of this eighteenth-century Frenchman, who despite many obstacles was able to play such an important role in the discovery, description, and cataloging of the botanical riches of North America.

# Acknowledgments

Two women whose names do not appear in our work must first be recognized for their understanding, patience, and years of support for this project. Karin Taylor, wife of Walter Taylor, and Lydia Williams, wife of Charlie Williams, have been supportive throughout the long period of research and writing for this study. Without their encouragement and cooperation there could be no book.

This study of André Michaux is the product of research by the three author-editors working together as a team for more than a decade. Individual studies of Michaux by each of us were begun many years earlier, and the present work is the culmination of our investigations. Our work builds on the efforts of many predecessors beginning with Asa Gray, who was captivated by an unnamed specimen that he found in Michaux's herbarium in 1839. This little plant, which he subsequently named *Shortia galacifolia* (now commonly known as the Oconee bell), sparked Gray's lifelong quest to find where Michaux had discovered it, and, because of Gray's prominence in nineteenth-century American botany, rekindled interest in Michaux and his work long after the French botanist's death.

The present work is an outgrowth of the 2002 André Michaux International Symposium (AMIS), the keystone event of the five-day Michaux Celebration sponsored by a consortium of Belmont Abbey College, Daniel Stowe Botanical Garden, and Gaston Day School, assisted by the Southern Appalachian Botanical Society. The twenty-eight invited and contributed AMIS presentations provided the source material for the symposium proceedings published two years later under the editorial direction of Michael J. Baranski: *The Proceedings of the André Michaux International Symposium* (2004).

A member of the Michaux family in France has provided us with much appreciated direct research assistance in French archives. During the course of our long study, Michaux's great-great-great grandnephew, the French journalist Régis Pluchet, published studies that resolved the previously disputed death date of Michaux, provided unknown details of the botanist's family genealogy, and offered fresh insights and details concerning his participation in the Baudin expedition and his death in Madagascar. Monsieur Pluchet followed in 2014 with a full-length book about Michaux's Middle Eastern expedition: *L'extraordinaire voyage d'un botaniste en Perse*. A careful researcher, he has provided us with facsimiles and transcriptions of many documents, including previously unknown letters from Michaux to correspondents in France that were important primary sources for our study as well as a wealth of information on France in the eighteenth century.

We are also grateful for the assistance of a large number of scholars, archivists, and librarians in the United States, France, Canada, the United Kingdom, and Sweden. Especially helpful among this group were historian of science James E. McClellan III, who shared with us the applicable parts of the research in France that led to his 2011 book (coauthored with François Regourd), *The Colonial Machine: French Science and Overseas Expansion in the Old Regime*, and Laurence J. Dorr, who made available to us his unpublished paper about Michaux in Madagascar (2004).

We greatly appreciate the assistance of Rita Dockery, Roy Goodman, Marion Crist, Mary Cross, Scott DeHaven, Elizabeth Carroll-Horrocks, Alison Lewis, Eleanor Roach, and Valerie Lutz at the American Philosophical Society Library, who, numerous times over the last three decades, provided assistance with the wealth of Michaux materials in that archive. The other librarians and archivists who have helped with our research would fill an auditorium, and not all can be listed here, but they include Bill Burk, Ron Gilmour, and Jeffrey Beam, University of North Carolina at Chapel Hill; Pascale Heurtel, Mediathèque centrale, Muséum national d'histoire naturelle, Paris; Graham Duncan, Henry Fulmer, and Bev Bullock, University of South Carolina's Caroliniana Library; Lisa Reames and Mary Jo Fairchild, South Carolina Historical Society; Frances Pollard, Virginia Historical Society; Gayle Alvis, Kentucky Department of Libraries and Archives; Sara Gillespie Swanson, Davidson College Library; Tina Wright, University of North Carolina at Charlotte Library; Lois Sill and Susan Hiott, Clemson University Libraries; Doug Holland, Missouri Botanical Garden Library; Steven Sinon, New York Botanical Garden Library; Lisa De Ceasare and Judith Warnement, Harvard University Botanical Library; Dan Lewis, Huntington Library; Lynn Catanese,

Hagley Library; Maria Asp Romefors, Royal Swedish Academy of Sciences; Håkan Halberg, Uppsala University Library; Anne-Marie Allison, Lailia Miletic-Vejzovic, Winnie Tyler, Deidre Campbell, Joanie Reynolds, and other staff, Interlibrary Loan Department, University of Central Florida Library; Joel Fry, Historic Bartram's Garden; Debbie May, Nashville Public Library; Celine Arsenault, Montreal Botanical Garden Library; Caroline Lamothe, Literary and Historical Society of Québec; Carrie Hogan, American Swedish Historical Museum; Gail Benfield, Burke County Public Library; Jean Krause and Patty Holda, McDowell County Public Library; and Shelia Bumgarner, Valerie Burnie, John Camenga, Tom Cole, Tammy Cooper, Barbara Gynn, Peter Jareo, Jane Johnson, David Waters, and others, Charlotte Public Library.

Other libraries, archives, and historic sites visited for unique materials and helpful insights whose staffs provided much appreciated assistance include the Bartlett Tree Expert Company Library, Charlotte; Western Kentucky University Library, Bowling Green; the Filson Club Library, Louisville; the McClung Collection of the East Tennessee Historical Society, Knoxville; Tipton-Haynes Tennessee State Historic Site, Johnson City; the New Hanover County Public Library, Wilmington; Grandfather Mountain, Linville; Augusta-Richmond County Public Library, Augusta; Bryan-Lang Historical Archives, Woodbine; South Carolina Department of Archives and History, Columbia; and the Charleston Library Society Library, Charleston Public Library, Drayton Hall, Middleton Place, Magnolia Plantation, and the Huguenot Society of South Carolina Library, all in Charleston.

We have shared drafts of sections of this work with botanists who specialize in the various geographic areas that Michaux visited. Jacques Cayouette, with Agriculture and Agri-Food Canada, reviewed our chapters on Canada. Arthur V. Gilman, author of *New Flora of Vermont* (2015), reviewed the sections around Lake Champlain. Michael Vincent of Miami University and Brett Jestrow of the Fairchild Tropical Botanical Garden examined the botany of Michaux's Bahamas journey. Edward Schwartzman of the North Carolina Natural Heritage Program reviewed a composite chapter for the Southern Appalachians. Julian Campbell of the Nature Conservancy and Ronald Jones of Eastern Kentucky University assisted with identifications of the plants of Kentucky. Steven Hill of the Illinois Natural History Survey, formerly with Clemson University, brought his expertise to bear on a wide swath of Michaux's travels from Georgia to Illinois. T. Lawrence Mellichamp of the University of North Carolina at Charlotte reviewed North Carolina materials, and, with James Matthews and Catherine Luckenbaugh of the James F. Matthews Center for Biodiversity Studies in Charlotte, assisted with several botanical identifications. Tom Jones and Mary Garland Douglass of the South Carolina Association of Naturalists examined chapters from their state and advised on both plants and animals.

George W. Williams, emeritus professor of English at Duke University, a lifelong student of André Michaux and proponent of Michaux research, kindly critiqued each chapter of our manuscript as it appeared. The late Joseph Ewan, legendary botanical historian, provided information on Louis Augustin Bosc and other French botanists; Kanchi N. Gandhi, curator at Harvard University Herbarium, provided assistance with botanical names. The late Gérard G. Aymonin, botanical historian at the Muséum national d'histoire naturelle, was very helpful during our visits in Paris. His colleague Christian Jouvanin, ornithologist and expert on Empress Joséphine and her interest in natural history, sent us documents that revealed the existence of Merlot, Michaux's black helper; Gabriela Lamy, researcher at the Jardins de Trianon, Versailles, provided us with material about Louis Claude Richard and his family. Michel Jangoux at the Université libre de Bruxelles helped to clarify Michaux's position in the Baudin expedition.

Benjamin Mudry, Tom Brewster, and Kimberly Brewster of Asheville and Karen Goldsmith of Charlotte provided technical assistance with the taxonomic index that is part of our work.

The color map of Michaux's North American travels was originally prepared by the Catawba Lands Conservancy's first geographic information systems (GIS) specialist, Jack Drost, and has been revised and enhanced by his successors Ashley Conine, Scott Bodine, and Sean Bloom. We owe a special thank-you to Donald Cresswell and his staff at the Philadelphia Print Shop, who allowed us to spend the better part of a week studying the wealth of period maps in their inventory, permitted us to use their extensive map reference library, and encouraged us to reproduce portions of several of their old maps in our book. Many of the other period maps we reference in the text are available digitally on the web.

We are grateful to Jacques Cayouette and Marie-Ève Garon Labrecque, who generously provided us with beautiful photographs of Canadian plants. We also thank Lenny L. Lampel, Catherine Luckenbaugh, and other staff members of the James F. Matthews Center for Biodiversity Studies, who allowed us to use their database of plant photographs, and to Richard Jackson, who drew Michaux's projected plant-carrying case.

In almost all cases, Eliane Norman translated Michaux's journal and letters, but we have benefited from some other unpublished translations. A few translations of documents found in the papers of Michaux's biographers Henry and Elizabeth Savage are cited in our work. Mary Radford's 1964 translation of the journal, made available to us by curator Wendy Zomlefler at the University of Georgia Herbarium, and the 1998 translation of Michaux's expense account by young Daniel Stowe Botanical Garden intern Marie-Eve Berton Tomlin also formed part of the rich fabric of our study.

Many other people, some of whom are now deceased, made a variety of important contributions over the course of this study: C. Ritchie Bell, Hugh Morton, Mary Coker Joslin, Edgar Love III, Jack C. Moore, Bill Steele, Martha Jane K. Zachert, Robert Zahner, George A. Rogers, Fred Friday, Mike Moore, Alan May, Jamey Donaldson, Dennis Werner, William S. Bryant, Brad Sanders, Joyce Handsel, Sara Grissop, Alfred Rhyne, Mae Tucker, Jim Love, Jim Green, Morgan McClure, Lucile Maclennan, Mary Miller, Marlene Ward, Lish Thompson, Michael J. Heitzler, Dee Dee Joyce, Mary Boyer, Wayne Adkisson, Bill Logan, Joe McLaurin, Bruce Fraedrich, Allein Stanley, Suzanne Barber, Susan Pannill, John Blythe, Bert Pittman, John Garton, Kathy Gunter Sullivan, Eli Springs IV, Tim Lee, Patrick McMillan, Ben Sill, Judy Purdy, David Rembert, Penny McLaughlin, Carol Brooks, Françoise Winieska, Robert Tompkins, Mike Stacy, Mike Bush, Ken Moore, Carla Vitez, Nancy Crockett, Lindsay and Louise Pettus, Ronny and Donnietta West, Jimmy and Mary Lou Avery Furr, Kim and Dianne Caraway, Perry Deane Young, Michael Taylor, Randy Johnson, Walter Judd, Dan Ward, Ingrid P. Shields, Edward Weldon, Sheherazad Navidi, Charles P. Tingley, Virginia Farmer, Peter Spyers-Duran, Bertil Grovers, Xylander Kroon, Paul Hiepko, Pedro Gonzales Garcia, and many others not named here whom we hope will fondly remember positive encounters with us as we met them during our pursuit of the authentic André Michaux.

A special salute of appreciation must go out to Donald Beagle, director of the Abbot Vincent Taylor Library at Belmont Abbey College, who has served as secretary of the André Michaux International Society (AMIS) since its 2002 founding. His valuable assistance and unwavering support have been indispensable.

We also wish to thank Kathryn Holland Braund and two reviewers who chose to remain anonymous for their helpful insights with our manuscript and our editor Elizabeth Motherwell, who deftly guided us through the publication process.

APRIL 22, 2018

# Abbreviations

| | |
|---|---|
| AAE | Archives des Affaires etrangères, Paris |
| AAS | Archives de l'Académie des sciences, Paris |
| AM | André Michaux (as the author or recipient of a letter) |
| AN | Archives nationales, Paris |
| APS | American Philosophical Society |
| BDUSC | Biographical Directory of the US Congress, 1774–2005 |
| BM | British Library, London |
| BNF | Bibliothèque nationale de France |
| CLL | Clements Library, University of Michigan, Ann Arbor |
| *DAB* | *Dictionary of American Biography* |
| *DCB* | *Dictionary of Canadian Biography* |
| FAM | François André Michaux |
| HA | Hagley Library and Archives, Wilmington, DE |
| HU | Huntington Library, San Marino, CA |
| IDC | Inter Documentation Company, Zug, Switzerland (herbarium on microfiche) |

| | |
|---|---|
| KG | Royal Botanic Gardens, Kew, UK |
| LC | Library of Congress, Washington, DC |
| MABG | Archives Real Jardin Botánico, Madrid |
| MNHN | Bibliothèque centrale du Muséum national d'histoire naturelle, Paris |
| NYBG | New York Botanical Garden Library and Archives, New York |
| RRCHNM | Roy Rosenzweig Center for History and New Media, George Mason University, Fairfax, VA, Papers of the War Department: 1784–1800 |
| RSAS | Royal Swedish Academy of Sciences, Stockholm |
| SCHS | South Carolina Historical Society, Charleston |
| SCL | Caroliniana Library, University of South Carolina, Columbia |
| UP | Archives Uppsala University Library, Uppsala |
| USGS | US Department of the Interior, Geological Survey |
| USDA | US Department of Agriculture, PLANT Database |
| VERS | Bibliothèque municipale de Versailles, Versailles |

André Michaux
*in North America*

# Introduction

## *Biographical Sketch*

ANDRÉ MICHAUX WILL BE REMEMBERED as long as North American plants are studied or cultivated.[1] The botanist spent almost eleven years in North America. He arrived in 1785, before the federal constitution was adopted, and departed in 1796 near the end of George Washington's presidency. A bold explorer and intrepid traveler, he was often the first scientist to visit frontier areas. His life story is an engaging one; from modest beginnings in the landless French middle class, he worked his way into the front rank of eighteenth-century science. In two letters of introduction, Lafayette noted both the esteem that scientists had for Michaux and his humble origins.[2]

### FAMILY AND EARLY LIFE

Michaux's story opens on a farm in the shadow of the great wealth and power of the French kings at Versailles and blossoms into a career built on intelligence, opportunity, and hard work. His father, also named André, managed a farm at Satory, a royal domain acquired by Louis XIV in the park of Versailles. It was said to be about 370 acres (150 hectares) in some sources and 500 acres in others.[3] His paternal grandfather, Toussaint Michaux, was a farmer of the Pontoise region north of Versailles, and his wife, Marie Gosse, came from Gérocourt near Pontoise. Michaux's maternal grandparents, Claude Barbé and his wife, lived at Les Granges de Port Royale near Versailles. Although these families did not own the land that they farmed, they were considered well-off because they owned other property, including farm animals and tools. The farm at Satory that Michaux's father cultivated was one of thirty farms on the royal domain, coveted because they produced good incomes for their tenants.

André's mother, Marie-Charlotte Michaux (née Barbé), was the young widow of François Lespart, the previous manager of this Satory farm. By marrying the widow, André's father gained the farming rights to the property. The couple had at least seven children.[4]

André was the firstborn of André and Marie-Charlotte Michaux. He was baptized the day of his birth, March 8, 1746.[5] His brother André-François was one year younger, followed by his sisters, Marie-Victoire, Gabrielle, and Agathe. Two other younger siblings, Jean-Baptiste and Augustine, apparently did not survive to adulthood. At least four of the Michaux children married, and there are many descendants of the Michaux family today through André-François.[6] Michaux's formal education was brief and grounded in the classics; his biographer Deleuze[7] reported that when André was about ten years old, he and André-François were sent away to boarding school where André demonstrated an aptitude for languages. Many years later, André related to colleagues that one of his favorite books was the biography of Alexander the Great in Latin by Quintus Curtius Rufus, a first-century Roman writer. Young André enjoyed the descriptions of the exotic countries that the Macedonian king conquered as he swept into Asia, and he dreamed of visiting those lands one day.

Their father withdrew the boys from school after fewer than four years. He anticipated that his sons would follow him in managing the Satory farm and wished to instruct them in agriculture and accustom them to the routine hardships of rural life while they were young. André was an apt pupil of agriculture, exploring nearby gardens, making experimental plantings, and learning all he could about growing plants. He soon combined the farmer's skills with his curiosity and enthusiasm for natural history.[8]

The Michaux brothers shared the management of the farm at Satory after their father's death in 1763. Their mother died three years later. During this time André continued his studies in Latin and Greek.[9] At age twenty-three, André married twenty-year-old Anne-Cécile Claye, who was from a wealthy and cultured family living at Tremblay le Vicomte, near Chartres. Several members of her family were prosperous cereal growers, one brother and two of her uncles were priests, and another brother was a merchant in Paris.

Prospects seemed bright for the young couple, but their happiness was short lived. Anne-Cécile died in October 1770, less than two months after the birth of their son, François André, probably from complications of childbirth. Michaux was overcome by grief and mourned his wife's loss deeply. Louis-Guillaume Lemonnier (1717–99), physician to Kings Louis XV (1710–74) and

FIGURE 1. Jardin du Roi, 1794, by Jean-Baptiste Hilair (1753–ca. 1822)
(Courtesy of Bibliothèque nationale de France, Paris)

Louis XVI (1754–93) as well as professor of botany at the Jardin du Roi, took an interest in the sad young widower, encouraging and tutoring him in botany. The young man responded enthusiastically. From this early relationship, Lemonnier became Michaux's patron, mentor, and lifelong friend.[10]

## APPRENTICE BOTANIST

André Michaux soon left the management of the Satory farm to his brother and became an apprentice botanist under Lemonnier. He had the idea that when his training was complete, he might eventually become a botanical explorer in the service of his country, visiting foreign countries and bringing their best and most useful plants back to France. In 1777, with Lemonnier's recommendation, André began to study at the Trianon gardens at Versailles with Claude Richard (1705–84) and Bernard de Jussieu (1699–1777). Two years later, he moved near the Jardin du Roi, Paris, and continued his studies with the prominent botanists there, especially André Thouin (1747–1825), the chief gardener of that establishment,[11] and Antoine-Laurent de Jussieu (1748–1836), nephew of Bernard.[12]

His training continued with a journey to England in 1779 or 1780, where vast numbers of exotic plants were being brought into cultivation.[13] The French

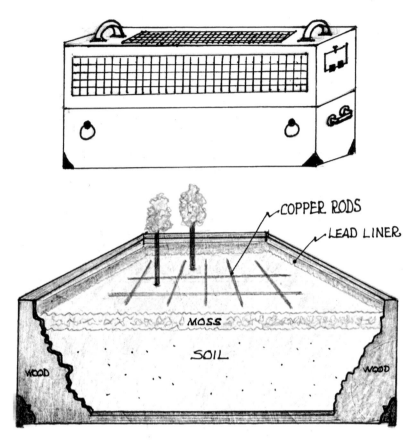

FIGURE 2. Michaux's proposed plant transport case
(Drawn by, and courtesy of, Richard Jackson)

alliance with the American revolutionaries and war with Great Britain were
not insurmountable obstacles to his visit, nor was his inability to speak English. Michaux's correspondence indicated that he met Sir Joseph Banks (1743–
1820), head of the Royal Society, who as a young man had circumnavigated the
globe as Capt. Cook's chief naturalist. While in Britain, Michaux also met
nurseryman James Lee (1715–95) of the renowned Kennedy and Lee Vineyard
Nursery, Hammersmith, famously rich in trees and shrubs. Perhaps this nursery was a source of trees that Michaux brought back to France and planted in
the gardens of his mentor Lemonnier and that of the latter's own patron and
friend Maréchal (Louis) de Noailles (1713–93), an avid amateur horticulturist
and forester.[14]

Two more short journeys soon followed. On the first, a visit to the mountains of the Auvergne region of France, Michaux accompanied several other botanists, including Jean-Baptiste Lamarck (1744–1829) and André Thouin. With the support of Count de Buffon (1707–88), the celebrated *intendant* of the Jardin du Roi, Lamarck had recently (1778) published his innovative *Flore française*. With the passing of Bernard de Jussieu in 1777, these men were among the most prominent of the younger botanists; Michaux impressed them with his zeal and energy. Deleuze notes that Michaux collected not only plants but also birds, insects, and minerals. Michaux followed the Auvergne journey with another trip, over the Pyrenees Mountains into Spain, where he collected seeds that were distributed to several gardens.[15]

Soon after completing these journeys, Michaux asked Lemonnier to recommend him for some possible botanical exploration overseas.[16] During this time he also considered the practical difficulties involved in successfully shipping live plants long distances on seagoing vessels and designed a model for a special transport case that Lemonnier and his colleague Louis Jean Marie Daubenton (1716–1800) shared with their associates at the Académie des sciences.[17] Michaux's personal financial situation seems to have been reasonably secure at this time since he had the funds to invest in a shipment of arms to the American revolutionaries.[18]

## Middle Eastern Mission

In 1782, Lemonnier obtained a commission and a small sum from the comte de Provence, the king's brother, often referred to as Monsieur, who later became King Louis XVIII (1755–1824). This appointment and stipend allowed Michaux to accompany Jean-François Rousseau (1738–1808), cousin of the philosopher, as he traveled to take up his new post as consul at Basra in modern Iraq. Michaux could explore that region for plants useful in France. This Middle Eastern journey lasted from March 1782 until May 1785. Michaux traveled widely in today's Syria, Iraq, and Iran and collected plants and seeds as well as other objects in the lands he had dreamed of visiting as a youth.[19]

Michaux's stellar performance on the Middle Eastern journey won him the respect of influential officials in France.[20] He was prepared for his mission and informed enough about the plants in the region to know what was new or could be useful in his homeland from the abundant floral riches that he encountered.[21] He was a well-organized and disciplined person. Even though he encountered a variety of difficult and even some life-threatening situations, Michaux remained calm, resourceful, and focused. There is a suggestion in one of his letters to his son that his attitude came from religious beliefs. He explained to young François André that "God has twice rescued me from the greatest danger when no number of men could have."[22]

Michaux returned to Paris in June 1785. His initial wish was to return to the Middle East and travel on to South Asia to continue his research for valuable plants in that region. Instead, he was chosen to travel to America on an economic mission. The origins of this mission may be traced in part to a 1784 report sent to correspondents in France by St. John de Crèvecoeur (1735–1813), then consul of France in New York.[23] Crèvecoeur had just published his sympathetic portrayal of America in his 1782 book *Letters to an American Farmer*; French relations with the United States were very good. In addition, it was known that there were plants of potential value to France to be found in the United States. Sending Michaux to the United States was the most cost-effective course and the one that France chose to follow. French forests had been depleted of the best timber for shipbuilding. Replenishing this vital national resource with species from America was a major objective of Michaux's mission.[24]

## Royal Botanist in North America

In July 1785, Michaux received an appointment as Royal Botanist to King Louis XVI and detailed instructions on how to carry out his mission in the United States. His son, François André, age fifteen; a young servant named Jacques Renaud; and an experienced gardener trained by Thouin, Pierre-Paul Saunier, age thirty-four, accompanied the botanist. Michaux was pleased by the confidence shown in him and excited at the prospect of what he might accomplish in North America. Two quotations from letters the botanist wrote while waiting to sail to America reveal some of his thoughts. To Monsieur, Michaux wrote "I imagine myself setting out to the conquest of the New World." To Count d'Angiviller (1730–ca. 1809),[25] the government official who had given him his instructions and would supervise his American mission, the botanist wrote that he preferred to work in countries with temperate climates because of the advantages of naturalizing the plants in France from those countries. Michaux's closing line in this letter reveals his ambition: "I shall have nothing to fear so much as leaving discoveries to be made by those who shall come after me." Before his ship even had wind in its sails, André Michaux's aspirations were far more than to collect American seeds and plants, many of which were already known in Europe. He wanted to be the one to discover new plants on this continent.[26]

After waiting in Lorient for a month for fair winds, Michaux's ship, the

*Courrier de New York*, finally sailed into the Atlantic. Michaux and his companions arrived in New York Harbor on November 13, 1785, after a stormy voyage lasting forty-seven days. There Michaux met Louis-Guillaume Otto (1754–1817), the French chargé d'affaires.[27] Otto had been frustrated in his attempts to procure American plants and seeds desired in France. Not knowing that Michaux was already on his way, he had recently requested that a French botanist be sent to do this work. Otto's letter announcing Michaux's arrival, written when the botanist had been in America only two weeks, described Michaux as intelligent and reported that he could read but not yet speak English and would begin to travel in the interior in a few months. In his second letter, Otto suggested that a garden in the southern states would be needed as well as one in the New York area. The diplomat believed that trees from the American South would thrive in the southern regions of France.[28]

After a month and a half of enforced idleness at sea, Michaux was anxious to get to work, and a whirlwind of activity followed during his first weeks in the United States. Despite multiple obstacles, including some of the coldest weather in years in that part of the country, he immediately supervised the collection and preparation of a substantial shipment of plants and seeds for the *Courrier de New York*'s return voyage, which took place less than four weeks after his arrival. His first letters convey his pleasure and excitement at beginning his task, and the descriptions of the plant material he sent are remarkably detailed and perceptive.[29]

Within a few weeks, assisted by Otto, Michaux had located and purchased a 29¾-acre property across the Hudson River in New Jersey to develop as a holding garden for his collections. A special act of the New Jersey legislature was required for Michaux, a foreign national, to own the land.[30] Michaux was pleased by the easy access to the port of New York for his shipments and noted that the site provided a variety of habitats ranging from a wooded upland to a constantly inundated white cedar swamp. Michaux promptly cleared land for planting, constructed a four-room dwelling, and added other improvements that enabled him to increase the pace and efficiency of his work.[31]

Evidence of Michaux's activities and future plans appear in his expense records and in his letters. He purchased English dictionaries, maps, and an atlas of the United States. After three months he dispensed with the bilingual laborer who had served as his interpreter, apparently confident of making himself understood. He noted that Americans of different national origins retained the strengths and weaknesses of their nationalities even when transplanted to America. Michaux observed that America was a rich country

FIGURE 3. Michaux's New Jersey house
(Provided by James E. McClellan III;
reproduction courtesy of Archives nationales, Paris)

and that Americans did not need to work hard.[32] In fact, the work ethic of free people frustrated the French botanist, who was apparently accustomed to laborers who worked hard for long hours in poor conditions at low wages. He found that Americans demanded higher pay to dig trees from frozen ground and even then did not give him what he considered a full day's work for a full day's pay. We may conclude that André Michaux was both a tireless worker and a demanding taskmaster.

Among the letters from his first months in America is a remarkable ten-page document dated January 19, 1786, sent to his colleague at the Jardin du Roi, André Thouin. After less than nine weeks in the United States, and before he had completed purchase of the land for the New Jersey garden, Michaux wrote to Thouin that he felt that his truly important work in North America would be in the Carolinas, Georgia, and Florida. Michaux's letter revealed his disenchantment with New York. He had failed to find other men in the New York area interested in plants and reported that everything was frozen and that the people were only interested in "whatever would bring in money," an observation that echoed the sentiments of Finnish-Swedish botanist Peter

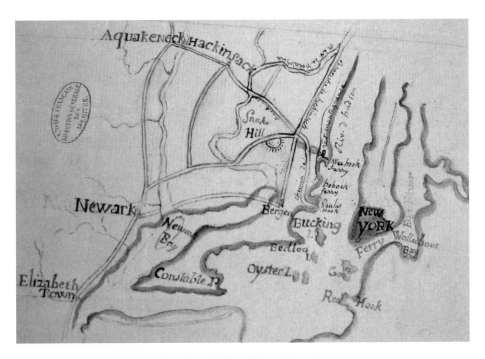

MAP 2. Michaux's New York–New Jersey map
(Provided by James E. McClellan III;
reproduction courtesy of Archives nationales, Paris)

Kalm (1715–79) more than three decades earlier.[33] Michaux had also learned that there were planters in the Carolinas who had naturalized trees of Asian origin. One section of this long letter concerns a request for specific species that Michaux wanted to "multiply in America." The plants listed include some species that have long been associated with Michaux in Charleston: mimosa, *Albizia julibrissin*; ginkgo, *Ginkgo biloba*; camellia, *Camellia japonica*; and the tea plant, *Camellia sinensis*.[34]

In June 1786 Michaux made his first extended journey beyond the immediate environs of New York City. He traveled south on an official trip to Pennsylvania, Maryland, and Virginia. In Philadelphia, he visited Benjamin Franklin and Bartram's Garden.[35] Michaux then continued south, passed through Baltimore, and entered Virginia at Alexandria where he visited George Washington at Mount Vernon.[36] He noted in a letter that he also reached Winchester (he referred to it as Frederick Town) and the Shenandoah River.[37]

During the first week of September 1786, Michaux made a second, brief visit to Philadelphia specifically to visit the Bartrams and William Hamilton (1745–1813).[38] Bartram's Garden was a magnet for the French botanist. John

Bartram Jr. (1743–1812) had inherited the property from his father, John Sr. (1699–1777), and he and his brother William (1739–1823) had rescued the garden from neglect and restored their father's plant and seed business.[39] William Bartram had also traveled extensively in the South and was generous with information useful to Michaux. The very wealthy Hamilton lived near the Bartrams, and his estate, the Woodlands, featured a magnificent garden containing many exotic trees.[40]

## CHARLESTON

Michaux then sailed from New York to Charleston, South Carolina, arriving on September 20, 1786, with his son and Renaud. Saunier was left in charge of the New Jersey garden and would continue to make shipments of plants and seeds from New York to France. Charleston was at that time a city whose people were primarily British, but with a French flavor added by the many Huguenots, and a large, mostly enslaved black population. Michaux found a warm reception in this community and promptly began his work. The need for a garden outside the city quickly became apparent, and he first rented, and later purchased, a 111-acre plantation in the Goose Creek section ten miles north of the city. Goose Creek was a neighborhood of large plantations and great wealth.[41] As with the New Jersey garden purchased earlier in the year, the botanist evaluated the situation and acted swiftly to purchase this property on his own authority without waiting for formal approval from France.[42]

Located on the main road into the city from the interior where Michaux needed to travel to collect plants, the Charleston garden also had easy access to the seaport for shipments to France. To lower his expenses, Michaux worked to develop the property as a self-supporting plantation. About eighty acres of the property were in forest, while another fifteen-acre section was devoted to producing food for the Michauxs, their workers, and the farm animals. There were wetlands and a permanent stream; he built a new road across the marshy lands and a bridge across the stream. He engaged four workmen, including a skilled carpenter, to construct a new dwelling, repair an existing one, and build cold frames for his garden, fences, and crates for shipments to France. As he had done in the north, Michaux reported to Count d'Angiviller that labor was expensive. At the same time, he collected plants and seeds and made several substantial shipments to France before the first cold season ended.[43] Four years later, the US Census report of April 1790 noted two white men and one white woman living at Michaux's Charleston garden. Because Michaux's son, François André, had embarked for France in February 1790,[44] the white men at the garden must have been André Michaux and his servant Renaud, who eventually became his gardener. The white woman, who was not named, was not reported elsewhere.[45]

As was common, the larger plantations nearby were worked by enslaved black men and women. Following the local economy, Michaux also used enslaved black people, although it is unclear how many of them usually lived and worked at the Charleston property and under what arrangements. The census report listed twenty-four enslaved persons on the property, a figure that seems high considering the financial constraints of Michaux's expedition and the size of the cultivated land, but there may be explanations. Michael J. Heitzler, the modern historian of Goose Creek, points out that the census report did not necessarily mean that Michaux owned twenty-four enslaved persons. At this time, after the decline of rice and indigo culture, but prior to the rise of cotton, large enslaved populations had become an economic liability, and plantation owners often rented out their enslaved workers. These people would then be counted where they were working, not on the plantations where their owners lived.[46] Beginning in January 1787, we have indications in his expense summaries that Michaux recorded both wages to workers and payments for the rental of enslaved people.[47]

Little is known of Michaux's attitude toward black people, but having observed Africans in positions of responsibility in the Middle East early in his career, he did not appear to see them as inferior.[48] This background did not prevent him from acquiring enslaved black people with his personal funds. Michaux's journal indicates that he and François André were accompanied by a black assistant on journeys in 1787, 1788, and 1789; this man was certainly enslaved. Michaux also reported selling three of the enslaved people that he personally owned to finance his journey to Canada in 1792.[49] Nonetheless, the most revealing insight that we have about his attitude toward Africans comes from the Swedish abolitionist Carl Bernhard Wadström (1746–99), who visited Michaux in Paris in November 1797. Wadström reported: "I had the satisfaction to see a young African 12 or 14 years old at Michaux's home that appeared to have a very happy disposition. He devotes himself especially to drawing and botany. Citizen Michau [sic] who doesn't spare his care or expense to train him, plans to encourage him to do research in the interior of Africa, after he has acquired the necessary knowledge."[50] The young man was Merlot (Merlo), the young boy whom Michaux had brought from America and who, at Michaux's request, would later join the Baudin expedition as a gardener and then accompany Michaux to Madagascar.

In April 1787, with the Charleston garden functioning, and having made several shipments of plants and seeds from there during the cold season, Michaux began using the garden as his base for a series of explorations of the interior. His extant journal, which contains detailed reports of all his journeys, opens on April 19, 1787, with the beginning of a major journey of exploration. Accompanied by his son and the Scottish botanist and nurseryman John Fraser (1750–1811), whom he had met in Charleston, Michaux headed south on horseback along the coastal road toward Savannah, Georgia, with a cart for plant collections. His initial idea was to see whether it would be possible to continue south into Florida, but he also had a contingency plan to explore the southern mountains if the journey to Florida proved impractical.

At Sunbury on the Georgia coast south of Savannah, he turned north once again and traveled overland to Augusta, Georgia; between Savannah and Augusta, Michaux parted company with Fraser. From Augusta, Michaux proceeded into the most remote settlements in the South Carolina Piedmont and then plunged into the mountain wilderness at the headwaters of the Savannah River.

Although Michaux was an experienced traveler and plant collector who had often encountered danger on his Middle Eastern expedition, this journey was his initial foray into the North American wilderness. In the Middle East, the major threat to his personal safety was the lawlessness among some of the people and the conflicts that he encountered. In the wilds of North America, the greater threat seemed to come from the land itself. Michaux traveled through a virgin wilderness so rugged and wild that a few days beyond the settlements, his Cherokee guides became lost. Rain-soaked, ill-fed, and bone-weary, Michaux wrote about the fear he felt traveling in this gloomy wilderness, but he overcame his fear, focused on plant collecting, and returned safely. He had found several new species, and he was eager for many more discoveries.

This initial exploration in spring 1787 set the pattern for the next two and a half years as Michaux traveled from Florida to New York by paddling canoes, walking, and riding horseback; he also took a ship to the Bahamas. National boundaries proved to be no obstacle to his botanical explorations. He met with the Spanish governor of East Florida and the British governor of the Bahamas during his single visit to each of these territories outside the United States. In each locale, with the cooperation of local officials, Michaux gathered plants and seeds for shipment to his Charleston garden.

After initial journeys scouting the country in warm weather, the botanist would return to the mountains of South Carolina and then to the higher mountains in North Carolina to collect large quantities of plants in the difficult cold season. Because the plants were dormant at this time, the chance of survival was improved.

While Michaux worked industriously in America, the Old Regime of France that he served as Royal Botanist was beginning to collapse, and the chaos of the French Revolution was becoming imminent. For many months, he had found it increasingly difficult to obtain funds from his government to pay expenses.[51] The regular packet boat service between France and New York that he had used for his early shipments was suspended, and delivery of his mail from France was disrupted and delayed. Not many months later, he was directed to cease his botanical journeys and to economize his operations.[52] In addition to these problems, his son suffered an accidental eye injury in late 1789 and returned to France in early 1790.[53]

Throughout 1790 and the early months of 1791, Michaux appears to have remained in South Carolina principally at the Charleston garden, attempting to obtain the funds to continue his operations. He did not keep his journal during 1790, but lists of seeds he planted in the Charleston garden on specific dates confirm his presence there.[54] Despite the order to curtail his travels, in spring 1791 Michaux sailed from Charleston to the coast of Georgia to explore the territory along the St. Marys and Satilla Rivers and the neighboring islands.

The continuing disintegration of the French royal government in the 1790s forced the botanist to make painful decisions. Michaux did not choose to take personal possession of the Charleston property when Count d'Angiviller offered it to him. When he received an order to sell the Charleston property, the botanist was reluctant to do so. He realized that a sale of the property to strangers meant loss of the plants that he had collected and grown, often at great risk and with great exertion. The plants represented a treasure that he wished to preserve for France, even though at the moment, France, absorbed in overwhelming political difficulties, placed little value on Michaux's American plants. Therefore, Michaux arranged for his friend John J. Himely, a Charleston watchmaker, to purchase the property at public auction, and then he leased the land from Himely.[55]

Michaux wished to travel and to study and collect the American flora, but not to be burdened with the costly and time-consuming maintenance of the Charleston garden, a drain on his personal resources. Living among wealthy planters with a keen interest in adorning their own gardens with the exotic plants that Michaux had obtained from his correspondents in France, the Bartrams, and other sources, Michaux came to an arrangement with these local landowners. In return for their assistance in caring for the Charleston garden, he made available to them treasures that he described as "the fruit-trees of Europe and the old Continent which I had naturalized in the garden." He went on to say that "I reserved for myself the property in all those [trees] of North America, which I destined for, and for which I considered myself accountable to my country. To this proposition, I added a memoir upon the advantages of naturalizing several exotic trees, chosen from those of China and the South Sea islands, which vegetated imperfectly in the hot-houses of the botanic gardens of Europe."[56] These men in turn made Michaux an honorary member of the Agricultural Society at Charleston.[57]

### EXPLORING AMERICA WITHOUT A GOVERNMENT'S SUPPORT

In spring 1792 Michaux knew that he could neither depend on the organizational and financial support that he had received from the Old Regime nor count on instructions from its successors. Michaux first used his own resources to embark on an exploration of an unknown territory in Canada, and then he sought financial help to undertake an even longer journey across the entire North American continent. The first journey, into northern Canada, was the longest journey that he had thus far undertaken. One of the things that motivated him to make this long voyage was the desire to learn the distributional limits of the many species of North American trees. Michaux made his Canadian journey primarily to collect information for his projected book on North American flora rather than to collect plants and seeds for France, although he did send collections to France from this journey.[58]

Michaux first traveled to New York and then to Connecticut where he visited Peter Pond (1740–1807), a veteran fur trader who had lived in the vast Canadian wilderness north and west of the Great Lakes.[59] Discovering that it was too late in the season to accompany the fur traders going west, Michaux resolved to undertake his own expedition. He followed the same general route up the Hudson River and along Lake Champlain into Canada that Linnaeus's student Peter Kalm had followed four decades earlier.[60] Unlike Kalm, however, Michaux continued traveling north and west by birch bark canoe deep into the vast unknown interior region north of the Saint Lawrence at the latitude of Hudson Bay, an area never before visited by a botanist. Michaux was treated well and assisted by British authorities and private citizens in Canada. Accompanied by three native guides, an interpreter, and a dog, he paddled, portaged, and waded hundreds of miles across this watery wilderness to the latitude where he found little plant diversity. Then, on the advice of his guides, he turned back three days short of Hudson Bay itself, to escape being marooned in this forbidding landscape by swiftly approaching autumn storms.[61]

Entries in Michaux's journal from the Canadian wilderness show that his years of almost ceaseless travel on the American frontier had seasoned him as a wilderness explorer. Describing the perils of one journey among many on a white-water river, he wrote: "These voyages are frightening for those who are not accustomed to them and I would advise the little dandies of London or Paris to stay home."[62] Returning from Canada to Philadelphia, now the seat of the national government as well as the leading center for science in the United States, Michaux sought a way to continue his explorations. He found a possibility in the aspirations and dreams of Thomas Jefferson, who desired information about the lands west of the Mississippi River. Michaux offered his services to Jefferson and the American Philosophical Society (APS) to explore the vast Spanish territory that stretched from Saint Louis to the Pacific Ocean. Jefferson drew up a detailed plan for this trip that was to be a model for the one drawn up a few years later for the voyage of Lewis and Clark. The APS was a private scientific society founded by Benjamin Franklin, John Bartram, and others. Its members were some of the most accomplished and prominent men in the United States. Jefferson, the vice president of the APS, was also secretary of state of the United States. Planning and preparations for the expedition were still underway in spring 1793, when a new minister representing the French Republic, Edmund Charles Genet (1763–1834), arrived in Philadelphia and claimed the botanist's services for France.[63]

## In the Service of the Republic of France

Michaux's next adventure was not a botanical exploration but a political mission on behalf of the new French Republic. Michaux had been uneasy with some of the provisions of his arrangements with the APS concerning the western exploration. Jefferson's instructions indicated that Michaux would be an employee of the APS rather than a French civil servant acting independently for the benefit of both the Americans and his own country. Michaux was a proud Frenchman and valued his independence as a collector.[64]

Although the botanist had faithfully served the royal government all his life, he had also been exposed to revolutionary ideas and was aware of the changing political currents in France. While visiting Canada he may have either met or learned of the activities of Henri Mézière (1771–after 1819), a young

Canadian who wanted to foment revolution in Quebec. In Philadelphia, Michaux joined a group of French émigrés who favored revolution, La société française des amis de la liberté et de l'égalité de Philadelphie. After Mézière fled Montreal and arrived in Philadelphia in May 1793, it was Michaux who proposed him for membership in this society.[65]

Michaux delayed his departure on the western mission for the APS to meet with Genet, who asked him to undertake a political mission to Kentucky. He then acted as Genet's secret courier to George Rogers Clark, the American general who had proposed that France help launch a bold military strike against Spanish outposts in the Mississippi Valley and New Orleans to open the Mississippi River to free navigation.[66] France and Spain were at war; Clark's plan fit French objectives, and Genet readily accepted. Michaux made the journey to Kentucky, met with the general, and faithfully attempted to carry out his instructions from Genet. After his conferences with Clark, Michaux realized that the financing Genet had provided for Clark was inadequate and returned to Philadelphia to seek additional funds, only to find Genet replaced as minister and the military plans abandoned. Released from his Kentucky political commission, Michaux soon returned to Charleston and resumed his botanical pursuits, but his political sympathies were now aligned fully with the French Republic.[67]

Arriving in Charleston in March 1794, Michaux worked to restore his garden until July, when he made an almost-three-month journey to the high mountains of northwestern North Carolina. He had already made two visits to this area, but this time he remained there longer, explored more fully, and climbed several of the highest peaks, including Grandfather Mountain, which he believed to be the highest mountain in North America. On the summit of Grandfather, he demonstrated his love for his country by singing the new national anthem, "La Marseillaise." Ten days after returning from this journey, he experienced one of several attacks of fever that periodically marked his American sojourn, and on this occasion he reported that it was six weeks before he fully recovered from the illness.[68]

As spring arrived in 1795, Michaux embarked on another ambitious exploration, a journey financed with the last of his personal resources. Leaving Charleston on April 19, he did not return until April 11 of the following year. He traveled through the Carolinas over the mountains into Tennessee and Kentucky, then ferried across the Ohio at Louisville and continued west across Indiana and Illinois, reaching Cahokia, Illinois, the old French settlement across the Mississippi from Saint Louis, on September 6. After three months exploring Illinois, down the Mississippi to its junction with the Ohio, and upstream on the Tennessee and Cumberland in Kentucky, he began his return journey to Charleston.[69] In all, he traveled more than three thousand miles in 358 days on this last American journey.

Michaux had learned of the war's end between France and Spain while in Illinois, and he had inquired about guides to accompany him up the Missouri as he and Jefferson had planned in 1793, but he simply did not have the financial resources needed for the journey. During the summer of 1796, he assembled a collection of plants that he had been growing in the Charleston garden and packed his baggage. In August he sailed for France. There he hoped to gain recognition for his discoveries and government support for a return to America, where he knew that he could add to the luster of French science with additional discoveries.[70]

## RETURN TO FRANCE: DISASTER, DISAPPOINTMENT, AND RENEWAL

On October 9, 1796, after an uneventful Atlantic crossing, a violent storm in the English Channel drove Michaux's ship ashore on the Dutch North Sea coast near the village of Egmond. Michaux survived, but he lost personal possessions, the first notebook of his journal, a small trunk of papers and manuscripts, and a heavy box filled with bird and mammal skins. He saved almost all his plant collections, although they were damaged by seawater. Repairs to his collections became his highest priority, and he worked ceaselessly for six weeks restoring his herbarium. Finally, he returned to Paris and was reunited with his son, friends, and colleagues whom he had not seen in years.[71]

Michaux received a warm and honorable welcome from his colleagues, but heartbreaking disappointments awaited him. Of the sixty thousand trees and ninety cases of seeds that he had collected and packed with much effort and sent to France between 1785 and 1791, only a fraction had survived. Many perished in transit, and some were diverted to Marie Antoinette's Austria or to private gardens during the Old Regime, but the losses in the wave of revolutionary destruction directed at royal properties were probably the most difficult to bear.[72] He also soon learned that the republic would neither make good on the promises of its predecessor to pay him what he felt he was owed nor finance his return to America. Michaux had imposed on himself the task of publishing a flora of North America, but now he found himself financially ruined. Even though he was acutely disappointed by this turn of events and lacked the means to publish his discoveries, he nonetheless began to work on his books, a monograph on the oaks and a comprehensive flora of North America. He soon suffered an additional loss when his lifelong friend and mentor Louis-Guillaume Lemonnier died at the age of eighty-two. It is said that in helping his friend during his fatal illness, Michaux put aside his own difficulties and disappointments and once again became positive about his life and prospects.[73]

Both of Michaux's books are important landmarks in American botany. His *Histoire des chênes de l'Amérique* (Oaks of North America; 1801) was the first book published on an important genus of North American plants, with descriptions of twenty species and nine varieties of oaks that Michaux recognized. It was written in French, with short Latin descriptions for each entity, an indication of synonyms, common names in English and French, and distribution and uses, and it was illustrated. He finished the text on the oaks before he left on the expedition with Baudin and wrote the preface while he was on the voyage.[74] In this preface, he also mentions his next book, which was originally called *Flore de l'Amérique* and eventually titled *Flora Boreali-Americana* (1803), and says that he will include in it a résumé of his travels to that country. The *Histoire des chênes* was not published until 1801 because the illustrations were not ready. In this book, Michaux established his authority as a botanical writer with a combination of his own observations in the field and in nurseries in America and careful study of previous work on the genus by eighteen European botanists who wrote between 1601 and 1790, as well as Thomas Walter's 1788 *Flora Caroliniana* and William Bartram's 1791 *Travels through North and South Carolina, Georgia, East and West Florida*. *Histoire des chênes* was the work of a skilled, confident professional at the height of his powers of observation and exposition.[75]

Preparation of the North American flora was a far more difficult undertaking. No comprehensive American flora existed, although European plant collectors had been working on the continent for over two hundred years and although Linnaeus had included many species from eastern North America in his works. We do not know precisely why Michaux did not go beyond studying the plants in his own collection in preparation for his *Flora Boreali-Americana*.[76] Clearly, a period of study in the British capital would have been useful for him to examine other important botanical collections from North America, however his lack of funds as well as the political situation rendered such a journey almost an impossibility.

In preparing his *Flora*, Michaux used *Species Plantarum* (1753) by Linnaeus, but he did not always agree on the interpretation of certain species found in that book. He also knew and possessed a copy of Walter's *Flora Caroliniana*.

John Fraser, who had traveled with Michaux on his first southern field trip, was an associate of Walter's and had added to *Flora Caroliniana* many new species found on the journey with Michaux that should have otherwise been Michaux's discoveries.[77] Perhaps his distrust of Fraser was one reason Michaux did not accept many of Walter's names. Linnaeus's natural history collections were in London, having been purchased from his widow in 1784.[78] John Clayton's collections for the 1739 *Flora Virginica* were also in London in the care of Michaux's correspondent Sir Joseph Banks.[79] The collections of Walter and Fraser were also there, although Fraser was in Russia for much of 1797 and 1798.[80]

We speculate that Michaux wrote the initial manuscript for the *Flora* in a style similar to that used in his book on oaks—that is, it would have been written in French, with a minimum of Latin, with common names and information on usage, ecology and distribution. François André explained in his introduction to his father's *Flora* that when his father's manuscript was shown to specialists (probably Antoine-Laurent de Jussieu and René Desfontaines), they judged it unwieldy.[81] Before the *Flora* could be completed to everyone's satisfaction, Michaux received an offer to accompany an expedition to Australia. He left the unfinished *Flora* in the hands of his son and the botanist Louis Claude Marie Richard (1754–1821), who had probably already been selected to assist him, many months earlier, by de Jussieu. Richard and de Jussieu were close friends, and the latter was the most powerful botanist at the Muséum national d'histoire naturelle in the 1790s and early 1800s. We concur with Reveal (2004) and others that Richard reworked Michaux's text, so that the final product was entirely in Latin and lacked all mention of common names or usages. It was also missing the narrative of his travels. Even though both Michaux and Richard had been students of de Jussieu and his natural system of classification, the *Flora* followed the Linnaean sexual system. The most likely reason for Michaux's and subsequently Richard's use of that system was that Michaux closely followed Linnaeus's *Species Plantarum* as he made his collections, and he subsequently organized his herbarium according to that methodology. *Genera Plantarum*, de Jussieu's book, did not appear until 1789. It would have been difficult and time consuming to rearrange his collection according to de Jussieu's system.[82] The Linnaean system was much easier to use, and the groupings were distinct. Richard was not adverse to the Linnaean system, as some botanists have claimed; he is known to have formulated a revised version of it, perhaps for teaching purposes at the École de médecine.[83] In addition, Michaux showed interest in the natural system in various instances. For example, in *Histoire des chênes*, he organized the species into two well-defined

FIGURE 4. Antoine-Laurent de Jussieu
(Courtesy of Muséum national d'histoire naturelle, Paris)

FIGURE 5. Louis Claude Marie Richard
(Courtesy of Institut de France, Académie des sciences, Paris)

groups, the white and the red oaks, based on morphological characters of the leaves as well as growth patterns of the acorns. Similarly, he attempted to classify the genus *Vaccinium* into groups based on several morphological criteria while in Charleston in summer 1796.

*Flora Boreali-Americana* went forward, and when it appeared the only author's name on the title page was that of André Michaux. Speculation about the magnitude of Richard's role in the book's preparation began almost immediately. Michaux's original draft, mentioned by his son in the introduction, has never been found. Like Michaux in North America, Richard suffered severe hardships while collecting in Africa, South America, and the Caribbean. He also had the misfortune to return to France bankrupt. The two men were likely kindred spirits, and we speculate that Richard chose to give Michaux the full credit for authorship.

## BAUDIN EXPEDITION

In October 1800, Michaux left France on an expedition to explore Australia. It was led by Capt. Nicholas Baudin (1756–1803), a Breton who had impressed André Thouin and others with his ability to bring live plants back from long sea voyages. Baudin was a seasoned commander not awed by the group of mostly young scientists on this voyage, and a rift soon arose between the captain and the scientists. Michaux and several others chose to leave the expedition when Capt. Baudin prepared to depart the island of Mauritius, a French outpost in the Indian Ocean.[84] Michaux remained for more than a year on Mauritius, where he found old friends, the company of talented naturalists, and an interesting island natural history to study. In late spring 1802, he left Mauritius for the French settlements on the larger, and botanically much less well-known, island of Madagascar, about five hundred miles to the east.[85]

## MADAGASCAR

On Madagascar, Michaux worked with Louis Armand Chapelier (1778–1806), a young gardener sent there by Thouin. One of Michaux's intentions was to introduce useful European and tropical plants on the island, and he worked very hard to develop a garden for this purpose. His confidence in his physical powers and his work habits, however, were misplaced in this disease-ridden region. For several months he suffered no ill effects, and then he contracted a tropical fever and died on October 11, 1802.[86] As had happened with so many other explorers before him, and with so many others who would follow, it is likely that a mosquito brought to an end the illustrious career of André Michaux.

# LEGACY

André Michaux's legacy was secure. His son stood at the threshold of a distinguished career in botany and forestry that would carry the family name forward in science for another generation. Furthermore, André Michaux's *Histoire des chênes de l'Amérique* was already published, and his *Flora Boreali-Americana* appeared less than six months after his death. His colleague J. P. F. Deleuze (1805) would soon recount the story of André Michaux's life of service to science and to France. Deleuze's memorial was quickly translated into English and printed for British and American audiences.[87] It would be useful to know what the botanist looked like, but no portrait or other image of André Michaux made during his lifetime has ever been found. His son's portrait was painted by both Rembrant Peale of Philadelphia and an unknown artist in France. Late in his life, François André was also the subject of a daguerreotype.[88] Nonetheless, all we know about his father's physical appearance is drawn from a description of the personnel who accompanied the Baudin expedition in 1800. André Michaux was a man of medium size with gray hair.[89]

# CHARACTER

Most of the records that André Michaux left were not personal in nature but rather products of his work. He kept a journal with notes probably meant only for his own eyes, and he sent many letters to colleagues, but the only two explicitly personal letters known to survive were addressed to his son. Nonetheless, we can form a picture of this active, intelligent man who began life on a lower social rung and with less formal education than many of his peers and who therefore strove to excel by working harder than anyone else. He was proud of having been selected as a member of the new Institut de France at the age of fifty, and he attended its meetings regularly after his return to Paris in 1796. With admission came the opportunity to meet regularly with the scientific elite of France, a group that Michaux had aspired to join for most of his adult life.[90] Michaux did not seek to become rich or powerful, but he sought the admiration and respect of accomplished men.

What we can observe in the life and botanical career of André Michaux is a person who made the most of his opportunities. He learned early in life the virtues of simplicity, the rewards of hard work, the value of education, the importance of persistence, and other basic lessons that served him well throughout his life. He matured strong in body, strong in mind, and strong in spirit. It is not surprising that at times he showed a defensive pride with people

FIGURE 6. *Michauxia campanuloides*, rough-leaved Michauxia.
The genus name of this Middle Eastern plant honors André Michaux.
(Photo by Charlie Williams)

of higher status and more education, especially if he felt that they looked down on him. In studying his journal, we observe that he learned from his experiences as he worked and traveled, and he was remarkably centered, able to overcome his fears, focus his energies, and accomplish his work in the most difficult of circumstances. He was an honest and upright man whose dealings with others are characterized by humanity and fairness. He probably had a sense of humor, but unfortunately it appears in a very few places in his journal. He appreciated good-looking women. He treated his Native American guides with respect and shared their hardships. Régis Pluchet described Michaux as a scientist of the Enlightenment who was eager to leave absolute power behind and who analyzed the world in terms of science and reason, but not always without prejudices.[91]

André Michaux was a man of good judgment who was self-reliant and cautious when caution was called for, but bold when boldness was required. His constant companions on every journey were hardship, fatigue, hunger, and danger. He feared nothing so much as wasting his time, and he rarely complained. A dangerous fall from a horse with injuries sustained when far from the nearest settlement was described chiefly in terms of time lost, not as the life-threatening close call it must surely have been. It has been easy for some writers to deride the lack of literary merit of his journal. Some have misconstrued his motives, while others have romanticized him, but he lived his life to its fullest potential. Interpreters of his legacy have, sometimes unwittingly, fashioned legends about André Michaux, but when his life and career are closely studied, the remarkable flesh-and-blood botanist is found to exceed the imagined botanist of legend. As the late journalist Charles Kuralt remarked in 1994 when commemorating the two-hundredth anniversary of Michaux's climb to the summit of Grandfather Mountain, Monsieur Michaux was "a man for the ages."[92]

CHAPTER ONE

# Arrival in New York, November 1785, and Relocation to Charleston, September 1786

SUMMARY: ANDRÉ MICHAUX'S FIRST NOTEBOOK covering his initial months in America was lost in the shipwreck during his return voyage to France in 1796. The record of his first eighteen months in North America is reported from other documents, primarily letters. In these documents we learn that Michaux was welcomed to New York, then the capital of the United States, by Louis-Guillaume Otto and other officials in the French embassy. With Otto's assistance, Michaux was able to quickly begin his work collecting and making shipments of plants to France and to establish a plant nursery in New Jersey. After a few months in the New York area, where he gained familiarity with spoken English, he made a journey to Philadelphia, Mount Vernon, and the northern Shenandoah Valley delivering his letters of introduction from French officials to Benjamin Franklin, George Washington, and other Americans who welcomed him and promised assistance. In early September 1786, leaving his gardener in charge of the New Jersey nursery, he sailed to Charleston, South Carolina, with his son and a servant. There he soon purchased a plantation where he established another nursery garden that became his new base of operations, and he began collecting and shipping plants from Charleston to France. In April 1787, seven months after arriving in Charleston, he undertook the first of a series of journeys in search of plants that his surviving journal notebooks report in detail.

## PART 1. NEW YORK, NOVEMBER 1785–AUGUST 1786

### Louis-Guillaume Otto to Count d'Angiviller, from New York, November 25, 1785[1]

I received the letter that you honored me by writing to recommend M. Michaux, botanist sent by His Majesty to ship to France shrubs and seeds from this country and to gather a collection of all the curiosities from the vegetable kingdom. I just conferred with this person who seems to be most fitted to fulfill the wishes of the government. He arrived here only day before yesterday, and he will leave directly for [New] Jersey where he will still have time to collect a large quantity of seeds before the departure of the [packet] boat. I gave him the instructions that are added to the letter that you honored me with, as well as the statement that M. l'Abbé Nolin[2] had given me before my departure from France. All shipments from now on will be made by M. Michaux; he will send them to me in New York, and I will take care to ship them punctually. I cannot tell you, Monsieur le Comte, the satisfaction that the arrival of this botanist has given me; I did not have any hope of sending anything this year, and all the requests that I had made to nurserymen of my acquaintance were received with so little interest that I had almost given up on successfully fulfilling the mandate that you had been willing to give me.

We have time, M. Michaux and I, to select during the winter in the neighborhood of New York a piece of land suitable for the establishment of a king's garden. The purchase of this piece of land will be an outlay whose value will double in a few years; however we will attempt to be as frugal as possible. M. Michaux already knows English; he will learn to speak it this winter, and next spring he will be ready to travel successfully in the interior of the country.

Independently of the garden that M. Michaux proposes to establish here for productions of the northern portion of America, you will think perhaps appropriate, Monsieur le Comte, to give orders to M. de Châteaufort, consul in Charleston, to establish an analogous garden in the Carolinas for trees and plants that live only in a warmer climate. I have reason to believe that most of the products of the two Carolinas and Georgia will be able to thrive in the south of France, and M. de Châteaufort will have the advantage of dealing with a very intelligent botanist who will be on the spot and who has already made several interesting shipments to l'Abbé Nolin. We do not have this facility in the northern states, and without the assistance of M. Michaux, I do not see that we will ever obtain from these states the great benefits for the kingdom's culture.

## AM to Louis-Guillaume Lemonnier, from New York, December 9, 1785[3]

We arrived in New York on the thirteenth of November, after 47 days crossing with almost continuous contrary winds.

On the 17th I set out to visit the woods of New Jersey. I found there an abundance of *Liquidambars*, sassafras, tulip trees, in addition to the oaks of various species. Almost all the trees are found in moist places and even in places submerged during the winter, with the exception of *Liquidambar asplenifolium* recognized today as that is found in sandy and dry places. There is a chest addressed to Monsieur, the brother of the king, containing besides other trees about 150 at least of this little *Liquidambar*. There is also in the bottom of this chest a small shrubby tree whose wood is blackish, very remarkable for its scarlet fruit with which it is loaded, and of all the shrubby trees that are cultivated, I have seen none whose effect is as remarkable.[4] It is found in very humid places and very rarely in dry ones; you will recognize it as it is tied with three little knots of hay. I cannot express to you the problems and difficulties experienced for this shipment that we have been obliged to rush so as not to miss the departure of the packet boat, the day having been set. Fourteen days have been spent in collecting seeds and digging up trees both on Long Island and in different spots in New Jersey. We have not been able to obtain tree moss that is found in the woods, so necessary for packing the trees. The woods that we visited only had moss that is short and scattered; it is not possible to make any use of it. We have been obliged to use sod or turf from the land, which is already frozen, making the work painful, and the packing will not be as well done as it would have been if we had had the facilities. I hope that Monsieur le Comte d'Angiviller will make allowances for this first shipment.

Besides the box of trees addressed and marked thus, M #, to Monsieur, brother of the King at Versailles, which is to come to you directly, there is in one of the boxes addressed to Monsieur le Cte d'Angiviller a box that is almost square on which I forgot to put your address, because while I was busy this morning writing letters, the gardener, having finished filling the chest, closed it and carried it to the ship, so that it is now impossible to remedy this.

*Louis Wm. Otto.*

Pub. by Janes & Bumford, 1808.

Institut Royal
de
France.
Acad.ⁱᵉ des sciences.(Économie rurale.)

THOUIN,
(André)

Membre de la Légion d'honneur.

Né le 10 Février 1747, au Jardin du Roi, élu à l'Académie en 1786, à l'Institut en 1795.

FIGURE 7.  (*Opposite, top*) Portrait of Louis-Guillaume Otto
(Courtesy of the New York Public Library)

FIGURE 8.  (*Opposite, bottom*) Portrait of Count d'Angiviller
(Courtesy of the Metropolitan Museum of Art, NY)

FIGURE 9.  (*Above*) Portrait of André Thouin
(Courtesy of Muséum national d'histoire naturelle, Paris)

The chest in which is enclosed the box for you is the one marked R number 4. It contains very interesting seeds, most of which I have not been able to recognize and that I have gathered in the woods.

We have experienced difficulties that I cannot begin to express to you, the bad weather, freezes or snows in turn, the distance from where it is necessary to carry what we have collected, and not having one single convenient spot in which to [deposit] our collections. I am sending five strong boxes addressed to Monsieur le Comte d'Angiviller—one rich in trees, the others with seeds. There is one whole box of a *Vaccinium* here called cranberry, the fruit of which is as large as a cherry and is used for making jam here; in this same box there is a bottle filled with this fruit in water so that it will be preserved in its natural state.

There is also in the same box some sweet potatoes called Carolina potatoes that are eaten here; they are subject to freezing. In the little box that is meant for you and on which the address is omitted, there is a well-corked bottle of this fruit.

In the box addressed to Monsieur, you will find above the *Liquidambars* two little shrubs wrapped and bound with a little creeping *Vaccinium*, and near the top of this box I have put some Carolina potatoes.

I have the greatest hopes of making interesting shipments, the woods are filled with plants that seem to me never to have been described, and next spring I hope that you will be very pleased. Not being settled in permanent lodgings because it is difficult to find rooms to remain in this city, I have placed my belongings in a very small room, which prevents me from finding again a very extraordinary plant, found in the marshes. It has some relationship with the *Droseras* I presume; its basal leaves and other leaves resemble the flower of the digitalis in shape and size, rather hairy at the entrance, somewhat recurved; and furthermore there is an extension of one side only from the expansion toward the base of the petiole; on only one of these plants did I recognize a dried peduncle with fruit. This plant [probably a pitcher plant] is found among the moist areas where the *Sphagnum palustre* grows.

I have one more little creeping plant with red berries, heart-shaped leaves, and some ferns that I cannot send at this time.

You will recognize in all the haste and confusion that has overwhelmed us, it is now one hour after midnight, no fire: I was told this evening that it would not be surprising if the rivers are frozen tomorrow morning, and that happens often here.

I ask your indulgence; the gardener that M. Thouin procured for me is a treasure for his good will, enthusiasm, eagerness; he is intelligent and very industrious.

We have been to see a nurseryman who has advertised trees for sale for this climate. [Him] not being acquainted with us, and [us] asking to buy fruit trees so that he should not be suspicious of our curiosity, he led us into his nursery; we saw there only six hispid *Robinia*, one Weymouth pine, and a spruce. The *Magnolias*, he says he is obliged to have sent from Carolina.

Immediately after the departure of the packet boat, I plan to go to Philadelphia and, from there, farther south for there is nothing to do here for several months. It is probably not possible to cultivate here *Magnolia* and other interesting trees, even those from Virginia. Perhaps it would be convenient to establish a garden near Williamsburg. I am assured in advance that the land will not be expensive there because it will not be necessary to choose it near a great city like the one we must have near New York on account of the proximity of the packet boats. When I shall be free, I will send an account [report] to you and to Monsieur le Comte d'Angiviller of all that I learn concerning this establishment.

## AM to Count d'Angiviller, from New York, December [9?], 1785 (excerpt)[5]

We did our best to send you a package of the most important [plants], but we had some major difficulties and repeated accidents at each stage. The distance, the bad weather, and what is more disappointing, the poor attitude of the inhabitants of this land; we can't pay them enough money to make them work. In many places in New Jersey, where we would find some farmers sitting idly around a fire, I was often offering them wages of one piastre [dollar] and sometimes as much as two per day, and [yet] I was obtaining only two or three hours of work. It is in this province where we did our harvesting; going as far as Elizabethtown. Near Newark, a few miles from here, is a very interesting place because of the quantity of trees and shrubs of different kinds [that] I recognized. We dug up four to five thousand of them, which are now deposited in a garden and which are not ready for the departure of the packet boat because of the difficulties that we encountered getting things done promptly here.

I have been told that I will find more resources among the Germans and French, who are more widespread in the other provinces. I am certain that, if I was being helped by farm workers like we have in France, I would be able to send several ships completely filled with interesting productions every year.

In addition to the collecting trips in New Jersey, Michaux's expense report for this period mentions other collecting trips to Long Island and Harlem in New York.[6] After sending the shipment to France in mid-December, Michaux and Otto quickly decided on a suitable parcel of land for a plant nursery and botanical garden in New Jersey. Michaux described this property in New Jersey and gave reasons for its purchase in a letter to Count d'Angiviller.

### AM to Count d'Angiviller, from New York, January 18, 1786 (excerpt)[7]

Since I arrived here, Monsieur le Comte, I realized the necessity and the need of a nursery, not only for trees whose seeds don't survive the journey, but also for those that form taproots and can't be pulled out without cutting the roots. We will therefore have to sow those species abundantly and send them the second year.

I found land very convenient for the establishment of a nursery. The main advantages are: proximity to diverse woodlands for gathering seeds and pulling out young trees (important consideration and that will save time and expenses), the proximity of New York and of the [Hudson] River for communication with the packet boats, and finally, the land's quality. The quantity is twenty-nine acres on a gentle slope, [of] which the highest part is of clay and substantial soil, the middle part is of gray sand, and the lowest part is a woodland of aquatic cypress. Under the shadow of the cypress, the *Magnolia* and *Rhododendron* are found. This land forms a long square at one end of which are large woodlands that border on the Hudson River, at the other extremity is a little branch of the Hackensack River, and on both sides there are some private residences. The road to New York and from Bergen to Hackensack crosses this land, and when going through the woods in a straight line, we arrive at "Bull-ferry," i.e., the ferry of the bull situated four miles from the Hoboken Ferry. You can see this location exactly on the map detailed and translated by Le Rouge, Quai des Augustins, Paris. I will send you this map and its survey calculated into the French measure.

The price is three hundred pounds. . . . This transaction is not finished as the owner didn't bring all of the land titles needed. There are also some problems linked to laws and traditions as only naturalized Americans can possess lands; I didn't want to act in this affair without M. Otto's advice, and we are waiting for the answer from the New Jersey state governor.

There would be some costs of renovation, but I will make them with economy. . . . Due to repairs that are not practiced down here, I can assure [you] that the price of this acquisition would more than double its acquisition and renovation price in 4 or 5 years.

Count d'Angiviller did not receive this letter before he sent letters that must have questioned the need for acquiring the New Jersey property. By the time that the count's letters reached Michaux, however, the purchase was complete, and the development of the property was well underway. Because of the slow communications, we skip ahead almost four months for Michaux's spirited defense of his actions.

### AM to Count d'Angiviller, from New York, May 13, 1786[8]

I fear [I] have been too hurried in the acquisition of land, seeing by your letters of the 23rd and 25th of January last that your orders were not then determined in this regard. I have however confidence that there will be a great deal to gain beyond the expenditures when one will wish to rid oneself of it as much because of the price, which is not beyond the current ones, as because of the improvements of which this place is susceptible. The greatest of all the advantages and the one on which we congratulate ourselves every day is to have in our proximity all the productions that can be found from Canada to Virginia, so that one can go make abundant collections and return to the house at bedtime. This advantage is very considerable because I presume that for four months at least it will be necessary to be occupied with the collection of seeds, and that remoteness would have used precious time and increased expenses for carrying our collections in trees and seeds and then sending them to the packet boats. The situation of this land at the foot of a mountainous forest will produce for us abundantly all sorts of trees that we can desire. The extremity is terminated by a moist land composed of cypress, *Thuya*, *Magnolia*, and *Rhododendron*. These kinds of woods are called here Cedar Swamps [and] are almost impenetrable, and we have six acres covered with this kind of trees.[9]

The object that I have principally in view, Monsieur le Comte, and the true economy, is to make expenditures only as far as they can produce evident advantages, and those expenditures being recognized as necessary, not to defer them; this is why we have this first year beyond the acquisition of the land and the construction of a building in which to deposit our collections, made an enclosure of garden and lands and purchased two horses, plough, and other utensils for working and preparing the earth, because beyond the shipping of the trees and seeds, it would be necessary to acquire [tools] for sowing them [on] the most land possible.

The purchase of the land has cost, as I have had the honor to inform you, the sum of 750 dollars. . . . I have a price for the construction of the building, which is already well advanced and will be finished in 3 weeks, at 800 dollars.

. . . The building is composed of four rooms, two on the ground level, two above, and a granary. I would have wished, Monsieur le Comte, not to make this expense and to find a house to rent; this place, which is so advantageous in other respects, is inhabited by 10 or 12 families whose houses or cabins have no floors above the ground; a storehouse in the ground for preserving the trees and a granary for seeds, absolutely necessary, would be about two thousand pounds, and I have preferred to have a house built. With the exposure I have given it, its pleasing design and the group of improvements of which the place is susceptible will multiply a great deal its value. Furthermore, I have consulted MM. Otto and La Forest as to finding other means, but the expense in the city or near the city, a year that would have been lost for the gardener without occupation, the fear of sowing and other operations on the land of another, have decided me. I dare to hope that with the good will of the gardener this establishment will be also very interesting for America and will bring honor to France by the facility of sending there a great number of trees that are unknown here. In the summary of instructions that M. l'Abbé Nolin has sent to me from you, he mentioned to me that I would be able to offer to the Americans the useful trees of which we are in possession in France, and I have shown the usefulness of this in the petition presented to the Assembly of the State of New Jersey to obtain the liberty of purchasing without being forced to [do] the formalities insisted on with regard to foreigners.

Having recognized that the gardener could not feed himself easily in this country with the sum granted for his wages . . . I have permitted the gardener [Pierre-Paul Saunier] to cultivate some acres of land on condition that I should have your agreement for this advantage, that the employed workers should be nourished by the products of the land as well as the horses that one would be obliged to have for the cultivation. The gardener's helper is a young man [Jacques Renaud] whom I have brought because of his good will, who has cost only his maintenance up to now and who has no right to demand steady wages because he is more fortunate here than at home [and] because it has been expensive to bring him here, and one will be able to reward him in the future if his good will continues.

The next evening, after his letter to Count d'Angiviller outlining the advantages of the New Jersey garden, Michaux wrote to his friend and colleague André Thouin. This ten-page document is one of the most interesting letters from Michaux's stay in North America. Michaux and Thouin were only a year apart in age and knew each other well.

## AM to André Thouin, from New York, January 19, 1786 (excerpts)[10]

I arrived in the worst period, having stayed 27 days in Lorient and 47 days at sea.[11] Winter had already started, and snow covered the plants a few days later. Immediately after our arrival, we went to New Jersey, and we collected seeds and dug up trees. In this way, by return boat we were able to send five cases of seeds and trees. By this shipment, I sent to M. Comte d'Angiviller 12 heavy boxes of trees and one to M. Lemonnier; of the 12, there was one for you and one for M. de Malesherbes.[12] It is not without adversities and difficulties with the season so advanced. When snow started to fall we had just discovered the moss *Sphagnum palustre*, which we gathered and which became a block of ice overnight. We had to melt it with the heat of a furnace, and it finally was available as packing material for our trees. But one has to forget about the hardships once they have gone by. At first I was really concerned about the supply of moss, as one does not find it as easily as in European forests. Here the mosses in the woods are scattered and very small, but fortunately *Sphagnum palustre* is found in abundance, and I feel confident at being able to send much material and in good state. Please let me know in what condition your trees arrived as the moss might have been a little too wet. You will find a plant that I think is *Sarracenia purpurea* in among the moss, and if you empty the box in a wet area and shake the moss, the trees and shrubs will loosen. I have observed many interesting things, but I want to finish the letter so that it can be sent on the boat, and I would like to write to our friends, Cels, l'Héritier, Dantic.[13] There are many birds, very interesting because of their beauty and rarity. The first time I killed some, I did not have the time to stuff them; I had to abandon this enterprise and will wait till the departure of the ship to start again as I want to send a shipment to M. Daubenton,[14] but it is not here that much of the work will take place but rather in the Carolinas, Georgia, and Florida. That is where I would like to have the company of M. Dantic to travel through the forests, cross rivers and swamps, [and] penetrate through dark forests where trees of enormous size have fallen perhaps as many as 100 years ago and are reduced to a powder when you walk on them. There is probably no better country to collect insects. People here who have little knowledge or interest in these things speak to me of things that astonished them. The high temperature and humidity of this countryside cut up by streams and rivers, stagnant water, all multiply the abundance of plants from which other living things obtain their sustenance. Many people want to frighten me concerning the unhealthful air, but I am an old soldier in this regard. You cannot imagine what one calls swamps here; the swamps here are full with *Cupressus thyoides*, found in the wettest areas, and so low that in high tides there is in some parts one or two feet of water. These trees are so close together that it is often difficult to penetrate. The trees do not have strong lateral branches but are very tall and straight; the summit of these trees is evergreen, but the rays of the sun do not penetrate in the interior of these forests. It is here in the shade that the most beautiful *Rhododendron* obtain their freshness. The *Magnolias* have a most honorable situation and are found where there is air and shade. *Prinos glaber* are always found at the edges and never in the most shaded areas; they extend beyond the forests and form plants of such a vivid green that the perspective is interesting. A moss (*Sphagnum palustre*) about one foot high covers the whole area, and the roots of the shrubs (*Rhododendron*) do not penetrate much beyond. *Gaultheria procumbens*,[15] *Vaccinium repens*, *Sarracenia purpurea* (wonderful plant), etc. etc. are here in their native element. When these swamps are closer to higher sandy ground, the soil is covered by *Kalmia angustifolia*, etc. It is often necessary to wear boots to go through these swamps, and the odor of the stagnant water can be very bad. People say that it is dangerous to go in the swamps during the summer without boots because of snakes. I forgot to tell you that *Prinos verticillatus* is found among the *Prinos glaber*, and their branches are loaded with red fruits looking like coral. You can also find this shrub in areas that are alternatively dry and wet. The Weymouth pine is also found among the cypress in areas that are often submerged.

I beg you to tell our friends MM. l'Héritier, Lamarck, Cels of my good wishes and hope to transmit to them before the end of the year my heartfelt thoughts. I am taking to the Carolinas the work of l'Héritier;[16] there is no amateur here, nor expert; people are all ice-cold here and indifferent to everything except what will bring money. People give me the agreeable prospect of finding in the Carolinas horticulturists who have plantings from eastern India that are naturalized there.

I thank you infinitely for your offers, and if you and our friends, especially M. l'Héritier could put together for me a small herbarium of the *Kalmia* family, I would be infinitely obliged. It is not necessary to send fine samples because it is only for making comparisons. It is not necessary to have *Rhododendron*, but I would like to have *Vaccinium*, *Gaultheria*, a leaf only, *Arbutus*, [and] *Andromeda*, but not *Erica* as it is not found in America. Some seeds from eastern India that might succeed in the Carolinas, such as *Pandanus odoratissimus*, *Lawsonia*, or henna from Egypt, *Kaida* from Madagascar, seeds of *Diospiros ebenum*, ebony wood from Isle de France, etc etc. This small collection of seeds should be given to M. Desaint or M. Dantic, who know the name of my correspondent in Lorient. I would like to ask M. l'Héritier for a bushel of seeds of female cypress and seeds of *Thuya* from China, and also of Lebanon cedar and also mountain ash of the birds and *Paliurus*, and I would like to know what would interest him from this country. I will try very hard to make his work known in the Carolinas where I will make every effort to go to as soon as possible. Please excuse all the mistakes in my letter. The cold weather, which increases daily, is already at 13 degrees F [Fahrenheit; –10.5°C]; the houses are made of wood, but the cold penetrates in spite of the fire. It is actually midnight, and I still have much to do before the departure of the boat. . . .

Michaux followed the six-page body of this letter with four pages of postscripts and plant lists. We have reproduced his special instructions and the lists of plants that he wanted Thouin to send so that he could "multiply them in America."

This boat [*Courrier de New York*] will arrive around the 10th of March and will remain five weeks in L [Lorient]. The season will be very favorable to make this shipment. I ask you to take plants in excellent form from Sèvres for the ones that are starred and that you be reimbursed by M. Desaint, assuming that each one will be worth two louis.

Excuse all the work involved.

They must be shipped with quite a large amount of moss. Seal them, and then enclose them with two circles of thin wire. I ask that M. Desaint does this last operation.

Summary of all the objects that I asked for previously either from M. Desaint or from our friends MM. Dantic, Thouin, Cels, and l'Héritier

Immediately upon receipt of my letter,

First case

1 Detailed maps (in which are found even the areas around New York and the city of New York), translated by Le Rouge, Quai des Augustins, and with the title *American Atlas*[17]

2 Dictionary of English words[18]

3 English method by Siret, two samples[19]

4 Work of Mutis, a Spanish botanist (if it is easy for M. Dantic to find)[20]

5 By the same shipment, some seeds of female cypress from which the cones are dry, a cone of cedar of Lebanon

Michaux included a somewhat enlarged list of the plants he wanted to "multiply in America" in the second case of this shipment.

Second Case

6 By a second shipment a few days after the return of the [packet] boat (*Courrier de l'Europe*)

   (a) The plants indicated with an (a) can be asked from l'Abbé Nolin

   (b) Those marked with a (b) must be bought from a nurseryman from Sèvres, and M. Desaint will reimburse for the amount

   (c) Those marked with a (c) and others can be provided by MM. Cels and Thouin

   (a) *Arbutus unedo* [*Arbutus unedo* L., strawberry tree]
       [*Arbutus*] *andrachne* [*Arbutus andrachne* L., Greek strawberry tree]
       *Arundo donax* [*Arundo donax* L., giant reed]
       *Buxus balearica* (cuttings) [*Buxus balearica* Lam., balearic boxwood]
   (b) *Camelia japonica* [*Camellia japonica* L., camellia]
       *Cerasus azarero* (grafts) [*Prunus lusitanica* L., Portuguese cherry-laurel]
   (b) *Croton benzoin* [*Styrax benzoin* Dryand., Benjamin tree]
       *Cupressus fastigata* [*Cupressus sempervirens* L., Italian cypress]
       *Heliotropium peruvianum* [*Heliotropium arborescens* L., grey leaf heliotrope]
   (a) *Larix orientalis* [*Larix kaempferi* (Lamb.) Carrière, Japanese larch]
       *Lawsonia inermis* (cuttings) [*Lawsonia inermis* L., henna]
   (a) Mimosa el abrisin [*Albizzia julibrissin* Durazz., mimosa]
   (b) *Olea odorata* [*Olea europaea* L., olive]
       *Populus fastigiata* (cuttings) [*Populus nigra* L., Lombardy poplar]
   (c) *Robinia chamlaga* [*Caragana sinica* (Buc'holz) Render, Chinese pea shrub]
   (a) *Sorbus aucuparia* [*Sorbus aucuparia* L., European mountain ash]
   (a) *Sophora sinica* [*Sophora japonica* L., Japanese pagoda tree]
   (b) *Thea viridis* [*Camellia sinensis* (L.) Kuntze, tea plant]
       *Thuya orientalis* (two individuals) [*Platycladus orientalis* (L.) Franco, Oriental arborvitae]
   (b) Ginkgo [*Ginkgo biloba* L., ginkgo, maidenhair tree]
   (a) *Quercus suber* [*Quercus suber* L., cork oak]
       *Ulmus* said to be from Siberia [*Ulmus pumila* L., dwarf elm]

7 By this second shipment, ask M. l'Héritier for a bushel of female cypress, and have it added in a second case with rye grains that my brother from Satory [André-Francois Michaux (1747–ca. 1808)] will send to M. Desaint

Add also the little herbarium of the *Kalmia* family to be provided by MM. l'Héritier and Cels.

These three cases that will be sent to M. Desaint, he will seal them and will give orders that they are not to be opened at Lorient. M. Desaint will reimburse for all the expenses of purchase and anything else. After they are sealed, put two circles of wire around the second case. They should not be opened.

Parisian printer and bookseller Jean Charles Desaint, husband of André Michaux's sister Marie-Victoire, was a trusted family member.[21] The specific instructions and the extra security requested for this shipment of plants are unique in the known correspondence of Michaux. Clearly, obtaining this shipment of plants, seeds, and books was a very important matter for the botanist. Noteworthy among the species mentioned on his list are the mimosa, ginkgo, tea plant, and camellia.[22] Michaux has long been credited with introducing these species, as well as several others, into the southeastern United States through his garden near Charleston. Likewise, both Michaux and his gardener Saunier have been credited with spreading the Lombardy poplar from the king's garden in New Jersey.[23]

Direct evidence that the Old World plants available in France were being sent to Michaux with the blessing of his superiors is found in Michaux's letter to Count d'Angiviller of May 13, 1786, reproduced earlier in this chapter.

In another letter written long after his return to France, dated January 25, 1799, Michaux wrote that the state of Carolina had benefited from the garden near Charleston by the introduction of trees from China, India, and Europe.[24] We have already noted in the introduction how Michaux's 1792 arrangement with the Agricultural Society of Charleston made all the nonnative plants growing in the Charleston garden freely available to this influential group of local citizens.[25]

An act of the New Jersey legislature was necessary for Michaux to complete the purchase of the garden property. Otto was well connected to assist Michaux in this matter since a niece of New Jersey governor William Livingston

(1723–90) would become his first wife.[26] Once he had completed the purchase of the property, Michaux wasted no time contracting to have a house built. In the next few weeks, other expenses of equipping the New Jersey garden appear on his expense report, such as the purchase of farm tools, 150 window panes for his garden cold frames, and a team of horses.[27] Young Jacques Renaud, who had come to America as a personal servant to Michaux, became Saunier's helper until he (Renaud) moved with Michaux to Charleston.

In mid-March André Michaux dispensed with the services of the bilingual laborer who had served as his interpreter. He traveled, renting horses, but the duration or specific destinations of these overnight trips cannot be established from the known letters and expense summary. His detailed description of the manners of the citizens of Connecticut reported in a letter to Cuvillier, Count d'Angiviller's secretary, confirms his travel in that state.

### AM to Cuvillier, from New York, May 12, 1786[28]

You have asked me, Monsieur, for a description of the characters of the Americans. I began some details, but time does not permit me on this occasion because this description is long and is nevertheless incomplete. It is necessary to be acquainted with and to know and observe before describing, and that is difficult in this country. The individuals of the different nations of Europe who have come to settle in this country have brought with them the customs, costumes, talents, industry, virtues, and vices characteristic of each of them; one recognizes among the inhabitants of Bergen and of the neighborhood of New York, that of activity, diligence; and a continued solicitude, an exquisite cleanliness extending to the smallest things is noticeable on entering into the houses of those inhabitants from Holland; they still speak the language. The Germans who are in great number especially in the state of Pennsylvania are very industrious and hospitable; the Americans for several generations are idle, do not take the trouble to cultivate the land, which nevertheless, yields a month [in crops] 12 for one, and sometimes 20 [times] or more the amount planted. The abuse of rum, which one brings from the islands and which one sells at a low price, is the principal cause of this, combined with several others. The multitude of game that one hunts in the forests for food or furs, the fish with which the rivers abound, and which cut across the country on all sides, distract the inhabitants from diligent work. In Connecticut chiefly when one enters the houses, one is received with an air of gravity, seriousness, and reserve, which lasts several minutes, which are used in examining the stranger. They then become more friendly and the usual questions are: Who are you? What is your business? What is your religion? When their curiosity is satisfied, they treat you with all the hospitality possible, but one must keep oneself from annoying them in their way, even in not imitating them, for they are the best friends and the worst enemies.

The observance of Sunday begins on Saturday a little before sunset; all work ceases at this moment, they prepare in advance food for the meals, one must speak only for the most necessary things and in a low voice. The farmers of Connecticut are prudes; they do not allow themselves to read comedies, nor even to speak of whist, the quadrille, or the opera, but [they] speak willingly on the subjects of histories, geography, etc. They are great moralists loving to talk religion, and many possess Latin. If a stranger travels on Sunday on horseback he is likely to be stopped during the hour of the service.

Michaux had come to America with letters of introduction from French officials to Benjamin Franklin in Philadelphia, to George Washington in Virginia, and to others. Now the arrival of warm weather and the opportunity to examine the plant life on his route coincided with Michaux's growing confidence in his spoken English and his son's as well. He had sent François André to live with an American family in order to learn English. He purchased maps of the western frontier of the United States and of Virginia, and he no doubt sought travel advice from French officials in New York as well as from other people he met.

The first journey outside New York and New Jersey that can be reconstructed in any detail is the June 1786 journey to Philadelphia, Maryland, and northern Virginia. Three letters as well as entries in his expense report give us some highlights of this journey.[29] George Washington's diary provides information about Michaux's visit to Mount Vernon and his gift of seeds to the future president. Michaux's letters of introduction to Franklin and Washington also add useful insight.

In the first letter written from Philadelphia, Michaux informed Count d'Angiviller that he was collecting plants on the journey and had made his first visit to Bartram's Garden.

### AM to Count d'Angiviller from Philadelphia, June 11, 1786 (excerpt)[30]

I cannot [now] give you longer details as to what I collected and saw at Bartram's Garden. There is only one tree that is interesting and new, *Franklinia*.

I leave tomorrow for Baltimore and Virginia and then to the Appalachian Mountains, but nevertheless I propose to return to New York in less than two months' time as it is more advantageous to go to the Carolinas at the time when seeds are mature.

Upon my return, I will tell you and M. Lemonnier about my travels.

Michaux's expense summary informs us that he traveled by stagecoach from New York to Philadelphia on June 5 and remained in Philadelphia until June 11. During this time Michaux called on Benjamin Franklin, where he presented the letter of introduction quoted below.

### Count de Vergennes, French minister of foreign affairs, to Dr. Franklin in Philadelphia, August 9, 1785[31]

This letter, Sir, will be presented to you by Mr. Michaux, botanist, associated with the king's nurseries, under the order[s] of Monsieur le Comte d'Angiviller, director of his majesty's [properties]. Mr. Michaux having been well regarded by his work and his discoveries on the culture [of plants] on his trip of several years in Asia, the king has requested that he continue his research in other climates starting with North America; he has received the order to go there and establish a collection of trees, shrubs, plants, and seeds from the New World, little known or unknown in France and of which His Majesty wants to introduce the cultivation on His Properties. Our botanist, having received his instruction[s] from the director general of properties, is getting ready to leave on the new trip, and I hope that he will find that you have arrived in good health in your country. I expect, Sir, from your love of science, because of your affection for France, who misses you, and for your friendship toward me, that you will under these circumstances help in the king's wishes and in the success of Mr. Michaux's mission, both by your wise counsel as well as by that of bright agriculturalists with whom he must be in contact. I pray you to help him in ways that will depend on you, independently of the assistance that will be forthcoming from M. Otto, to whom I have likewise written.

Receive, Sir, the new assurances of my wishes for the general prosperity of the United States and for your happiness in particular.

Franklin welcomed his French visitor and promised help, including letters of introduction for persons in the Carolinas, where Michaux informed Franklin that he would soon travel. Continuing his journey south through Baltimore and Alexandria,[32] Michaux arrived at Mount Vernon, Virginia, on June 19. There he presented George Washington with letters of introduction and gifts of plants and seeds. Washington recorded the visit in his diary, quoted below.

### George Washington, diary entry, Monday, June 19, 1786[33]

A Monsr. André Michaux, a Botanist sent by the Court of France to America (after having been only 6 weeks returned from India) came in a little before dinner with letters of introduction and recommendation from the Duke de Lauzen[34] and Marqus. de la Fayette to me—he dined and returned afterwards to Alexandria, on his way to New York from whence he had come; and where he was about to establish a Botanical Garden.

The letter from Lafayette to which Washington refers is excerpted below.

### Marquis de Lafayette in Vienna to George Washington in America, September 3, 1785 (excerpt)[35]

This letter has been requested of me as an introduction for M. André Michaux whom for many reasons I am very happy to present. In the first place I know you will be glad to know a man whose genius has raised him among the scientifick people, and who, as a botanist, has at his own expense travelled through countries very little known. He is now sent by the King to America, in order to know the trees, the seeds, and every kind of natural production whose growth may be either curious, or useful, and for them the King will set up a nursery at a country seat of his which he is very fond of. I am the more pleased with the plan as it oppens a new channel of intercourse and mutual farming good offices betwen the two nations. I beg, my dear General, you will patronize this gentleman, and I much want it to be said in France that he has been satisfied with his reception in America.

Washington's diary also includes details about planting the seeds that Michaux brought to him as gifts.[36]

### George Washington, diary entry, Thursday, June 29, 1786

Planted in one row between the Cherokee Plumb and the honey locust, back of the No. Garden adjoining the green House (where the Spanish Chustnuts [Chestnuts] had been placed and were rotten) 25 of the Paliurus, very good to make hedges and inclosures for fields—Also in the section between the work House and Salt House, adjoining the Pride of China Plants, and between the rows in which the Carolina laurel seeds had been sowed, 46 of the Pistatia nut in 3 rods—And in the places where the Hemlock Pines had been planted and were dead, Et. & Wt. of the Garden gates—the seeds of the Piramidical Cyprus 75 in number—all of which with others were presented to me by Mr. Michaux, Botanist, to his Most Christian Majesty.

### George Washington, diary entry, Saturday, July 1, 1786

Planted 4 of the Ramnus Tree (an evergreen) one on each side of the Garden gates—a peg with 2 notches drove down by them (Pegs No. 1 being by the Pyramidical Cyprus)—also planted 24 of the Philirea latifolio—(an evergreen shrub) in the Shrubberies by Pegs No. 3—and 48 of the Cytire—a tree produced in a cold climate of quick growth, by pegs No. 4—all these plants were given to me by Mr. Michaux.

After he returned to New York from this journey, Michaux reported a few details about his trip south in two letters dated July 15, 1786.

### AM to Count d'Angiviller, from New York, July 15, 1786 (excerpt)[37]

I visited Pennsylvania, Maryland, and a part of Virginia as far as Frederick Town [Winchester] and the Shenandoah River; I gathered a great many interesting plants, but I did not remain long because the essential purpose is to gather seeds, and because I would have made expenditures for which one could not extract the full advantage, which indisputably can be hoped for in two months. I have been very well received by Dr. Franklin and all the persons to whom I have had introductions. Dr. Franklin has promised me letters of recommendation for Carolina and Georgia. Gen. Washington has offered to allow me to send my collections to his home on deposit, and I will profit on all the occasions by the offers that have been made to me.

I shall set out before the 15th of next month for Carolina and Georgia so that by arriving soon enough for seeds, I can reap every advantage.

### AM to Cuvillier, from New York, July 15, 1786 (excerpt)[38]

Last month I visited Pennsylvania, Maryland, and some parts of Virginia. Philadelphia deserves to be compared to the best cities of Europe after Paris. The first impression is very pleasing because of the regular arrangement of its streets and its disposition; the merchandise of all parts of Europe are there in large quantity. There are more enlightened and scholarly people than in any other areas of this continent. The foreigners complain of the few amenities that are found in the communities because the Quakers who form the largest portion of the inhabitants are very religious and there is no theater in this city. One judges the wealth of the inhabitants of the province by the richness and the strength of the carriage, horses, and carts that bring provisions to the city. In the countryside, the good soil that is clayey and substantial, added to a warm and humid climate, contributes to a great fertility. The inhabitants of this

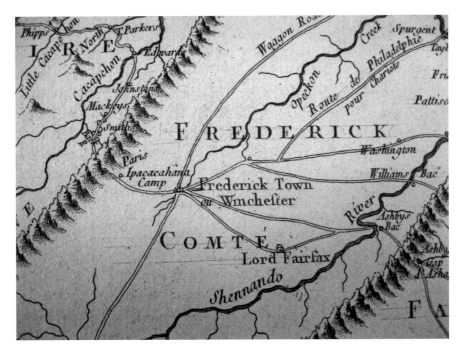

MAP 3.  Detail from French version of Fry-Jefferson Virginia map, Paris, 1777 (Courtesy of the Philadelphia Print Shop, Philadelphia, PA)

province, who are German in origin, well deserve these advantages as there is no group of people that are more hardworking among the different nations that came to form the United States. Maryland is probably the most sterile of all the states. The soil is sandy, and rye can hardly grow there. The regions of Virginia that I saw are likely to be fertile, but they are inhabited by owners of very large properties. They are not very active or hardworking, and the earth is only cultivated in proportion to the number of slaves [that they have].

Michaux noted in his expense report that he visited Baltimore for several days and botanized in Maryland and Pennsylvania on his return trip from Virginia. During the month of July he was occupied with other botanical trips in New Jersey and improvements for the king's garden. He also made the final payment on the new house that he had constructed at the garden and paid for fencing an enclosure.

One indication of Michaux's future plans is seen in the expense entry for a book he described as a "complete atlas of America." At the end of July, he was making final preparations for the journey to Carolina that he had contemplated for several months. It appears that he initially planned to travel overland, because he purchased horses for the trip, but he changed his mind and traveled by sea. This last-minute change of plans no doubt prompted the quick trip to Philadelphia by stagecoach to visit the Bartrams and William Hamilton that is mentioned in his expense report.[39]

It appears from the following letter to Count d'Angiviller that Michaux did not initially plan to change his base permanently from New Jersey to Carolina.

### AM to Count d'Angiviller, from New York, July 18, 1786 (excerpt)[40]

I shall set out within eight days for Carolina and Georgia: this journey, which is long and will cause an absence of more than six months, has obliged me to draw on M. Dutarte for the sum of 5,000 livres, M. Otto being absent. I have had the letter of exchange endorsed by M. De la Forest, Vice-Consul. Having still some funds, I would not have wished to draw now for this sum, but M. de La Forest has assured me that I must not count on finding any money in Carolina, that there is only paper money, which has no value from one state to another, and that service would be lacking. . . .

The confidence with which you have been willing to honor me, Monsieur le Comte, will be a more than sufficient motive, if it were possible, to have more enthusiasm and gratitude. I confess that I cannot desire a situation more fortunate than that of having an immense country to visit, immense collections to make, and for which I would devote all my life, my time, my fortune, and not withstanding that of being furnished with all the assistance possible. I experience with the greatest satisfaction the fatigues that result from this, and the least botanizing procures for me collections for which botanists would go from one end of France to the other. But this satisfaction is often much altered, seeing that I cannot practice all the economy [that] I had flattered myself I could, and that I cannot gather trees and seeds the tenth part of that which I would be able to do with the same means if the inhabitants of this country had the dispositions of those of France. It is manual labor that is lacking among the few inhabitants, and the indolence of those who do live here. There is even in these thirteen states a spirit of migration [among] the inhabitants who have no love of work, and it is the greatest number [who], seeing that the forests furnish less game than formerly and that the lands are better beyond the Appalachian Mountains and the Alleghany [Allegheny] chain, abandon their habitations in crowds to establish new ones in these regions. Those who are naturally industrious are attacked by the same epidemic for going to acquire great possessions because land is sold there for a shilling an acre. In Kentucky there is almost no more to sell. Five years ago one counted only seventeen hundred habitations there, and now, one estimates at thirty thousand the population of that state.

Michaux's initial thinking concerning the development of a southern garden may be discerned in this excerpt from a letter to Count d'Angiviller.

### AM to Count d'Angiviller, from New York, August 19, 1786 (excerpt)[41]

Someone had told him [the consul in Charleston] that I was thinking of establishing, under his orders, a garden in Carolina, but for several months I have realized the inconvenience of an establishment when there is no one trustworthy on the place itself and that is too difficult to meet with. All that one can hope for will be a small garden at the consul's with a French gardener.

In what was likely his last letter before his departure for Charleston, we again have insights into Michaux's work ethic as he describes to his friend and colleague Thouin at the Jardin du Roi in Paris the personnel difficulties involved in establishing and maintaining the gardens in North America.

### AM to André Thouin, from New York, August 1786 (excerpt)[42]

I announce my departure for the Carolinas. Sorry that I was obligated to stay here until today and did not have the opportunity to go there earlier, but nothing is lost for collecting for this year, and I will stay there until February to have [plants] dug up all winter. You will find a little package of seeds in this letter. Pressed on all sides to finish needed tasks before my departure, pressed also by the vessel's departure, I cannot give you all the details that you might wish for. M. de Chateaufort, having arrived here and ready to leave for France, asked me whether I was going to establish a garden in the Carolinas. I told him about the difficulty of finding someone trustworthy to direct these establishments. This would keep me from obtaining all the advantages that it might offer. However if the young Archibal, whose merit you are aware of, were resident of the one in New York, and Saunier, the one in the Carolinas, to assemble the productions of Florida, Bahamas, Georgia, etc., the project would be practicable. I explained this viewpoint to the comte d'Angiviller. Saunier is active and hardworking, he even means well, but since he is obliged to look after and be in charge of the garden, I have often had to reproach him because he forgets the most essential matters, watering, conserving, examining where things are. He collects seeds that are often precious but allows them to get moldy or eaten by rats. I often remonstrated with him but I saw that there was danger in expecting absolutely everything that needed doing and that it could not be remedied, and I was afraid that he would become discouraged, which would be worse, as he is infinitely superior to what could be found here. . . .

The inhabited lands [here] are often poorly maintained because the inhabitants are not very hardworking or [do not] have [sufficient] concern with their husbandry. Hunting, fishing, idleness, time used for drinking; people are so arrogant that the poorest believes himself dishonored to be working for another man. They say that this is only for Negroes and that we should do our own harvest as well as [the] digging of trees. . . .

Michaux closed his letter with a note to let Thouin know that he would actively support his colleague's efforts to enrich the plant collections at the Jardin du Roi.

I will send you at the first possible occasion a list of all the plants that I have collected for my herbarium and will continue to do so, so that you might be able to ask for things that have been forgotten in the shipments.

Various establishments and individuals in France desired American plants from Michaux, although his primary task in America was to send trees and shrubs to the royal nurseries at Rambouillet. He was originally instructed not to share with others, but this was difficult for Michaux who had many friends who desired novelties from North America. Antoine-Laurent de Jussieu would later unfairly chide him for not doing enough for the Jardin du Roi.

At the end of the first week of September, Michaux, along with his son, François André, and his helper, Jacques Renaud, embarked on the brig *Mercury* at New York and sailed to Charleston.

## Part 2. Charleston, September 1786–April 1787

At the beginning of his period of residence in Charleston, Michaux lived in one or more inns in the city, but he neither identified nor located these establishments. He mentioned meeting wealthy people in the Charleston community and their offers of help with his work, but he neither identified any individuals by name nor described any visits to the many gardens in this city. His letters of this period were business communications reporting on the issues that he encountered in carrying out his mission.

### AM to Count d'Angiviller, from Charleston, September 30, 1786[43]

I have the honor to announce to you my arrival here on the 21st of this month, having set out from New York on the 6th. I congratulate myself on having made the decision of coming by sea, for M. Duplessis, brigadier of the king's armies, who arrived the same day as I, to go to Georgia, had set out from New York the 3rd of August. Because the floods made roads impassable and destroyed bridges, he was obliged to embark from Wilmington to come here.

Since my arrival, Monsieur le Comte, I have not lost time; I have given one day to visiting the consul and the governor of this state, and all the other days I have visited the woods where I already recognized a great number of the most interesting trees and a great number of other [plants].[44]

The day that I visited M. Petry,[45] consul of France, because of the absence of M. de Chateaufort, we discussed at length the advantage and our desire that a packet boat should come here, saying that M. Le Jay, the American minister, had requested it of the French minister,[46] but we believed that M. Otto and M. La Forest, residing in New York, were opposed to it. As for me I would find it a great advantage for my shipments, and, if I were notified in advance, I could prepare 200 boxes of trees and seeds and even more if the ship could take them. I would hold ready also several kinds of interesting birds. But as there is no appearance that this can be done this winter, I will profit by the best occasions. The consul of New York and the one from Charleston have advised me to make my shipments by the ships that will be here this winter destined for France. I will be perhaps obliged to take this decision for one fears that the packet boat between New York and Charleston may be interrupted.

The following letter to Count d'Angiviller accompanied Michaux's first shipment of plant material collected in Carolina and includes his initial description of the Goose Creek plantation that became the southern botanical garden and plant nursery.[47] It also reports on the botanist's continuing efforts to collect wild animals for shipment to Rambouillet, site of a royal hunting preserve as well as royal plant nursery.

### AM to Count d'Angiviller, from Charleston, November 12, 1786 (excerpt)[48]

It is difficult to perform my work while staying in Charleston, because the forests are three leagues away at least, because often it is necessary to go ten leagues' distance for certain species of trees: the high prices and the excessive expenses of living here, where one spends ten times more than in the country, in a word, the loss of time, has obliged me to rent a house in the country at three leagues' distance from the city, in the midst of the forest but, however, in the neighborhood of better people, which is important. Along with the house there are also one hundred and eleven acres of land (of which) more than eighty [are] in forests, the remainder in pastures and land for cultivating maize and necessary things for the subsistence of the farmer. If I had not feared to go beyond your orders, I would have purchased it, since this plantation is for sale for only one hundred guineas, and it would have cost me more than a hundred guineas if I stayed in the city with my son and the two horses that I have. However the assistance and the horses are so necessary to me that I do not know how I would have been able to fulfill the purpose for which you have sent me [without them]. If I place my son with your agreement on this plantation, the subsistence for him and some laborers will be taken from the produce, and I am assured of making numerous shipments for you without cost other than the most extraordinary disbursements. The very boxes for the shipments will be taken from the place because it will cost only the manual labor to know how to make the planks, an object of great economy since one could have a box for 40 sous, while if we had it made in Charleston, it would not cost less than 7 to 8 livres French silver.[49] It is the same with all the other expenses. The price is up to six dollars for sending a wagon with two horses three leagues from here and [is as high] for all sorts of [other] expenses. It is necessary to be in a position to do without help from others. That can come about only in the country. There is an infinity of other advantages that the house combines where I am going to establish the home of my son, the domestic that I have brought, and myself when I shall not be traveling into other parts of this continent, from that of having a spring of pure water at a distance of 20 toises [120 feet or 40 meters] from the house, of being able to raise there in the domestic state the summer ducks, which merit more than any other kind of bird being naturalized in France (not only for their beauty and usefulness, but for the ease), [of having] wild turkeys, dwarf partridges, the American deer, and finally the other species, which I do not yet know enough [about] but which will be susceptible of it, and [accomplishing] that [goal] without increasing expenses and, above all, [gathering] collections more abundant than one can imagine.

I have never had enough presumption to dare to flatter myself with something with which I am not certain, and I would not have the temerity to promise you, Monsieur le Comte, all these advantages without having examined the difficulties and the obstacles that could oppose them. I can assure you the completion of a good shipment if you could send a packet boat here. Being informed soon enough, I could pack it with not only 200 boxes but even many more. There is also the plan concerning samples of wood. I just received an instructive memoir from M. Daubenton on this matter, and I have not forgotten what you recommended to me.

The country here is most interesting for the beautiful woods that can succeed in temperate or cold climates. They bring here from Georgia and even from the forests of Carolina that are located on the rivers, some glossy woods with different colors. While waiting for [your] orders, [I will] take advantage of some opportunities that will not be expensive, and as early as this winter I will prepare for you some *Magnolia grandiflora* wood with which to make furniture, a table, a desk, or some others. There is also an interesting animal to become acquainted with for a nation that concerns itself more than all others with extending knowledge. It is the American buffalo. I am as well placed as can be for obtaining it, either from St. Augustine or from North Carolina; it will be easy to tame by retaining it in an enclosure and then breeding it to send to Europe.

When I speak of placing my son on this plantation, it is because I do not recognize in him either the energy or the disposition for any kind of knowledge, and that, being obliged to work for a living, he will learn the necessity for it; in addition, I will not supply him with any help except to place him here, since in this region there are no inhabitants at all who are not very well off.

As for what concerns the service of the king for which you have sent me, Monsieur le Comte, I have the greatest interest in performing it well, to have a depot in the center of the country, the most interesting area of all of America; that is to say, Georgia, South Carolina, and North Carolina, without speaking of the mountains that border these provinces, [and] Florida, where we can go often from here in twenty-four hours by sea. It is near the sea that these regions have not yet been visited; one speaks however of a new species of *Gordonia* or *Franklinia*, a beautiful tree that I have seen in Philadelphia,[50] a new *Nyssa*, [and] an azalea with a scarlet flower. These trees are found on the river Ogeechee and on the Savannah River.

Not only the advantage of uniting easily all the productions of this country, but that of being able to increase them, to cultivate them, and to remain here as many years as the service of the king requires without any increase of expenditures and without any expense for maintenance of laborers, these advantages are beyond what I had hoped. Most of the rich inhabitants of Charleston have made me most welcome as they all [also] have homes in the country; they have invited me to visit them and have promised for the most part all assistance, either [by] getting [me] acquainted with local areas or by having their Negroes accompany me into the inhabited regions. For my part, I have interested them by pointing out to them the many resources that this beautiful country has and discussing with them all aspects of agriculture.

Finally, Monsieur le Comte, I am determined to take advantage of every circumstance and means to acquit myself well on the commission that has been given to me.

### AM to Count d'Angiviller, from Charleston, December 2, 1786[51]

In my last letters I have had the honor to announce to you my arrival at Charleston. I soon recognized that this southern part of the United States is the richest and the most fertile in different productions, and that the weather for collecting seeds is without interruption from the month of May until the month of December. The quantity of seeds that we gathered from [the time of] our arrival and the different kinds that require different care for their conservation made me feel the necessity for a convenient place to store them. Several species are fruits or berries long in drying; others must be put in sand; there are others of which it is necessary to extract the juice such as grapes, etc. I could not perform all that is required in this regard while residing in Charleston because, beyond the excessive price for all kinds of expenses, such as lodging, living, workmen, etc., it was not even possible to find a temporary residence convenient for my purpose.

In addition to the care necessary for the conservation of seeds, it is necessary that one must be on the lookout for collecting those that ripen successively, which I could not even do myself while residing in Charleston. The forests are full of a multitude of birds of different kinds at a proportion of a hundred to one in Europe. They take away seeds and fruits from certain kinds of trees, even before maturity, so that three or four days' delay cannot be made up for the rains and the floods to which this region is susceptible. But the stay in Charleston, because of the distance from the forests, brought new difficulties, such as my two horses becoming sick from fatigue and the rains they endured in places remote from residences, despite all my care for them. I was fortunate to know how to treat them myself in this country, but to make up for lost time it was necessary to buy another one. The necessity to go into remote forests, to return from them daily loaded with trees and plants has often made me feel the necessity not only of [having] a companion, but even of having two horses to increase my collections and continue these hardships. It was with this view that in setting out from New York for Carolina, I had brought my son and put my two horses on board ship.

I desired to find a lodging for some months, as much within reach of Charleston as within the proximity of the forests in order to make my collections and return daily, to give the care necessary for their conservation. All the southern regions of the United States are a great deal less populated than the others, are composed commonly of great plantations whose proprietors do not wish to share lodging. In a word, the searches and the difficulties for a little lodging made me find, and obliged me, to take the one where I am now. It is what they call here a plantation, of which I have rented the house and the land that contains one hundred and eleven acres because one cannot find here any house either for rent or for sale without taking the land that belongs to it. I wrote to you, Monsieur le Comte, in my last letter that this plantation was for sale for one hundred guineas and that I would wait for your orders to make the purchase of it, but subsequently the proprietor, who needed money, threatened to sell it to another; because several purchasers presented themselves and in order not to lose the advantage that would result from this, I resolved to make the purchase. The fear of not having your agreement brought me to ask to pay it [the balance] in six months, of which I paid one hundred piastres on account in order to take this expense out of my fund. But this acquisition is so advantageous that I am persuaded of having your approval. The good bargain of this plantation, situated nine miles from Charleston and between two navigable rivers, is a small advantage in comparison with those [advantages] that will result from my plans. The base of my actions is to make abundant collections with the greatest possible economy. In traveling directly without having a residence [for] five or six months, I would have spent more than a hundred guineas and had no place to deposit my collections [and] could not even find carriage for transport. Through the difficulties or disloyalty that one experiences, my collections would be limited to some samples from which one could not even have profited entirely because there are some seeds that are likely to spoil at sea [and that] could be sent only when they become young plants. In the present situation, I will derive every advantage and economy possible from my travels because I will have a carriage follow me to transport my collections, my instruments, and provisions for sustenance to sites far away from houses. This method will make it much cheaper than any other way.

Michaux's first report of hiring workers at the plantation near Charleston is found in his December 1786 expenses. Four men were paid for work performed during the preceding month and a half. Other expenses included food for the men and horses, as well as the cost of tools and household implements such as dinnerware, cloth, nails, and boards. Michaux moved his horses to this plantation on December 12, and he and his son followed on December 16.[52]

**AM to Count d'Angiviller, from Charleston, December 26, 1786 (excerpt)[53]**

I have already finished the fences in the place of cultivation that I announced to you, Monsieur le Comte. I have worked unceasingly, and despite the expenses of a new establishment, which consist of repairs, purchases of work tools, living expenses for men and horses, which the following years will be taking out of the products of the land, it will be nevertheless a great deal more economical than if I had remained in the city. It is true, Monsieur le Comte, that despite the solitude and effort that I have borne, the quantity of my collections will not equal this year those that I can make the following years, but if I had remained in the city they would have been fewer and more expensive.

The excess labor to which I have been subjected since my arrival, without being assisted except only rather weakly by my son, had obliged me to speak poorly of him to you. He has been very aware of it, and from that time on the firmness and strictness that I have been obliged to use has produced good results. I am happy with him for he seems to do all that he can and is interested in our affairs, and this is what I principally desired.

In addition to the seeds of this shipment, there are several others that I could now send you but for the difficulty of drying them; you will receive them, Monsieur Le Comte, only at the end of spring. . . . The work with which I am entrusted is very interesting, but there is a great deal to do, for every day I recognize something new, which will not be neglected because I will take note of everything so as to profit by it when circumstances permit.

In January 1787 we find that Michaux not only paid wages to four workers, but he also paid for the rental of two enslaved black persons. The following two months, February and March, he possibly rented an additional enslaved person. Some of the wages were paid to carpenters, and the highest paid workman was a skilled free black carpenter. Michaux had a new frame house built on the plantation, installed fencing and cold frames in his garden, and had boxes and crates constructed for his shipments of seeds and plants to France. Other items listed on the expense report indicate that Michaux equipped the plantation as a working farm, purchasing farm animals and tools.[54]

While he was making repairs and developing and equipping the plantation, Michaux continued to make collections and shipments to France. The letter accompanying a shipment of six cases in February mentions also that he had found the flowering trees *Styrax* and *Stewartia*. Despite the botanist's efforts, accidents and problems did occur with his delicate cargoes. A very large shipment of twenty-one cases of plants destined for Bordeaux in March 1787 came to grief when the ship ran aground attempting to leave Charleston Harbor. The shipment was not lost, but it was damaged despite Michaux's exertions to protect it.[55]

Michaux's March 1787 expense report includes an item for "brandy, ham, cheese, and provisions for the trip to Georgia," indicating preparations for his upcoming journey. It is also likely that his expenses listed in March for powder and lead shot as well as a barrel of biscuits were part of his travel preparations. His travel plans were flexible; depending on the conditions he encountered, he was ready to change destinations and thus not waste valuable time.

### AM to Count d'Angiviller, from Charleston, April 15, 1787 (excerpt)[56]

I will leave today for my trip to Georgia and will do the best that I can to go as far as Florida; however, if there are problems, I will continue my trip toward the mountains that border Georgia and the Carolinas.

I hope to be in New York in July and return here as soon as possible unless you send me other orders.

The problem that Michaux refers to was Georgia's ongoing conflict with the Creek Indians, which would lead him to travel to the mountains bordering the Carolinas and Georgia, where the peaceful Cherokee lived, rather than attempt to travel along the Georgia coast to Florida where the Creek Indians were to be feared.

Between April 2 and April 15, 1787, Michaux dispatched at least six letters on three different vessels to either Count d'Angiviller or his secretary, Cuvillier. In one letter he reported to Cuvillier that he had not received replies to any of the letters he had sent from Charleston since September 1786. Other letters outlined the care of the ducks he was sending to France, his problems obtaining funds from the French consul Petry at Charleston, his actions after the ship ran aground in March, and of course his plans to travel to Georgia. Perhaps the most useful of these letters, however, is the following, which he directed to Count d'Angiviller in April 1787. In this letter he repeats his reasons for purchasing the Charleston garden property and lists some of the improvements he had made. Michaux's enumeration of the improvements allows us to form an idea of what this property was like in the early days of his occupation of the site.[57]

### AM to Count d'Angiviller, from Charleston, April 2, 1787[58]

I have the honor to address to you, here attached, the account of my expenses from my departure up to the end of March. I have abridged as much as possible the details so as not to bore you. It is my desire that you, Monsieur le Comte, recognize the necessity and the circumstances that have often obliged me to spend more than I would have wished.

My unceasing purpose has been to respond to the confidence you have honored me with, Monsieur le Comte. In all my expenditures or undertakings I have considered seriously and for a long time the advantages that should result, and I have acted only on the certainty of [the advantages].

You will see, Monsieur le Comte, by the statement of expenditures that the expenses were rather considerable during the first three months after my arrival at Charleston. They have been infinitely less during the three following months if one considers that there were always at least four workers employed, that one pays them very dearly here, more dearly even than in the French colonies.

The work has been, first, to repair the old building and to construct a four-room house.

Second, the enclosure of an acre-and-a-half garden in rough boards and an enclosure for wild turkeys, dwarf partridges, summer ducks, etc. The enclosure of fifteen acres of land to sow in maize and potatoes for the nourishment of horses and men.

Third, a new road across the marshes and a bridge over a continual stream.

Fourth, the collection of seeds and trees for which it is often necessary to go great distances away from the place to dig up. The care for putting seeds and trees in condition to be shipped, despite the vagaries of the rain and the seasons, [and] lack of lodgings there. To make the shipments by all possible occasions.

Fifth, the cultivation and even the clearing of several acres of land for the nursery. The working and sowing of 15 acres of land for the subsistence of workmen and horses on this habitation.

All these projects have been carried out with less expense than if we had resided, my son and I, with two horses in the city during this space of time.

If we had traveled, it would have cost more, for everyone agrees that one cannot travel in these southern regions without having three persons, and if one were always obliged to sleep in inns, it would cost no less than five or six piastres a day. On account of these difficulties and the advantages of being able to now go to make abundant collections and to make journeys taking carriages and horses from one's own home, carrying provisions in a vehicle so as to sleep in the forest as do the inhabitants in remote places; this is what has decided me to look for a residence that could procure these advantages. There will be economy through the decision I have taken in buying this land, not only because a house for one year's rent in the city costs often more, but also because the value of the lands situated in the proximity of Charleston increases every day.

Before this winter the maintenance of the horses has been the most expensive item because it was necessary to purchase everything. From now on, the maintenance [of the horses] will be taken from the produce of the land and will cost nothing.

The detail of my expenditures, which I shall send to you every quarter, will prove to you that, beyond the advantage of gathering a great number of plants, economy will have been my purpose in this acquisition.

With his travel preparations complete and his correspondence now up to date, André Michaux was ready to begin his first extensive collecting journey in the south.

## CHAPTER TWO

# Initial Journeys from Charleston, 1787

SUMMARY: SIX AND A HALF months after his arrival in Charleston, Michaux began a plant-collecting journey that he recorded in detail in his journal. Heeding the advice of local residents against traveling in the backcountry alone, Michaux set out with a party consisting of his teenaged son, François André; Scottish nurseryman John Fraser; and an enslaved black man. They headed south from Charleston toward Savannah, Georgia, along the route of modern US Highway 17. This was a prosperous and long-settled part of South Carolina, home to many great plantations and some of the best roads in the state. Michaux reached the town of Savannah after ten days and continued the journey to Sunbury, a seaport on the Georgia coast south of Savannah. He made inquiries about the route to St. Augustine, Florida, but found that travel to Florida was unsafe because of hostilities between Georgia and the Creek Indians. Michaux then returned to Savannah where he began his journey to the mountains that border Georgia and the Carolinas.[1]

From Savannah, Michaux traveled to Augusta on roads near the Savannah River. Before reaching Augusta, he lost his horses and parted company with John Fraser. A few miles north of Augusta, he crossed the Savannah River into South Carolina and soon passed through the French Huguenot settlement of New Bordeaux. Continuing toward the mountains, he met Gen. Andrew

Pickens, one of the leading men of the district, then resumed his journey north to the Cherokee village of Seneca located at the present site of Clemson University. At Seneca he employed Cherokee guides to lead him up the Keowee River and its tributaries into the wild, roadless mountains of North Carolina. After crossing the four-thousand-foot Highlands Plateau, Michaux's party wandered westward to the Little Tennessee River near the modern border of Georgia and North Carolina. They then returned to Seneca on Cherokee trading paths leading through the northeastern corner of Georgia.

Following this wilderness journey, Michaux returned to Gen. Pickens's plantation, then continued to Charleston along a direct route across South Carolina following roads generally paralleling the south fork of the Edisto River. Soon after arriving in Charleston, he took a ship to Philadelphia where he continued to New York by stagecoach. He briefly visited the New Jersey garden and met with French officials in both New York and Philadelphia before returning to Charleston from Philadelphia by ship. Michaux spent the remainder of 1787 either in Charleston at the garden or on short collecting trips in South Carolina.

### Notebook 2 begins

THURSDAY, APRIL 19, came from Charleston to the plantation. I left the plantation and went to sleep at Ashley Ferry: 10 miles [away].[2]

FRIDAY, APRIL 20, *Styrax angustifolia* and [*S.*] *latifolia* in bloom, *Nyssa aquatica* in flower, and *Sarracenia lutea* and other species.[3] I came within one or two miles of Parker's Ferry, 32 miles from Charleston.[4]

APRIL 21, on the river's side entering the woods, on the left, before coming to the ferry called Parker's Ferry on the Edisto River, found a *Gleditsia* . . . three species of *Mespilus*;[5] and a little shrub with milky juice, flowers in clusters, not [yet] expanded, and several fruits from the previous year, resembling *Tythymalus* (*Stillinga*); and a pine with two needles.[6] While continuing on the road to pass by Ashepoo Ferry, I found several pines with two needles. We came to spend the night two miles beyond Ashepoo Ferry.[7]

SUNDAY, APRIL 22, I passed the ferry named Combahee Bridge that is found 10 miles [south] of the preceding on the Combahee River.[8] A little before arriving at this ferry between [it and] Mr. Dais's plantation,[9] I collected a bulbous plant, flower with usually 2 spathes calyx tubular divided equally in six segments, 6 stamens whose filaments being very long are exserted beyond

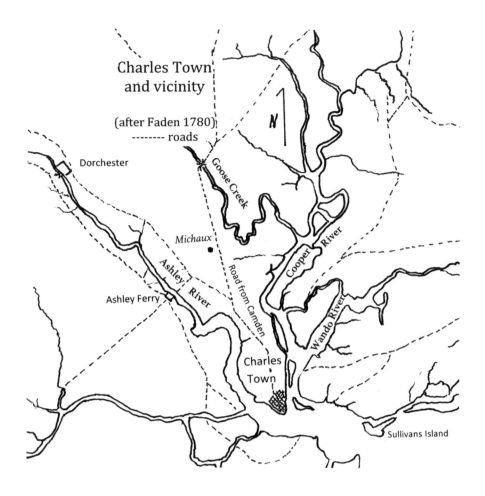

MAP 4. Charleston area in the 1780s (Drawing by
Charlie Williams, based on 1780 Faden map)

a type of corolla (white) nectariferous, long pistil, ovary inferior.[10] After having passed the ferry by several miles, we found abundant *Nyssa* with large dentate leaves.

APRIL 23, we walked 13 miles, and we passed . . . We went through several prairies continuously washed over by the sea, which produced only rushes.[11]

After arriving at the camping site, I gathered *Verbena** with long flower, calyx 5 parted, laciniate, subulate. Corolla almost irregular, long tube, mouth of the corolla velutinous and the tube under the stamens also velutinous. Stamens 4, of which 2 are shorter. Pistil 4-angled, style the same length as the tube. Flowers in spikes; leaves opposite, pinnatifid.[12]

* Michaux's footnote: *Verbena aublatia*.

April 24, our hike was 12 miles long, and we camped seven miles from Two Sisters.[13] We walked in the woods all the time; we only saw 3 plants found at some distance from the road and in low, less sterile areas. In general, one can't travel in an area that is more sterile. The forests are composed of pines. In wet places, I saw *Nyssa aquatica*, *Cupressus disticha*, and *Gordonia lasianthus*.

April 25, we traveled eight miles, and we lodged at the ferry house on the left bank of the Savannah River, situated equidistant between Purrysburg and Abercorn.[14] For five or six miles the ground continued to be dry, with only a few pines growing, and cypress in the wetter areas.

Three miles before reaching the ferry [the soil] appeared clayey, and then we found sandy soil—but uneven and hilly ground with many plants that I had not seen previously. I collected two species of lupine, namely, *Lupinus perennis* and *L. pilosus*;[15] two species of verbena, namely, *Verbena . . .* and *V. carolinana*. Two species of *Asclepias*. Several species of *Tythymalis*. In the swamps, *Nyssa* with dentate leaves, *Stillingia sylvatica*, a species of *Annona*, etc. Upon arriving on the river's edge, I saw *Sideroxylon tomax*, a *Ligustrum* (*monospermum*?),* two species of *Annona*,[16] *Magnolia grandiflora*.

In submerged sites, *Betula papyrifera*,[17] *Platanus*, *Gleditsia*, *Nyssa*, etc. etc.

In the morning I killed a very beautiful snake with yellow bands, black and red; these snakes were active and well marked. I killed three snakes of the type called moccasins; one was 3 feet 9 inches [long] and 8 inches in circumference. My son killed another snake called black snake. This is the enemy of the rattlesnake, and although it is not venomous, it is successful in killing him uniquely by the speed of its course (if we can call a *course* the progressive action of an animal that has no feet). The rattlesnake drags himself heavily and rarely goes far from his hiding place, which is ordinarily cavities formed by the roots and soil of a rotten or fallen tree. When the black snake meets his adversary, it moves speedily over his body and beyond. It comes back with the same speed and continues until the rattlesnake, by its repeated efforts to bite his enemy, bites himself and thus gives himself the kiss of death by the venom of his bite.[18]

April 26, after having looked around for one mile on this river, we crossed by boat to the other side; we went for about 4 miles to an area bordered by hills and covered by woods where I collected *Dirca palustris*,[19] [and] *Kalmia* that differs a little from [*K.*] *latifolia* by the flower color. An *Azalea coccinea*, whose flower color is dark red throughout. Although this color is not very

* Michaux's footnote: *Olea americana*.

vivid, this shrub will be a most agreeable addition to the ornament of gardens.[20] It seems to be related to *Azalea nudiflora*.[21]

I collected *Silene virginica*; I saw many *Chionanthus*, a *Magnolia* with flowers of the size and form of those of *M. tripetala*, very agreeable odor instead of which in *M. tripetala*, the odor of the wood is pleasant but not that of its flowers; it differs from [that] *Magnolia* by its leaves, which have 2-inch-long petioles and are cordiform at the base; leaves are long and apex is 3 angled.[22] A very large shrub that I think is *Andromeda arborea*; it was not in flower, but the cluster of fruits from the past year and the acid taste of the leaves allowed me to conjecture that it might be *Andromeda arborea*.[23] We passed the ferry at around two o'clock in the afternoon, and we found the roads so bad that we only traveled 2 miles in five hours. One had to pass through sludge up to our horses' knees and sometimes in water. In a spot where the bridge had collapsed, the horses had to swim across.

APRIL 27, we found the soil rather dry, except in the ravines or streams that do not run continuously. I collected the *Azalea* with the color of fire. The color of this *Azalea*, that is of uniformly dark color, corolla, stamens, and pistil, is that of *Hemerocallis fulva*, but in areas more open and less shaded, this color is even more pronounced. After having walked . . . miles we arrived at a poor hamlet called here a town, composed only of 4–5 houses. This area is called Ebenezer.[24] One mile from this spot, always following the road to Savannah, in the low areas covered with *Betula papyrifera* and near a river named . . . I collected *Gleditsia* [with a] one-seeded indehiscent oval capsule; I collected several remarkable plants, a species of *Asclepias*.* I collected in this site another small climbing shrub having much in common with *Bignonia sempervirens*.[25] A *Polygala rosea*? An *Astragalus*,[26] etc. *Arethusa divaricata* and *Arethusa ophioglossoides* and another species that I named *Limodorum*.[27] (Note: having found *Arethusa bulbosa* near New York, in addition to the three species indicated by Linnaeus, I have acquired a fourth). We slept in a plantation inhabited by a Dutch woman who furnished us with some provisions and permission to visit the woods, where I found a variety of *Halesia* [that] someone has named *diptera*.

APRIL 28, we traveled 12 miles.

SUNDAY, APRIL 29, we traveled nine miles, and we camped near Savannah.[28]

---

* Michaux's footnote: *Apocynum cannabinum*.

APRIL 30, we stayed in Savannah. In the morning I botanized; I saw a species of palm, different from the *Chamaerops* of Carolina,* having a stem instead of the one near Charleston that has no stem;[29] its leaves come out from the inner growth, as does the stem, which bears the fructification, the same as in *Osmunda cinnamomea*. The leaves also differ, and I will speak of them later. I soon returned to town and spent the rest of the day making visits.

MAY 1, 1787, the day was spent obtaining provisions necessary to continue our journey. The town of Savannah is composed of about one hundred and fifty houses situated near the river with the same name, on a hill formed by the sands accumulated by the winds. The town is arranged regularly, but the few houses built around this layout do not allow the perception of this regularity of which the inhabitants boast. The streets are very wide, and the ground, which is formed by moving sand, adds to the heat and the discomfort that one feels in a climate that is always very hot.[30]

MAY 2, we traveled 12 miles, and the rain obliged us to sleep in a small, uninhabited house that was near the road. I saw more *Magnolia grandiflora* than I had seen earlier. *Nyssa* [with] leaves [of] acute dentate and a *Tradescantia umbellata* with flowers pink.[31]

MAY 3, we walked 16 miles. We passed the Ogeechee Ferry[32] situated on the Ogeechee River, and a mile before arriving at this ferry I found *Nyssa ogechee* of Bartram. This tree could be considered a shrub if it did not differ in height in other locations. It is closely related to *Nyssa* with leaves acute, dentate, but the leaves are oval and completely entire [lacking teeth or lobes], velutinous on the lower surface. On the border of this river, in the inundated areas and amid the willows, I collected *Zizania palustris*: 6 stamens in the male flowers, and the female flowers are separate although on the same plant. At the same time I collected *Pancratium mexicanum* among the willows, in very wet and sometimes even submerged areas.

MAY 4, we traveled eight miles, and I saw nothing of great interest. Rain from the previous night had delayed our departure.

MAY 5, we traveled six miles, and we found in abundance an *Andromeda* that I will name *ferruginea*, a *Kalmia* (?) creeping, whose leaves are hairy. The flowers are very late. After searching I found a kind of flowers whose stamens differ from those of other *Andromeda*. I collected an *Arum* with spotted stem,

* Michaux's footnote: *Chamaerops recurvata caule.*

but the spathe was as white as a fleur-de-lis, and [I] collected another plant of the *Annona*.*

May 6, we stayed at Sunbury, and we tried to determine the ways to go to St. Augustine, but we returned to 6 Mile House.[33] That same day my son left with a servant and an English traveler to visit the borders of the Altamaha River,[34] and I went to stay in an inn at 6 miles from Sunbury because of a problem with my leg that had worsened for several days. This trouble was caused by the bite of an insect that these woods are full of, and the continued friction of the horse against this part [of my leg] produced an abscess and considerable inflammation.

May 7, I visited the surroundings on foot. I busied myself describing several plants for which time had not been available in preceding days. May 8 was occupied in the same way.

May 9, same occupations, I took count in writing of the number of plants collected. I organized my herbarium by orders.[35]

May 10, I left for Augusta, and we walked 25 miles. We passed the Ogeechee River.

May 11, we walked 25 miles, and we stayed overnight at Fifteen Mile House: 15 miles distant from Savannah.[36]

May 12, we traveled six miles, and we camped at 21 miles from Savannah and about four miles from Ebenezer. A little river that passes this spot at the base of the prairie where we camped allowed me to obtain a collection of *Halesia diptera*, of which I had always been doubtful until now; I collected *Populus heterophylla*, a small tree with opposite leaves; the ripe fruit [that] had fallen for the most part resembled that of *Viburnum*. A *Mespilus*?, a very large treelet with red fruits on the hill that borders this river; *Zizania palustris*, *Chelone glabra*, *Gleditsia aquatica*, *Vinca* lutea? Near nightfall on a creek that borders the house of a Dutch widow, I saw several *Halesia diptera*, a large shrub, and in this creek many *Zizania palustris*.

Sunday, May 13, we traveled 15 miles, and we camped on top of the chain of hills that borders the Savannah River across from the ferry called Two Sisters. I found at this spot *Andromeda arborea* ready to bloom.

* Michaux's footnote: *Annona lanceolata*, calyx 3 segments, petals 6, 3 inside nectariferous, stamens numerous, pistils 5.

MAY 14, our journey was nine miles. We stopped with Capt. Prevott, the son of an old Frenchman.[37] He took me into a part of the woods that abounds in *Annona*, from which he made a rather strong rope by beating the bark.

MAY 15, in the morning, we noticed that our horses had been stolen an hour earlier. According to custom, whenever we found a good prairie too distant from houses, we camped near a well, and we put a bell on each horse. I had put into effect all these precautions. Apart from this, it was customary for me to get up several times during the night. I saw them at 3 o'clock in the morning, and at 4:15 they had disappeared.

We looked for them all day, and we sent for information in all areas. The inhabitants of the region told us that they had been stolen. We encountered two individuals who were running after a certain captain known in the area for stealing horses.

MAY 16, we were occupied with the same searches and went to sleep only four miles farther at an inn.

MAY 17, we sent letters to different parts of the district, particularly to the captain, Maj. Revots, and to Savannah. Finally, I resolved to continue on foot with my son, and we slept only three miles from the inn. The owner of the inn where we spent the night promised us, in return for remuneration, to make every effort to recover them if they only went astray, and the 18th we spent part of the day searching for them. We stayed the night four miles from where we started.

MAY 19, we traveled 14 miles, and we camped near a bridge on Beaver Dam Creek River.[38] A little before arriving at Beaver Dam, I collected on the road, at around 60 miles from Augusta, a *Rumex* shrub that I will name *Lapathum occidentale*, a large shrub 15 to 30 feet high. It is found near the Altamaha River, where my son had brought it to me several days ago.[39]

SUNDAY, MAY 20, we walked four miles because of the rain and went to bed in a little house found near the road. The soil is very sandy and sterile.

MAY 21, we traveled 10 miles, and we camped near an inn found 45 miles from Augusta. The soil changes at this site; it is clayey mixed with sand, in some parts ferruginous. There are several hills on which I recognized *Calycanthus* and *Robinia hispida*. We became certain that our horses had been stolen; an individual from the area from which they had been taken had also lost two of his; [he] ran after a certain captain known in the neighborhood to steal horses. He caught and killed him. His accomplice, who had taken ours, got away and took the road toward the Creek Nation.[40]

MAY 22, we hiked 10 miles not counting the hikes that we were obliged to take away from the main road whenever we saw hills or swamps or other soil variations that harbor different plants.

MAY 23, we traveled only two miles as going down a hill a wheel broke from the cart that served to transport our collections and our provisions.

MAY 23 and 24, were employed in visiting several hills of the district, and I recognized at this site *Trillium cernuum* and [*T.*] *sessile*,[41] *Cypripedium calceolaria*—yellow flower, *Calycanthus . . .* , *Zanthorhiza* or *Marboisia* and . . .

MAY 25, we traveled 12 miles toward Augusta. We saw dry sandy soil, except for a portion that was very wet, and we were obliged to cross water that went up to our knees, and the remainder was a creek that we had to traverse on a medium-size tree just above the water level at the risk of being attacked by the alligators that were numerous in this site.[42]

MAY 26, we traveled ten miles, and we passed a little river whose bridge had been broken by the overflowing water, it was necessary to work in the water to repair it so that we could pass with our cart.[43] We finally arrived in Augusta. Alligators or caimans are in abundance in the rivers, torrents, and swamps of Georgia and even in the Carolinas. We did not see any here and would have been very troubled because we had to spend over 3 hours in the water to repair the miserable bridge where we had to pass.

MAY 27, we stayed in Augusta. They are so strict in America that one does not dare to go out or even take a walk on Sunday in the big cities.

MAY 28, I went to visit Col. Leroy Hammond, whose home is at 3 miles from Augusta in South Carolina, as we are in Carolina as soon as we have passed the Savannah River on which Augusta is situated. When I came back [later] the same day the colonel was not at home, although I received all kinds of courtesies from his wife.[44] There were also two young ladies, his nieces, who were very polite, and this house appeared very distinguished in all ways for its good manners, its riches, and its elegance. A lawyer from Ninety-Six took it on himself to give me a letter of recommendation for the District of Keowee, where I proposed to go.[45] I continued to follow the river to return to Augusta, and I collected a *Pavia* (*spicata*),[46] a new *Vaccinium . . .* , *Aquilegia*? *Tilia*, *Annona*.

Augusta is one of the most agreeably situated cities of all North America, but with few houses. Three years ago there were only a dozen, and now there are 120; the most essential supplies are lacking because the inhabitants get supplies only for themselves.[47] Most inhabitants are idle, playing cards and imbibing rum, of which inhabitants of all ages and of all ranks in America drink to excess.[48]

Some English businessmen have warehouses or stores for commerce for things necessary to inhabitants living in remote portions of the Carolinas and Georgia.

MAY 29, rain obliged us to stay all day without being able to leave Augusta. We were informed in Augusta that a certain Mr. Fraser, a Scot sent to collect ornamental shrubs for English nurseries, had lost his two horses. This man had left Charleston with me and had sworn to follow me everywhere that I went. I had accepted his company because, [his] being English, I had hoped that he would have more resources to obtain necessary items in these southern underpopulated areas. But his lack of knowledge in natural history, in areas he wanted to collect, especially plants and insects, made him collect in large quantities objects of little value and already well known such as *Prinos glaber, Ceanothus . . . , Styrax*. He lost a lot of precious time that he could have used to collect more interesting items if he had been able to recognize them. I was tired of his continuous questions and of his ignorance that in addition to his low confidence led him to collect endless numbers of abnormal productions that are found in larger numbers in America compared to the old continent because of the humid climate. I always traveled with him on good terms, but having lost my horses 12 days before arriving in Augusta, I took advantage of the circumstances to tell him that since I planned to look for my horses, he should not wait for me and [should] continue on his trip. From that moment we separated.[49]

MAY 30, we left Augusta, and we only walked five miles because it rained all day. I had no business in Augusta, but the difficulties of finding provisions had obliged us to stay there three days. There is only one baker, but it was not possible to persuade him to make us bread for several days because he was afraid that he wouldn't have enough. He did not even want to sell us flour. We were unable to find any [bread] at other merchants, nor was it possible to purchase corn for the horse or cornmeal as provision for us.

MAY 31, we traveled 12 miles by a new road full of stumps through the woods. We saw several plantations at which we asked to purchase cornmeal because we could not obtain bread. An honest planter five miles away from Augusta let us have approximately six quarts. His name was Mr. Pece, and he received us very civilly because we were French. He treated us to free milk and other provisions. He told us that the large number of farmers who had arrived from Virginia, Maryland, and other northern parts to establish themselves in these isolated parts of Georgia had made the price of corn so high that famine was feared. In effect this staple is of primary need as no wheat bread is available. The inns are few, and one is obliged to sleep in the woods.[50]

June 1, we walked nine miles.

We passed Scot's Ferry on the Savannah River, situated at 21 miles from Augusta.[51] After crossing the river, we walked 5 miles without seeing a single house, and the road through the woods was little used.

The soil is clayey and reddish, and one frequently finds blocks of pure quartz; there are also pieces of mica, and once I recognized clay schist. Two miles from the river, the soil is wet and often submerged, but no alligator is seen.

June 2, we traveled twelve miles without seeing a single house or finding water. The road was hardly cleared. I killed two black squirrels and two birds: one was a woodpecker, and the other appeared to me to belong to the chaffinch, large beak, yellowish plumage; I found in its gizzard debris of scarabs.[52]

I found no new plants. The woods were composed of pines with 2 needles, black oak,[53] white oak, *Diospyros*, etc.

In a wet place I found *Andromeda arborea*; a small stream nearby prompted us to have our supper and to pass the night there.

Sunday, June 3, we traveled 10 miles. The difficulty that we encountered while passing through a creek delayed us more than an hour and a half because we had to empty the cart and [because] we had to transport all our belongings, books, and dried plants by horse.[54] We passed by the French establishment called New Bordeaux. The houses are far one from another, and I only visited one.[55]

The French people of this area are generally well regarded for their honesty and their proper lifestyle. The soil here is good; it is generally clayey, reddish in color, and we can find blocks of quartz adhering to the soil, unlike in those areas where I was yesterday, [where] they were isolated and not an integral part of the soil. In the streams one can find only quartz and mica. I found on the borders of the streams *Dirca palustris* and *Andromeda arborea*.

June 4, we traveled 16 miles; we saw a rather sparsely inhabited country, and even two plantations were abandoned; we were nevertheless rather happy to find a woman in a poor house who sold us 3 pounds of butter, treated us to milk, and made us bread with the corn flour that we had. She added wheat flour and yeast; thus we received a very good bread. At night we approached a place with more people called . . .

The soil was generally rich in iron and clay and did not produce grass that made the horse suffer a lot; the woods had been burned everywhere that we passed. We finally arrived at a place with grass and a source of water. We met several inhabitants who were returning from church.[56] They told us that we would find a country more inhabited, that we shouldn't worry about our horses in these parts, the inhabitants of this area having all the principles of integrity, good morals, and religion, 300 of whom had on this very day received sacramental communion, and [who] did not allow strangers or adventurers who lacked good morals to move in among them. There was among them a rich planter named Squire Coohm who was much respected by others.[57] I found near the water source many *Andromeda arborea,* and I measured one to be 2 feet 6 inches in circumference at 3 feet above soil level.

June 5, we arose at three o'clock in the morning to pack up and seek shelter under the cart from a thunderstorm and heavy rain. The weather improved around noon, and we traveled four miles. We arrived at the plantation of Gen. Andrew Pickens, for whom I had a letter from Col. Leroy Hammond near Augusta; he welcomed us very courteously, and we spent the night at his house.[58]

June 6, we traveled 17 miles. We stopped at the home of Capt. Middle, seven miles from Gen. Pickens.[59] I ordered some wild turkeys from him, and he promised me that after making some inquiries, we would come to terms about the price on my return.

We spent the night at the home of a planter named Th. Lee near the Rocky River.[60] The soil was clayey, and the stones or rocks I found were quartz. Sometimes I found granite composed of quartz, mica, black tourmaline, and iron ore.[61] I noticed an owl of the large species typical for the Carolinas, and I killed it; it plunged down with a black snake, a species of coach whip.

June 7, we traveled 15 miles, and we spent the night at Deep Creek. The area was getting more hilly.[62]

June 8, we traveled [a total of] 15 miles and arrived in Seneca. After ten miles we crossed a creek.[63] On its banks I recognized *Epigea repens, Kalmia latifolia, Panax quinquefolia*. I took a walk on the same evening on the river-bank near Fort Seneca, actually Fort Rutledge.[64] This river is called Keowee; it is deep in several places, and in other areas rocks reach the surface of the water; I collected *Hydrangea arborescens*, and I noted *Cornus alternifolia, Kalmia latifolia, Zanthorhiza* or *Marboisia, Panax quinquefolia*.

June 9, we went with a Frenchman named Mr. Martin,[65] who was an established planter in this area, to hire two Indians to accompany me in the mountains that separate the state of Carolina from the Indian nations of the Cherokees, Creek, Chickasaw, etc.

The Indians were very reluctant to agree to accompany me; not only was their price exorbitant, but also they wanted a horse for the two of them. It was even more difficult to get an interpreter, and I decided to go with one young man and the two Indians that I wanted. I arranged a meeting for the next day to conclude the agreements, and to persuade them to keep their word, I promised them half a gallon of rum. I passed by a place abandoned by the Indians and that had been the location of the town called Seneca. I noticed *Gleditsia*, with which they nourish themselves, peaches, and wild plums. I collected a black oak that I had not seen previously in the Carolinas or Georgia.

Sunday, June 10, the Indians came with their chief and several others of the Nation. After I had made them understand that I wanted to visit the sources of the Keowee River and the Tugaloo River that unite, forming the Savannah River;[66] [and that I wanted to visit] those [streams] that form the Tennessee that empties into the Ohio; and that I wanted to go as far as Tennessee; each of them asked for a blanket and a petticoat, [at] the price of six dollars for each for the 12 days that the journey was to take. I agreed, but half had to be paid in advance because many other white men had deceived them. I promised them that if I came back satisfied with the journey, I would fill their bellies with rum. They were very satisfied and told me that they would be waiting for me the next day whenever I was ready to leave.

June 11, several honest residents of the area who took an interest in my journey provided me with provisions; one had bread made for me and had cornmeal ground, another sent me corn, lent me a horse and mount, etc. I left with a young man who had lived with the Indians for five months and went to the rendezvous we had arranged, and at noon we set out with the Indians

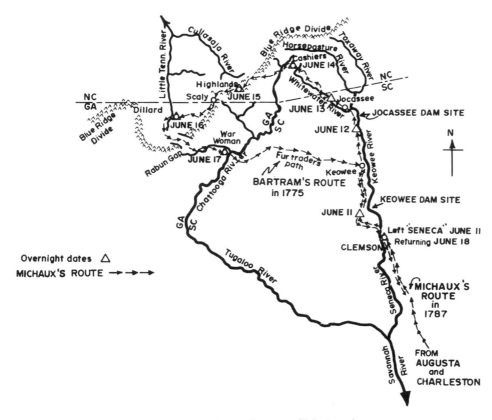

MAP 5. Zahner and Jones's map of Michaux's 1787
wilderness journey, reproduced in *Castanea*, 1983
(Courtesy of the Southern Appalachian Botanical Society)

to whom I had given powder and lead. They guided me alternately through
mountains and torrents, which are called creeks. This same day we passed
through very steep places, and we crossed a small river called Little River; it
is very swift, and I became fearful when I saw that we had to pass on rocks
that were one foot or sometimes even two feet below the level of the water.[67]
The current was so swift that anyone but an Indian would have been carried
away. These rocks were sloping and covered with slimy moss. I feared the fall
of one of our horses, but there was no other way, and the Indians would not
indulge in listening to reflections that I might make under the circumstances.
The deep creeks and the riverbanks were covered with *Rhododendron maximum*. Our day's journey was twelve miles. The Indians took advantage of the
remaining daylight to go hunting, but since they did not kill anything, all the
bread that had been baked for us was eaten that very day.

June 12, the Indians went hunting at daybreak, and because they didn't get anything, we ate some of the cornmeal boiled in water. At noon we took a short stop to rest the horses and to drink from a brook whose water was the purest and best that one can drink in America. Following the example of my two Indians, I soaked the cornmeal in that water, and that was our dinner. The poor food and the bad roads didn't bother me as much as the lack of success in finding any interesting plant since May 8, and I was preoccupied with the unpleasantness of such a journey without results.[68] That day we traveled fifteen miles through mountains full of rocks and had to pass through deep brooks, through marshy areas filled with *Smilax* with horrible spines that wrapped itself continuously around the face, the body, or the legs. I saw along the river very fertile plains. At three different sites the Indians showed me the location of abandoned villages whose names they told me.[69]

June 13, shortly before crossing the Keowee River, which we had followed on our right, going upriver, one of the Indians killed a wild turkey, and at 10 o'clock I found a dioecious shrub whose fruit is in the shape of a pear, a superior calyx with 5 very short sepals; it [the fruit] had not formed yet but was developed enough to see one stone inside. I said that this shrub is dioecious because I saw some of these plants whose flowers had gone where only the raceme remained. The female individuals bore their fruits in 4s and 5s also on similar racemes.[70] I saw several *Magnolia acuminata*; it was the first time that I saw this tree in America.

The Indians killed a deer, and while they were dressing it, I visited the creeks where I saw in abundance *Kalmia latifolia* and *Rhododendron maximum*.

That day we traveled nine miles, and we were all too starved to continue, especially since we had made such a great capture.[71]

June 14, we continued, still keeping the river on the right, and we had to go over rocks and alternately climb over huge trees that had fallen over thick bushes where we had trouble finding our way because of the dense thicket, the high mountains we were approaching, and the darkness produced by an overcast sky and the fog that seemed to envelop us like night. The trouble and confusion were increased by the waterfalls of this river falling over rocks, and those of several creeks that we had to ford with water up to our knees. The speed with which the two Indians crossed the torrents, sometimes in the water, sometimes over trees, which hampered our passage made it difficult to follow them. The young man and I had to lead horses; we were obliged to abandon our horses for one of us had to run ahead and determine what had become of the Indians, for in these parts there is no passage except those made by bears or sometimes by Indians. To the added worry of stepping on snakes, I had an increasingly horrible fear of stepping on huge trees that were so rotted that they gave way underfoot, and we became half buried in bark and the leafy plants that surrounded them. Having finally arrived at a spot where the river was only one and a half feet in depth on a rocky slope, we crossed it, and I recognized *Pinus strobus* on the banks, pine or sapinette,[72] a new *Magnolia* that I will name *Magnolia* (*hastata*), a large *Aristolochia scandens*. We finally arrived at the source of the Keowee River. This site resembled a bay, being a plain of more than one mile, surrounded by higher mountains, extremely swift [water], with very regular contour. We stayed there more than two hours to rest our horses and eat strawberries that were abundant there. Our day's journey was ten miles, and rain forced us to camp under a cabin made of bark abandoned by Indians who had come in this area to hunt, which we recognized by bones of animals that they had killed and eaten and from the scaffold that they had used to smoke-dry the meat.[73]

June 15, the Indians led us through high mountains, not dangerous for the horses, and despite continued rain, we arrived at the Tugaloo River.[74] I recognized in several creeks a new species of very large *Clethra*, the stem 4 inches in circumference; [and] a violet with hastate leaves from which I had the happiness of obtaining some seeds.[75] Frequently I saw *Magnolia* (*hastata*), [and] several new plants past flowering whose genus I could not identify. Our journey was twelve miles or more, and we camped at four o'clock between mountains so steep that they obscured the daylight.[76]

June 16, we traveled over several mountains whose creeks flow into the Little Tennessee River and found nothing in these areas but *Magnolia hastata* and a new *Vaccinium* (or *Arbutus*) that is a favorite of the bears; the Indians pointed out this fondness to me from [the look of] their droppings.[77] Despite the rain that had fallen for three days, I decided to go on to the Little Tennessee River while avoiding all the branches that make up the river, and we traveled about eighteen miles that day. We camped near the river, which in this spot flows through rocks that divide its course into three or four parts. In this way we could cross it, but above and below [these rocks] the river was about 60 feet wide. I found a large number of the shrub with pyriform fruits and an *Azalea* with yellow flowers.[78]

SUNDAY JUNE 17, the young man who knew a little of the Indian language told me that they themselves did not know the way and that it was impossible to continue through those mountains crossed by this river. We decided to go to some of the Indian villages to buy some flour, because we were tired of eating only meat without bread. Having happily found the path of the fur traders, we decided to return, and we passed through much less steep mountains, always covered by the *Arbutus* loved by bears. Our hike was 15 miles long. We had for the first time good weather, and the clear air made the scene of some of the mountains beautiful.

JUNE 18, we traveled twenty-seven miles through rather flat and easy terrain except for the creeks swollen by the rain. We passed through an Indian village of about 60 families,[79] and at night we arrived at Seneca, exhausted with fatigue. Five miles before arriving in Seneca, I recognized *Magnolia acuminata* on the banks of a creek called Cane Creek.[80]

JUNE 19, I rested and prepared to leave for Charleston since it had been exactly two months since I had left.

I collected herbarium specimens of *Zanthorhiza*, and I noted many *Annona* behind the house of Col. Henderson.

Michaux wrote the paragraph that follows later, but he indicated that it should be placed after June 19.

Not only did I have the displeasure of finding few new plants in these mountains in comparison with those collected recently in Georgia, but I didn't see a single bird of interest. The rocks that we see in these mountains are made up of quartz, and we can find granite made of quartz, mica, and red clay. In the portion of the mountains that belong to the Indians, the soil is always better. Ten miles from the Keowee River, the border has been established between that Nation [the Cherokee] and South Carolina, but several villages were vacant, and I saw the remnants of five villages in the small area that I visited. This Nation is one of the largest after that of the Creeks who inhabit the area between Georgia and Ohio. I learned upon my return of the hostilities begun between the Nation of the Creeks and the Georgians. The inhabitants who live in the country around Seneca had assembled to construct a fort and to retreat there. It was feared that the Cherokees, who no longer have a chief, except those who govern each village, would join themselves to the Creeks.

June 20, I left Seneca for Charleston; we traveled twenty-two miles.

June 21, we traveled twenty miles by the same road that we had taken previously.

June 22, we traveled five miles since one of our horses had gone astray in the morning, and we slept at Gen. Pickens's.*

June 23, the same horse was still gone, and we left the house of the general and slept only 3 miles away (hope to arrive before the 10th of next month in Charleston).

Sunday, June 24, we spent the night at Hard Labor Creek, 14 miles away by a new road through the woods;[81] [we] saw several *Magnolia acuminata*.

June 25, we traveled seventeen miles, passing through Turkey Creek.[82]

June 26, we traveled nine miles and were surprised by a storm.

June 27, we traveled nineteen miles, and we left the hills whose soil is clayey and rocks with quartz.

June 28, we traveled twenty-one miles through sandy terrain without water.[83] We camped near several springs where one can find *Sarracenia tubifolia*.

June 29, we traveled eleven miles through a sandy and humid area and encountered often *Sarracenia tubifolia* and *Cupressus disticha*.

June 30, we traveled 15 miles; rain was continuous, and we camped one mile from the Edisto River.

Sunday, July 1, we traveled sixteen miles. I saw along the river swamps full of cypress and *Nyssa*; they are near the road to Augusta, and there is such a large number of young plants that I plan to return next winter as the distance is only 80–100 miles from Charleston.[84]

July 2, we traveled eighteen miles.

July 3, we traveled sixteen miles.

July 4, we traveled twenty-one miles.

* Michaux's footnote: Capt. Vedle promised to supply me with wild turkeys by telling all the inhabitants of the district. ["Capt. Vedle" is Capt. Moses Liddell; see note 59]

JULY 5, we traveled ten miles, and we arrived at the plantation [i.e., the garden near Charleston].

JULY 6, I arrived in Charleston having traveled ten miles.

JULY 7, I visited the plantation and looked over the work of the gardener.

JULY 12–15, I was obliged to stay in Charleston to complete the business of acquisition of the king's property in Carolina and also to get ready for a trip to New York. I was obliged to take passage on a boat for Philadelphia.

JULY 16, I went on board.

JULY 27, I arrived in Philadelphia.

JULY 28, I visited the French consul.[85]

SUNDAY, JULY 29, I was busy writing.

JULY 30, I visited Bartram and dined with the consul.

JULY 31, I traveled by stagecoach to New York.

AUGUST 1, I arrived at the king's establishment in New Jersey.[86]

AUGUST 2, I arrived in New York.

AUGUST 3, I was occupied with making a list of seeds brought from Carolina.

AUGUST 4, I dined with the French chargé d'affaires and . . .[87]

SUNDAY, AUGUST 5, I visited the king's nursery with Mr. Roland, a marine engineer.[88]

AUGUST 6, I spent the day in New York to receive money and do accounts with M. de La Forest.

AUGUST 7, I visited the garden [in New Jersey] and took count of the fruit trees and American trees to be sent to France.

AUGUST 8, I packed my books and herbarium to send them to Carolina, looked at accounts, and paid the gardener.[89]

AUGUST 9, I wrote to Count d'Angiviller, Abbé Nolin, M. Lemonnier, M. Thouin (to M. Desaint to let him know of a draft of 1,200 to the order of M. de La Forest).[90]

AUGUST 10, I left New York and . . .

AUGUST 11, I arrived in Philadelphia and the same day embarked for Charleston, and [I] withdrew from M. Dutarte the sum of three thousand pounds to the order of M. de Marbois, the French consul in Philadelphia.

SUNDAY, AUGUST 12, we went by Chester.

SUNDAY, AUGUST 19, we went beyond Cape Hatteras.

AUGUST 20, we had winds strong enough from the northeast that at night all the sails were taken down, even the uppermost ones from the two masts; during the night we were subjected to almost continual rain, accompanied by thunder, lightning, and a furious wind.

AUGUST 23 and 24, we were becalmed.

AUGUST 24, at 5:00 a.m. the thermometer exposed to the air registered 21¾ degrees [Réaumur; 81°F, 27°C], seawater was 21 degrees [79°F, 26°C], weather very calm; at noon 23 degrees [84°F, 29°C]; and at night 18 degrees [73°F, 23°C].

SUNDAY, AUGUST 26, calm as in preceding days. At 3:00 p.m. a breeze arose, and we hoped that we would soon be on our way.

AUGUST 27, variable winds.

AUGUST 28, we arrived in Charleston having taken eighteen days from Philadelphia to Charleston.

AUGUST 29, I obtained word of the receipt of [21] boxes of trees sent by Capt. Clark that arrived in Bordeaux on May 20.[91] On the same day, I wrote to Count d'Angiviller to announce the draft of M. Dutarte. I wrote to Abbé Nolin, to M. de Marbois, and to Saunier.[92]

AUGUST 30, I continued writing and made several visits.

AUGUST 31, I received several visits and at night left for the plantation.

SEPTEMBER 1, I remained there and took note of the different collections made by my son.

SUNDAY, SEPTEMBER 2, I bought a horse.

SEPTEMBER 3, I collected plants around the plantation and made grafts.

SEPTEMBER 4 and 5, I traveled beyond the Cooper River to look for palm trees, found in abundance *Sideroxilon tomax*, *Ligustrum monospermum*, and *Magnolia grandiflora*.

SEPTEMBER 5, I bought 7 sheep to have manure for the garden.

SEPTEMBER 6, I sowed seeds of *Laurus benzoin*, [L.] *aestivalis*, etc.

SEPTEMBER 7, I sowed several different kinds of seeds, dried, and looked over all the seeds that had been gathered.

SEPTEMBER 8, I worked and sowed.

SEPTEMBER 9, I sowed.

SEPTEMBER 10, I went to Charleston, and I rented a different room. I received letters from New York; I wrote to Philadelphia.

SEPTEMBER 11, . . . [no entry]

SEPTEMBER 12, I removed boxes that came from New York and returned to the plantation.

SEPTEMBER 13, I collected seeds of *Gleditsia triacanthos* and began the construction of a loft for the seeds.

SEPTEMBER 14, I gathered *Cassia chamaecrista* and *Cassia nictitans*, *Cacalia atriplicifolia*.

SEPTEMBER 15, I sowed seeds.

SEPTEMBER 16, I sowed.

SEPTEMBER 17, I prepared a box with seeds to be sent to France.

SEPTEMBER 18, I went to Charleston.

SEPTEMBER 19, I returned to the plantation.

SEPTEMBER 20, I saw to plowing.

SEPTEMBER 21 and 22, I sowed seeds from the Carolinas.

SEPTEMBER 24–26, [I] went to town. I sent two boxes of seeds to Bordeaux. I wrote to d'Angiviller, Abbé Nolin, Lemonnier, Th. M. Nairac, and Saunier. Put on board the 8 ducks.[93]

SEPTEMBER 27 and 28, I was busy at the plantation.

SEPTEMBER 29 and 30, I went with my son to harvest *Sideroxylon tomax* on the Cooper River.

OCTOBER 1 and 2, I went with my son beyond Dorchester to harvest *Gleditsia aquatica*.

OCTOBER 3, I went to Charleston.

OCTOBER 4, I went with my son and a black man to collect *Magnolia grandiflora* near the Cooper River.

OCTOBER 5 and 6, I plowed and planted seeds.

OCTOBER 7, I collected a large quantity [of seeds].

OCTOBER 8, I sent my son to Charleston and received letters via New York from M. d' Angiviller and Abbé Nolin.

OCTOBER 8–15, I sent my son to collect *Magnolia grandiflora, Cyrilla, Juniperus, Quercus phellos, Liriodendron,*[94] and I extracted from my journal all my expenses.

OCTOBER 18–30, my son and I had a fever. I was also in pain with rheumatism.

OCTOBER 25–31, I looked over the seeds that had been collected and started to prepare to ship some.

NOVEMBER 1–4, I continued to fill the boxes with seeds.

Eight of the daily journal entries from October 30 through November 6 consist only of temperature readings.[95]

NOVEMBER 4 and 5, I sent my son to Strawberry Ferry to get *Nymphaea nelumbo* and the vine supplejack.[96]

NOVEMBER 6, I stayed in town to write letters, ship 7 cases of seeds and a cage of 8 summer ducks.

NOVEMBER 7–12, I was busy with the above shipment and writing letters.

NOVEMBER 12, I observed in Watson's garden a *Crinum rubrum* said to be from Mississippi.[97] Spathe 2 leaves, 2–3 flowers. Corolla tubular with 6 parts. A plant with aggregated flowers, calyx . . . , corolla tubular 5 stamens inserted on the corolla. Pistil stigma single, capsule hairy with only one seed.[98]

NOVEMBER 13–14, I looked over the seeds and paid the blacks who had been busy in the preceding days in seed collection.

NOVEMBER 15–16, I went beyond the Cooper River for *Olea americana*.

NOVEMBER 18 and 19, I went to town to receive a box of trees from New York.[99]

NOVEMBER 20–28, I was busy planting the trees and the several species of seeds.

NOVEMBER 29 and 30, I traveled to Moncks Corner for *Olea americana* and *Sarracenia tubifolia*.[100]

DECEMBER 1, I planted the trees that I brought back and prepared the site of an enclosure for the dwarf deer.[101]

DECEMBER 2, I looked over the seeds and prepared to ship some.

DECEMBER 12, I gave the package to a packet boat going to New York.[102]

DECEMBER 15, I traveled to the interior of Carolina for *Gleditsia monosperma*, *Stewartia*, etc., in order to have a full consignment for a ship announced to go directly from Charleston to Le Havre [Havre-de-Grâce]. From this day until December 27, I was busy with digging up trees, packing them, packaging seeds, and writing letters, and . . .

DECEMBER 27, I gave the cases and the summer ducks to the ship destined for Le Havre and entrusted to M. Limousin, a merchant.[103]

DECEMBER 28, 1787, I worked to take account of all my expenses and settle with those persons and workmen to whom I owed money.

Notebook 2 ends here. In his letter of November 6, 1787, to André Thouin, the botanist explained the difficulty of supervising his gardener in New Jersey, defended himself against criticism, and revealed his own personal ambition.

### AM to André Thouin, from Charleston, November 6, 1787[104]

I have the honor to announce the shipment of a little package of seeds in a box addressed to Monsieur Lemonnier. I made two other shipments to you on 10 August and 25 September of this year. I was very upset to learn of the poor success of the shipment to Count d'Angiviller through Bordeaux. This shipment had been made on 15 February to coincide with the departure of the vessel. It deferred its departure, then it was disabled, the boat was emptied and the boxes underwent another delay of three weeks and were exposed during this time to cold and dry winds.

Abbé Nolin complains of the way that I have executed the wrapping of trees, he reprimands me for having enveloped roots of trees with pine needles. One would have to be without sense to work in such a fashion. I used the pine material to separate areas between branches and to prevent fermentation that could happen in the ships if one were to put as much fresh moss in the middle of the box and among the branches as we must put around the roots. In the bottom of the boat where the boxes are usually placed, fermentation and heat often make the trees develop their buds and leaves.

Abbé Nolin reproaches me even more about the shipments that the gardener [Saunier] made last winter. Before leaving for Carolina, we selected together a quantity of the most interesting trees and shrubs, I sent the duplicate of this list to the Abbé Nolin and he approved it. I have placed in this list fewer specimens of each species than the gardener himself promised to send so that he would not complain that I demanded more than he could furnish. In addition to the list I asked him to send nuts of *Juglans nigra* that I bought myself from a private individual, about 2 miles from the Nursery on the way to New York. All he had to do was to obtain the amount needed when the nuts were ripe; this should not have been a pretext for his not collecting other seeds. I made people gather moss two months before my departure. We worked on this task every morning because the snakes are dangerous during the day. A quantity more sufficient than necessary was collected and dried. I had made a number of boxes in order to be ready at the time of shipment. I looked over his herbarium and I named the plants so that he would know the plants that are the most interesting. I told him that I would make interesting shipments and that we would compare my shipments with his and that he should force himself to do better than I . . .

I recommended that he cultivate the area with a spade or cultivate a portion of the lower terrain known to be very favorable to seeds that we proposed. He can hire workmen by the day as there is money deposited for his needs at the consulate in New York. When I arrived this year in August I did not see any seeds, he told me that everything that he sowed died. He used all of the land and probably all his time for cultivating corn, buckwheat, barley, and potatoes. He had a good harvest. I had told him to use for his profit all the land that was not occupied by seeds of trees and I would have seen his harvest with pleasure as I gave him all possible encouragement. He lives in a new house; he finds wood for heating on the King's land without even being obliged to cut for several years. He has pigs and cows which run in neighboring forests like that of other inhabitants and thus without expense they furnish a part of his food. He takes hay for the maintenance of his animals in winter on the lower portion of the King's land. I had recommended to him seeds that would have given him much honor because it shows the effect of work and that specimens of young plants are worth much more than those that are taken in the woods. Even if he had only worked 30 toises [190 feet or 60 meters] with the flat shovel and sowed about 2 meters with each kind of trees, for example 2 meters of tulip poplar, 2 meters of *Liquidambar*, another in *Magnolia glauca*, another with oaks, etc., this would have furnished more than two or even three hundred for each 2 meters if we suppose the success which rarely is lacking if the land is well prepared and if one immediately sows the tree seeds. During my stay in New York I did not cease to give him all the possible encouragement. These encouragements only served to increase his avarice and they rendered him blind to his own interest, as he thinks only of making an immediate profit. At first when I arrived in America he seemed very hard working and sober I had been moved to tell you good things that I noticed, but time has shown his character. If I recommend something to him, he shouts and makes a sermon that I can trust him that he will do the work and I should not even think about it.

I did not complain against him but made only vague remarks to justify my position concerning what he said about the collection of walnuts that he had said kept him from obtaining other seeds. He flatters himself to have your esteem, but as everything that you have written concerning him had as its object the good of the project, I pray please tell me what to do. Knowing the confidence that he has in you and the fear of losing your esteem, I am persuaded that your recommendations and your reprimands will have all the possible success.

These unpleasantries, but more particularly those stemming from undeserved reproaches from the Abbé Nolin, would be more than enough to cause me to withdraw if they were to continue because I do not work as I do for the honorarium which is given to me, I do not work either like good souls who do not realize their merit but are still useful. I placed my reward in the honor and merit of having executed my mission, to have put into it the success of not letting any discoveries [if possible] be made by those who come to visit these lands after me. The approbation of Count d'Angiviller is not the least reward of which I was desirous, but also my being named correspondent of the Académie des Sciences which is well deserved and which I have already asked for. If thus the hope of these rewards is lost because of the reproaches of Abbé Nolin who indicates that I have worked poorly and have lost the esteem of the Count, will I be compensated by the continued honorarium and by the profits of cultivation done upon the counsel of Abbé Nolin in the King's nursery in Carolina? I could not acquire these profits of which I would be ashamed if they were against the interest of my mission.

The observations of my trip to Georgia and the Appalachian, the descriptions of the plants which I collected (they are with my herbarium which I have not had a chance to visit since my return), the probability of obtaining more rich natural historical discoveries in the next year, have given me the right to aspire for this title of corresponding member of the Académie for the publication of my discoveries which are increasing daily. I can cite seven species of *Ludwigia*, three *Rhexia* that do not coincide with those of Linnaeus. The descriptions of *Laurus aestivalis*, *L. borbonia*, *L. indica*, etc. must be redone and are not at all recognizable.[105] I would like that France beat the English in publishing these discoveries. There are here two ignorant men sent by nurseries, who pick up everything and send some also to Mr. Banks.

I also had remarks about a collection of birds, for the most part new, which I addressed to M. Daubenton during my trip to New York this year in August. I will increase by many this winter the collection of wood samples. I would like to present my request to M. Lemonnier to have his approbation regarding this title of correspondent to the Académie, but I do not even know if he is receiving my frequent shipments. I think also that he puts me in the rank of those people who are not aware of the worth of their work.

Excuse my sincerity and believe in my affection,

Sir and dear friend,

Your very humble and very obedient servant,

Michaux

The prohibition against sending seeds to anyone has kept me from sending those I had promised to Messieurs Cels and l'Héritier.

Wouldn't it be possible to profit from the garden that I have in this climate, which is so favorable for the growth of plants, for increasing seeds of which you are desirous? I would send them to you in abundance and this would only be between the two of us.[106]

# CHAPTER THREE

# Exploring Florida, 1788

SUMMARY: MICHAUX'S THIRD NOTEBOOK TELLS of his journey to Spanish East Florida in 1788. Michaux, his son, and a black assistant traveled from Charleston to St. Augustine, Florida, by ship. There, the botanist met with Spanish officials, including Governor Vizente Manuel de Zéspedes y Velasco (1720–94), and obtained approval for exploring Spanish territory.[1] Michaux made two botanical trips in a dugout canoe that he purchased. First he explored the rivers and lagoons from St. Augustine south to the vicinity of Cape Canaveral and then returned to St. Augustine. Michaux's party then went overland on horseback to the St. Johns River, where they paddled south, exploring this waterway and its lakes, tributary streams, and springs, to within a few miles of Lake Monroe in present-day Seminole County, before returning to St. Augustine. They then traveled by horseback and dugout canoe north to the St. Marys River, the border between Florida and Georgia. Crossing into Georgia in Michaux's canoe, they proceeded to Savannah along the waterways between the barrier islands and the sea. In Savannah, Michaux embarked on a ship and returned to Charleston. Upon reaching Charleston, he quickly set to work planting what he had collected in Florida. Short local collecting trips in the Charleston area, the preparing and shipping of plants to France, and work in the garden occupied the next several weeks. In August, a botanical expedition north toward Camden, South Carolina, was thwarted by an attack of fever.

## Notebook 3 begins

THURSDAY, FEBRUARY 14, 1788, I prepared myself to embark. I bought a young black man for the price of 50 pounds, and I rented another one for one shilling a day. I went on board at 12:30 p.m. for St. Augustine in Florida.[2]

FEBRUARY 15, calm weather and contrary winds; we stayed anchored inside the Charleston Inlet.

FEBRUARY 16, a strong wind came up during the night; several ships were dragging their anchors. A schooner came up and struck our ship, but without any damage. We managed to separate them. Then came the rain, and we hoped that the wind would turn from the south to the north, but it continued, and that evening we obtained shelter from the wind below Sullivan's Island, in view of Charleston.

SUNDAY, FEBRUARY 17, we stayed at anchor, and I went to look for plants on Sullivan's Island. I recognized only a few plants that were worthy of being noticed, because this little island, besides being sterile due to its exposure to the winds, is burned every year according to the custom of the Americans who annually set fire to all the forests. The English, during the last war, cut all of the *Chamaerops*; all that's left are young ones that do not have fruits.[3] I also noticed a shrub whose fruit indicates it to be a *Croton*, and a grass.

FEBRUARY 18, the wind calmed down, but it still was not favorable.

FEBRUARY 19, we raised anchor, and we went beyond the inlet, but the contrary winds forced us to return.

FEBRUARY 20, we sent a canoe to the city, and I took advantage of the occasion; I went there not only to renew our provisions consumed during the bad times spent due to the contrary winds, but also with the hope of hearing some news from France by way of New York that should have already arrived. There was a schooner destined for New York, and I regretted all the more bitterly the eight days lost due to the contrary winds; if I had stayed in Carolina, I would have been able to prepare a shipment for February 24, which was the day the schooner was to depart; and likewise, if the wind had been favorable to go to St. Augustine, I would have been able to send a very interesting shipment by the schooner, whose departure date from Charleston had been established for the 24th of February and which could have easily arrived before the 10th of next March in time for the departure of a packet boat headed for France.[4]

FEBRUARY 21, we still remained anchored, and in the evening there arose a heavy wind accompanied by rain.

FEBRUARY 22, the sea's turmoil and the wind having stopped, we were hopeful of having the north wind that would have been favorable for us.

FEBRUARY 23, the wind was very favorable, but we spent all day trying to pull up our anchor, which had become so stuck that we were about to abandon it, but toward the evening, with the help of another ship that was stronger, we were able to pull it up.

SUNDAY, FEBRUARY 24, we set sail with a very feeble wind, but favorable enough.

FEBRUARY 25, we encountered a south wind that was contrary; it lasted until the next morning.

FEBRUARY 26 and 27, we stayed at sea, and finally toward the evening, we observed the Florida coast.

FEBRUARY 28, we entered the port of St. Augustine and disembarked at 1 o'clock in the afternoon.[5]

There came on board some officers of the government who asked what I came to do and whether I had brought any merchandise: I responded by saying that I came only to observe the natural history of Florida and that I had already obtained permission from His Excellency the governor. They told me immediately that it was necessary to present myself to him. I told the governor that I had no other object than natural history and [that] when I would be ready to visit the different parts of the country, I would inform His Excellency, and I would give him acknowledgment of the most interesting discoveries.

He [the governor] told me that I was welcome and that all the services he could render, he would give me. He showed me every courtesy and had a message sent to the place where I had taken lodging so that I would be given the best of care.

FEBRUARY 29, the day was spent in visits.

MARCH 1, I went to look for plants, and I recognized new species of *Andromeda*, no. 1, 2, and 3.[6]

SUNDAY, MARCH 2, we went to church, and we heard mass at which His Excellency the governor was present.

MARCH 3, the thermometer was at 9 degrees Réaumur above zero [52°F, 11°C] at 6 o'clock in the morning. We went a distance of 5 miles, but a storm accompanied by thunder and lightning went through and completely soaked us, and we came back without picking up any interesting plants.

MARCH 4, the wind accompanied by rain lasted all night; the thermometer was at 5½ degrees [44°F, 7°C]. The storm was a little less violent during the day; we went a distance of more than 6 miles and saw only the interesting shrubs found on March 1, namely, no. 2 and no. 3. I also collected an unknown shrub that looked like *Andromeda* no. 4 but which differed totally in its fruits.

MARCH 5, the wind from the northwest, thermometer 2½ degrees above zero [38°F, 3°C] in the morning. The day was used to read the description of Florida and to verify this description with a map that had been lent to me.[7]

MARCH 6, I consulted several inhabitants on ways of going to look for plants in the south of the province.

MARCH 7, I bought a small canoe and hired two men who could maneuver it.[8]

MARCH 8, I bought some provisions for the voyage and a lot of powder and lead to be able to kill some game because the areas where I was going to visit were uninhabited and frequented only by Indians.

SUNDAY, MARCH 9, I settled all things for the voyage. The thermometer in the morning was at 5 degrees [44°F, 6.5°C].

MARCH 10, the thermometer in the morning was at 5½ degrees [44°F, 7°C]. Wind from the northwest. A worker was occupied sewing the sail of the canoe and making some repairs.

MARCH 11, the thermometer in the morning was at 4¾ degrees above zero [43°F, 6°C]. The wind was from the northwest. Because the sail and other parts of the canoe were not ready, I went to visit the property of a local person so that I could establish a depot for trees there.

WEDNESDAY, MARCH 12, we left St. Augustine in our canoe with five persons, namely my son and I, two oarsmen, and the black man whom I had brought from Charleston. The wind was favorable, but the tide that was against us formed waves that came into our canoe, and we decided to stop at the house of a respectable old man established here for 52 years on the isle of St. Anastasia. This man, who was the most hard working and the most

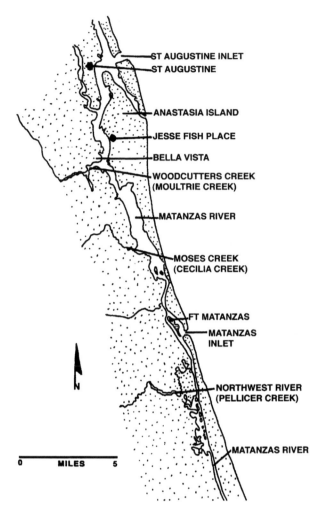

MAP 6. East Florida, St. Augustine vicinity (Drawing by Walter Kingsley Taylor, first appeared in *André Michaux in Florida* [Taylor and Norman, 2002])

industrious in all of Florida, had turned his place into a paradise despite the several pillages by pirates to which he had been exposed and the revolutions that he had twice experienced due to the changing of ownership of this province to the English, and once again to the jurisdiction of the Spanish.[9]

MARCH 13, thermometer in the morning was at 12 degrees [59°F, 15°C]. We went around St. Anastasia Island; we stopped at around 14 miles from St. Augustine, and I recognized on the shore two species of trees . . . called mangrove by the English and, in some parts of this island, the *Zamia* . . .[10]

We arrived in the evening at Fort Matanzas situated on this island. I used the few hours remaining to me to look for plants within a short distance from this fort.

MARCH 14, we tried to go past the Matanzas Inlet, 20 miles from St. Augustine, where the island of St. Anastasia ends, but the wind that came from the sea formed waves that filled our canoe; we decided to stop at the house of a Minorcan who lived at the mouth of the North West River, 3 miles distant and 24 miles from St. Augustine.[11]

MARCH 15, the wind, still coming from the sea, kept us at the house of the Minorcan. I visited the surroundings, and I recognized only the same plants known to me from Carolina and from Georgia, namely: *Magnolia grandiflora, Quercus phellos, Pinus taeda, Myrica cerifera, Bignonia sempervirens, Juglans hiccory.*[12]

SUNDAY, MARCH 16, the thermometer in the morning at 14 degrees [63°F, 17°C]. We took a horse and a guide to go up the North River, called the North West River. We traveled 22 miles and noticed only plants common to Carolina and Georgia, such as the *Magnolia grandiflora, Gordonia lasianthus, Acer rubrum, Laurus borbonia, Cupressus disticha, Myrica cerifera*, etc. etc. Apart from these trees, I also saw along the river, which should not be called a river but [called] a stream, *Andromeda arborea, Zamia pumila, Chamaerops repens*, and a leguminous shrub with ternate leaves no. 17, and another unknown shrub no. 18, a *Halesia tetraptera* with small flowers, and two species of *Annona*, etc.[13]

MARCH 17, we continued to follow this river; in a short distance I saw *Viburnum cassinoides, Ziziphus scandens, Lupinus pilosus*, blue flowers.

I gathered a lot of seeds from shrub no. 17 and a new *Andromeda*. Finally, seeing that the soil was still arid without interesting plants, I decided to turn back.

MARCH 18, I collected no new plants, but I recognized on the bank of the North West River and along the Matanzas River an *Andromeda* with almond-shaped leaves around 10 to 12 feet in height; it formed hollow stems that were very straight, which the Indians, so they say, use for their pipes. I didn't see it flowering, but I believe it is the one that Bartram indicated to me under the name of *Andromeda formosissima*.[14]

MARCH 19, the two oarsmen whom I had sent with my black man not having given the signal on which we had agreed, I decided to go to that place, and there I learned from a soldier of Fort Matanzas that they had found the wind to be favorable to cross the inlet and that the tide had forced them to leave without having the time to light the fire, the signal we had agreed to give. Coming back I went by a place abundant with oranges, and two miles farther I found several interesting shrubs.

March 20, the Minorcan, at whose house we had been staying, gave me three horses so I could go join the oarsmen because the sea was so rough at the Matanzas Inlet that it would have been imprudent for us to pass with our baggage.

We left at 7 o'clock, and we walked until 6 o'clock in the evening without stopping. I saw the most arid countryside in Florida during this walk, with the exception of one plantation at which we arrived at 5 o'clock in the evening that had belonged to Governor Moultrie at the time when the English possessed Florida. Finally at 6 o'clock, we arrived at the mouth of Tomoka Creek, and we camped on the bank of the lagoon. This is a canal formed by the islands that are strung out along the coast of America. When these islands are interrupted, the sea comes in and breaks against the river's edge, and navigation by boats is dangerous when the wind comes from the sea. One can go by little boats from the Carolinas until Cape Florida, and this is called inland navigation, and the different arms of the sea formed by the islands that are strung out in such a manner are named lagoons. These have different names according to the places and the islands that contain them. We fired a gunshot, and our oarsmen responded quickly by firing another gunshot. They had arrived the evening before without any danger except that their canoe had been overturned twice by the waves, and thus they got wet, but they were very experienced.[15]

We were then about 40 miles away from St. Augustine in a straight line and a mile from the mouth of the Tomoka Creek.

March 21, we passed over to the left bank of this lagoon where there was an abandoned residence. I saw orange trees covered with fruits, and I picked up several interesting shrubs. We camped that evening on the Island of the Orange Trees, 4 miles from the abandoned plantation of Mr. Penman. In the interval, we visited also several houses that had been abandoned and that were numerous enough to have been called a village.[16]

March 22, we had a considerable rain, which had started during the night and which lasted until midday.

Our navigation was about 6 miles, and we camped on dry ground at 4 miles from the mouth of Spruce Creek. There I found *Carica papaya*.[17]

EASTER SUNDAY, MARCH 23, the wind was favorable enough, and we camped between the New Smyrna Inlet and the ruins of this town, which had been founded there in the time of the English. This establishment had been built by Dr. Turnbull at the expense of the company of which he was the director. More than 1,200 people—men, women, and children, the greater part of them from Minorca—had been enticed from their country. The harshness and the oriental despotism with which this barbarian led his colony was still the subject of conversation of the inhabitants of St. Augustine when I was there. This place is designated in a new map of Florida published in London several years ago by the name of Musketo Shore (*côte des mosquites*).[18]

MARCH 24, the thermometer was at 7 degrees [48°F, 9°C]; the wind from the northwest was very noticeable.

We came to camp on the ruins of New Smyrna; there I noticed more than 400 houses destroyed; there remained only the chimneys because the Indians, who came each year for the oranges that survived despite the annual fires, destroyed the wood, from which these houses were built, to warm themselves.

MARCH 25, thermometer at 5 degrees [43.5°F, 6.5°C]: white frost.[19] I visited the humid places and the surroundings of this establishment that had flourished in the time of the English, but I didn't notice any other plants besides the ones that had interested me the preceding days. We were now 75 miles from St. Augustine.

MARCH 26, our navigation was 12 miles, and we stopped on the ruins of a plantation that had belonged to Capt. Besy in a place that was very fertile and that made me want to visit the swamps.[20]

There I found only one species of *Pancratium* and an annual plant 12 feet in height that had dried out and from which I collected several seeds.[21]

MARCH 27, we still navigated between the isles of mangroves, and we dined at the foot of a hill named Mount Tucker. I collected several shrubs and plants of the tropics.[22] In the evening we came to camp on the ruins of the residence of Capt. Roger.[23]

MARCH 28, I crossed the swamp that once formed this habitation on which sugarcane had been cultivated, and finally toward midday, we came to the Indian River and for some the Aïsa Hatcha, that is to say, River of Deer, and called by the Spaniards Rio d'Ais. This property was the most southern one that the English had established in Florida. We camped 4 miles farther.[24]

MARCH 29, our navigation was around six miles because the contrary wind was very strong; the oarsmen, despite much effort, were not making much progress. Moreover my son and I went on the west bank to try to discover the narrowest place between the Indian River and the canal where we were. Around 11 o'clock from above the trees one could distinguish easily the two arms of the sea; that is to say, the one where we were, called by the English . . . and the Indian River, thus named by the English, which is not a river at all but a very narrow arm of the sea, like all the others made so by a chain of islands that extends from north to south from the Carolinas to the Cape of Florida. Our two oarsmen came ashore, and we walked around the whole territory so as to try to find a less difficult passage to transport the canoe. Around four o'clock in the evening, we came back to camp with the hope of being able to transport the canoe. We wanted all the more to be near dry ground because, since our departure from New Smyrna, we had had only brackish water. The provisions of rum for the oarsmen had been consumed, and they were just as eager as we were to leave this place where we were being devoured by mosquitoes. As for me the place presented alternately only considerable stretches of rushes and saw palmettos with saw teeth, with leaves acute, dentate, rootstock prostrate.[25]

However, I found among the trees that made up a part of these woods situated on the Indian River, a fig tree of entire and oblong leaves, a new *Sophora*, and two other shrubs that were unknown.[26] This increased my hopes for the expeditions that I prepared to undertake the following days on this river.

SUNDAY, MARCH 30, we have been occupied all day rolling our canoe on the ground the space of one mile across rushes and spiny plants. It was necessary to cut trees, but the greatest difficulty was when we had to cross an area of 600 feet all covered with *Chamaerops* with saw teeth that not only cut our boots and our legs, but their strong stems resisted the good tools that we had. In effect, a very skillful worker whom I had hired for this trip said that he would much rather cut a 60-foot-tall cabbage palm than one of these shrubs because the sprawling stem is often interlaced with other stems or branches of the same size passing one over the other. Finally, toward evening, we got the canoe across and all of our baggage transported to the bank of the Indian River.[27]

MARCH 31, we were ready to leave at daybreak. But the place where we were was a kind of gulf that (to the judgment of our oarsmen) formed with the river a stretch six miles in width. The wind was against us, and there was so little water in this part of the gulf that our canoe could not advance even though my son and I had traversed more than four miles in the water that only came to halfway up our legs. When the water became too deep, we would get back in the canoe, but then the waves came into the canoe, and by midday we stopped near a swamp full of mangroves. Not being able to camp at this place because of the slimy soil, we returned to the place from which we had left, but it didn't take long before our canoe was almost submerged because of the large amount of water that came into it, and all our provisions became wet.

TUESDAY, APRIL 1, 1788, the same wind from the south that had brought us also kept us at the same place. It blew with more violence than the preceding day. Our oarsmen profited from the occasion to dry the rice and the biscuits that had been soaked the preceding day. They went fishing and brought back two fishes that weighed more than 18 pounds each. I looked for plants after having dried my baggage, which had also been submerged the day before, and I collected *Pteris lineata* and *Polypodium scolopendroides*, which commonly grow together on stems of the large *Chamaerops*. I found also *Acrostichum aureum* in very humid places and even among the mangroves that border the immense swamps of this river. We saw aquatic birds of several species, and my son killed more than 12 that day by repeated firing of his shotgun. We cut cabbage palms so as to save the bread that was diminishing, and we allotted two biscuits per day for each of the five people.[28]

APRIL 2, we took advantage of a calm wind to cross the river to the side with firm ground. It was at least six miles distant, and toward midday we reached shore. The wind, which had gotten considerably stronger, impeded us from continuing our route after midday. I found on dry ground an abundance of *Sophora occidentalis*, a beautiful shrub, and I collected an abundance of seeds and a beautiful stalk of its flowers, which confirmed that this was a *Sophora*, of which the flower is very attractive. I picked up several other plants that the night impeded me from describing. . . . A new species of *Spigelia* and another plant that has an affinity with. . . . Our walk was judged to be twelve miles long.[29]

APRIL 3, our walk was fifteen miles, and the hope of finding interesting and new plants excited me to overcome the obstacles (as I was still traveling by foot to relieve the oarsmen, who had had contrary wind). Instead I only found the trees and shrubs of Georgia and of Carolina, *Magnolia glauca*, *Gordonia*, *Acer rubrum*. However, I collected two *Annonas*, one of which was a new species with very large white flowers and leaves. The expanse of the canal, which was 4 to 8 miles wide in several places, frightened our oarsmen, and they judged that it was more convenient to take advantage of the wind to return, so we decided that we would benefit from the calm wind, which comes every day just before sunrise until 9 o'clock in the morning. In effect, on the 4th, having embarked before sunrise with a favorable wind, we had the good fortune to traverse the deepest part before 8 o'clock, and in the evening we found ourselves on the eastern bank of the Aïsa Hatcha River.[30]

Every evening we saw from our camp the fires that the Indians made on the other bank of this river, but since our departure from St. Augustine, we had not encountered any directly; our oarsmen advised us to avoid any encounter with them because of the demands to which one is exposed on their part in order to get rum, for which they are at least as passionate as our oarsmen could be, who by the way, were among the most sober I have yet seen among this sort of people.[31] Our navigation was estimated to be 24 miles.[32]

APRIL 5, was entirely used up in transporting the canoe and rolling it in the same way as we had done on the preceding Sunday.

Toward the evening I profited from a small time interval to collect several shrubs and trees that I had noticed on the shore of this river that I had not seen before. I packed them in such a way as to transport them to Charleston, so I could plant them there, and everything was ready to return to St. Augustine the next day.

SUNDAY, APRIL 6, before leaving this most southern part of Florida to which I had been able to advance, I decided to visit an island where I saw trees different from those (other than only the mangroves) found commonly on these islands, and I did not waste any time picking up the *Guilandina bonduccella*, the mangrove with fruits like those of Catesby's fig tree. . . . An unknown tree and a *Phaseolus* or a *Dolichos* with large fruits.

Our navigation was . . . , and we came to camp on the ruins of the residence of Capt. Roger. This settlement was the most southern that the English had had in Florida. There they had cultivated sugar, but the Indians have destroyed all of the canes.[33]

APRIL 7, the wind had blown from the south for several days and was very favorable for our return and pushed us all the way to New Smyrna of which there is nothing there but ruins, as I have already remarked. Our navigation was . . .

APRIL 8, we came to sleep on an island ten miles distant of . . .[34] We were at the latitude of . . .

APRIL 9, we had the wind behind us, and despite our many stops we covered 24 miles.

We camped at the mouth of the Tomoka Creek, latitude of . . .

APRIL 10, we went up the Tomoka River, which is truly a river, although it was named a creek by the English who didn't get to know Florida very well during the time they possessed it. The wind was very favorable, and toward the evening we found an island covered by woods. We camped a little past the island, and our navigation was around 18 miles at best.

I collected an *Annona* with large, white flowers that I believe to be the *Annona palustris*, *Annona glabra*, which seems to me to be a variety of [*A.*] *triloba*. The other plants on this river are: *Acer rubrum*, *Cupressus disticha*, *Fraxinus* . . . , *Magnolia grandiflora*, and [*M.*] *glauca*, *Pinus* with two needles.[35]

APRIL 11, we went up around five miles, but the river, which was filled with trees, prevented the canoe from passing, so I decided to eat here and to look for plants while the meal was being prepared and to leave immediately after. That evening we came back to camp at the mouth of the Tomoka River.

APRIL 12, one man left to go look for some horses so we could transport the baggage that could not be transported in the canoe, in order to cross the Matanzas Inlet.

SUNDAY, APRIL 13, the man whom I had sent to the home of the Minorcan to get horses arrived in the evening, and he brought the provisions that we lacked.

I had spent the preceding day and this one in visiting the surrounding woods and the swamps where I was, but there were no interesting plants in this very disagreeable place, because of the presence of caimans and serpents, which were abundant, and the mosquitoes that tormented us and didn't let us rest during the night.

APRIL 14, we started walking at daybreak and didn't arrive until very late in the evening due to the detours that we were forced to take several times across saw-toothed *Chamaerops*, which covered the surface of the ground, because the woods are very open. We were required to take a considerable detour because the woods had been on fire the preceding days. They were still burning, and the wind that was blowing toward us brought the fire at a rapid pace. One has no idea in Europe of the considerable amount of woods that are annually set ablaze in America either by the Indians or by the Americans themselves. They have no other motive, either one or the other, than to have new grass come up without the dry grass of the preceding year and to hunt deer more easily and to feed the cattle. I am persuaded this is the principal cause of the decline of the North American forests.[36]

APRIL 15, we waited for the oarsmen who had gone by sea to cross Matanzas Inlet.

I went to look for plants in the woods, and I noticed the *Andromeda* that I had seen before as being truly a new species, having enough resemblance with the *Andromeda arborea*, but different in several regards particularly in the disposition of its flowers and . . .

I also recognized an *Annona* and *Stillingia silvatica*.[37] I collected all the rare shrubs and trees to finish a box that I had decided to take with me to Charleston, at some risk since the season was then very advanced.

APRIL 16, we left this place to return to St. Augustine, and we camped two miles from Fort Matanzas.

APRIL 17, we set out at two o'clock in the morning, and we arrived in St. Augustine (the wind having been favorable) at midday.

APRIL 18, I went to visit the Spanish governor, and I visited Mr. Leslie, agent for Indian affairs, to discuss with him the means of traveling in Indian land.[38]

APRIL 19, I was invited to dinner by Mr. Leslie.

SUNDAY, APRIL 20, I received a visit from the governor, who came to see my plants and other collections that I had made during my trip, birds, etc. I was invited to have dinner with him, and the afternoon was spent in the gardens of His Excellency with the hospitable ladies of his family.

APRIL 21, 22, and 23, I looked for plants in the area around St. Augustine, and I sent a man to the St. Johns River to reserve a canoe to shorten our trip by keeping us from having to enter the river at its mouth.

APRIL 24, 25, and 26, I wrote Count d'Angiviller to give him an account of my trip to the south [and] my collections and to announce to him the draft of 2,000 francs to the order of M. de La Forest drawn on M. Dutarte.

I wrote to l'Abbé Nolin to respond to his letter received here and to make some observations on the plants that I am sending him.

Furthermore I asked him for the *racine de disette* and for seeds of the male *véronique* for Capt. Howard.[39] I have written to M. de La Forest to send him the bills of exchange drawn on M. Dutarte in triplicate to employ funds for the establishment near New York.

I also wrote to M. de Marbois, consul of France in Philadelphia, to register the package addressed to Count d'Angiviller. This week I described several grasses and *Carex*, *Scirpus*, and other plants that grow in the area around St. Augustine.

SUNDAY, APRIL 27, I wrote out the lists and the descriptions of the plants collected since my arrival, numbering 40 species whose genera and species are well known to me.

The second notebook contains 36 whose genera are well known to me but [whose] species are doubtful or unknown. The 3rd notebook contains 29, the greater part of which are unknown or could not be determined because I have not seen them in flower. In all, 105 trees or plants collected since March 1 until this day.[40]

APRIL 28, I bought provisions and prepared to leave to go visit Lake George, situated beyond the St. Johns River.

Gave the letters written earlier to Capt. Hudson who was to leave to go to St. Marys to get his ship and go to New York making a stopover in Savannah. I wrote at the same time to M. Ferry Dumont.

I addressed the package to M. de La Forest as well as the bills of exchange on M. Dutarte. I observed at the home of Sieur Roquet an abundance of *Annona grandiflora*.

APRIL 29, we left for the St. Johns River.[41]

APRIL 30, we arrived at the residence of Mr. Wigin found on this river, 40 miles from St. Augustine by land.[42]

THURSDAY, MAY 1, 1788, I looked for plants in the surroundings, and I collected an *Andromeda formosissima* in flower. The canoe being ready [on] May 2, we embarked and we passed by the store established for commerce with the Indians situated 10 miles away. We camped farther away, and we had traveled sixteen miles on this river.[43]

MAY 3, we traveled between 14 and 16 miles, still having contrary wind, and we camped in a place called *camp des Indiens*, which seems to have been cultivated at one time. I recognized *Sapindus saponaria*, some orange trees, and a pretty *Convolvulus dissectus*?[44]

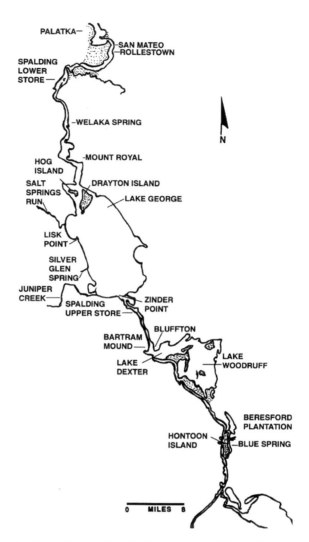

MAP 7. St. Johns River, Florida (Drawing by Walter Kingsley Taylor, first appeared in *André Michaux in Florida* [Taylor and Norman, 2002])

SUNDAY, MAY 4, we traveled only four miles, and we camped on an island at the entrance to Lake George on the east bank opposite a place named Alligator Point.[45] The wind, which was contrary and very strong, forced us to stay in this place, where I recognized *Erythrina*, again woody, and *Sapindus saponaria*. The woods were filled with sour oranges.

MAY 5, we saw as we entered Lake George a big deep bay to our left, which is to the west, and after having directed our route straight ahead, we entered a river that one cannot see before arriving, at a distance of only 120 feet. The mouth (at latitude 29° south) of this river is so filled with sand that it was necessary to drag the boat the distance of 150 to 200 feet. Then one finds it over 15 feet in depth. The water is salty and more disgusting than that of the St. Johns River and that of Lake George. After having gone upstream for more than three miles, we found the spring that comes out of the ground and forms bubbles that rise up over half a foot on the surface. One can see the bottom, greater than 30 feet in depth. Around the basin formed by this spring, we recognized *Illicium*.[46] The ground was composed of sand darkened by plant debris and shells.

The other trees that abound in this place, as well as everywhere where one finds *Illicium*, are *Magnolia grandiflora* and [*M.*] *glauca*, *Ilex cassine*, *Olea americana*, and *Laurus borbonia*. This river abounds so prodigiously with fish that they would hit our canoe as we traveled along. Our course was five miles up to the mouth of this river.[47]

MAY 6, we continued upriver, again following the shore, and as I was walking on the sand while the canoe continued, I recognized at a mile away from the place where we had left, that is, from the mouth of the Salt River, a spring of water, the purest and the best that I had yet drunk in Florida. We stopped there to eat because we were all thirsty and disgusted with the bad water that we had been drinking for several days. One mile farther I again recognized *Illicium*, and it was found in abundance at the southern point of the bay.[48] After having passed the bay (latitude 29°3'), we camped at the hill of oranges to shelter ourselves from a furious storm that was about to descend on us. At the bottom of this hill is the mouth of a fairly wide river, the water of which is not as good as the preceding water. I went up this river about two miles, and I recognized in the woods *Sapindus saponaria*, a species of *Coffea* that I had observed before at Mosquito Shore, and two other trees that I had seen there, but which had remained unknown to me. I also saw *Crinum americanum*. Our journey was judged to be 15 miles long.

MAY 7, 1788, our navigation was eight miles. We passed Lake George, and we entered the river that lies above, and we camped in a place abundant with orange trees. We arrived early enough to construct a shelter of leaves of the wild palm *Chamaerops* . . . so as to be safe from a storm.[49]

May 8, our navigation was 10 miles, and we endured a storm more severe than the one of the preceding day. We saw a place frequented by Indians. There was a boat that belonged to them on the bank of the river and a cooking pot. I had someone place some biscuits, beans, and some sweet oranges in this pot and we continued on our way. We heard two gunshots; that proved that the Indians were hunting nearby. We passed through a place so abundant with oranges that I traveled more than one-half mile through the interior of these woods without seeing any other trees. This place was over one mile long. We camped on a hill where I recognized *Rivina humilis*, an *Asclepias* shrub, and *Gleditsia monosperma*. The bottom of the hill and the summit were covered with orange trees.[50]

May 9, our travel was estimated at only 12 miles, although our oarsmen had worked all day long, but since our departure we were going up the river against the current, and the wind was always contrary. For more than eight miles, we saw nothing on either side of the river but rushes and a few trees; the ground was always muddy. The river was bordered on both sides by alligators or caimans of an enormous length and size with horrible faces. We approached them from 6 to 10 feet away. Their form is that of a lizard, but they are black and armed the whole length of their back with big points that they bristle up when they are angry. One cannot kill them but by charging the gun with balls and aiming at the lower neck. The nose is more turned up than that of a pig, the head flattened and two feet and four inches in length, sometimes even longer. The eyes are very close to the top of the head. They have 72 teeth in the jaw. They easily swallow dogs, pigs, and young calves, but at the least movement of a man, they jump in the water with a big commotion. They are amphibious and came every morning to visit us to get the leftovers of the fish with which we were well furnished on this river. We were also regaled with their music, the sound of which resembles a stronger and more continuous snort than the bellowing of a bull situated in a valley one mile away. The Indians eat the lower part of the alligator sometimes, but only when they lack other game.[51]

May 10, our navigation was 15 miles; we continued up to the source of a river that came out of the ground. The water was salty and gave off an unbearable odor, although one could see the bottom of the river at more than 15 to 20 feet in depth. We had much difficulty passing over trees that covered the bottom and sometimes obstructed the surface. There were no habitations more remote from the time of the English than these ruins where we ate lunch. I found a species of wild squash at the most secluded place to which we had advanced.[52]

SUNDAY, MAY 11, we traveled 11 miles still continuing to go upstream against the current of the river that seemed more and more obstructed and losing itself in marshes covered with rushes. I picked up an *Ipomoea*, the flower of which was perfectly white and the tube six inches long. This plant appears to me to be an annual and grows in moist places; the leaves are entire and heart-shaped. Seeing little success in continuing my voyage, I made the group turn back, and we returned to sleep at the same place from where we had left that same day.[53]

MAY 12, the wind was favorable for our return, and we traveled 27 miles. We camped at the hill of orange trees.[54]

MAY 13, the wind and the current were again very favorable, and we arrived at the shore of the stream whose water was so agreeable and beautiful. It is situated only one-half mile from the Salt River, the water of which is just as bad as the water of the little river is good. I experienced furthermore the satisfaction of collecting *Illicium* only 150 feet farther. It should be remarked that this shrub is found in places where *Magnolia grandiflora*, *Annona grandiflora*, *Olea americana*, *Ilex cassine*, etc. etc., grow, but more particularly where one also finds *Aralia spinosa* and a grass called "Canes" that grows to ten feet in height, which always indicates a good but sandy and cool soil. Our trip for the day was 18 to 20 miles.[55]

MAY 14, our navigation was . . . , and we arrived at the home of Sr. Wiggins . . .[56]

MAY 15, we started back on the land route to return to St. Augustine.

MAY 16, we arrived at St. Augustine at two o'clock in the afternoon.

MAY 17, I went to visit His Excellency the governor, etc.

SUNDAY, MAY 18, I wrote down [notes] and put my collections in order.

MAY 19, I was invited to lunch at Capt. Howard's house.[57]

MAY 20 and 21, I went to look for plants at the extremity of St. Anastasia Island.

MAY 22, was Corpus Christi, and I was present at the procession.[58]

MAY 23, I took leave of His Excellency the governor and several other people of distinction from whom I had received a cordial reception.[59]

MAY 24, I gave the government an account of the observations made in Florida during my stay.[60]

SUNDAY, MAY 25, I left St. Augustine for the Post of St. Vincent, and we slept at Twenty Mile House.[61]

MAY 26, our horses having strayed during the night, we searched for them the following morning. The sergeant of this post who had taken charge of our horses had us led by two soldiers with two other horses to the Post of St. Vincent, situated 40 miles from St. Augustine.

MAY 27, we embarked in our canoe that had come by sea to wait for us at the Post of St. Vincent as we had already taken advantage of a small ship that had set sail for this part of Florida.[62]

MAY 28, 1788, we navigated between islands of rushes, and we camped opposite Nassau River Inlet.[63]

MAY 29, we arrived at the mouth of the St. Marys River, which separates Florida from Georgia, and we camped in the territory of Georgia. The place where we crossed the river was about two miles wide.

MAY 30, we went around Cumberland Island, which is more than . . . miles in length, and we camped on the island. Because of the considerable detours that we were obliged to make in the canal between the continent and this island, we only arrived at the camping area at 9 o'clock at night. The land in Georgia opposite the island is named Camden County.[64]

We saw several houses on this island belonging to inhabitants of Georgia who had taken refuge here to elude the ravages of the Creek Indians who had killed their livestock, burned their houses, and killed several from among them.[65]

MAY 31, we continued our route in the canal that is found along this island, and at eleven o'clock we passed St. Andrew Sound, which is five miles across at its narrowest point. Several rivers have their mouths there.[66] We then continued our way to St. Simons Island, and at 10 o'clock at night we went across the St. Simons Sound.[67]

SUNDAY, JUNE 1, 1788, we left at 2 o'clock in the morning, and we arrived around ten o'clock at Frederica.[68] I delivered letters to different people, and I dined with my son at Mr. Spalding's, where there were ladies of the family of Gen. MacIntosh and several others of considerable reputation.[69]

JUNE 2, we came to the southern point of the island called Little Sapelo Island and we camped after passing the sound, called Frederic Sound.[70]

JUNE 3, we crossed two sounds, and we camped on St. Catherines Island.

JUNE 4, we crossed St. Catherines Sound at 7 o'clock in the morning. The day was calm; the width of the sound is more than four miles, and we found four very swift currents that, despite the calm that prevailed, gave our rowers a great deal of trouble and exposed us to the danger of not being able to overcome these (currents) or of being submerged by the least amount of wind that might develop.

JUNE 5, we navigated approximately 22 miles and arrived in Savannah toward evening.

JUNE 6, we remained in Savannah.

JUNE 7, we left by ship destined for Charleston.

SUNDAY, JUNE 8, we arrived in Charleston, and I stayed there until the next day.

JUNE 9, I went to the house.

JUNE 10, I visited the garden and the nurseries.

JUNE 11, I returned to Charleston to remove my belongings from the ship and had the boxes of plants transported to the house.

JUNE 12–14, I planted the trees brought back from Florida.

JUNE 15 and 16, I was busy planting seeds brought back from Florida and a large number of other species.

JUNE 17, I went back to Charleston.

JUNE 18–20, the workers were busy weeding the nurseries.

JUNE 21, I went collecting and obtained *Fothergilla gardeni*.

SUNDAY, JUNE 22, I returned to the house.

JUNE 23–25, I worked in the garden.

June 26, I went to Charleston.

June 27, I returned home.

June 28–30, I worked in the garden and continued collecting *Fothergilla gardeni* with several black men.[71]

Tuesday, July 1, the harvest of *Fothergilla* amounted to four bushels.

I wrote to M. le Comte d'Angiviller and sent seeds from Florida. I also wrote to l'Abbé Nolin by way of M. de Leyritz.[72]

July 2, I returned from town.

July 3, I went with my son to look for *Stewartia*.

July 4, I was busy with both the garden and with different trips toward the Santee and Cooper rivers, etc. I was obliged to go to Charleston several times until the end of the month.

I noticed *Zizania palustris* at a small distance from Moncks Corner.

August 2, I gave a box of seeds for the count to Capt. Elliot by way of New York.[73]

August 3–9, I was busy at the garden as I had not been satisfied with the previous gardener.

August 10–14, traveled toward Moncks Corner and beyond, toward Camden.

August 15, I was attacked by a fever.

August 20, I withdrew from M. Dutarte 2,000 pounds for the functioning of the New York nursery in a letter of exchange from M. de La Forest, consul of France in New York.

The fever continued, and I decided with the advice of several people to go back to Charleston to be close to the doctor and necessary help.

September 7, 1788, since I had not had a fever for several days, I returned to our house in the countryside.

September 13 and the following days, the fever returned and I was obliged to return to town and stayed there till the end of the month. During the course of the month, I made several trips to the house, particularly to gather seeds of chinquapin, *Styrax*, etc.

October 7, I returned to the house.

October 8, rain all day.

October 9, the rains continued.

October 10, we went to gather *Stewartia* and noticed a *Populus heterophylla* in the plantation of a person named Willimon.[74]

October 11, I prepared a shipment of seeds by way of New York for the use of the establishment.

Sunday, October 12, I continued working on the shipment and writing letters.

October 13, I went into town to deliver the boxes to the ship. I wrote to the count, the Abbé, M. de La Forest, M. Saunier. I received a box of trees from Philadelphia.[75] I bought planks.

October 14, I had to remain in town.

October 15, I returned home after finishing my affairs in town.

October 16, I planted the trees that I received and sowed the chinquapins.

October 17, I traveled to Dorchester to collect *Gleditsia aquatica*.[76]

October 18, I planted seeds of *Magnolia glauca*, *M. tripetala*, *Chionanthus*, *Stewartia*, *Alaterne* from Carolina,[77] *Zanthoxylum*, *Styrax*, *Halesia*, *Fothergilla*, *Magnolia acuminata*, *Viburnum dentatum*.

Sunday, October 19, I pruned the trees of the garden and prepared the cold frame of the little garden; I replaced the glass.

October 20, I made a shelter to protect the *Illicium floridanum* from the north winds and the rains from the northwest.

October 21, I sent a case to Capt. Marshall so that he could bring me trees from St. Augustine in Florida.[78] Wind from the north, and temperature in the morning 10 degrees [Réaumur; 55°F, 13°C].

October 22, the temperature in the morning 9 degrees [52°F, 13°C], sowed in a special enclosure chinquapin, persimmons, *Fothergilla*, *Magnolia glauca*, *Styrax*, *Juniperus*, etc.

OCTOBER 23, in the morning the [Réamur] thermometer showed 6½ degrees above zero [47°F, 8°C]. I harvested *Pinus palustris* and *Fraxinus palustris*.

OCTOBER 24, 1788, like the previous day, I harvested pine seeds and [observed that] the seeds of several trees had already fallen. Although the year has been more productive than ordinarily, a tree of 1.5–2 feet in diameter produces only about one peck or at most a half-bushel of cones.

OCTOBER 25, I collected pine seeds and put earlier seed collections in order.

SUNDAY, OCTOBER 26, collected pine seeds and put my earlier seed collections in order.

OCTOBER 27, my son accompanied the blacks to harvest pine seeds, and I worked with the gardener to make a ditch to divert water around the *Illiciums*.

OCTOBER 28, I went to Charleston, and I had to stay overnight to obtain hard currency from paper money.[79]

OCTOBER 29, I went back home.

OCTOBER 30, I collected *Baccharis* and several other species.

OCTOBER 31, I sowed seeds.

Notebook 3 ends here.

CHAPTER FOUR


# Exploring in the Carolinas, Georgia, and the Bahamas, 1788–1789

SUMMARY: THE FOURTH NOTEBOOK OF Michaux's journal begins with the French botanist in Charleston preparing for a late fall plant-collecting expedition to Georgia and the southern mountains that he had visited in 1787. To begin this journey, he first traveled in the opposite direction along the route that he had followed from the South Carolina mountains to Charleston in June 1787. Near White Pond in Aiken County, South Carolina, he left this route and continued to Augusta, Georgia. Upon reaching Augusta, he turned south on the road toward Savannah and collected plants in modern Richmond, Burke, and Screven Counties before returning to Augusta. After making arrangements to ship his collections of plants and seeds down the river to Savannah and then on to Charleston on commercial transport, he began the next part of his journey toward Seneca (modern Clemson, South Carolina), the Indian village that he had visited the year before. In 1787, Michaux had explored the South Carolina counties bordering the Savannah River north of Augusta, and now he traveled through the Georgia territory on the opposite bank. After reaching Seneca he again obtained Indian guides and continued his journey following the Keowee River upstream toward the higher mountains in North Carolina. After collecting plants in the mountains along the border of North and South Carolina, he returned to Seneca. There he arranged to ship a part of his collection by wagon, while keeping another part with him as he returned directly to Charleston on horseback along essentially the same route that he had followed the year before.

Michaux ended 1788 and began 1789 in Charleston, preparing and sending shipments to France, including cases containing some of his freshly gathered mountain plants. In February 1789, he sailed to the Bahamas, where he met the governor and made extensive plant collections on the islands; he returned to Charleston in April. While working in his Charleston garden in early May, he began preparations for a new journey to the high mountains of northwestern North Carolina.

## Notebook 4 begins

NOVEMBER 1, I gathered seeds of *Bignonia sempervirens* and covered the shrubs from Florida to protect them from the freezes of the winter.

SUNDAY, NOVEMBER 2, I gathered seeds of *Nyssa* with large fruits and prepared for my trip to Georgia; [I] received a note from M. Petry to recommend that I do not go to Georgia south of Savannah as some Indians have again begun their devastations.

NOVEMBER 3, I went after a collection of *Bignonia crucigera*, *Andromeda nitida*, *Clethra*, and dug up *Spigelia marylandica* that had especially been asked for in the last letters of Abbé Nolin.

NOVEMBER 4, I sent to Charleston [for information] concerning the arrival of several ships that had arrived from New York.

NOVEMBER 5, I left home for Augusta and slept at Givhan's Ferry while going through Dorchester.[1]

This day I traveled 36 miles while calculating this journey as if I had left from Charleston itself.

NOVEMBER 6, I dined at Stanley House, 26 miles, and slept at Peoples's house near the D'Antignac Ferry: 35 miles.

NOVEMBER 7, I breakfasted at Bruton House 6 miles away, having completed half the distance between Charleston and Augusta. I slept at Chester House: about 30 miles.[2]

NOVEMBER 8, I had dinner at Robertson House on White Pond: 15 miles.

Here the route of Long Canes joins that for Augusta.[3] From Roberts to . . . house: 10 miles. Total 25 miles.

SUNDAY, NOVEMBER 9, I went across pine barrens and lunched at 12 miles distance and finally arrived in Augusta after walking 10 [additional] miles, about 22 miles. Total distance 148 miles.

NOVEMBER 10, I visited several persons to whom I had been directed; it rained all day.

NOVEMBER 11, I went to the home of Col. Stallion,[4] and [I] recognized on the riverside *Kalmia latifolia, Rhododendron . . . ,*[5] *Padus sempervirens, Halesia . . . , Annona . . . , Acer . . .*

NOVEMBER 12, I returned to Augusta.

NOVEMBER 13, I went to gather *Pavia spicata*, small white flowers, new species, and [also] found on the banks a new tree with opposite leaves observed the previous year in Georgia along riverbanks.[6]

NOVEMBER 14, I went 8 miles from Augusta to pick a shrub that has the aspect of an *Erica* and also brought back two hundred *Epigea repens.*[7]

NOVEMBER 15, 1788, I left Augusta to go on the road to Savannah taken last year. I observed several rare plants, especially *Lapathum occidentale*. Dined at the Widow Brown's house located between two ponds, 27 miles, and bedded at Mr. Lambert's, 37 miles, found *Calycanthus* near his home.[8]

SUNDAY, NOVEMBER 16, I went by the house named Bel Taverne at 42 miles. Then found in a pine barren 12 miles in diameter *Ceanothus floridanus* and a shrub with large spreading roots of the *Euphorbia* [family] with an affinity to oaks. I found these two shrubs near the home of the Freemans: 54 miles.

Continued my journey until Beaver Dam Creek, 60 miles from Augusta, and went back to sleep near the home of Mr. Bel.[9]

Note: Last year, the wheel of the cart broke in the hills 25 miles from Augusta.

NOVEMBER 17, I returned to sleep in Augusta, and I gathered all the most remarkable plants. My trip during these three days was 120 miles long.

NOVEMBER 18, I packed the plants obtained during my short trips around Augusta.

November 19, I went to dig up young plants of a new species of *Rhododendron* and a *Kalmia* that is closely related to *Kalmia latifolia*.

November 20, I went to pick plants of *Andromeda arborea* and of *Annona triloba*. Then in the afternoon, I packed these plants; I gave the cases containing eleven hundred sixty-eight trees or plants to Mr. [name illegible] to send via Savannah to Charleston.[10]

November 21, I left Augusta and went by way of Bedford, a town consisting of 4 or 5 houses at 3 miles from Augusta.[11] Five miles farther as I continued along the road of Wilkes County, one finds several houses near a creek, and close to the creek one could dig up several thousand plants of *Calycanthus*.

Mr. Grays's house is 15 miles from Augusta, and one can stay there overnight. I slept at the Widow Marchall, whose house is 20 miles from Augusta.

November 22, I was bothered by pain in the kidneys, and I only traveled 12 miles. I went across Little River and four miles farther. I spent the night at Col. Graves's, a Virginian.[12]

Sunday, November 23, I arrived in the town of Washington, located 46 or 48 miles from Augusta. Note: Washington is the capital [seat] of Wilkes County.[13]

November 24, I went to see a French doctor settled in the country.[14] He gave me remedies and ordered that I rest. I recognized near Washington *Magnolia acuminata*, which I had not seen on my journey since my departure from Charleston.

November 25 and 26, the fever in addition to another indisposition kept me from continuing the journey that I had planned on the Broad and Tugaloo Rivers.

November 29, I felt better and left Washington. I visited in Washington an esteemed Frenchmen, M. Terundet.[15] I stayed at Col. Stablerfield's.[16] I went to bed at Col. Gains's, whose house is situated on the Broad River at 20 miles from Washington.[17]

Sunday, November 30, I was unable to see Mr. Meriwether who lives close to Col. Gaines, and I crossed the Broad River.[18] In this area there were rocks in the river that made it difficult for the horses, especially after the rains. There is a ferry called . . . over the Savannah River, 5 miles away.

Meriwether is said to be a botanist; he attempts to know all the plants of the region, and I am sorry that I was unable to see him. I turned my trip toward the Tugaloo River, and I stayed the night at Capt. Richardson's at 15 miles' distance from the crossing on the Broad River. On the way I had dinner at Esquire Tets.[19]

DECEMBER 1, I crossed several creeks, the first, Beaver Dam, situated one and a half miles from Capt. Richardson's. Another creek, Coldwater Creek, is found 5 miles from the first near the home of Col. Cuningham;[20] I passed Cedar Creek at 8 miles from the third and slept on Lightwood Log Creek at the home of Mr. Freeman. I was very well received by the lady of the house, whose husband was absent.[21] This woman was young and very pretty, but very devout, and [she] was endlessly preoccupied with the different ways of thinking among the Methodists, Anabaptists, and Quakers. The conversation on this subject lasted from 7:00 to 10:30; I began to be weary of it, and despite the honesty and attractiveness of this woman, I went to bed. The creek on which this house is found meets the Savannah River in this area, at 15 toises [90–100 feet or 30 meters] below the house. The journey today was 20 miles long.

DECEMBER 2, I left the junction of the two rivers Tugaloo and Keowee to go up the Tugaloo 19 miles, and I spent the night at [the home of] Mr. Larkin Cleveland, Esq.

DECEMBER 3, I crossed the Tugaloo River at the only site for such a passage. It was so hazardous that two of our horses were in danger of drowning. I had lunch at John Cleveland's on the other side of the river.[22] They told me that there were no more settlements, and I passed through country covered with forests similar to those of the southern provinces, but in addition it was very hilly, and I arrived at sunset at Seneca after a trek of 19 miles.

DECEMBER 4, there was frost. We found ice the thickness of a fingernail or more [2 millimeters]. At dawn I went to look at the banks of the river and recognized *Zanthorhiza*, *Rhododendron* new species, *Kalmia latifolia*, *Hydrangea* (*glauca*), *Abies* spruce,[23] *Acer negundo*, *Carpinus* in fruit . . . ,[24] *Annona triloba*, *Halesia tetraptera*, *Cornus alternifolia*, *Calycanthus*.

DECEMBER 5, I continued my explorations while my black man was busy digging up trees that I had shown him. I was looking for an interpreter and a Cherokee Indian to go into the mountains inhabited by that nation.

December 6, I left for the mountains, and I slept with my guide in an Indian village. The head of the village received us courteously. He told us that his son who was due back from hunting that night would lead us in the mountains to the source of the Keowee. But he did not return, and this old man, who appeared to be about seventy, offered to accompany me. This man, who had been born in a village near the source of this river, knew the mountains perfectly. I hoped that his son would not return. For supper he served boiled fresh deer meat and cornmeal bread mixed with sweet potatoes. I ate with my guide, who because he spoke the Indian language acted as an interpreter. The chief ate with his wife at another bench. Then the mother of his wife and his two daughters, one married and the younger one fourteen or fifteen, came to sit around the pot in which they cooked the meat. These ladies were naked to the waist, each having no other clothing but a skirt.[25]

Sunday, December 7, the lady of the house roasted corn in an earthen pot with hot ash that had been sifted. When it was a little more than half roasted, it was removed from the fire, and the ashes removed. Then it was carried to the mortar, and after crushing, it was put into a fine sieve to separate the fine flour that was put into a sack for our provisions. When one is tired, one places approximately three spoonfuls in a glass of water, and often raw or brown sugar is added. This very pleasant drink is restorative and renews strength immediately. The Indians never go on a trip without a supply of this meal that they call . . .[26]

We walked about fourteen miles from 7:30 in the morning until 6 o'clock at night. We stopped only for one hour for dinner. We camped on the banks of the Keowee River at the base of the mountains among two species of *Rhododendron*,[27] *Kalmia*, and *Azalea*, etc. etc.

December 8, as we approached the source of the Keowee, the paths became more difficult. Our trip was . . . , and two miles before arriving there I recognized *Magnolia montana*, which was named *M. cordata* or *auriculata* by Bartram.[28] A family of Cherokee Indians lived in a little cabin at this site. We stopped there to camp, and I rushed to explore. I dug up a new shrub with toothed leaves that grew up the mountainside not far from the river.[29] The weather changed, and it rained all night. Although we were sheltered by a large *Pinus strobus*, our clothing and blankets were soaking wet. In the middle of the night, I went into the Indians' cabin, which could hardly contain the family made up of eight people, men and women. In addition, six big dogs increased the dirtiness and discomfort. The fire was in the middle without an opening at the top of the cabin to let the smoke out, however there was enough of an opening to allow rain to percolate through the roof of the house. An Indian offered me his bed made up of a bearskin and took my place near the fire. But finally I was so annoyed by the dogs that fought continuously for a place by the fire, that, the rain having stopped, I returned to the campsite. The site, known as the source of the Keowee, is thus improperly named. It is the junction of two other rivers or large torrents that join at this point and that have not been named except as branches of the Keowee.[30]

DECEMBER 9, we left with my Indian guide to go to the highest mountains and ventured to the source of the torrent that seemed to be most precipitous. We had to cross precipices and torrents covered by trees where ten times our horses sank and were in danger of perishing. We climbed up to a waterfall where the noise of falling water sounded like distant muskets.[31] The Indians said that at night one can see fires from this place. I wanted to camp there, but the unexpected snow and wind were so cold that we looked for an area lower down the mountainside, one less exposed to the cold wind and with more forage for our horses. The night was terribly cold, and there was no pinewood to keep the fire that burned poorly because of the intermittent snowfall. Our snow-covered blankets were stiffened by ice soon after they had been warmed.

DECEMBER 10, I went to several mountains on the slope, and in the lower regions we dug up *Magnolia cordata*. The day was spent specifically in search of this tree.[32]

DECEMBER 11, there was a considerable freeze, and the air was clear and bracing. I noticed a line of high mountains that ran from west to east and where the freezing weather was little felt because of the sun's orientation.[33] I collected a creeping juniper that I had not previously seen in the southern part of the United States, but it must be observed that I saw several trees of the northern parts such as *Betula nigra*, *Cornus alternifolia*, *Pinus strobus*, *Abies* spruce, etc.[34] We crossed an area of about three miles among *Rhododendron maximum*. I returned to camp with my guides at the head of the Keowee and collected a large quantity of the shrub with saw-toothed leaves that I had found on the day of my arrival. I did not find it on the other mountains. The Indians of the area told me that the leaves taste good when chewed and smelled good when crushed, which I found to be the case.

Directions for finding this shrub:[35]
The head of the Kiwi [Keowee] is the junction of two largish streams that cascade from the highest mountains. This junction occurs in a small

plain where previously a town, or rather a Cherokee village, existed. Coming down the junction of these two streams, keeping the river to the left and the north-facing hills to the right, one finds at 100 to 300 feet [60–100 meters] of this junction a trail made by Indian hunters. It leads to a brook where one can recognize the remnants of an Indian village by the peach trees that survive in the middle of the brush. Continuing this trail one swiftly arrives on the hills where we find this shrub that covers the ground, along with *Epigea repens.*

December 12, while returning, I visited the southern-facing hills as provisions were low and we had had a very meager breakfast. I collected many *Magnolia cordata* in better condition than those of previous days.

We followed the river and saw several flocks of wild turkeys. Our Indian guide fired at them, but the rifle failed repeatedly as it had not been possible to protect it from the rain in preceding days. Thus our supper consisted of a few chestnuts that our Indian had received from another tribe.

Our trek was eighteen miles long. The weather was very clear; frost was felt at nightfall, and after asking my Indian guide the names of several plants in their language, I wrote in my journal by moonlight.

December 13, I attempted at daybreak to shoot wild turkey. They were abundant in this area. I was unsuccessful, and we broke camp without breakfast. We left famished on our route toward the camp of Indian hunters, and although the hills became less steep, it was one o'clock in the afternoon before we arrived, after a six-hour journey of about fifteen miles. They cooked for us bear meat cut up into small pieces and fried in bear fat, and although it was very greasy, we had a good dinner, and even though I ate a large amount of the fattest portion of this meat, I was not uncomfortable. Bear fat has no taste and resembles good olive oil; it has no smell even if used in roasting several different foods. It becomes solid only if it freezes. After dinner we walked sixteen miles, and we arrived at night at Seneca.

Sunday, December 14, I was advised that a wagon was due to leave for Charleston the following morning.[36] I sent for the two wild turkeys that I had purchased three miles from this area, and I collected several tree species, *Rhododendron . . . ,*[37] *Nyssa,* Montana [illegible word], *Mespilus* of the mountains,[38] etc. etc.

December 15, I paid the Indian who had accompanied me in his tribal land; I worked at packing my trees; I dug up some more and had seeds collected. I collected *Pavia lutea* (?), *Quercus glauca,* etc. etc.

December 16, I worked all day packing trees and collected several species that I had recognized in the vicinity.

December 17, I finished packing the trees, settled the account of my expenses during the trip, and prepared everything for my departure.

December 18, I left Seneca; one of the wild turkeys that I had bought died two miles from where we left, and the other died as we arrived at camp. We walked only fifteen miles because we were obliged to stop several times to adjust the cages, which were on a horse and kept falling from one side or the other because of the birds' struggles. We camped in the forest since there was no settlement.

December 19, we ate the turkey that had last perished when we arrived in camp since we had thrown out the one that died first, and not having had any dinner or supper the preceding day. I spent the night at Rocky River 26 miles from Seneca, and I traveled only 12 miles because of bad weather.[39]

December 20, the cold was penetrating, and I stayed the night at the plantation of Gen. Pickens, 45 miles from Seneca.[40] That day I only traveled 20 miles since I visited the Little River to look for *Magnolia acuminata*. I recognized *Magnolia tripetala*, *Annona*, and *Magnolia acuminata* near the Pickens plantation in clayey, reddish brown soil.

Sunday, December 21, it was still very cold. We had to cross more than twenty large creeks, and I spent the night at Turkey Creek, at an American Tory's home, who told me when I arrived that he would kill me if I stayed overnight at his house. I replied that I did not fear this as I wasn't fat enough nor was my purse. He wanted to joke about my country, but I had plenty of answers for him, and he was content to make me pay a hefty sum for the lodging. I did 29 miles that day.*

December 22, the cold continued, and toward the afternoon a cold rain fell. I stayed overnight at Capt. Randel's.[41] There were two horse thieves there. The inhabitants of the neighborhood were assembled to take legal action against them. They released one of them, and the other was beaten. On this occasion they were all drunk with rum, and all night long I was troubled and tired by this disagreeable company. My trip that day was 28 miles long.

* Michaux's footnote: A little before crossing the creek is the house on the right of Col. . . . , where it is preferable to stay.

I observed *Epigea repens* in abundance on a hill whose soil is calcareous and clayey. It is rare to encounter calcareous soil in the lower part of the Carolinas.

DECEMBER 23, I left this spot and had breakfast two miles away on the right side of the road at a very honest man's house.* Then we had to pass through a sterile forest (pine barren) 18 miles long, and I arrived at the Robertson House.[42] I still did 12 miles and a total of 32 miles for this day. I slept at the Walkers'.

DECEMBER 24, I passed by Chester House situated 4 miles away and bedded at the house of Mr. People[s].[43] This journey was 34 miles long.

DECEMBER 25, I passed by Stanley House, 9 miles distant, and I slept at Givhan's Ferry.†

DECEMBER 26, I left Givhan's Ferry, and I went to sleep at the house.[44] The distance between this ferry and Charleston is 35 miles.

DECEMBER 27, I planted the trees from the collection that had been brought by horse.[45]

SUNDAY, DECEMBER 28, I went to see the seeds that had been collected during my absence, etc. I went to Charleston on December 29.

DECEMBER 30, I learned of the destination of a ship for Le Havre, and I returned to the nursery to prepare a shipment of trees and seeds.

DECEMBER 31, I packed several species of seeds, and I sent to Charleston for the three cases of trees that I had obtained on my last trip and that had arrived by way of Savannah.[46]

JANUARY 1, 1789, I opened the cases and found the trees in good shape, but a little of the vegetation had developed buds; the latter had grown, and to prevent the cold and even a little frost, which occurred every night, I removed the trees from the moss with which they were enveloped. In the middle of the day I dipped them immediately in a bucket of water that I had near me,

---

* Michaux's footnote: Sleep in this spot on another trip.

† Michaux's footnote: Between the ferry and the house I found 10 miles beyond the road to Seneca, several ponds where there is abundant *Ilex* with narrow and small leaves. Travelers can spend the night. (A site about 45 miles from Charleston.)

and after a good immersion, I held them under the wetted moss until the moment of planting: I also covered with moss those that had breaking bud.

JANUARY 2, I sent someone to inquire whether the trees that I had awaited had arrived.

I continued to plant the trees received from Georgia that I had reserved for the garden in order to send them off later.

JANUARY 3, I sent the five prepared cases to Charleston. I went there also and returned the same day.

SUNDAY, JANUARY 4, I continued packing seeds.

JANUARY 5, I packed seeds for the botanical garden in New York in order to benefit from a ship going to that port. I wrote to M. de La Forest and also to M. Saunier.[47]

JANUARY 6, I continued the same work regarding seeds.

JANUARY 7, the same.

JANUARY 8, I sent to Charleston to determine the day of departure of the boat and learned that the shipwright did not want to load any freight on his boat.

JANUARY 9, I went to Charleston, and I was allowed to send 10–12 cases, with M. Petry's help.

JANUARY 10, I completed the shipment of trees and seeds.[48]

JANUARY 12, I sent 13 cases to Charleston, and I went there that very day. I stayed until the 15th to provide hoops for the cases and to make sure that they got on board.

JANUARY 15, I returned home.

JANUARY 16, I prepared a shipment of seeds for the king's garden in New York.

JANUARY 17, I continued the same work.

SUNDAY, JANUARY 18, I prepared a shipment of birds for M. Dantic sent to the address of the Baron d'Ogny;[49] I put in order the collection of different *Vaccinium* and sent it to M. l'Héritier.[50] I sent duplicates of letters to Count d'Angiviller, etc etc.

January 19, I went to Charleston, and I shipped the birds and the *Vaccinium* on a boat going to Nantes.

January 20, I completed the shipment for Abbé Nolin that had been deferred by the ship's captain.

January 21, storm and rain, and I worked on the same shipment. I wrote to Mr. Beaudin, Mr. Plane, and Mr. Bartram in Philadelphia.

January 22, I went to Charleston to send off this shipment.

January 23, I sent someone to bring two dwarf deer to the nursery and worked to send a shipment to Le Havre, having been informed the preceding day of a ship going to that port.

January 24, I received my collection of trees from the mountains, amounting to six cases and a package of trees.

Sunday, January 25, I continued to work on the shipment for Le Havre and planted a portion of the trees received.

January 26, I went to Charleston to put on board several cases, and I wrote several letters.[51]

January 27, I returned to the house.

January 28, I planted the trees that had been obtained in the mountains and wrote several letters to France and a letter of exchange to M. Desaint to the order of M. Petry.[52]

January 29, I went to Charleston. I gave my letters to the captain of the ship, etc. I returned to the residence on the same day.

January 30, I sowed seeds.

January 31, I collected flowers from *Alnus*, first male catkin imbricate, scales 3-flowered . . . , corolla minute, 4 parted; female catkin, imbricate, pistil with 2 styles, sometimes 3.[53]

Sunday, February 1, I packed trees and seeds for the New York nursery.

February 2 and 3, I did the same sort of work.

February 4, I went to Charleston and spoke to the captain of the ship concerning going to the isles of Bahamas.

February 5, I came back to the house.

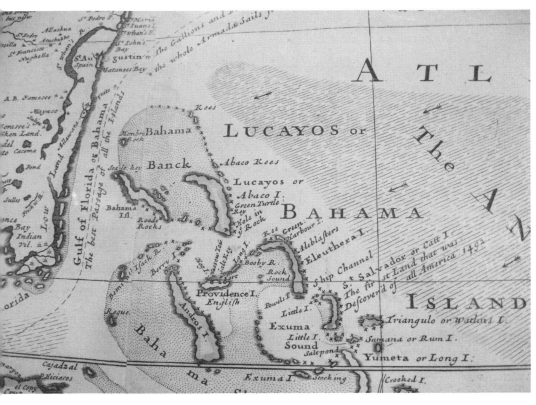

MAP 8. Detail of the Bahamas, Herman Moll, *Map of the West Indies . . .* ,
1715–20 (Courtesy of the Philadelphia Print Shop, Philadelphia, PA)

FEBRUARY 6, I had work done in the garden.

FEBRUARY 7, I continued the same work.

SUNDAY, FEBRUARY 8, I sowed seeds of trees and shrubs.

FEBRUARY 9, I went to Charleston.

FEBRUARY 10, I returned to the house to prepare for the trip to the
Bahamas.

FEBRUARY 11–13, I sowed seeds and put things in order for my departure.

FEBRUARY 14, I went to Charleston.

SUNDAY, FEBRUARY 15, I stayed in Charleston [and] waited for favorable
wind.

February 16, I embarked on the schooner the *Hope*, Capt. Weeks, for the Bahamas.

I stayed 10 days at sea and went ashore February 25 at New Providence.[54] I made several visits on that day.

February 26, I visited My Lord Dunmore, governor of the Bahamas.[55] He welcomed me graciously and asked me to give him some seeds and samples of plants to be sent to Mr. Banks such as: cedar, ebony, etc. *Elathera cortex.*

February 27, I went botanizing and recognized the following plants: *Vinca lutea,*[56] *Annona glabra, Laurus persea, Laurus indica* leaves evergreen; and also *Cornus,* leaves like laurel willow, acuminate, white flowers, fruit like Catesby's sassafras;[57] *Calceolaria?* leaves entire, *Psydium, Tamarindus indica, Catesbaea spinosa, Bursera gummifera, Coccoloba,* etc.

February 28, I continued my botanizing.

Sunday, March 1, botanizing continued.

March 2, botanizing continued; I recognized a *Gardenia* known popularly as the seven-year apple.

March 3–5, I continued.

March 6–8, I dug up trees to send to Charleston.

Sunday, March 8, I completed my collection of eight hundred and sixty trees among which are found: *Amyris elemifera, Winterania canella, Croton cascarilla, Gardenia* new species,[58] *Chrysocoma* new species,[59] *Annona glabra, Annona muricata, Annona . . . , Catesbaea spinosa, Bignonia pentaphylla, Passiflora cuprea,*[60] *Anacardium?,*[61] etc. etc.

March 9, I wrote to the count, to Abbé Nolin, to M. Petry, and to Mr. Robinet,[62] and to my son.

March 11, I botanized on the isle of New Providence until the 14th.

March 15, I looked over my herbarium and my seeds.

March 16–19, I went botanizing on the small islands called Keys.

March 20, I accompanied a coastal pilot to the Lucayas Islands.[63]

March 21, I had a bout of fever because of my having slept outside at the seashore.

SUNDAY, MARCH 22, I have collected seventy-five different species of seeds since my arrival here.

SUNDAY, MARCH 29, I prepared to leave for Charleston but the boat only set sail the . . .

THURSDAY, APRIL 2, with a favorable wind, we lost sight of the island of New Providence that same day.

APRIL 3, calm and . . .

APRIL 4, we recognized a little island named . . .

APRIL 5, I recognized the Bahama island that is more than 15 leagues long.[64]

APRIL 6–8, we were becalmed and still in sight of the Bahamas.

APRIL 9, thunder, cloudburst, and contrary wind.

APRIL 10 and following days, bad weather. I arrived and disembarked the 20th, all the provisions used up, and very tired by the bad weather.

APRIL 21, I stayed in Charleston.

APRIL 22, I went to the house.

APRIL 23, my trees and seeds gathered in the Bahamas arrived at the nursery—more than 900 trees.

APRIL 24, I prepared land and planted trees.

APRIL 25, 26, 27, and 28, I continued the same work.

APRIL 28, I shipped seeds to Count d'Angiviller and to Monsieur, brother of the king, and to the Jardin du Roi.

APRIL 28 and 29, I continued to work on my shipment and write my letters.

MAY 1, I sowed seeds brought back from the Bahamas.

MAY 2–5, I worked in the garden and continued to sow and plant, etc. I began to prepare myself to travel to the mountains.

MAY 6, I went to Charleston.

MAY 7, I approached the consul of France, M. Petry, several times to obtain money for my trip, and I returned to the house without having completed the [negotiations].

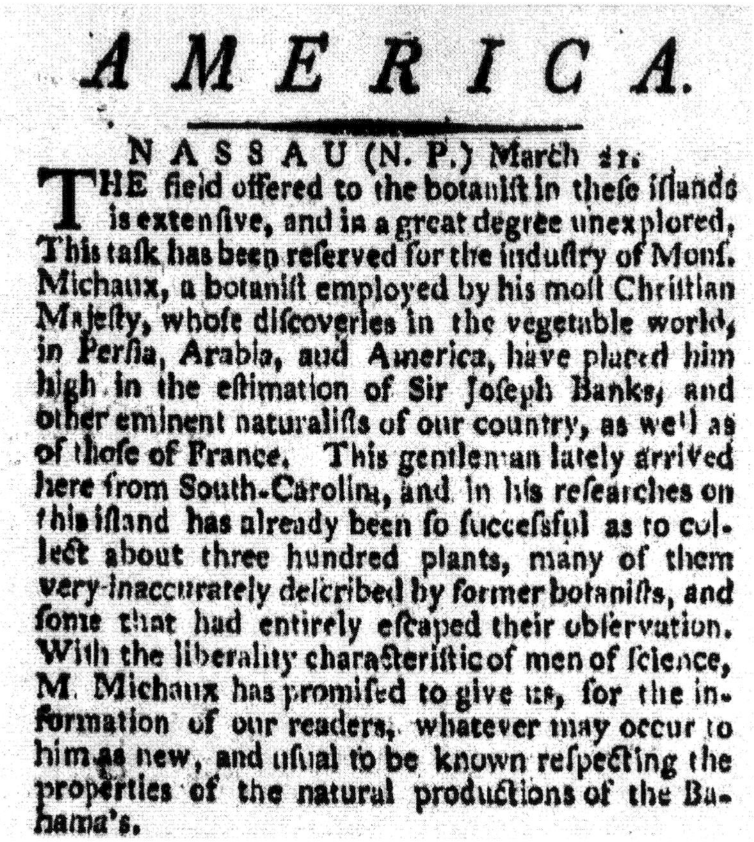

## AMERICA.

NASSAU (N. P.) March 21.

THE field offered to the botanist in these islands is extensive, and in a great degree unexplored. This task has been reserved for the industry of Monf. Michaux, a botanist employed by his most Christian Majesty, whose discoveries in the vegetable world, in Persia, Arabia, and America, have placed him high in the estimation of Sir Joseph Banks, and other eminent naturalists of our country, as well as of those of France. This gentleman lately arrived here from South-Carolina, and in his researches on this island has already been so successful as to collect about three hundred plants, many of them very inaccurately described by former botanists, and some that had entirely escaped their observation. With the liberality characteristic of men of science, M. Michaux has promised to give us, for the information of our readers, whatever may occur to him as new, and useful to be known respecting the properties of the natural productions of the Bahama's.

FIGURE 10. Philadelphia newspaper report of Michaux's 1789 visit to the Bahamas (Courtesy of Newsbank)

MAY 8, I continued several jobs essential in the garden concerning the plants from the Bahamas.

MAY 9, I made a little shelter to protect the trees from the big showers.

SUNDAY, MAY 10, I described a dioecious *Spiraea*, pistil with 3 styles, brought back from the mountains.[65] I wrote in and organized my papers.

Notebook 4 ends here. In a letter to Count d'Angiviller, Michaux offered a brief report on his trip to the Bahamas and explained the reasons behind the trip that he now planned to the high mountains of North Carolina.

### AM to Count d'Angiviller, April 28, 1789 (excerpt)[66]

I have the honor of sending you seeds gathered in my trip to the Bahamas. I recognized there all the trees cited by Catesby, with the exception of only two or three species. A very large number of these [species] have been described by Plumier, Jacquin, and Sloane in the Antilles, and several new species.[67] I brought back more than 1,500 trees that I planted, waiting for the right season to send them to you. I used only two months for this trip so as not to miss the season to go to the mountains of North Carolina. These mountains are the highest of all those that make up the Allegheny and Appalachian, and their products can succeed better in France that those of the other southern parts of the United States. The temperature there is so moderate that I found in the mountains of South Carolina trees of Canada and of New York state.

After writing to Count d'Angiviller and others, Michaux sent the following letter to Sir Joseph Banks, president of the Royal Society, giving highlights from his recent trip to the Bahamas. Michaux had met Banks when he visited Great Britain in 1779 or 1780 and had also written to Banks from Persia in 1784 and from New York in 1786. Banks replied to this letter about plants from the Bahamas, and Michaux responded in 1790 sending him a small selection of plants and offering to send him others.

### AM in Charleston to Sir Joseph Banks in London, May 20, 1789 (excerpt)[68]

I went to America three years ago to become acquainted with its natural productions. I established in Charleston my principal residence and there set up a garden, to place my collections in order to send them later to Europe under the best of circumstances. It is with much pleasure that I would send you objects that would be of interest to you.

I recently took a trip to the Bahama Islands, and My Lord Dunmore told me that you desired to have, sir, samples and seeds of *Juniperus* or cedar, ebony, and Eleuthera. I recognized a *Juniperus* that appears to me *virginiana*, and I collected male and female specimens, but they did not have seeds at that time.[69] I recognized the shrub Eleuthera-bark; it did not have any flowers or seeds. I did not find the tree or shrub, ebony. I will send you specimens of the *Juniperus*, and I will keep you informed of the shrub, Eleuthera-bark or *Croton cascarilla*.[70] I brought it back, and it is doing well in my garden. If you desire, I will send you a live one. I was only two months in the Bahama Islands because I did not want to miss the opportunity to go to the mountains west of the Carolinas.

# CHAPTER FIVE

# Charleston to New York, 1789

SUMMARY: ANDRÉ MICHAUX'S FIFTH NOTEBOOK includes the record of two journeys. Only five and a half weeks after he had returned from the Bahamas, on May 30, 1789, he began what would be his longest overland journey so far. This trip would take him to Philadelphia, the New Jersey garden, and New York. By this time, the botanist had become aware of the political turmoil in his homeland that would turn into the French Revolution, and he realized that he might soon be recalled from America and his mission ended. He lost no time beginning a fresh journey with the goal of crossing the high mountains in North Carolina and continuing west into Kentucky. His companions were his son and an enslaved black man whose name is unknown. The second journey recounted in the fifth notebook was a late-fall collecting trip on the same route as his first, as far as the high mountains of northwestern North Carolina.

Michaux's party first traveled north through central South Carolina and passed through the town of Camden. After entering North Carolina near the village of Charlotte, the party turned northwest across the gently rising Piedmont toward the new town of Morganton, gateway to the high mountains of northwestern North Carolina. There the botanist met Col. Waightstill Avery, probably the most important citizen in the district. Two days' journey beyond Morganton, Michaux secured a local guide and made a five-day reconnaissance of the Black Mountains, a roadless wilderness topped by summits of over six thousand feet. Continuing northwest, he explored the lofty Roan Highlands along the border of North Carolina and Tennessee, as he followed a difficult route to Elizabethton, Tennessee. Two days later the botanist altered his original plans when he arrived at Anderson's Block House in Virginia, near the Tennessee border.

Michaux's initial plan had been to continue into Kentucky along the Wilderness Road route through the Cumberland Gap. After learning of a recent Indian massacre of a party of travelers along that road, Michaux turned northeast and followed well-traveled roads through the Virginia mountains and Shenandoah Valley. After crossing the Potomac River at Harpers Ferry, he traveled through Maryland and eastern Pennsylvania to reach his initial destination, Philadelphia. Michaux then made a quick round-trip to New York City to meet with French consular officials as well as to inspect the king's garden in New Jersey and meet with his gardener.

For the return journey from Philadelphia to Charleston, Michaux traveled rapidly overland along a more coastal route. Coming south from Philadelphia through Wilmington, Delaware, and Baltimore, Maryland, the botanist's party crossed the Potomac River into Virginia at Alexandria. Continuing south through Virginia, Michaux and his party traveled through Fredericksburg, Richmond, and Petersburg before exiting Virginia near Halifax, North Carolina, eight days later. They then followed a direct route south to Wilmington, North Carolina, where they followed roads near the coast into South Carolina and on to the Charleston garden, 110 days after their departure.

Three days after the party's return from this long journey, François André suffered a serious eye injury. Michaux sought medical assistance, and his concern for his son's welfare is evident. Nonetheless, in less than three weeks, André Michaux, possibly once again accompanied by François André, and almost certainly accompanied by one or more enslaved black men, who would dig up plants, set out on a fall collecting expedition to the North Carolina mountains. He retraced the route of his June journey and returned to the wild country of the Black Mountains in North Carolina. Despite encountering cold, snowy weather conditions, he succeeded in gathering twenty-five hundred trees and shrubs for his garden near Charleston and eventual shipment to France. The last entry in his fifth notebook is for December 9, 1789, when he passed Morganton on his return journey to Charleston.

The chapter concludes with a letter sent to Michaux by his colleague André Thouin in Paris that shows how his work in North America was being appreciated in France. The letter also indicates how optimistic the scientists at the

Jardin du Roi Paris were about the French Revolution, then in its early stages. It is not known whether Michaux actually received a copy of this letter.

### Notebook 5 begins

MAY 30, 1789, I left Charleston.*

JUNE 6, we arrived in Camden, a little town found at 120 miles from Charleston.[1]

JUNE 10, [we] went by Charlotte in Mecklenburg County,[2] 80 miles from Camden.[3]

Saw a *Magnolia cordata* 18 miles from Charlotte. This *Magnolia* seems to differ from the *M. cordata* discovered a few years earlier; the leaves were glaucous with a well-marked bluish tint underneath.[4]

A little before arriving at the ferry on the Catawba River, I saw an unknown shrub with neither flower nor fruit; it resembled *Calycanthus* in certain aspects.[5]

Saw the same shrub near the Burke Courthouse.[6]

JUNE 13, 1789, we arrived at Burke Courthouse, 80 miles from Charlotte. Visited Col. Avery and left Burke the 14th.[7] We were at 298 miles from Charleston and saw *Magnolia cordata* at the foot of several rather high mountains,[8] noted the clay soil and the quartz rocks.

JUNE 15, we arrived at the home of Col. Waford,[9] found between high mountains. This spot is called Turkey Cove.[10] The distance from Burke to Turkey Cove is 30 miles, 310 miles from Charleston.

JUNE 16, Lodged at Capt. Ainsworth's [house],[11] one mile away, and [who is] related to Col. Waford.

JUNE 17, we left for Black Mountain at . . . miles from Turkey Cove.[12]

Our collecting on this mountain lasted until June 22. I recognized a new species of *Azalea, Andromeda* . . . , *Vaccinium* . . . , *Viburnum* . . . , and several other plants that the loss of my journal keeps me from describing, but my herbarium will be proof that these plants are new.

JUNE 22, we arrived again at Capt. Ainsworth's.

* Michaux's footnote: My journal having been lost on July 1, a large number of interesting observations since May 30 until this date will be abridged.

JUNE 23, 1789, [we] left for Yellow Mountain.[13]

JUNE 24, we arrived at night at the foot of Yellow Mountain at 30 miles from Turkey Cove. This mountain is considered in North Carolina and in Virginia as the highest mountain in the whole of North America.

I recognized . . .

There is a five-mile ascent to the top of the mountain. Before arriving one has to hike for several miles on the chain of the highest mountains, called Blue Ridges.[14]

JUNE 28, we arrived at Capt. Farkinson's, the first house after leaving Turkey Cove.[15] The trail is narrow, steep at several points; it is often necessary to go on foot; several times we were obliged to cut branches of *Kalmias* with a tomahawk, which one must always carry when traveling in the forests, the so-called Wilderness. The distance from the summit of Yellow Mountain to Capt. Farkinson's residence is 15 miles.

JUNE 28 we slept at Maj. Carter's [residence] at 20 miles from the summit of Yellow Mountain.[16]

JUNE 29, 1789, we crossed the river,[17] and we slept at . . . , 4 miles from Block House. The Block House is a site well known as a rendezvous for travelers going to Kentucky.[18] The distance from Maj. Carter's house to [the] Block House is 25 miles, 390 miles from Charleston.

We learned that the preceding week several travelers were killed by Indians as they returned from Kentucky, and I decided to abandon the trip to Kentucky and continue my plant search in the mountains of Virginia.[19]

JUNE 30, I continued my route toward the mountains and the same night entered the territory of Virginia.

JULY 1, we arrived at Washington Courthouse [Abingdon], the first town* of Virginia that one finds on the western side of the mountains when leaving North Carolina. From Block House to Washington is 35 miles.[20]

JULY 2, we slept at 30 miles from Washington, 65 miles [from the Block House].

JULY 3, at 35 miles from Washington, noticed a plant whose solitary fruit on

---

* Michaux's footnote: first town, if we can call that a town composed of 12 houses (log cabins). In this town one eats only corn bread. There is no fresh meat or cider, but only bad rum.

FIGURE 11. Maj. Landon Carter's home, Carter Mansion, Elizabethton, Tennessee. André Michaux spent his first night in Tennessee in what was likely the finest home in the territory, this dwelling on the Watauga River in present-day Elizabethton. As the first trained botanist to visit Tennessee, Michaux would spend a hundred more nights there and meet many other important settlers of the generation that carved the state out of the wilderness. (Photo by Charlie Williams)

a stalk had the form of a pipe mounted on its tube. The leaves are double on a single petiole.[21] Observed the fruit of the ginseng: umbel undivided, involucre of subulate leaflets, 5-parted, sepals minute toothed, petals oblong recurved, stamens 4 as long as the corolla, anthers inclined downward, ovary subcompressed, inferior, styles 2, stigmas recurved.

JULY 3, we slept at Stone Mill, found 93 miles from Block House.[22]

JULY 4, [we] passed by Montgomery Courthouse, named also Fort Chiswell, and slept 2 miles farther on. Our trek was only 13 miles because of the rain.[23] [Another small sheet inserted in the manuscript by Michaux indicates: "slept two miles away from Fort Chisses and traveled 26 miles."][24]

SUNDAY, JULY 5, we walked 28 miles and slept at the ferry of New River.[25]

JULY 6, at half past noon, we went by the mountains called Appalachian and entered those named Alleghenies. Our hike was 36 miles long. As soon as we passed the eastern side of the mountains where the rivers run toward the sea (whereas west of the mountains the rivers are supposed to run into the Ohio and the Mississippi),[26] [we] saw *Diospyros*, *Cephalanthus*, *Annona*, *Styrax*, and several other trees that I had not seen before.

July 7, 1789, noted a *Pavia lutea* 3 feet in diameter, and on the Roanoke River, *Thuya occidentalis* among the steep rocks that border this river on the north. Our trek was 34 miles.

July 8, we visited a natural arch 300 feet in height. Our journey was 25 miles.[27]

July 9, left Lexington, little town whose commerce is kept alive by the enterprises on the western rivers (Western Waters settlements) thus named.[28]

The trail while out of the high mountains was broken by hills covered by rocks and streams. The rocks are made up of calcareous blackish material substance, but in much smaller amounts than clay. Hike of 24 miles.

July 10, we went through Staunton, a very commercial small town in the mountains.[29] A mile and a half before arriving, I noticed in a prairie along streams a *Spiraea* whose flowers, disposed in panicles, are pink as are the pedicels. Petals 4, somewhat rounded, angled, clawed, each with linear pedicel; stamens 32, inserted on calyx, filaments very long, anthers subrotund erect; ovaries 6, oblong styles short, recurved, stigmas capitate; sepals 4, laciniate, small, reflexed, persistent.[30] We walked only 15 miles because of the rain.

July 11, we went across the North Branch River[31] and continued our journey across an uneven, hilly territory having the Blue Ridge on our right and the Alleghenies on our left; our journey was 25 miles long.

Sunday, July 12, I noticed in a prairie on the edge of a stream the same *Spiraea* found two days before. This one was close to a property where the house was the most decorated of all the houses that I had seen in this part of Virginia. The man showed me everything that he had done to enhance the cultivation of his farm and to make it even more attractive. He had a breed of cows recently imported from England, very large and fat pigs of a breed different from those found in the district. He regularly fertilized his land. The fruit trees were well taken care of, etc. Our trek was 30 miles long, and we passed by a little village called New Market.[32]

July 13, 1789, went by Strasburg [then Stoverstown], another township located 40 miles from Winchester. I noticed *Thuya occidentalis* a little before arriving in this small market town in the foothills of the mountains, along the river. Our journey was 27 miles long.[33]

July 14, we passed Winchester, a little town whose commerce with the Kentucky settlements is done overland. Merchandise arrives from Philadelphia, Alexandria, and especially Baltimore. Our trek was 25 miles long.[34]

July 15, we passed through Charles Town, a little town made up of 6–10 houses, found 22 miles from Winchester.[35] Afterward we crossed the Potomac River to enter the state of Maryland, 30 miles from Winchester. The Shenandoah and Potomac become joined at a site called Harpers Ferry. High and steep mountains covered with rocks are close to this spot. I observed several European plants before leaving Virginia, such as *Hypericum perforatum*, *Arctium lappa*, *Echium vulgare*, *Trifolium lagopus*, *Verbascum album* and *V. nigrum*, *Veronica officinalis*, etc. etc.[36]

July 16, 1789, we went through Frederick, a well-built small city in Maryland; the houses are made of brick, and commerce is quite flourishing; [our journey was] 25 miles.[37]

July 17, nothing special; the land was flatter, the quartz rock often very pure but sometimes combined with ferrous compounds. I also saw some hills whose rocks were calcareous and the soil clayey, as found in large parts of Virginia. In the areas of Maryland where limestone is combined with clay, cereals that were still on their stalks appeared to be in better condition, and the vegetation on uncultivated land, stronger, more robust, the trees greener, herbaceous plants more vigorous, and farm animals more hardy.

In the whole of Virginia from north to south beyond the Blue Ridge, the soil appeared to me to have generally this combination of clay with lesser amounts of calcareous matter, the countryside is rich, producing much grain, the farm animals are numerous and well fed all year long, the horses are vigorous, and the inhabitants enjoy very good health. A farmer from this district told me that generally wheat yield was 15 bushels per acre, but very often 12 bushels, and rarely 20 bushels. Our journey was 31 miles long.

July 18, went by York, rather pretty town at 59 miles from Frederick. The countryside appeared to me to be better cultivated in these parts. The inhabitants are German as in Pennsylvania.[38] They are generally hard working and industrious. Soils in these parts of Maryland are alternatively clayey [and] calcareous and sometimes containing iron. The stones and rocks are made of quartz, iron-bearing schist. In several areas one finds rocks of primitive

FIGURE 12. Bartram house, Philadelphia, Pennsylvania
(Photo by Charlie Williams)

calcareous substance mixed with veins of quartz. Our journey was only 24 miles long.

SUNDAY, JULY 19, we passed through Lancaster, little town of Pennsylvania, peopled by Germans ([in] the morning we crossed the Susquehanna River).[39] Our journey was 21 miles long.

JULY 20, our journey was . . . miles.

JULY 21, we arrived in Philadelphia after having traveled more than . . . miles since our departure from Charleston,[40] not counting the journeys into the mountains that are off the main road.

JULY 22, I visited M. de Marbois, French consul.[41]

JULY 23, visited the garden of Mr. Bartram, botanist near Philadelphia; noted in his garden a new species of *Prinos* whose acuminate leaves are not denticulate [toothed]. Saw a monoecious *Zanthoxylum* from the northern regions of America.[42] *Hydrastis* . . .

JULY 24 and 25, I was busy with visits.

SUNDAY, JULY 26, I visited several gardens around Philadelphia.

JULY 27, I sent my horses to the countryside to lower the expense, which is considerably higher in a large town.

JULY 28, I continued activities of the previous day to stock up on items that are not available in Charleston and that are necessary for my garden in the Carolinas.

JULY 29, I left for New York.[43]

JULY 30, [I] arrived in New York.

JULY 31, I visited M. de La Forest, who was about to leave for Albany. I asked to see M. le Comte Dumortier, but M. de La Forest told me that he was out in the country, as was M. Otto.[44]

AUGUST 1, I visited the garden near New York and found it in quite good order. The gardener had sowed much grain and planted many young shrubs that would be sent to France after they developed a good root system.

SUNDAY, AUGUST 2, 1789, I settled accounts with the gardener, and I advised him on the trees and shrubs that should be sent next winter.

AUGUST 3, I visited Mr. Willet, grandson of Dr. . . . and left the same day to return to Philadelphia.

AUGUST 4, I arrived in Philadelphia.

AUGUST 5, 6, 7, and 8, I was busy with depositing a letter of exchange to obtain funds that I needed to pay travel expenses and [to pay] for my return to Charleston.

SUNDAY, AUGUST 9, I visited different gardens, particularly that of Mr. William Hamilton.[45]

AUGUST 10, an accident occurred to one of my horses that resulted in a cut on the shoulder and knee caused by a fall over a rock, obliging me to replace him by securing another horse.

AUGUST 11, I remained to have the horse's wounds dressed.

AUGUST 12, I went to visit Mr. Le Coulteux, who sold me a horse for 70 dollars.

AUGUST 13, 14, 15, and Sunday, 16, were spent in finishing my affairs in Philadelphia.

August 17, [I] left Philadelphia to go and get my horses, which I had sent to the countryside at Mr. Bartram's.

August 18, rain kept me from leaving.

August 19, rain continued all day.

August 20, I left very early in the morning and slept at Wilmington, small city in the state of Delaware situated 30 miles from Philadelphia.[46]

August 21, I passed by Christiana Bridge [and] Elk River.[47]

The soil in the state of Delaware is not as good as that in Pennsylvania, less clayey and more sandy. I noted *Magnolia glauca* and also *Chionanthus* more frequently at 52 miles from Philadelphia. Today our journey was 27 miles long.

August 22, [I] crossed the Susquehanna River and entered Maryland; the soil is dry, sandy, and ferrugineous. Noticed that *Fagus pumila* (Chinquapin) is found in abundance. Journey was 27 miles.

Sunday, August 23, we arrived in Baltimore, capital of the state of Maryland.[48] We traveled 24 miles.

August 24, 1789, I visited M. Le Chevalier d'Annemours, consul of France.[49]

August 25, left Baltimore, our journey was 34 miles.

August 26, we went through Bladensburg and Alexandria, first city of Virginia, whose commerce was languishing despite its favorable location on the Potomac River.[50]

This town is the homeland of Gen. Washington. His residence is 8 miles south of this city on the bank of the river.[51] Our journey was 28 miles long.

August 27, went by Colchester, a little unremarkable town.[52] Dined at Dumfries, small town of shopkeepers with 8–10 houses and a total of approximately 30 families.[53] The soil is clayey but cold and with poor fertility in this area of Virginia. Our journey was 30 miles long.

August 28, we went by Fredericksburg, a rather agreeable little city on the southern bank of the river.[54] Our trip was 27 miles.

August 29, we walked 30 miles.

Sunday, August 30, 1789, we arrived in Richmond; our journey was 27 miles long.[55]

August 31, we stayed in Richmond.

September 1, we left Richmond and went through Petersburg, a small but commercial place;[56] the soil was sandy between Richmond and Petersburg. *Mimosa . . . , Hopea,*[57] etc. Many plants of the Carolinas. Trip of 34 miles.

September 2, the soil continues to be sandy; trip of 29 miles.

September 3, went by Hick's Ford [now Emporia], last courthouse in Virginia;[58] soil arid and sandy; the houses are poor, and the inns very bad; arrived in Halifax, first city in North Carolina.[59] Trip of 35 miles.

September 4, went by Enfield Courthouse; soil sandy, long succession of woodlands and nonagricultural land.[60] Trip of 21 miles.

September 5, went by Dorchester Bridge on Swift Creek, by Duncan Lemon's Ferry: 30 miles.[61]

Sunday, September 6, traveled by Peacock's Ferry on the Contentnea Creek: 31 miles.

September 7, went by Whitfield Ferry: 31 miles.[62]

September 8, went through Rockfish and Washington Township;[63] noticed at several sites the plant *Dionaea muscipula* in sterile, sandy, humid sites: 28 miles.[64]

September 9, 1789, went across the Northeast Cape Fear [River] and arrived in Wilmington: 34 miles.

September 10, visited M. Ducher, vice-consul of France, and left in the afternoon.[65] After having crossed three rivers, saw at two and a half mile[s] from the city *Kalmia angustifolia, Dionaea muscipula,* and a new species of *Andromeda.* Slept at Town Creek, 10 miles away.[66]

September 11, went by Lockwood Folly, and went to bed at Little River on the border of North and South Carolina: 40 miles.[67]

September 12, went by east end of Long Bay and slept in a small house bordering the sea: 25 miles.

Sunday, September 13, went by west end of Long Bay and stayed overnight on the border of the River Santee: 32 miles.[68]

September 14, the wind was so strong that we were unable to cross the river, which is 5 miles wide in this area. The wind died down the following night, and we crossed to the other side. I paid $3 for the crossing of three horses and $5 for the dinner, supper (hot water) for two persons, and the expense of three horses and a black man. We went only 4 miles.

SEPTEMBER 15, our journey was 32 miles long.

SEPTEMBER 16, the distance traveled was 28 miles.

SEPTEMBER 17, crossed the Cooper River and arrived home; 5 miles.[69]
    A total of 190 miles from Wilmington to Charleston.

SEPTEMBER 18, 1789, we spent the day resting at home and resting our horses.

SEPTEMBER 19, I went to Charleston, where M. Petry handed me letters received during my absence.

SUNDAY, SEPTEMBER 20, my son while walking on the path was injured in the eye at the base of the pupil by someone who was firing on a partridge.

SEPTEMBER 21, he was bled from the arm on the advice of the doctor.

SEPTEMBER 22, the white of the eye was considerably swollen, and I decided to take him to Charleston to be within reach of assistance.

SEPTEMBER 23, the pain continued to increase until Sunday, [September] 27. During this interval, I made several visits to town and returned to our house to look over the different tasks in the garden, which I had found in rather poor shape at the time of seed collection. I gathered seeds of *Illicium*.[70]

SEPTEMBER 30, there was hope that things would improve; the doctor having made an incision, the eye was less inflamed, and the swelling went down after I applied a cooling compress. The grief to which he surrendered indicated that the pain continued to increase.[71]

OCTOBER 1, 1789, rain that had started the day before allowed me to prepare trees by placing them in pots to be put into the greenhouse, and we spent time around the garden doing essential tasks.

---

SUNDAY, NOVEMBER 8, 1789, left and slept at Moncks Corner, 32 miles from Charleston.[72]

NOVEMBER 9, ate breakfast at Jackson's Tavern, 9 miles.[73] At 7 miles saw *Ilex angustifolia*. The trees that are most abundant are: *Quercus alba, Q. nigra, Q.*

MAP 9. Eastern South Carolina locations in Michaux's narrative
(Map by Brad Sanders)

*nigra aquatica,*[74] *Q. salicifolia,*[75] *Q. rubra, Liquidambar styraciflua, Nyssa aquatica, Crataegus . . . , Nyssa dentata, Cupressus disticha.* Slept at Eutaw Springs,[76] 19 miles from Jackson's [Tavern] and 28 miles from Moncks Corner.

NOVEMBER 10, crossed the river Santee 2 miles from Eutaw Springs and had dinner and slept at Capt. Deauty's, 22 miles away.

NOVEMBER 11, went through sterile sands named High Hills of Santee and had lunch with a man named. . . .[77] Saw the philosopher Le Fevre. Observed in the sands *Andromeda glauca*; spent the night at 16 miles on the other

[south] side of Camden at [the home of] someone named Willow (nice looking girl): 22 miles.

November 12, 1789, had lunch in Camden, visited Dr. Alexander,[78] and
slept at Capt. Nettle's,[79] 6 miles from Camden: 22 miles.

November 13, breakfasted 4 miles away, and we slept near Bear's Creek at
[the home of] someone named Johnson, 29 miles, and 7 miles beyond Hanging Rock.[80]

   Note: Five miles before arriving at Johnson's there is an abandoned house
at the base of which the trail forks; the left branch of this fork also leads to
Charlotte according to Maj. Bartley, but it is 80 miles from Camden by this
road.[81]

November 14, 1789, I left at 6 o'clock from the Johnson house and arrived
at a creek above which there is a house, 6 miles distant from Johnson's. Saw
*Triosteum*. Four miles farther there is a plantation to the left and a creek to
the right of the trail.[82] Near this creek, [I] saw on the raised bank of the creek
an unknown *Viburnum*, whose dried leaves seemed to be trilobed. This *Viburnum* is 2–3 feet high and with very thin stems. It is 7 miles from this creek
named . . . to the plantation of John Cry.[83] Between this large creek and the
plantation, [I] saw another little creek near which was a type of pear, unknown small tree. This day's journey was 17 miles.

Sunday, November 15, 1789, went by a plantation 8 miles away; and 9
miles before arriving in Charlotte, saw *Triosteum*, *Clematis erecta*; soil alternatively yellow or red clay gravely; rocks of granite and very often of very
white hard quartz, as well as iron-bearing flint: Red oak with long petioles,
oak with long petioles [and] tomentose leaves, and black oak are the most
common; cultivated soil produces wheat, oats, and corn. Along the Catawba
River, things are very good; the herbaceous plants are a little better than in
South Carolina, but the sheep are not very good looking and the other farm
animals are not plump.

   Arrived that night in Charlotte in Mecklenburg County in North Carolina: 25 miles. Two hundred miles from Charleston.

November 16, 1789, crossed the Catawba River at the place called Tuckasege Ford, 14 miles from Charlotte;[84] 2 miles before arriving at this ford, we
found an unknown shrub with opposite leaves, and we went to sleep at the
home of Peter Smith;[85] one or two miles before arriving, saw near a creek on
the banks on which grew *Kalmia* and *Ilex* a *Magnolia* leaves glaucous, very

long and cordate, globose fructifications, branches whitish appearing silky. This *Magnolia* is of lesser stature than other known species.[86] This journey was 26 miles long.

November 17, 1789, we went by Lincoln Courthouse,[87] 12 miles, and we spent the night at Henry Watner's [house],[88] 16 miles from Lincoln; in all 28 miles.

November 18, considerable white frost. Found the countryside mountainous and the rocks made up of black tourmaline, quartz, and mica, but more often quartz or iron-bearing flint and clay in the softer stones. Arrived at Burke Courthouse. Two miles before arriving, I saw the unknown shrub from the Catawba River: 29 miles.

November 19, left Burke and went by the home of Col. Avery, whose house is on the Catawba River at 3 miles from Burke. Just before arriving there, found in the creeks a new *Astragalus* and a *Menispermum* with black fruits; slept at 12 miles from Burke.

November 20, we had breakfast 6 miles farther and saw *Magnolia cordata*, *Juglans oblonga*, and then we arrived at Turkey Cove. On the road noted *Epigea procumbens* and *Gaultheria procumbens*. It is 15 miles from where we slept to Turkey Cove.

Turkey Cove is the place from which one can go to different sites on the high mountains.

November 21, went on the northern branch of the Catawba River.[89] Saw an *Andromeda arborea* 43 inches in circumference.

Sunday, November 22, collected and picked up acorns of the glaucous oak in the high mountains.

November 23, left for the high mountains. Saw an *Andromeda arborea*, 49 inches in circumference.

November 24, I moved into the Blue Ridge of North Carolina.[90]

November 25, I arrived in the lower portions of Black Mountain and collected *Azalea fulva*, a new Azalea, etc.

November 26, I collected *Magnolia cordata*, *M. acuminata*, etc. etc.

November 27, I arrived at the southern cataracts of the Toe River and collected a new *Viburnum*.[91] Frost and snow.

November 28, thaw and rain all day.

Sunday, November 29, I came back to the home of Mr. Ainsworth.

November 30, I collected *Kalmia latifolia* and *Rhododendron*.

December 1–5, went to several high mountains and then packaged my collections, which approximated 2,500 trees, shrubs, and plants; in total, 7 boxes.

(Remember to call at Capt. Smith's, the 2nd house below Mr. Seagrove, and get letter directed to Capt. Stafford)[92]

December 9, went by Burke Courthouse [Morganton].

The . . . arrived in Charleston.[93]

Notebook 5 ends here. Few letters describing the receipt of Michaux's shipments in France are known. The letter quoted below provides useful information in this regard as well as about the situation during the opening stages of the French Revolution.

### André Thouin to AM, from Paris, November 21, 1789[94]

I received last February 28 the letter that you gave me the honor to write the preceding January 5, and a while later Abbé Nolin gave me seeds that you mentioned in this letter. These seeds appeared to be excellent choices and had perfect maturity. I sowed them with care, and the majority grew during the year. I hope that those that did not germinate so far will do so this coming spring and that we will have important results from this shipment. You write that M. Lemonnier will give me *Illicium floridanum* from a shipment destined to Monsieur. I did not receive them [because] either these shrubs arrived in poor condition or for other unknown reason. In any case, many thanks to you for this interesting shipment.

The material sent from Charleston on April 28 has still not arrived. I worry about its fate as so much time has elapsed without sign of it as well as the troubles that made Count d' Angiviller leave France several times. However it is rumored that within four days he will be back in Paris and that his administration will take up its normal course.[95] As soon as he returns, I will make all the necessary inquiries to recover this shipment, which interests me greatly, [both] from the fact that it comes from you and because it contains valuable species.

It is true, sir, that I do not receive as often as I might wish news of you and that the shipments that you make do not arrive in intact condition. But it is an unhappy situation for which I cannot find a remedy because, on the one hand, since you are always on the go, it is not possible to write us more often, and on the other, the shipments go through so many hands that it is not unlikely that things get lost. In spite of that, I assure you that we are perfectly satisfied with your correspondence and that we earnestly wish it to continue.

You have probably learned that botany has a new genus that bears your name (*Michauxia*). It is your beautiful *Campanula*, which you brought back from the Near East and which we cultivate in our gardens. Here is a note from M. l'Héritier, who tells you this and would request from you live *Vaccinium* and dried specimens of *Geranium*. This botanist works on both genera and would like to complete his work as much as is possible. If you could fulfill his wishes, you would give him pleasure as well as render a service to science.

M. Mauduit asks me to send you a note in relation to the promises that you made him regarding demands for birds and insects that he wants.[96] You will respond depending on the circumstances that offer themselves. The thing that made him remember you is the superb shipment that you made to the royal garden and that he saw at M. Daubenton's. All the birds appeared to be well preserved with many new ones. They were stuffed and placed in the royal collection. No doubt M. Daubenton will have acknowledged this beautiful shipment.

The work of M. de Jussieu has finally been published in the last few weeks. I rush to send you a copy, and I ask you to accept it as a keepsake and souvenir. You will find all the genera, ancient as well as modern, described with care and arranged according to a natural order, as well as many good observations on natural families and a new viewpoint on botany. I hope that this work will be of interest, that it will be useful and pleasing to you.

Since your departure, everything has changed in France. The nation has taken back its indefeasible rights, [and] it has shaken its fist at the ministerial despotism. The nobility has lost its aristocratic privileges. The vast property of the clergy has just been declared property of the state. Finally the courts have been reduced to only the functions of the judges. France will acquire a constitution that will become in the future a source of its happiness as well as its glory. But all these large changes could not occur without making many people unhappy; this results in plotters who cause troubles and several slight disturbances. Nevertheless we got off at a good price because this memorable revolution was accomplished with hardly any loss of blood. When you return to your country, especially if several years elapse before this takes place, you will not recognize anything; everything changes and thank God to the satisfaction of the good patriots.

I wish you good health, much happiness in your research, and [I] ask you to remember me sometimes and our garden . . .

# CHAPTER SIX

# Charleston Interlude, 1790–1791

SUMMARY: ANDRÉ MICHAUX DID NOT keep a journal during most of the year 1790; the first entry in notebook 6 is dated December 30, 1790. Nonetheless, there is evidence for his activities in letters and notes indicating that he made no journeys outside South Carolina until April 1791.

One important personal circumstance changed during 1790. André Michaux no longer had the assistance and companionship of his son, François André. The young man sailed for France in February carrying a collection of seeds and a letter for André Thouin. He arrived sometime before April 30, 1790, when Abbé Nolin reported his arrival to Count d'Angiviller. The servant Jacques Renaud, who had been trained as a gardener, remained at the garden near Charleston. Visiting the property in 1790, the US Census taker also reported the presence of one white woman, not mentioned elsewhere, and twenty-four enslaved black people. There is no information as to how many of the latter were adults and under just what arrangements they lived and worked on the property (see discussion in the introduction).

One source of information that we have about Michaux's activities during these years is the notes he made in a notebook concerning seeds planted in his garden. These seed lists and his brief accompanying notes were not included in the 1889 published transcription of his journals and are published here for the first time. These lists not only confirm Michaux's presence at this garden but also reveal that he was obtaining seeds from around the world. He planted seeds from China, the Cape Region of South Africa, the Caribbean, and Central and South America. Some seeds were planted in a cold frame, and others in a greenhouse, confirming that the Charleston garden had both of these facilities in 1790. A note in the 1790 seed lists and a 1791 letter to John Bartram Jr. confirm Michaux's continuing exchange of seeds and plants with the Bartrams in Philadelphia. A September 1790 letter to Sir Joseph Banks in London, responding to a letter that he had sent to Michaux, confirms that Michaux had sent a small shipment of plants to Banks.

The other primary sources of information for Michaux's activities in 1790 are the surviving letters that he sent to Count d'Angiviller, a letter André Thouin wrote to Michaux from Paris, and a few sentences in a report that Michaux drafted in 1797, after his return to France. We learn from Michaux's letters to Count d'Angiviller, that, at the direction of Consul Petry in Charleston and M. de La Forest at the consulate in New York, he had ceased his distant journeys and was trying to reduce the expenses of both gardens. He also urgently needed to draw funds to continue their operations. Saunier continued to work at the New Jersey garden, and Michaux was having difficulty supervising him from Charleston.

The difficulties that Michaux faced in continuing his mission during the early stages of the French Revolution are apparent in three letters he sent from Charleston that we have reproduced below.

### AM to Count d'Angiviller, February 5, 1790[1]

I had the honor of informing you last September 29 that I had received upon my return from beyond the mountains two letters that you did me the honor of writing, one dated November 9, 1788, concerning the grass of Mr. Walter. The second was a duplicate of the one from July 28, 1788, concerning *Panicum altissimum*. Since the letter that I had the honor of writing to you then, I have only been able to obtain three pieces of turf of Mr. Walter's grass. I am sending you two, having divided the third into pieces to multiply it to send it to you in greater abundance later.[2] As to [the] *P. altissimum*, which I brought back in April from the Bahamas, [it] multiplied prodigiously during last summer, but at the beginning of winter, it was not able to survive the least amount of cold, and it died as early as November, while sugar cane [and] guinea pea that were also covered were able to survive the winter cold, which this year did not go below 5 degrees [43°F, 7°C].[3]

I would wish, Monsieur le Comte, that the letters and the orders that you give me would come from you or from M. l'Abbé Nolin, thus having a clearer direction. The last two letters were only duplicates, not having been received the first time, dated from 9 month[s] previous. Abbé Nolin announced in his last letter that he is writing in detail about the shipment and that he is enclosing his remarks with one of yours. I received neither one. In addition Abbé Nolin wrote that I sent two cases with chinquapins without announcing them. This shipment was part of a collection of seeds, placed in a third case, and M. de La Forest wrote October 28, 1788, that he had sent the 3 cases to Amsterdam. On August 2, 1788, I sent a box of seeds of which Abbé Nolin never acknowledged.

M. Petry demands that I inform you that he is no longer able to endorse letters of credit; M. de La Forest had made [note of] the same difficulties when I was in New York, and the difficulties have been such that these men have asked that I stop my traveling. I have done everything to work under the circumstances and only to spend money for the present shipments. In addition I reproached M. Saunier concerning [his] recent high expenditures for which he did not give details. I was obliged to threaten him [with] placing the keys to the house [of the New Jersey garden] in the hands of M. de La Forest, and I [would] give a detailed account of this affair to M. l'Abbé Nolin.

I have the honor of sending you 14 cases of trees and seeds from Carolina by a ship going to Le Havre. In addition I have received 15 cases from New York because M. Saunier had been told not to send large shipments to Bordeaux.

### AM to Count d'Angiviller, February 23, 1790[4]

I have the honor of sending you two cases of trees and a cage which contains two dwarf deer. Since a ship from St. Valery sur Somme is in port, I thought that I should take the opportunity to profit from the occasion as Paris is very close to this port.

I waited until this moment to draw [money] from the department. The numerous trips that I made in the preceding year had increased expenses. I will cease long voyages upon the receipt of your orders. I am obliged to withdraw today the sum of two hundred pounds at 30 days' *de vue* [sight draft] to the order of M. Petry. I pray you, M. le Comte, to order this payment. With the settlement of accounts that I have reached with the New York establishment, I hope that this sum will last a long time, without diminishing shipments by all occasions.

### AM to Count d'Angiviller, Oct. 4, 1790[5]

I have the honor to inform you that I obtained today from M. Dutarte a letter of exchange of three thousand pounds *tournois* at 30 days *de vue* [sight draft] by order of M. Petry, consul in Charleston. I pray that you will order the payment.

M. l'Abbé Nolin remarked that I should have obtained your permission, M. le Comte, but I received no letter since that concerning Walter's grass, November 1, 1788. I received it after ten months, and I answered it last year. This year I can send enough seeds of this grass to plant approximately one acre and a half.

To respond to your directions, M. le Comte, I cut back expenses in the New York establishment, regarding cultivation and shipments. Likewise, I did this here, and I stopped from going on voyages. Despite this I have the occasion to send you many new trees from Georgia, discovered three years ago, and with the end of the war with the Indians, this allows me to send them to you this year.

In February 1790, André Michaux sent a shipment of live plants directly to André Thouin at the Jardin du Roi in Paris. The same ship possibly carried his son, François André, who also had seeds and a short letter for Thouin. He received Michaux's shipment, but apparently he did not write a reply until March 10, 1792. It is not known when Michaux received Thouin's letter, if indeed he received it at all. Nonetheless, it offers us a report on the fate of the plants in one of his many shipments to France.

### Andre Thouin to AM, from Paris, March 10, 1792 (excerpt)[6]

I received, Sir and dear correspondent, May 11, 1790, the letter and shipment of living plants that you had the goodness to send me from Charleston on February 6 [1790]. Although packed with care and taken every care for its survival, it was not a success, the plants had developed roots on the stems during the crossing, the packing material had dried up; in addition all the growth decayed, and we were only able to retrieve *Fothergilla*, *Gleditsia monosperma*, *Spigelia marylandica*, and *Aristolochia macrophilla*; all the remainder died. I am so sorry especially since there were excellent items, such as *Magnolia acuminata*, [*M.*] *cordata*, *Illicium* new species, *Andromeda arborea*, *Styrax angustifolia*, [*S.*] *latifolia*, and *Gordonia*. If it were possible to replace this loss, you would render us a great service, and we would be under your most sincere obligation.

The transportation of live plants across the Atlantic to France and then overland to French nurseries was a difficult enterprise with the methods and equipment available in Michaux's time. Despite his best efforts, it is likely that much of the plant material that he shipped to France suffered the same fate as this unfortunate shipment.

Michaux's list of seeds planted in the Charleston garden in 1790 is found at the end of the small notebook that he used five years later as his ninth notebook. He recorded sowing seeds on eleven separate days from January 5 through June 30, 1790. For most of the entries, he provided only the genus. At other times he provided both the genus and species, and occasionally he gave only a common name or description. Some entries are followed by a question mark, and some lines in the list are entirely blank. We have reproduced the list, eliminating the blank entries and providing our interpretations of the modern plant names where there was sufficient information to make a determination.

Many of the seeds sown in January and February 1790 appear to be seeds that Michaux had obtained during his travels in the southeastern United States.

### List of Seeds Planted in 1790, from Notebook 9

Seeds sown in the garden in the Carolinas near Charleston

January 5

*Gentiana pumila* [*Gentiana pumila* Jacq., a European species]

*Astragalus* new species [probably *Astragalus canadensis* L., Canada milkvetch; see journal entry for November 19, 1789]

*Viburnum* [*Viburnum* sp., viburnum]

*Leontice thalictroides* [*Caulophyllum thalictroides* (L.) Michx., common blue cohosh]

*Clematis erecta*? [*Clematis ochroleuca* Aiton, curlyheads; see journal entry for November 15, 1789]

*Asclepias rubra* [*Asclepias rubra* L., red milkweed, purple savanna milkweed]

[*Asclepias*] *purpurea* [possibly *Asclepias purpurascens* L., purple milkweed]

[*Asclepias*] *linifolia* [possibly *Asclepias verticillata* L., whorled milkweed]

[*Asclepias*] *frutescens*? [*Asclepias* sp., milkweed]

Unknown plant from North Carolina

*Clematis semineparvo* [*Clematis* sp., clematis]

*Zizania* [most likely *Zizania aquatica* L., southern wild-rice]

*Menyanthes*? [*Menyanthes trifoliata* L., bogbean]

*Spiraea* new species [possibly *Filipendula rubra* (Hill) B. L. Rob., queen-of-the-prairie; see journal entry for July 10, 1789]

## January 6
First flat

*Magnolia glauca* [*Magnolia virginiana* L., sweet bay, most likely the southern variety, var. australis, in all Charleston plantings]

*M. tripetala* [*Magnolia tripetala* (L.) L., umbrella magnolia, umbrella tree]

*M. grandiflora* [*Magnolia grandiflora* L., southern magnolia]

*Stewartia* [probably *Stewartia malacodendron* L., silky camellia]

*Chionanthus* [*Chionanthus virginicus* L., fringe-tree, old man's beard]

*Cupressus disticha* [*Taxodium distichum* (L.) L. C. Rich., bald-cypress]

*Stillinga* [likely either *Stillingia sylvatica* Garden ex L., queen's-delight, or *Stillingia aquatica* Chapm., corkwood, water toothleaf; see journal entries for April 25, 1787, and April 15, 1788]

*Erythrina* [*Erythrina herbacea* L., coral bean]

*Sideroxylon* [possibly *Sideroxylon tenax* L., tough buckthorn, tough bumelia; see entries for April 25 and September 4–5, 1787]

*Gleditsia monosperma* [*Gleditsia aquatica* Marshall, water locust]

Mixture of *Magnolia* [Note that seeds of three Magnolia species were planted separately this same day, January 6]

*Fothergilla* [*Fothergilla gardenii* L., coastal witch-alder]

*Ilex* [*Ilex* sp., holly], etc. etc.

## February 18 in the nursery
1st and 2nd rows

*Magnolia grandiflora* [*Magnolia grandiflora* L., southern magnolia]

*Magnolia glauca* [*Magnolia virginiana* L., sweet bay]

*Stewartia* [probably *Stewartia malacodendron* L., silky camellia]

*Chionanthus* [*Chionanthus virginicus* L., fringe-tree, old man's beard]

*Stillingia* [probably additional seeds of the same *Stillingia* planted on January 6]

*Nyssa dentata* [*Nyssa aquatica* L., water tupelo]

*Halesia* [*Halesia* sp., silverbell]

*Sideroxylon* [probably additional seeds of the same *Sideroxylon* planted on January 6]

*Fothergilla* [*Fothergilla gardenii* L., coastal witch-alder]

*Zanthoxilon* [*Zanthoxylum* sp., prickly-ash, toothache tree]

*Gleditsia monosperma* [*Gleditsia aquatica* Marshall, water locust]

*Bignonia sempervirens* [*Gelsemium sempervirens* (L.) St.-Hil., Carolina Jessamine]

### February 22

First row

*Yucca aloifolia* [*Yucca aloifolia* L., Spanish dagger]

*Y. gloriosa* [*Yucca gloriosa* L., mound-lily yucca, Spanish bayonet]

*Y. filamentosa* [*Yucca filamentosa* L., curlyleaf yucca]

*Hopea* [*Symplocos tinctoria* (L.) L'Hér., sweetleaf, horse sugar]

*Evonimus* [most likely *Euonymus americanus* L., hearts-a-bustin'-(with-love), strawberry-bush]

*Calycanthus* [*Calycanthus floridus* L., sweet-shrub]

*Ilex angustifolia* [perhaps *Ilex myrtifolia* Walter, myrtle holly]

*Pinus foliis longissimus* [*Pinus palustris* Mill., longleaf pine]

*Liquidambar* [*Liquidambar styraciflua* L., sweet gum]

### February 24

In the northern cold frame

*Azalea* new species [*Rhododendron* sp., azalea]

*Azalea fulva* [*Rhododendron calendulaceum* (Michx.) Torr., flame azalea]

*Andromeda wilmingtonia* [*Zenobia pulverulenta* (W. Bartram ex Willd.) Pollard, Zenobia, honey-cups]

*Kalmia glauca* [probably *Kalmia polifolia* Wangenh., swamp laurel, bog laurel]

*Rhododendron* new species [possibly *Rhododendron catawbiense* Michx., Catawba rhododendron, pink laurel; see note for June 23, 1789 (chapter 5, note 13); but much more likely *Rhododendron minus* Michx. gorge rhododendron, punctatum; see journal entries for November 11 and 19, 1788, and December 4, 1788]

Beginning in March, the seeds that Michaux planted were generally not native to North America north of Mexico. Many were attractive ornamentals from warmer climates. We have little information about the sources of foreign seeds that Michaux planted. His biographer Deleuze wrote that during this period he obtained seeds from sea captains visiting Charleston. Michaux corresponded with individuals whom he had met traveling and sometimes asked them to collect for him. It is possible that these people also shared with him seeds that they may have obtained from outside the region. He also corresponded with botanists in France and elsewhere, often sending seeds to them.

## March 7

*Bixa orellana* [*Bixa orellana* L., lipstick tree]

*Chrysophyllon glabrum* ou grosse caünit [probably *Chrysophyllum argenteum* Jacq., bastard redwood, smooth star apple]

*Guilandina moringha* [*Moringa oleifera* Lam., horseradish tree, drumstick tree]

## April 2–10

In the greenhouse

*Guilandina moringha* [*Moringa oleifera* Lam., horseradish tree, drumstick tree]

*Quasia amara* [*Quassia amara* L., bitterwood, amargo]

*Mangifera*, 1 seed [probably *Mangifera indica* L., mango]

*Bixa* [*Bixa orellana* L., lipstick tree]

*Jatropha pinnatifolia* [probably *Jatropha multifida* L., coralbush, physic nut]

*Crotalaria laburnifolia* [*Crotalaria laburnifolia* L., rattlebox]

*Aeschynomene sesban* [*Sesbania sesban* (L.) Merr., Egyptian riverhemp]

*Aeschinomene grandiflora* [*Sesbania grandiflora* (L.) Poir., agati, hummingbird tree, vegetable hummingbird]

*Mimosa* [*Mimosa* sp.]

*Senna* (*cassia*) [*Senna* sp., cassia, sicklepod; or *Chamaecrista* sp., partridge-pea]

*Winterania* [*Canella winterana* (L.) Gaertn., pepper cinnamon, wild cinnamon]

*Passiflora* [*Passiflora* sp., passionflower]

Sapodilla from Saint-Domingue [possibly *Manilkara zapota* (L.) Royen, sapodilla]

Sapodilla from Curacao [*Manilkara zapota* (L.) Royen, sapodilla]

*Mimosa* [*Mimosa* sp.]

*Oriza aristrala* [probably *Leersia oryzoides* (L.) Sw., rice cutgrass]

Beginning with the planting of April 20, 1790, Michaux's notes about the seeds planted become less complete, and fewer determinations can be made.

about the identity of the plants. The frequent question marks and classifications as "unknown" are Michaux's.

April 20, seeds from China

Unknown

*Gledistia*? [possibly *Gleditsia* sp.]

*Balsamina* [perhaps *Impatiens balsamina* L., garden balsam]

Very hard fruit

Unknown

Seed with 2 compartments

Grenadier [perhaps *Punica granatum* L., pomegranate]

Malvaceous [undetermined]

*Rhexia*? [undetermined]

Aloca [undetermined]

*Laurus*? [undetermined]

Unknown

*Illicium* [*Illicium* sp.]

Umbellifer [undetermined]

Almond tree [*Terminalia catappa* L., tropical almond]

*Indigofera*? [possibly *Indigofera* sp., indigo]

*Rhamnus*? [undetermined]

*Hibiscus* [*Hibiscus* sp., rose-mallow]

*Convolvulus* [*Convolvulus* sp., bindweed]

*Citrus* [*Citrus* sp.]

Unknown

*Hibiscus*? [undetermined]

*Sida*? [undetermined]

Capsule with 3 chambers [undetermined]

Touchu? [undetermined]

Only one seed like a *Phaseolus* [undetermined]

Unknown

*Crataegus* [*Crataegus* sp., hawthorn]

Leguminosae [undetermined]

Little round seed

Unknown

*Fagara* [possibly *Zanthoxylum piperitum* DC., Sichuan pepper]

*Mimosa* [*Mimosa* sp.]

*Styrax*? [undetermined]

Unknown

Seed like those of calabash [undetermined]

May 18

Talipot [possibly *Corypha umbraculifera* L., talipot palm, native to southern India and Sri Lanka and having the largest inflorescence of any plant]

Agathy [probably *Sesbania grandifolia* (L.) Poir., agati, vegetable hummingbird, hummingbird tree]

Acacia de cayenne [probably *Acacia nilotica* (L.) Willd. ex Delile, acacia de cayenne, gum Arabic tree, two of the many common names of this plant]

Moraney peas [undetermined]

Mimosa gum Arabic [probably *Acacia nilotica* (L.) Willd. ex Delile, acacia de cayenne, gum Arabic tree]

*Lawsonia inermis* [*Lawsonia inermis* L., henna; requested in his letter to André Thouin of January 19, 1786]

*Acacia* from India [possibly *Senegalia catechu* (L. f.) P. J. H. Hurter and Mabb., khair]

Cashew nut [*Anacardium occidentale* L., cashew]

*Martynia* from Vera Cruz [*Martynia annua* L., una de gato, cat's claw]

*Crotalaria* [*Crotalaria* sp., rattlebox]

[The seeds listed above were] Sent by Mr. Gentilot [who is not mentioned elsewhere.]

May 18

Sent by Mr. Bartram from Philadelphia

Several *Phaseolus* [*Phaseolus* sp., beans] from India and a grass

June 23

Almonds? and Mangos [undetermined]

*Bixa orellana* [*Bixa orellana* L., lipstick tree]

*Mesembrianthemum edule* [*Carpobrotus edulis* (L.) L. Bolus, Hottentot fig]

*Aeschynomene grandiflora* [probably *Sesbania grandifolia* (L.) Poir., agati, vegetable hummingbird, hummingbird tree]

*Mimosa* [*Mimosa* sp.]

*Hedisarum* [*Hedysarum* sp.]

*Hedisarum gyrans* [*Codariocalyx motorius* (Houtt.) Ohashi, telegraph plant]

Red *Nymphaea* [*Nymphaea* sp., pondlily]

Asphodelle [probably *Aspodelus* sp., asphodel]

*Sapindus* [*Sapindus* sp., soapberry]

*Giroas* [undetermined]

*Hibiscus sinensis* [*Hibiscus mutabilis* L., cotton rose, confederate rose]

*Mimosa* and unknown seed [undetermined]

Phillant. Diadelph. [undetermined]

Unknown

*Protea argentea* [*Leucadendron argenteum* (L.) R. Br., silver tree]

*Protea conocarpus* [possibly *Leucospermum conocarpodendron* (L.) H. Buek,
  grey tree pincushion]

*Malacea dolichos?* [possibly *Pachyrhizus* (L.) Rich. ex DC., yam bean]

Cotton? [undetermined]

Bulbous plant

Nut with 4 compartments like *Palinrus*

*Guilandina?*

*Chrysoph?* and *Sapindus?* [undetermined]

*Phaseol* hairy [possibly *Phaseolus* sp., bean]

Three large round grains

*Annona* [*Asimina* sp., pawpaw]

Cereal grain

*Acacia* [*Acacia* sp.]

*Ricinus* [possibly *Ricinus communis* L., castor-bean]

*Protea argent.* [*Leucadendron argenteum* (L.) R. Br., silver tree]

*Protea conocarpus* [*Leucospermum conocarpodendron* (L.) H. Buek, grey
  tree pincushion]

Round grains, Malvaceous sp., Curcuma [possibly *Curcuma longa* L.,
  tumeric]

Canna? [undetermined]

Piper? [*Piper* sp., pepper]

Malanga [probably *Xanthosoma* sp., malanga]

Coix? [probably *Coix lacryma-jobi* L., Job's-tears]

etc. etc.

Extraordinary squash

These seeds came from the Cape of Good Hope

## June 30

Orange myrthe [possibly *Pimenta dioica* Merr., allspice]

Grapefruit [*Citrus x paradisi* Macfad., grapefruit]

Mixture

Michaux's avid interest in plants from all parts of the world is also exhibited in his correspondence with Sir Joseph Banks. This letter is in response to one sent to Michaux by Banks.[7]

### AM in Charleston to Sir Joseph Banks in London, September 20, 1790[8]

I only received your letter on August 15 that you gave me the honor to write March 6. If you write in the future, it is only necessary to put this address: André Michaux, Botanist, Charleston, South Carolina.[9]

I think, Sir, that you are satisfied in terms of cascarilla, while I saw in the Bahamas a shrub with leaves like rosemary that seems to me to be that designated by Sloane: Ricino affinis Rosmarini folio, and while it is surely not *Croton cascarilla*, nevertheless Linnaeus added this phrase of Sloane's in his treatment of *Croton cascarilla* in his *Species Plantarum*. No. 1 is a sample of this shrub. No. 2 is the one that I believe is really Eleuthera-bark of Catesby with leaves like *Eleagnus*. No. 3 contains samples of *Juniperus* from the Bahamas. I did not see this wood among those cut for export. I noticed that it was used in construction of houses, sloops, etc. I will take all information concerning it and will acquaint you with it.

I have added samples of several other plants as you will see from the enclosed list, but if I knew the species that interest you the most, I would send them with the greatest pleasure.

I thank you, Sir, for the offer that you have the goodness to make me. As the climate here is very favorable to vegetation, the summers long and winters short, if you or our friends have seeds that have trouble reproducing in Europe, I would be very happy to cultivate them and to send them in large quantity with the condition of taking account of who would be first to provide them; this condition seems justified.

I have the honor to be with respectful consideration, Sir, your very humble and very obedient servant.

Plant List.

No. 1 *Croton*, dioecious shrub Ricino (?) leaves like rosemary. Sloan. The odor is very pleasant [probably *Croton eluteria* (L.) Wright, cascarilla, Eleuthera-bark]

2. *Croton cascarilla*, Ricinoides (?) leaves of *Eleagnus* Catesb. P. 46, shrub 6–10 feet high, stems and branches gathered together and assembled

like those of *Cupressus fastigata* [*Croton cascarilla* (L.) L., Eleuthera-bark, cascarilla]

    3. *Juniperus* of the Bahamas [probably *Juniperus barbadensis* L., Barbados cedar, pencil cedar]

    4. *Lythrum lineare* [probably *Lythrum lineare* L., wand loosestrife]

    5. *Serratula* [possibly *Liatris garberi* A. Gray, Garber's gayfeather]

    6. *Hibiscus* [*Hibiscus* sp., rose-mallow]

    7. *Amorpha minor* [possibly *Amorpha herbacea* Walter var. *herbacea*, dwarf indigo-bush]

    8. *Thymbra* new species [*Thymbra* sp., mint]

    9. *Ceanothus* new species [possibly *Ceanothus microphyllus* Michx., redroot, littleleaf buckbush]

    10. *Polygala*, several species [*Polygala* sp., milkwort]

    11. *Hydrocotyle reniformis* [probably *Centella asiatica* (L. f.) Fernald, centella, coinleaf]

    12. *Illicium* flowers yellow [*Illicium parviflorum* Michx., yellow anise-tree, swamp star-anise]

As 1790 came to an end, Michaux resumed keeping his journal by beginning a new notebook, his sixth. This notebook principally recounts his activities from January through May 1791, and it confirms that he spent the greater portion of this period working at the garden near Charleston.

    Interestingly, however, the sixth notebook opens with a cryptic note about botanizing in South Carolina, during forty-one days across April and May 1790. Two 1791 journeys are mentioned in the sixth notebook. Early in January, Michaux mentioned a brief trip to the lower Santee region just north of Charleston, but he gave few details about this one-week trip. It is probable that Ephraim Mitchell, then the surveyor general of South Carolina, accompanied Michaux on at least a portion of this trip. The journey described in detail in the sixth notebook is a month-long return visit to the Georgia coast in April and May, after which the journal for 1791 ends. We have only a general idea of his activities during the last half of 1791, but there is no suggestion that he traveled during this period.

### Notebook 6 begins

APRIL 1790, botanized in South Carolina, from April 7 until May 18.

DECEMBER 31, 1790, the weather was very overcast; so much snow fell from four o'clock in the morning until five in the afternoon that the earth was covered with 6–8 inches in the countryside and 6 inches in town.

I worked a little packing seeds that I planned to send to France by the ship *Pennsylvania*, Capt. David Harding, destined for Le Havre.

January 1, 1791, I continued to prepare for the shipment of seeds. M. Godart, chancellor from the consulate in Charleston, had come to spend a few days with me at the house.[10] The frost that had lasted fourteen days continued, and it became even colder.

Sunday, January 2, I continued the same work.

January 3, M. Godart having left for Charleston, sent me information that the ship destined for Le Havre would not leave until the sixth. I received news that the Americans had sent troops, 1,453 men against the Miami Indians; there were approximately 100 Indians killed, but 183 Americans were killed, and 31 were wounded. I continued my work with seeds.[11]

January 4, I worked on packing of trees for the king and for Monsieur.

January 5, the snow disappeared. I wrote letters announcing the shipment and left at night for Charleston.

January 6, I had the shipment put onboard. I paid the captain for the freight. I finished my letters and returned home on the 7th at night.

Two of the letters that Michaux wrote accompanied packages of seeds that reached André Thouin at the Jardin du Roi in good condition. Thouin reported on the success of these seed shipments in a letter to Michaux the following year.

### André Thouin to AM, from Paris, March 10, 1792 (excerpt)[12]

I received from Abbé Nolin a package of seeds that you had placed in a box addressed to M. d'Angiviller of which you advised me in your letter of January 8, 1791. They arrived the following March 25 in very good state and were sown immediately.

I also received on April 3 the package of seeds that you asked M. Limousin of Le Havre to send to me. These two packages containing 57 species of plants or trees that are missing in our collection gave us the biggest pleasure. I planted them with care, and they gave us several new plants in the first year. I hope that those that did not germinate last year will do so this spring.

Please receive my thanks, dear sir, for this valuable shipment, and please continue all the opportunities that you will be able to.

**Notebook 6 continues**

JANUARY 7, I received word from Maj. Mitchell to accompany him to an area of the state where he had recognized a new plant.[13]

JANUARY 8, I left to visit the riverbank of the Santee from approximately Murray's Ferry until near its mouth.[14] The banks of this river are cultivated with rice for the most part. The greatest number of inhabitants live rather poorly, and among the richest planters I did not even eat bread but rather corn porridge (named . . .) and salt pork. My horses lived on fodder of peas or corn.

All week I was busy with this excursion and returned home January 16. The main result of this trip was the discovery of an *Andromeda* with glaucous leaves that is found at 38–40 miles from Charleston and only 30 miles from the home that I established in the Carolinas. To find it in abundance, leaving from Charleston one must pass by Strawberry Ferry and follow the road for Lenews Ferry (called Winningham) at a distance of 10 miles from Strawberry [Ferry];[15] while continuing on the main road, one can spot this *Andromeda* in the narrow swamps that one encounters frequently in the middle of sterile pine woods in the Carolinas. These piney woods extend over an immense surface, and because of the sterile soil, they only produce pines. Rain leads to the formation of muddy streams that carry water to the rivers during and after the rains. They contain slow-moving water since it is held back by leaves and other plant debris. In these areas, which are almost always moist, we find different species of *Andromeda*, *Laurus borbonia*, azaleas, *Magnolia glauca*, *Gordonia*, etc. etc.

JANUARY 17, I went to Charleston and received a letter from my son dated from April of the preceding year. I wrote to Abbé Nolin concerning the difficulties of obtaining money from my letters of exchange, and if the difficulties continue it will be necessary for me to return to France. I wrote to my son on the same transport with Capt. David Harding.[16]

JANUARY 18, and the following days until January 22, no work was done at the nursery as the blacks were obliged to work communally on the main road.[17]

JANUARY 22, I had the visit of Mr. Fraser; it seems that the good reception that he received in France made him more honest. He praised France abundantly. He wanted to know what new plants I had discovered. But knowing that his objective is to sell plants, I gave him nothing and limited myself to giving him the best welcome possible.[18]

SUNDAY, JANUARY 23, I was busy in grouping together all the species of *Andromeda* in the nursery.

JANUARY 24, 1791, I went to Charleston. There was no ship bound for France, and having bought some planks I returned the same day. There was a hard frost.

JANUARY 25, I had work done in the nursery as the wind was blowing south; we were primarily busy in repairing the enclosures.

JANUARY 26, same work in repairing the enclosures and assembling in the nursery a collection of *Andromeda* [for example:] *Andromeda arborea,* [*A.*] *paniculata,* [*A.*] *coriacea,* [*A.*] *mariana,* [*A.*] *nitida,* [*A.*] *racemosa,* [*A.*] *serrata,* [*A.*] *calyculata,* [*A.*] *wilmingtonia,* [*A.*] *polifolia,* [*A.*] *formosissima.*[19]

JANUARY 27–29, continued to repair the enclosures of the garden of the nursery.

SUNDAY, JANUARY 30, I studied *Betula alnus* and *Ulmus americana.*

JANUARY 31, worked on digging up trees from the garden and transporting them to the nursery.

FEBRUARY 1–5, I continued assembling trees of the same genus in the nursery.

SUNDAY, FEBRUARY 6, the blacks were busy helping a neighbor whose house was on fire.

FEBRUARY 7, I continued work on the tree collection in the nursery and answered M. de La Forest, from whom I had received a letter two days ago, as well as [sent letters] to Abbé Nolin and to my son.

FEBRUARY 8–12, continued the same work in the nursery.

SUNDAY, FEBRUARY 13, I grafted plum trees from Persia on common plum trees of this country.

FEBRUARY 14, I studied the flowers of the red maple of the Carolinas where the hermaphroditic ones have five stamens and the male flowers similarly have five stamens.

FEBRUARY 15, the scarlet plum tree from Persia flowered in my garden, as well as the Chickasaw plum. The night of the 15th there was a strong south to west wind and considerable rain.

FEBRUARY 16, 1791, the air cleared up, and the wind went from west to north. The night of the 16th there was a storm with furious winds, and several portions of the garden's enclosure fell backwards.

FEBRUARY 17, we worked to repair the enclosures. In the morning, frost, 5 degrees on the Réaumur thermometer [43°F, 7°C].

FEBRUARY 18, frost, 6 degrees [46°F, 8°C]; we discontinued planting to repair the fences broken down by the wind.

FEBRUARY 19, continued with repairs.

SUNDAY, FEBRUARY 20, I went to collect and study several plants.

FEBRUARY 21–26, we planted trees in the nursery.

FEBRUARY 27, I collected plants.

FEBRUARY 28, I went to Charleston.

SUNDAY, MARCH 1, I planted the trees in the nursery.

MARCH 2, I finished the task of assembling in the nursery the plants belonging to the same genus.[20]

MARCH 3, I organized and planted bulbous plants and different herbaceous ones from the mountains and other parts of the Carolinas.

MARCH 4, I started to put my herbarium in order and began with collections from New York, New Jersey, and Pennsylvania.

MARCH 5–6, continued the same.

SUNDAY, MARCH 6, rain; I sowed several types of seeds.

MARCH 7, I continued to put my herbarium in order. I received a letter from my son dated July 11 of last year. I received a letter from Mr. Bartram and one from Mr. Hamilton.[21]

MARCH 8, I prepared a case of plants for Monsieur, brother to the king, by way of Bordeaux, recommended to Capt. Bass and on board to Mr. P. Texier.

MARCH 9, I wrote letters to M. Lemonnier, to my son, etc.

MARCH 10–11, I worked on my herbarium.

SUNDAY, MARCH 13.

MARCH 14, I went into town.

On March 12 Michaux wrote a letter in English to the Bartrams in Philadelphia that accompanied a small shipment of plants and seeds.

### AM to John Bartram Jr., March 12, 1791[22]

I received the valuables seeds from China you was so kind to send me, I was very glad to make acquaintance with Mosas [Moses] Bartram. His conversation is very interessant [interesting].[23] I send you a box of some plants and [a] little parcel of seeds by the same conveyance but I was so busy since some days [ago] that I could not send you all that I wish to send. I am in a very few days going to Georgia and I pray to no[t] send my plants while I shall be abroad, but still you may inform [me] if you receive this [these] plants in good order.

On my return I will inform you, excuse my haste as I fear [I do] no[t] [have] enough time to send the plants to the vessel.

My compliments respectfull[y] to Mr. William Bartram

List of the plants [with number sent and modern names]

| Pavia alba | 2 | [Aesculus parviflora Walter, bottlebrush buckeye] |
| Sideroxylon | 8 | [probably Sideroxylon tenax L., tough buckthorn, tough bumelia] |
| Philadelphus inodorus | 4 | [Philadelphus inodorus L., Appalachian mock-orange] |
| Hydrangea | 2 | [Hydrangea arborescens L., smooth hydrangea] |
| Rhododendron minus | 2 | [Rhododendron minus Michx., gorge rhododendron, punctatum] |
| Magnolia auriculata | 3 | [Magnolia fraseri Walter, Fraser magnolia, mountain magnolia] |
| Illicium | 1 | [Illicium parviflorum Michx. ex Vent., yellow anise-tree, swamp star-anise] |
| Gordonia | 6 | [Gordonia lasianthus (L.) J. Ellis., loblolly bay] |

Approximately four weeks later, as he completed his preparations for the trip to Georgia, Michaux wrote a letter to Count d'Angiviller in response to a letter he had recently received from him. The letter was received in

FIGURE 13. William Bartram, after a painting by Charles Willson Peale in 1808 (Courtesy of the Library of Congress, Reproduction Number LC USZ62-38487)

France on May 31, 1791, but it was too late for any action by the count, who was stripped of all his property by the revolutionary government on June 5, 1791, and fled France shortly afterward.

## AM to Count d'Angiviller, April 15, 1791[24]

I just received the letter of October 4 that you honored me by writing.[25] I had spent several days before working on a draft to the order of M. Petry dated the first of April; I have just changed the address from M. Dutarte to that of l'Abbé Nolin. If M. Petry had not taken liabilities, I could have withdrawn this draft as, according to your orders, I did not anticipate having expenses to use all of it. Since the draft of October 12, there have been two, one with the date of February 20 and this last one of the first of the month. I used up little of that of February and did not need money at the time of the last one, but M. Petry [put] pressure on me as he said that he had an opportunity to cash it. In the last two years, the difficulties in placing drafts before these gentlemen has led to disputes and denials, while they said that if I did not give my drafts three month[s] in advance, they would not give me any funds.

Sir, I will prepare for my return; agreements having been made with workers, and expenses begun, the service of the department will be conducted with strict economy until the end of this year. I will select from among all the rarest that I have, of trees, plants, and seeds to ship them at that time because the temperature will be favorable. I will do everything in my power to obtain a favorable price for the New York property, but I do not count on more than that as there have been no improvements. Ever since I moved to the Carolinas, I never stopped reminding Saunier of his obligations; he did not take account of this. He trusted in the protection of the consuls; he never ceased to cultivate for his profit, potatoes, seeds, and hay. A distance of 1,200 miles does not allow me to go there often. He did not believe that he was under my direction; it is only when I gave him the choice of leaving the service or following my instructions that he acted in a better way. I would have taken this course much sooner if I had not feared that a momentary interruption in the service would have brought reproaches against me.

As to the Carolina property, the price of one hundred louis that it cost will always be easy to recover; one will regret the rather considerable collection of trees and new plants that is not appreciated here.

I will profit of all the time that remains until the end of the year to secure wild turkeys. I will send you the account of my expenses, and from the moment that I discharge the workers and when there is no longer work for the service of the department, I will account for funds that remain and will look on my mission as completed.

The goodness that you have toward me, Sir, in asking what my views are, leads me to explain to you that I would have the greatest aversion of remaining in America and that I would like to return to Persia; but there are circumstances that lead me to delay my return. Before the American war, I had an interest in a shipment of merchandise to the rebels, and I was to receive 28,000 pounds for my third of the sale made in Boston. Having here the titles to my property, I must take this step before returning to Europe. In addition, having taken the task to describe all the new plants that I saw and to compare the known plants with descriptions that are not always exact, this work in completing my herbaria will lead me until the middle or even the end of 1792. In all matters I will impose on myself to fulfill the obligations of your orders.

Your most humble and very obedient servant,

Michaux

Sunday, April 17, 1791, I embarked to go to St. Marys, Georgia (April has 30 days).

Memo. There is around the first pine after crossing the swamp with *Vaccinium repens*, a good quantity of *Vaccinium stamineum*. Near the enclosure by Dillon, many *Viburnums*. . . . *Andromeda axillaris* is found in abundance at the end of the field on the right side of Willman, approximately 200 toises [1,200 feet or 400 meters] before reaching the end.[26] *Magnolia tripetala* and a gynandraceous liana, opposite the wood shelter while coming from Ashley Ferry.

April 19, I arrived at night at Cumberland Island, across from St. Marys.[27]

April 20, I collected plants on Cumberland Island.

April 21, I went to St. Marys, known as New-town.[28]

April 22, I spent the day at the home of Capt. Stafford.[29]

April 23 I embarked to go and visit the banks of the Satilla River.

Sunday, April 24, I looked for plants near the property of James Moore.[30]

April 25, I remained in the same area.

April 26, I left to go to the higher areas of the Satilla River, a 17-mile hike.

April 27, the horses having been lost, I stayed with a man named Crawford at 3 miles from Mr. Right, who has a boat to cross the river.[31]

April 28, I walked 16 miles, leaving the dwellings behind; I camped near an encampment of Indian hunters.

April 29, I arrived at the store established for trading with the Indians and collected plants all day.[32]

April 30, I recognized *Nyssa ogechee* all along the St. Marys River and particularly at the home of . . .

Sunday May 1, I went down the river by boat and found *Sarracenia* new species.[33] Recognized at approximately 18 miles from St. Marys, *Pisonia baccifera*.[34]

May 2, I arrived at St. Marys, known as New-town, and collected nearby. At night I returned to Cumberland Island.

MAP 10. Coastal Georgia, from Carey's *American Pocket Atlas*, 1796
(Courtesy of the Philadelphia Print Shop, Philadelphia, PA)

MAY 3, I hired two men and a canoe to go on solid ground to collect an abundance of *Andromeda ferruginea*, *Kalmia hirsuta*, and *Befaria*, etc.

MAY 4, I botanized on the island and packed the rest of my collection.

MAY 5, the unfavorable wind kept the captain from setting sail. I recognized on Cumberland Island two areas where *Pisonia* grows.

MAY 6, the boat set sail for Charleston. At night there was a storm; thunder and lightning continued the following night; the wind having shifted several times, we found ourselves opposite St. Augustine in Florida.

MAY 7, after much difficulty and fatigue, we came back to Cumberland Island.

SUNDAY, MAY 8, I collected and studied plants from this area of Georgia.
(At Middleton's Place 3 miles from Dorchester, the cork-tree is to be seen. Inquire of the overseer.)[35]

MAY 13, we embarked again.

SUNDAY, MAY 15.

MAY 16, stopped in the Savannah River because of unfavorable winds.

MAY 17, we entered Savannah, and I looked for plants around this city.

MAY 18, I looked for plants in the countryside farther away, and I recognized a shrub that is related to *Mussanda*.[36]

MAY 19, I looked for plants along the canal built for the boats' safety, on the border of the sea.

MAY 20, the boat went down the river, and we were at sea.

MAY 21, we were becalmed at the entrance to the port of Charleston.

MAY 22, we arrived in Charleston and received letters from . . .

Note: I promised Mr. Belin half a bushel of rice and seeds of dry rice, sack of guinea grass; he promised to send me seeds of pawpaw. Promised to Mr. Bleym[37] seeds of *Pentapetes*. He promised to send me seeds of *Ipomoea* that were successful in the garden of a Mr. Clark, in the care of Francis P. Fatis.[38]

The person named Andrew . . . on the Crooked River promised me that he would gather seeds of palmetto at 2 shillings a peck,[39] and I have to send him a chest beforehand with moss in it.

---

*Saururus cernuus* is recognized as a very good remedy to bring to a head wounds with discharge and also diminishes inflammation. One brings to a boil the roots; they are crushed, and a little wheat flower is added to make a poultice. One also can use honey and corn flour to ripen wounds that tend to have discharge.[40]

Memo: don't forget to prepare several continuous shelves for planting the plants from Bahamas and Florida in such a way that they will overwinter. Prepare also a shelter to the north for the plants of the mountains. Assemble several species of *Viburnum* to graft on *Viburnum tinus* and particularly *V. cassinoides*.[41] Prepare immediately the frame for *Kalmia* and *Rhododendron*. Buy a barrel of tar. Make a straight enclosure with a trench behind it for the horses and cows. In addition to shade, put up long shingles to keep the rain from my shrubs on the north end.

---

From St. Augustine to Cow-ford Ferry on the St. Johns maintained by Pritchard . . . 36 miles

From Cow-ford to the home of one named Allen on the St. Marys . . . 46 miles

From Allen to the Brown Ferry on the Satilla River . . .

From Brown Ferry to ridge bluff on the Altamaha . . .

From the Altamaha to Savannah . . .

From Savannah to . . .[42]

These notes about routes and distances are Michaux's last entries in notebook 6. He did not begin notebook 7 for another ten months. We do have indications from other sources that he remained in Charleston. An item in the *Charleston City Gazette* of July 16, 1791, placed by James Bentham, a justice of the peace in the Charleston District, confirms Michaux's presence in Charleston. Mr. Bentham invited the owner of a bay horse that Michaux had reported having strayed onto his property on July 9 to contact him (Bentham) to claim the horse.

Michaux's urgent concern in 1791 and early 1792, however, was how to preserve the garden he had developed in Charleston with its collection of plants that he had gathered and nurtured for the past five years. We have Michaux's recollection of this period in a document that he drafted for the minister of the interior in February 1797, shortly after his return to France. In 1819 his son gave the French document to Francis Boott, an English friend, who then arranged for its publication in English.

### AM, report for the minister of the interior, February 1797[43]

The garden of Carolina was very flourishing, when I received, in 1791, an order from the intendant of the civil list to dispose of it. As this order was not officially communicated to me, I eluded it. I could not bring myself to the resolution of seeing this establishment pass into the hands of strangers, and of abandoning the rare collections which I had made; which were the more dear to me, as the acquisition of them had cost me so much fatigue, and exposed me to so many dangers. I saw that my country would be deprived of the fruit of the expenses which had been incurred for several years; and I determined, for the purpose of preserving it, to make myself lessee of the establishment. After I had observed, therefore, all the formalities prescribed by law, the public sale took place in March 1792.

From this period, the old government ceased to be charged with the expenses of the cultivation. I reduced them to 2000 francs from 3000, which had been the annual cost; and as my own fortune did not admit of bearing it alone, I proposed to the agricultural society of Charleston to share it with me, on the following conditions: I placed at the disposal of the society all the fruit-trees of Europe and the old Continent, which I had naturalized in the garden, and I reserved, for myself, the property in all those of North America, which I destined for, and for which I considered myself accountable to, my country. To this proposition, I added a memoir upon the advantages of naturalizing several exotic trees, chosen from those of China and the South Sea Islands, which vegetated imperfectly in the hothouses of the botanic gardens of Europe. The example of those which I had already naturalized in Carolina gave force to my observations. My proposition was accepted. I became a member of their society, and I consecrated the resources, which remained to me, to the continuation of my travels.

Hitherto I had not followed, in my journeys, that regular plan which I wished to observe, for a knowledge of the topography of plants; that is to say, by tracing them from the commencement of their existence toward the north, to the limits which nature had fixed to them towards the south; and to determine, within this space, those places where the same species constantly exist, those where they disappear and are replaced by others, in consequence of the diversity of soil, situation, and exposure. This manner of observation not having been practiced, I determined to make it the principal object of a journey in setting out for the north.

I was about to carry this project into execution, when I received, from the intendant of the civil list, the announcement that my mission was at an end. I refused the offers of the state of Carolina, which proposed that I should fix myself in America; and I set out without delay for those countries bordering upon the arctic pole.

The order to sell the nurseries had been opposed by some of Michaux's colleagues in Paris who were trying to take steps to save it, but in the confusion of the French Revolution the matter was not resolved before Michaux felt that he needed to act. We have here another excerpt from a letter that André Thouin wrote to Michaux that casts light on the difficulty these men encountered in their attempt to save the American gardens by rescinding the earlier order.

### André Thouin to AM, from Paris, March 10, 1792 (excerpt)[44]

Your son told me of the order that you had received to sell the nurseries that you established in America. I learned of the opposition that was forming against this sale. The National Committee of Agriculture is of the opinion that instead of selling these establishments, they should be conserved for the nation and become part of the Jardin des Plantes in Paris. Without a doubt this decision would be advantageous for the public good, and for me I would be very pleased because it would put me in more direct connection with you. I believe that for your part, you would not be displeased by this new arrangement as your hard work and hardships would benefit the nation more directly. But Abbé Nolin is occupied with many urgent problems, and I don't think that this business will be resolved for several months. I will do all I can to accelerate this decision either in putting pressure on the commissioner who is charged to make the report or in urging M. Broussonet to put pressure on him.[45] You can be sure of all my good will.

In the same month that Thouin wrote this letter, Michaux acted decisively on his own authority to preserve the garden near Charleston for his country. He advertised the property at public auction according to the legal requirements, and his friend John J. Himely, a Charleston watchmaker, purchased the property that he then leased to Michaux.

# CHAPTER SEVEN

# Journey to Canada, 1792

SUMMARY: ANDRÉ MICHAUX'S SEVENTH NOTEBOOK covers the almost nineteen-month period from March 27, 1792, through October 20, 1793. He made two long journeys during this period and did not return to Charleston between these journeys, but principally remained in Philadelphia or New York. He kept detailed journals for the two long journeys, the first to Canada and the second to Kentucky, but he made few journal entries during the seven months when he was not traveling. It is Michaux's longest notebook. This seventh notebook is unique for both its geographic sweep and its record of plants in regions where few naturalists had penetrated before. We have divided this long notebook into four chapters to make it more accessible to readers. Our first chapter covers his journey from Charleston, his important stop in Philadelphia, and his journey up the Hudson River Valley and across Lake Champlain into Canada, where he traveled down the St. Lawrence to the village of Tadoussac. Our second chapter follows his journey into the Canadian wilderness north of Tadoussac and his return to Philadelphia. The third chapter uses letters and the few journal entries that Michaux made in Philadelphia to illuminate the period when his proposed journey west of the Mississippi for the American Philosophical Society was under active consideration, and it ends with the arrival of Citizen Genet in Philadelphia. The fourth chapter combines the journey to Kentucky for Citizen Genet that

MAP 11. Northeastern North America map excerpt, Sayer and Bennett, 1775, London (Courtesy of the Philadelphia Print Shop, Philadelphia, PA)

concludes notebook 7 with the return journey from Kentucky to Philadelphia and Michaux's remaining weeks in this city found in the beginning of notebook 8 in order to combine his activities on behalf of Minister Genet into a single chapter.

### Summary of the first section: Charleston to Tadoussac

The initial journey described in notebook 7 began in Charleston with a sea voyage to Philadelphia, where Michaux continued overland to New York. After a side trip into Connecticut, where he learned from a former fur trader that he was already too late in the season to join the Canadian fur traders going west from Montreal, Michaux traveled up the Hudson River by ship to Albany, then overland from Albany to Whitehall at the south end of Lake Champlain. He continued the length of the lake and downstream on the Richelieu River into Canada by this water route to Montreal. He then proceeded down the St.

Lawrence to the city of Quebec. He was welcomed to Canada by officers of the British army regiments stationed there as well as by government officials and leading citizens.

At the city of Quebec, he gathered information for his proposed explorations and employed an interpreter who spoke the language of the native people. He continued downstream to the village of Tadoussac, where the deep waters of the Saguenay River entered the St. Lawrence. Tadoussac was the gateway to the northern interior region of the Province of Quebec, a region visited by few people of European descent and never before visited by a botanist.

André Michaux resumed the chronological entries in his journal on March 27, 1792. Opposite his chronological entries on the inside flap of this seventh notebook are three undated notes. The first is a centuries-old list of ingredients for making ink by combining oak galls, ferrous sulfate, and other ingredients (this is no doubt the method that he later used on January 2, 1796). The other two entries are medicinal and suggest that Michaux was affected by rheumatism and also in need of a remedy for skin rashes.

### Notes from the inside front cover of Michaux's seventh notebook[1]

Base for [making] ink
Gallnut [oak galls]
Green vitriol, known also as green copperas [ferrous sulfate]
Gum Arabic
Colored ink
Wood of Brazil or other color . . .

---

Tested remedy for those afflicted with rheumatism
Infuse red pepper, *Capsicum* . . . , cayenne pepper in strong rum or brandy
    and rub the afflicted area for several days[2]

---

General remedy for skin affliction
Rub the affected parts with liquor called catsup from China used by the
    English to season dishes.[3] John Hollingsworth

### Notebook 7 begins

MARCH 27, 1792, the property in the Carolinas was sold at public auction for the price of 53 guineas, the equivalent of 247 dollars.

April . . . , I passed the act of acquisition made by Mr. Himely.[4]

April 17, I wrote to M. de La Porte, minister to the royal household, to send him my itemized expenses and the sums that I have obtained since my departure for the United States. The same day, I obtained from Abbé Nolin a draft of 3000 livres tournois, equivalent to 555 dollars.[5] I wrote to my son in duplicate to let him know of my departure from Charleston for Philadelphia. This draft was given back to me by M. de La Forest,[6] who could not give me funds, and I gave him another 1,200 pounds on my family dated from Philadelphia on . . .[7]

April 18, 1792, I embarked from Charleston on the Charleston packet for Philadelphia and arrived Tuesday the 25th in the evening.

April 26, I visited the consul, M. de La Forest, who did not want to accept the draft from M. Petry.[8]

April 27, I visited James and Shoemaker.[9] Mr. Pinckney, Mr. Morphy, Mr. . . .

April 28, I finished the account of the sums with M. de La Forest of which he had taken charge, for the advances to be made to the gardener Saunier on the last draft of November 30, 1791. I visited M. de Ternan, minister of France in the United States,[10] Mr. De Brahm,[11] surveyor for the English colonies, and the Bartrams, botanists.

Sunday, April 29, I wrote several letters and drew up my accounts in order to send them to France.

April 26 [30 is meant], I visited Mr. Izard,[12] Dr. Benjamin Rush, physician,[13] Maj. Th. Pinckney,[14] Dr. Barton, [and] Mr. de Ternan, and [I] dined with Mr. De Brahm.

Unlike most of his contemporaries, Dr. Benjamin Rush has left us a report of a visit by André Michaux.

### Commonplace book (diaries) of Dr. Benjamin Rush (excerpt)[15]

May 1st, Thursday. This day Mr. Mecheaux [sic] a French botanist on a tour through the American woods, drank tea with me. He was recommended to me from Charleston by Mr. Bushe and Dr. Baron.[16] He had spent 14 months in Persia. He says that he found the Triticum spelt, the Lucerne, and clover wild in that country,[17] also many fruits, but the peach never. He spoke highly of the fruits of that country, that they were very saccharine and nourishing. He said that he once ate 120 nectarines for a breakfast, without being cloyed by them; that fruits composed the breakfast of rich and poor in Persia, rice and a small quantity of meat the dinner and supper. That musk melons were preserved from September til May upon high and dry shelves, and always retained a good deal of their flavor. This he ascribed to their being raised in a loose sand, and dry soil, and to the great quantity of saccharine matter in them. That water was scarce in Persia and brought 60 leagues in some cases to supply their towns and gardens by means of aquaducts. That he brought the seed of a plumb tree from Persia to Charleston which flourished there, although no European plumb had been known to thrive there. He said that the seeds of all plants declined the first, but thrived the second year after being transplanted. That the venereal disease was universal and incurable in Persia. That he once had pleurisy in Persia from drinking sour milk when he was very warm. He said that Chardin had published the best account of Persia.[18] The fig and the grape, he said, never rotted or became sour on the tree or after they fell but dried, became candied, and still retained their sweetness.

### Notebook 7 continues

TUESDAY, MAY 1, I wrote letters to Charleston.

MAY 2, I visited Mr. Hamilton.[19]

MAY 3, I visited Dr. Benjamin Smith Barton, physician in Philadelphia.[20]

MAY 4, I visited M. de Bauvois.[21]

MAY 5–26 . . .

SUNDAY, MAY 27, I left Philadelphia by the Amboy Road and arrived in New York on the 29th at 10 o'clock at night.[22] *Festuca* in the Carolinas and in New Jersey, Pennsylvania, etc., *Festuca* glume 2-valved, many-flowered, calyx 2-valved, valves lanceolate mucronate.[23]

MAY 30, I visited the New Jersey nursery, near New York, and botanized nearby.

*Celastrus*: calyx 5-parted, laciniate, oblong, obtuse, erect, petals 5, ovate, above reflexed, stamens 5, anthers oblong, erect, pistil small, immersed in receptacle, style none, stigmas 3. *Saxifraga pennsylvanica* flowers in panicles, *S. nivalis*.[24]

APRIL 31, 1792 [May 31 is meant], I continued botanical research.

FIGURE 14. William Hamilton's home, the Woodlands (Photo by Joseph Elliott for the Historic American Buildings Survey, courtesy of the Library of Congress, Reproduction Number HABS PA, 51-PHILA, 29–93 CT)

*Vaccinium hispidulum*,[25] leaves ovate, entire, hispid, flowers with epicalyx, pistil, 8-parted, inferior, fruit white, *V.* flowers single, axillary, short pedunculate, pistil inferior, at base with two small bracts ovate, opposite, calyx 4-parted, laciniate, ovate, apex of pistil persistent, corolla campanulate, 4-parted, laciniate, apex reflexed, stamens 8, filaments very short, anthers erect, pistil subrotund, inferior, style longer than stamens, stigma obtuse, pericarp of the berry snow white, subovate, navel-like, style persistent, seeds many; leaves ovate, entire, short petolate, alternate, below with rough hairs, rusty as are the stems. The stems prostrate, forming roots, filiform, fruit small, white. Lives in humid areas with cypress in Canada, New England, New York, New Jersey, etc. *Vaccinium* . . . , cranberry, with red edible fruit. *Hydrophyllum virginicum, Saxifraga nivalis,* [*S.*] *pennsylvanica, Trillium cernuum, Trientalis.*

JUNE 2, I left New York for New Haven in Connecticut, 98 miles from New York.

JUNE 3, I arrived at 10 o'clock at night.

JUNE 4, I left to visit Mr. Peter Pound,[26] traveler who lived 19 years in the interior of America, where he traveled west to Rainy Lake, Lake of the Woods,

Lake Winnipeg, Winnipeg River, Lake Manitoba. According to Mr. Pound, you must be in Montreal at the end of April in order to go with the Canadians who go trading.

June 5, left Milford and spent the night at New Haven.

June 6, I left at 5 o'clock at night and arrived in New York on the 7th at 4 o'clock in the morning.[27]

June 7, I prepared to leave for Canada.

June 8, at 8 o'clock at night, I left on board a sloop for Albany.

June 9, I collected 18 miles from New York while the sloop was at anchor because of unfavorable winds.

    *Cornus* branches punctate,[28] *Geranium . . . , Geranium . . . ,*[29] *Lupinus perennis, Verbascum blattaria.*

June 10, unfavorable winds.

June 11, in the morning, we passed between rocky mountains on which we can see the entrenchment of several batteries placed there during the war. The most remarkable site among these mountains is named West Point. These mountains are so close to each other at a point on the river that is narrow that the passage is closed with a chain that crosses the river.[30] At night we arrived at the level of Poughkeepsie.[31] Near this town, I saw *Juniperus europea? Thuya canadensis.*

June 12, the wind from the north was strong, and the cold was brisk. The Fahrenheit thermometer was at . . . at 5 o'clock in the morning. This same day we passed Kingston.[32]

June 13, the wind became favorable.

June 14, we arrived in Albany,[33] a distance of 164 miles from New York.

June 15, we left for Lake Champlain and slept at Lansingburgh.[34]

June 16 and Sunday, 17, I botanized on a high mountain near this site.[35] *Panax quinquefolia, Acer pensylvanica, Fumaria vesicaria scandens,*[36] *Mitella diphylla.*

June 18, we left again and arrived in Saratoga.[37]

June 19, we arrived in Whitehall;[38] 10 miles before our arrival, I observed:

*Linnea borealis,*[39] *Taxus . . . , Trientalis, Gaultheria procumbens, Helleborus trifolius.*

JUNE 20, we embarked on Lake Champlain; unfavorable wind. It is 60 miles or more long, very narrowed by the mountains that border the lake.

JUNE 21, at 4 o'clock in the morning, we passed by Ticonderoga in front of Fort Carillon; *Hyppophae canadensis.*

JUNE 22, the wind was unfavorable and calm. I botanized all day: *Arbutus acadiensis.*[40]

JUNE 23, we arrived in front of Burlington; on the right side we can see a very tall mountain situated at 20 miles from Burlington in the state of Vermont.[41]

JUNE 24, I botanized on the eastern side of the lake, which is part of Vermont. We arrived on the same day at Cumberland Head.[42]

JUNE 25, 26, and 27, I botanized while waiting for a chance to continue my trip. Plants observed around Lake Champlain:[43]

*Pinus abies canadensis, Pinus* with two needles, *P. strobus, Pinus* with few needles on all sides,[44] *Thuya occidentalis, Taxus monoica, Betula papyrifera, B. nigra,*[45] *Ulmus . . . ,* white elm,[46] *Carpinus . . . ,*[47] red elm, *Lonicera diervilla, Lonicera . . . , L. . . . , L. glauca,*[48] *Spiraea, Viburnum nudum, V. . . . , V. . . . , V. . . . ,*[49] *Fagus sylvatica americana, Hippophae canadensis, Actea spicata,*[50] *Vaccinium stamineum, V. corymbosum, V. resinosum, V. . . . ,*[51] *Arbutus acadiensis, Circea canadensis, Collinsonia canadensis, Iris coerulea, Carex,* Gramina,[52] *Cephalanthus occidentalis, Houstonia purpurea,*[53] *Galium . . . , G. album,*[54] *Cornus* 1, 2, 3 species; *C. herbacea, [C.] alternifolia,*[55] *Fagara . . . ,*[56] *Hamamelis virginiana, Cynoglossum . . . , C. officinalis, Symphytum officinale,*[57] *Lysimachia* 4-leaved?,[58] *Campanula . . . ,*[59] *Lonicera (chamaeceras),*[60] *Lonicera* (glaucous climbing), *L. diervilla, Verbascum thapsus, Rhamnus* (dioecious), *Ceanothus americanus, Celastrus . . . , Ribes cynosbati, R.* (miquelon), *Vitis . . . , Thesium umbellatum, Asclepias . . . , Asclepias . . . ,*[61] *Sanicula . . . , Rhus glabrum, Rhus . . . , Rhus . . . , Viburnum . . . , Sambucus . . . , Staphylea trifoliata, Aralia racemosa, [A.] nudicaulis, Lilium philadelphicum, Lilium canadense, Uvularia perfoliata, U. . . . , U. . . . ,*[62] *Hypoxis erecta, Leontice thalictroides, Convallaria polygonatum maximum, [C.] bifolia,*[63] *Prinos verticillatus, Medeola virginica, Trillium erectum, Trientalis . . . , Dirca palustris, Andromedu paniculata,*[64] *Epigea repens, Gaultheria procumbens, Arbutus acadiensis, Pyrola umbellata?, P. . . . , Helleborus trifolius, Pyrola*

. . . , *Mitella diphylla, Oxalis, Asarum canadense, Prunus* . . . , *Padus virginiana, Cerasus* . . . , *C.* . . . , *C.* . . . ,[65] *Crataegus* . . . , *Cr.* . . . , *Mespilus canadensis arborea, M. canadensis frutescens,*[66] *Spiraea* . . . , *Rosa* . . . , *Rubus occidentalis,* [*R.*] *odoratus,* [*R.*] *arcticus,* [*R.*] *hispidus,* [*R.*] *canadensis,*[67] *Potentilla* . . . , *P.* . . . ,[68] *Geum* . . . ,[69] *Actea spicata* . . . , *Sanguinaria canadensis, Podophyllum peltatum, Nymphea* . . . , *Tilia americana, Cistus canadensis, Aquilegia canadensis, Anemone hepatica,* [*A.*] *dichotoma, Thalictrum purpurascens,* [*T.*] *dioicum,*[70] *Pedicularis canadensis, Pedicularis* . . . , *Chelone glabra, C. hirsuta,*[71] *Scrophularia,*[72] *Linnea borealis, Orobanche virginica, Draba bursa-p.,*[73] *Lepidium, Geranium, Fumaria sempervirens, F. vesicaria, Polygala senega, P. viridescens, Hedisarum,*[74] *Trifolium rubens* . . . ,[75] *Hypericum,*[76] *Eupatorium, Gnaphalium dioicum,*[77] *Lobelia siphilitica, Viola* . . . ,[78] *Impatiens* . . . , *Cypripedium, Carex, Betula papyrifera, B. nigra, Urtica* . . . , *Sagittaria sagittifolia, Quercus* . . . , *Juglans oblonga, Fagus sylvatica americana, Carpinus* . . . , *Pinus* with needles in twos, *Pinus* with needles in threes, *Pinus* with needles in fives, *Pinus* needle apex shallowly notched, *Pinus* needles denticulate, *Pinus* needles in fascicles, *Pinus* needles inserted on all sides,[79] *Thuya occidentalis, Hippophae canadensis, Myrica gale, Fagara* . . . , *Smilax herbacea* . . . , *Populus balsamifera,*[80] *Menispermum, Juniperus virginiana,* [*J.*] *communis, Taxus monoicus, Veratrum, Acer rubrum, A. sacchariferum canadense, A. pensylvanicum, Fraxinus, Panax quinquefolia, Equisetum* . . . , *Osmunda.*

JUNE 27, we left Cumberland Head and went as far as Pointe au Fer.[81]

JUNE 28, we left in a little canoe and arrived on English territory at 5 o'clock in the evening.[82]

JUNE 29, we arrived and disembarked at St. Jean, and after dinner I rented a carriage to go to La Prairie, a small town on the St. Lawrence River.[83]

JUNE 30, I went by boat to Montreal. I visited several persons for whom I had letters of introduction.

SUNDAY, JULY 1, I botanized on a mountain near Montreal.[84]

JULY 2, I visited Capt. Hughes Scott of the 26th Regiment, amateur mineralogist.[85]

JULY 3, I botanized in the countryside and in the low prairies. I recognized two new genera. No. 1 genus intermediate between *Typha* and *Sparganium*, [this new] hermaphroditic plant with 3 stamens, spike-like inflorescence cylindrical, etc.;[86] No. 2 genus between *Moroea* and *Antholiza*, aquatic plant, 3 stamens, etc.[87]

July 4, I spent the morning with Capt. Scott; we talked about travels, botany, mineralogy, etc.

July 5, I botanized: *Alisma* . . .[88]

July 6, I dined at [the home of] Mr. Frobicher.[89]

July 7, dined at [the home of] Mr. Henry.[90]

Sunday, July 8, I botanized in the forest of Lachine:[91] *Dianthera* new species and *Hypericum* new species[92] in the area of about two miles going up the river.

July 9, I received several visitors.

July 10, I dined at [the home of] Mr. Frobisher with officers of the two regiments in garrison in Montreal: I took note of the merits of Maj. Murray of the 60th Regiment,[93] Capt. Robinson, etc. etc., and Capt. Scott.

July 11, I embarked.

July 12, wind contrary.

July 13, I got off at William Henry just before Sorel, a little town at the mouth of the Chamblis River.[94] In the evening the wind [was] favorable, and we crossed Lac St. Pierre.[95] I botanized near Sorel: *Andromeda calyculata*,[96] *Kalmia angustifolia*, *Vaccinium corymbosum*, *Vaccinium* . . . , *Calla palustris*, *Aralia* new species, *Vaccinium* creeping, stamens 8.

July 14, I botanized at [a place] known as Batiscan, about 20 miles from Trois-Rivières:[97] *Andromeda polifolia*, *Kalmia glauca*, *K. angustifolia*, *Azalea glauca*, *Ledum palustre*,[98] *Comarum* . . .[99]

Sunday, July 15, I botanized: *Triglochin*,[100] *Scheuchzeria* . . . ;[101] wind contrary.

July 16, I arrived in [the city of] Quebec.[102]

July 17, I visited Governor Clarke.[103] Botanized: *Oxalis* new species,[104] etc. etc.

July 18, I visited Judge . . . Dodd.[105] I botanized: *Lycopodium*, 5 different species. *Aconitum uncinatum*, popularly known as Tisavoyanne.[106]

July 19, I met Mr. Neilson, printer, a very learned man.[107] I collected several species of seeds. *Convallaria* . . . , *Cornus canadensis*, *Aralia nudicaulis*,

*Sambucus* (red fruit). Thermometer in the morning was 70 degrees [Fahrenheit; 21°C], an hour after noon 90 degrees [32°C].

July 20, Fahrenheit thermometer in the morning 67 degrees [19°C]; botanized: *Convallaria stellata*, [*Convallaria*] three-leaved, [*Convallaria*] two-leaved,[108] two other species, *Lycopodium* 6 different species.

July 21, thermometer in the morning indicated 51 degrees [11°C]; botanized: *Arbutus uva ursi*, *Arbutus* new species, *Sorbus aucuparia*,[109] *Narthecium calyculatum*,[110] *Euphrasia odontites*,[111] *Plantago maritima*, *Actea spicata* (fruit white), [*Actaea*] (fruit red), etc. etc.[112]

July 22, I visited Dr. Nooth.[113] I saw in his garden raspberries from the Cape of Good Hope.

July 23, I had breakfast at Dr. Nooth's [home] and saw a double bellows, his invention for maintaining a flame for the fusion of metals, glass for thermometers, etc.

July 24, Dr. Nooth showed me how to adapt the supports of a telescope to see small objects as with a microscope. There is nothing better for this purpose. The objects can be seen distinctly in stages more or less distant without fatiguing the eyes as happens with ordinary microscopes. If one looks at a flower, even a very small one, one can see as distinctly the inner surface of the corolla as the tip, etc. etc.

July 25, I made several arrangements to prepare myself for the trip into the interior.

July 26, I botanized at the Cascade de Montmorency;[114] plants noted: *Pinus balsamea*, *Pinus abies*,[115] Sapinette rouge [small conifer, red], Sapinette blanche [small conifer, white], *Thuya occidentalis*, *Larix*, *Betula papyrifera*, *Pinus balsamea*.

July 27, I lunched with Dr. Nooth.

July 28, I botanized in the woods to the right of the little river, St. Charles:[116] *Andromeda calyculata*, *Kalmia glauca*, *K. angustifolia*, *Ledum palustre*, *Sarracenia purpurea*, *Aralia* new species.

Sunday, July 29, I botanized at Lorette.[117]

July 30, I prepared for the trip to Lake Mistassini.

July 31, I left Quebec, went by way of Cape Tourmente and Cape Brulé,[118] the first 35 miles from Quebec, and the other 40 miles. I recognized around the mountains *Juniperus communis, Thuya* pines and spruces, *Epigea repens, Linnea*, etc. etc. The rocks are composed of the following minerals: quartz, mica, and tourmaline. In the evening we arrived at Baie-Saint-Paul: 50 miles.[119] We can see the Isle-aux-Coudres, approximately 52 miles from Quebec. At the entrance of the bay, I saw a wolverine and several porpoises; one was white as snow.[120]

August 1, at about one o'clock in the morning, the wind changed; and at 3 o'clock a heavy rain began that lasted until 10 o'clock. I botanized around the mountains: *Ledum palustre, Kalmia angustifolia, Populus balsamifera* . . . , *Potentilla nivea*,[121] *Calla palustris* in the swampy areas, as well as a *Vaccinium* creeping with white berry, *V. atoca*,[122] *Drosera*,[123] *Hordeum murinum, Galium album, Typha altissima*,[124] *Sparganium erectum*,[125] *Potamogeton*, etc.[126]

August 2, we arrived at Malbaie;[127] I saw *Cynoglossum* or *Pulmonaria maritima*,[128] *Glaux?*,[129] *Hippophae canadensis, Sisyrinchium bermudiana*,[130] *Galium album, Abies* with sparse needles on all sides, *A. balsamea, Pinus strobus, P.* with two needles, *P. larix, Pyrola uniflora*,[131] *Juniperus communis, Acer pensylvanicum, Populus balsamea; Juglans oblonga* is found in Quebec but ends here, as does *Abies canadensis*, while *Platanus occidentalis* ends at Lake Champlain, etc. etc.

From Baie-Saint-Paul, the falls and Malbaie are formed of clayey sands and rounded stones. Cape Tourmente [is] formed of quartz. On the rocks, just before entering the bay, there is a creeping shrub, leaves evergreen, small, oval, reflexed, glaucous underneath (this bloom only appears as a line as the leaves are narrower than those of rosemary), calyx 3-parted, corolla with 3 petals, stamens 3 whose filaments are very long, ovary superior, style 0, stigma simple, berry black, watery, with nine seeds, *Empetrum nigrum*.[132]

August 3, I remained at Malbaie.

August 4, I left and slept at the mouth of the Saguenay River.[133]

# CHAPTER EIGHT

# Journey into the Canadian Wilderness, 1792

Summary: When he reached Tadoussac, the trading post at the confluence of the St. Lawrence and Saguenay Rivers, Michaux had arrived at the gateway to the vast, interior region of northern Quebec, which had never before been visited by a botanist.[1] His geographic goal in exploring this region was to reach James Bay, the southern extension of Hudson Bay. Travel in this remote region was by canoe. Michaux was not the first European to attempt this journey. More than a century earlier, in 1672, the Jesuit priest Father Charles Albanel (1613/16–96) made the first successful round-trip to Hudson Bay by canoe. Father Albanel, who had lived for many years among the local native tribes, had distinct advantages. Accompanying Michaux were only a young interpreter and three native people in two canoes, while Albanel's much larger party included two other Europeans and sixteen native people in three canoes. Father Albanel also did not travel the entire distance from Tadoussac to Hudson Bay in one season as Michaux attempted to do, instead spending the winter at Lac Saint-Jean. This allowed the priest to travel the northernmost part of the route earlier in the season when the weather was more favorable.[2]

From Tadoussac, Michaux proceeded up the deep waters of the Saguenay River on his arduous forty-four-day journey deep into the interior by birch bark canoe. Michaux described this expedition in detail and made extensive lists of the plants that he encountered as he traveled north and west. After thirty-one days of paddling, wading through waters too shallow for the loaded canoes, and making dozens of portages, he reached roughly latitude 51° north on the Rupert River west of Lake Mistassini. There, instead of continuing downstream on this river to Hudson Bay, he turned back, heeding the advice of his guides who were apprehensive of being marooned in this wilderness over the winter. They backtracked on the same route and reached Tadoussac in less than half the time that the journey north had required. Leaving Tadoussac, Michaux began his return journey to the United States by retracing his route on the St. Lawrence at a slower pace, with extended stays of more than two weeks each in the cities of Quebec and Montreal.

The botanist returned to the United States with his desire to explore the great unknown lands of western North America still unrealized, but having sharpened his skills as a wilderness explorer in the Canadian North. Less than a week after his arrival in Philadelphia, Michaux approached the American Philosophical Society to discuss the organization's possible support for a new expedition west of the Mississippi River.

## Notebook 7 continues

SUNDAY, AUGUST 5, I arrived at 4 o'clock in the morning at Tadoussac.[3] I botanized: *Juniperus communis, J. sabina*? [It is] 135 miles from Quebec.

AUGUST 6, Fahrenheit thermometer registered 51.5 degrees [11°C] in the morning, wind from the east-northeast, noon 70 degrees [Fahrenheit, 21°C]. I retained three natives to go up the Saguenay River. From Cape Tourmente till Tadoussac the mountains stretch continuously along the north side of the St. Lawrence River and are principally of pure quartz, sometimes mixed with tourmaline. In several areas the base of the rocks is of black limestone.

AUGUST 7, thermometer in the morning was 52 degrees [Fahrenheit, 11°C]. I left in a canoe with three natives and the young man of mixed race whom I had hired in Quebec as interpreter, as he had lived three years with the natives.[4] Following the mountains, we covered approximately 25 miles. It was storming with thunder and rain.

AUGUST 8, the wind was contrary, and we rowed approximately 10 miles. On the rocky mountain I saw a shrub with berries (*Empetrum nigrum*), calyx 3, corolla 3-parted, stamens 3, *Arbutus* with leaves with wooly membranaceous margins, *Arbutus*? with leaves having glandular tip,* new *Aralia* having stiff bristly hairs. We slept near the cascade.

* Michaux's footnote: *Vaccinium vitis idaea.*

AUGUST 9, the weather calm, we went by a rock cut perpendicularly called the Tableau, estimated to be halfway between Tadoussac and Chicoutimi.[5] This post is situated on the Saguenay River where the tide ceases to rise. This river is said to be the largest of those that flow into the St. Lawrence. From the mouth or junction, one goes up 60 miles toward the northwest, then a large bay is seen in which probably another river empties, and at the entrance of this bay we continue on this river toward the north. The width of the river until the large bay is generally about 4 miles, narrowed by high mountains of perpendicular cliffs. There is no soil on these mountains, and the pines that grow there are only nourished by what is furnished by the mosses. They are generally composed of quartz mixed with tourmaline, sometimes on a calcareous base. But the limestone is hardly present after Cape Tourmente. I saw only one time some feldspar, a league before arriving at the large bay. After the large bay, the mountains are lower and less steep. We camped 3 miles beyond the large bay. Storm and rain prevailed.

AUGUST 10, the wind from the north was very violent. The mountains are covered by *Sphagnum*, *Ledum palustre*, *Andromeda calyculata*,[6] *Kalmia angustifolia*, *Vaccinium atoca*, *V. resinosum*, *Drosera rotundifolia*. The same day we camped five miles from Chicoutimi.

AUGUST 11, I recognized on entering the woods on the riverbanks *Swertia corniculata*, the same day we arrived at Chicoutimi.[7]

SUNDAY, AUGUST 12, we prepared to go to Lac Saint-Jean and the Mistassini Lake(s). We prepared two canoes with 300 pounds of flour, 155 pounds of salt pork, two wolverine skins for shoes, 100 pounds of biscuits, 50 pounds of bread, 10 pounds of salt, 5 pounds of gunpowder, 10 pounds of lead, 8 rolls of birch bark for tents, 3 guns, 6 aulnes [ca. seven yards] of rough wool material, three pairs of woolen socks, two pairs of woolen gloves. In addition to what we had already taken at Tadoussac; 26 pounds of salt pork, 50 pounds of bread cut in small pieces, blankets, shoes, gun flint, tinder box, etc., a large net, hammock, six blankets.

AUGUST 13, we transported the two canoes to the Chicoutimi River, leaving behind the Saguenay. Six native men and seven women were employed in carrying the provisions and the luggage. This portage is one of the longest; it being three miles or more from Saguenay to the beginning of the cascade. That day we had four other portages, mostly 400–1000 yards long, to move from the base of the cascade to the top. Often when the canoes arrive

FIGURE 15. Saguenay Fjord cliff face (Photo by Charlie Williams)

underneath the rapids or cascades, they are dragged by the violence of the water currents, which in these areas are often narrowed by enormous boulders. One must alternately row or forge ahead by hitting the bottom, then one takes hold of poles and struggles against the water. Despite the lightness of the canoes, the natives use all their strength, and they are skillful at avoiding the dangers of being carried away or knocked against the rocks or finally being overturned, which happens sometimes. There is rarely the danger of dying if one knows how to swim, because if you let go in the water current, you are carried immediately to an area where the water is calm and often less than two feet deep; then one must save whatever one can, canoes, baggage, and provisions. These trips are frightening to those who are not used to them, and I would counsel the little dandies from London or Paris, if there are any, to remain at home. I observed in the river and on its banks: *Potamogeton* . . . , *Nymphea lutea* calyx 3-sepals, petals 3, stamens numerous, and leaves cordate, *Nymphea lutea* leaves and flowers small,[8] *Ranunculus reptans* leaves narrow, stems creeping, *Chelone glabra* white flowers, *Fraxinus*, etc. etc.

AUGUST 14, there was considerable rain all day: botanized and collected many species of mosses, *Aster*, grasses, *Helleborus trifolius*, *Mitella aphylla*.

AUGUST 15, we canoed all day in a fine but continuous rain. We encountered two portages, one situated three miles from the other, and we ended the day by passing by Lake Kenogami,* which is 21 miles long and between one and two miles wide, sometimes bordered by mountains of rocks, sometimes by swamps. On the mountains, I recognized: *Juniperus communis, Abies,*[9] *Acer pensylvanicum, Potentilla nivea.* In the lower portions and swamps, I observed: *Myrica gale, Andromeda polifolia, Comarum palustre, Prinos verticillatus, Gentiana pneumonanthe, Mentha* stamens longer than corolla, *Triglochin palustre, Alnus glauca* stipules lanceolate, *Vaccinium atoca.* In the lake [I observed]: *Nymphea lutea major, N. lutea minor, Sparganium natans, Alisma subulata,*[10] *Potamogeton . . . , Polygonum . . . ,*[11] *Lobelia simplex, Eriocaulon . . .*

AUGUST 16, we had to have two portages in the morning, one about 320 feet, and the other 2 miles. At about noon we came to a river [Belle Rivière] that enters into Lac Saint-Jean. Navigating on this river, we made at least 7 miles per hour. We thus went approximately 25 miles, and we camped on the shore of the lake. On reaching the mouth of the river, we could see hills of sand where nothing else grows but *Artemisia crithmoides, Arundo arenaria.*[12] This lake resembles a sea because of its extent.

AUGUST 17, the contrary wind kept us from entering the lake in the morning, but after noon we paddled for four hours, always staying half a mile from land, and often the canoes touched bottom.

AUGUST 18, we arrived at about 4 o'clock in the afternoon at the post established by the company for the fur trade with the natives of Lac des Cygnes and Lake Mistassini. This post is occupied by two Canadian agents, the Panet brothers.[13]

AUGUST 19, a heavy fog kept us from crossing the lake to enter the river known as the Mistassini. Around midday a considerable wind was blowing. I botanized in the vicinity of the lake: *Nymphea lutea,* calyx 3, petals 3, stamens numerous; *N. lutea minor,* smaller, calyx 3, petals 2, numerous stamens, etc. . . . ; *Andromeda polifolia, Andromeda calyculata, Betula pumila, Arundo* glumes 2-flowered,[14] *Hyppophae canadensis, Eriocaulon . . .*

AUGUST 20, a gale lasted during the entire day.

AUGUST 21, we left the middle of Lac Saint-Jean where the trading post for merchandise exchange with the natives is located. It is the last post in this

* Michaux's footnote: lake of the bearberries.

wilderness. It was nine o'clock in the morning, and we entered the river said to be the Mitassini at 2 o'clock in the afternoon. We traveled on this river until eight at night. The mouth of the river is very shallow, and for 12–15 miles upriver one sees banks of quicksand that are more than a mile and a half long. *Thuya* stops at the lake, and I did not see any along the river.

I recognized *Abies balsamea, Pinus abies,*[15] *Pinus larix, Populus balsamifera, Ledum palustre.*

August 22, we continued upriver for an hour and then came to a portage. The portages are always caused by rapids across boulders, more or less steep. At the site of these first rapids, the river suddenly narrows, having been 3 to 4 miles in width. We encountered nine rapids, and consequently we had to pass through nine portages without leaving this river said to be Mistassini, although it does not come out of the Mistassini Lakes. It was approximately 45 miles from the post on Lac Saint-Jean to a large rapid where we arrived at 7 o'clock at night. Although one considers the rapids or cascades to be the result of nature, worthy of curiosity, it would be difficult to form an idea of the majestic perspective of this one. It is a natural amphitheater in whose bottom we can only see trees, as well as on its sides, and it enlarges at its base to about 1,600 feet and with a depth also of 1,600 feet. One can see countless rocks in the midst of the disturbed water, broken and reduced to fog like thick smoke. The river's bed at the base of these cascades forms a vast expanse of water on whose surface no rock is seen, but which is very turbulent and constantly produces large waves because of the rocks below the water and [because of] the smooth surface of the soil, narrowed by the hills that surround this base.[16]

The waters having crashed on the sides of these rocky hills, they come back again to mix and lose themselves at the base of the rapids and there form bands of quiet intervals between the rough waters from the different branches of the main waterfall. It is at this time that we are surprised by the natives' ability to know so well which alternative to take; either row at full strength or stop suddenly. Sometimes we found ourselves in a calm place or interval, while both sides were so rough as to send flakes of foam into the canoes. It was necessary to land between the branches of the waterfall in order to make portage and to place baggage and provisions on the rocks that were above the water's surface. The danger is that the rocks underwater are covered ordinarily by a species of *Byssus,* an aquatic moss that keeps one from getting a firm foothold. My guide, who wanted to jump from one rock to another barely one inch below the water, fell with his load that was a package of 50 pounds of flour and a bag with his old clothing. We camped near the Larges Rapides.

FIGURE 16. Mistassini River near Lac Saint-Jean (Photo by Charlie Williams)

Charles S. Sargent (1889) found the plant list below dated August 22 [1792] among Michaux's memoranda and inserted it at this point in the narrative.

AUGUST 22, I saw on the Mistassini River:

*Alnus glauca, Myrica gale, Gentiana pneumonanthe, Potentilla nivea, Linnea borealis, Epigea repens, Gaultheria, Ledum palustre, Kalmia glauca, K. angustifolia, Vaccinium corymbosum minus,*[17] *V. atoca, V.* white fruit, *Trillium* capsule angulate, violet, *Trillium* capsule ovate, red, *Narthecium . . . , Cerasus racemosa* petioles glandular,[18] *C. corymbosa* petioles glandular,[19] *C.* fruit black, petioles non-glandular, called *Cerise de Sable, Cornus canadensis, C.* stoloniferous red stems (red osier),[20] *C.* branches pitted,[21] *Convallaria*? berries blue,[22] *C.* 3-leaved, *C.* 2-leaved, *C.* alternate species,[23] *Lonicera camaecerasus* leaves tomentose, *L. diervilla, Lycopodium* sporangia in panicles, *Andromeda calyculata, Pinus larix, P. balsamea, P. abies alba, P. abies nigra, P. strobus, P.* needles twin, cone ovate, smooth, *P.* needles in twos, short,[24] *Salix sericea,*[25] *S.* stipules leaf-like,[26] *Arundo* glumes with one floret,[27] *A. . . . , Poa* glumes with 4 florets, *Ribes cynosbati,*[28] *Fraxinus* leaflets tomentose, serrate, *Betula alba seu* [or] *papyrifera, Ulmus . . .* white elm, *Rubus arcticus, R. occidentalis, Viburnum opulus* petiole glandular, *V. nudum, Taxus, Spiraea salicifolia,*[29] *Pteris, Oenothera,*[30] *Thalictrum dioicum,*[31] *Actea spicata alba,*[32] *Epilobium* stamens bent, *E.* petals split in two, *Aster.*

Aug 23, we had rain that began at 2 o'clock in the morning and that continued until noon. We stayed in camp all day.

August 24, we encountered two rapids and subsequently two portages. Our journey was estimated at 20 miles. We noted larches of a good size, while the other species of trees were of smaller stature in these latitudes.

August 25, we were obliged to use poles to fight against the river's current. This was made more difficult as a north wind was blowing hard, and we traveled about 17–20 miles.

Sunday, August 26, the wind was less violent, and we had to use the poles from the canoes only from 7 to 11 o'clock in the morning. We came across a cabin with natives, and we dined on boiled beaver meat, (*Vaccinium corymbosum*) blueberries, boiled to the consistency of jam, and also fresh blueberries. The mountains that have been burned in several places north of Quebec are covered by this shrub, and one can be satiated in less than half an hour and even in a quarter of an hour. This fruit is very tasty, and a large quantity never leads to discomfort. Our trek was 15–17 miles long.

August 27, we found the river much narrowed and the currents very swift, constricted by steep mountains on both sides. I found *Vaccinium* leaves with a glandular tip, namely *V. vitis idaea*. We estimated a course of about 20 miles.

August 28, the natives continued to use the poles to make way against the swift current, and at approximately two in the afternoon we made it to the Monte-à-Peine portage. It took us from 3 to 7 at night to climb this mountain and arrive at another small river on the other side. I estimate the height of the mountain to be 1,600–2,000 feet, and the river on the other side is 250–300 feet lower than the summit of this mountain, Monte-à-Peine.[33] The natives told me that this river has no name. The plants that I noted in the marshes from the summit were: *Ledum palustre, Kalmia angustifolia, Vaccinium corymbosum minus, V. niveum, Kalmia glauca, Betula . . . ,*[34] *Andromeda calyculata.*

August 29, in the morning I collected plants on the banks of the small river:[35] *Lycopodium inundatum, L. . . . , L. . . . , Andromeda rosmarinifolia, A. calyculata, Kalmia glauca, Ledum palustre.* We had to make four portages, during whose interval we traveled on two rivers that were no wider than 18 feet; the depth was sufficient for the canoes, but several times we had to lighten the canoes to lift them over the beaver dams whose lodges were on the banks. These lodges are always found on the banks of small rivers; they are constructed of wood and earth in the form of a three-to-four-foot-high mound on a six-foot-wide base. There is an entrance on land and an underwater exit so that during the winter freeze they can go out and eat the bark off the wood that they gathered in the water; they cut pieces the size of a thigh. The dams slow the water and raise its level, and thus it is less likely to freeze to its depth. Nevertheless the winters are so long and so severe that I have seen two-foot-deep holes in the ice. One cannot imagine the strength, the industry, the skill, and the patience with which these animals work to survive and protect themselves from the rigors of winter. When they cut down a tree, they make sure that it falls exactly where it suits their purpose, and if there are any lazy animals, they drive them out of the colony, and these live miserably and alone.

We reached Lac des Cygnes at about three in the afternoon. It is a large lake surrounded by lowlands covered with very small stunted trees. This area has a frightful aspect due to the soil's sterility added to the rigor and length of the cold season. The trees are birches, *Pinus* needles 2, *P. abies nigra*, *Ledum palustre*, *Kalmia glauca*, *K. angustifolia*, *Andromeda calyculata*, and *A. rosmarinifolia*. On entering Lac des Cygnes, I saw a new *Vaccinium* with straight stems, 1.5 feet high with quite a few branches, solitary fruits more acidic than those I tasted earlier in America, but this acidity is very pleasant. Apart from the natural habit of all *Vacciniums*, I take it to belong to this genus with 8 stamens and the fruit crowned by a persistent calyx. Its shape is that of a miniature apple, more elongate than round, but only the size of a pea. The fruit is bluish, and the leaves are entire and glaucous, related to *V. uliginosum*.

Lac des Cygnes is interesting because of the appearance of its surroundings, whose terrain is generally low but interrupted by hills of various shapes. The multitude of protrusions and recessions sometimes brings opposing shores closer together and sometimes distances them [by] more than two miles, sometimes making it very deep and other times too shallow for the canoes. I recognized *Potentilla fruticosa* at various points along the banks and almost submerged in some areas, as well as *Andromeda rosmarinifolia* and *A. calyculata*. The Indian who guided my canoe saw in a rather shallow area a beaver's head with no flesh on it, but with all the bones of the head and jaws intact. He gave it to me, but it was lost again in an accident that happened to us as we left the lake as we were climbing up over rocks around a rapid. We had decided to unload the canoes only partially, and while getting out, the Indian slipped on a rock covered with slippery lichen; as he still had a leg in the canoe, it made it tip, and immediately it was half full with water.

All my papers, plants, and other portions of my luggage were soaked, and all night was spent in drying and repairing in part the damage of this accident. [On] the 30th, I continued in the morning to dry my herbarium and my seed collection. My herbarium, which was wrapped in sealskins, had not suffered much as in some places the water penetrated only one inch.

August 30, we canoed through three lakes surrounded by low mountains that are connected by outlets between the hills. The terrain in all this area is intersected by mountains and hills with low places or valleys that are filled with water, forming many lakes mostly unnamed even by the Indians who often hunt there. Rather large areas are covered by *Sphagnum palustre*. We sink to our knees in it, and even during the height of the dry season we are wet to the knees. We made three portages and traveled 8–10 miles because of the difficulty in getting through these unpleasant marshes.

The marshes here are full of *Kalmia glauca* and *Andromeda rosmarinifolia*, *Sarracenia purpurea* and *Vaccinium atoca*. In areas with less moisture, we found *Andromeda calyculata*, *Ledum palustre*, *Kalmia angustifolia*, *Epigea repens*, *Pinus abies rubra*,[36] *P.* needles in 2s, short. *P. balsamifera* distribution stops at Lac des Cygnes; today I only saw three small shrubs of it, and the whole vegetation has the imprint of pygmies, stunted because of the poor soil and the severe cold. I saw a new *Vaccinium* with solitary fruits in the leaves' axils, fruit bluish, calyx with 5 segments, hardly woody in comparison with yesterday's *Vaccinium*, which was a well-shaped woody shrub. *Avena paniculata* spikelet? one-flowered, is the only grass that I saw today.[37]

August 31, we traveled for one hour and found a portage. The cold was bitter, the sky overcast for the last two days, and the rain like melted snow. We stopped for breakfast, but the cold took away our appetite. The Indians were shivering with cold, soaked not only from the rain but also from the wet shrubs through which they passed and the *Sphagnum* swamps that they had to go through and in which they sank to their knees in some places. Although I was better dressed, I still found the cold unbearable and had a fire made, and toward ten o'clock we set out again. We went through three lakes and a stream. *Narthecium calyculatum*,[38] *Epilobium* linear leaves,[39] *Kalmia glauca*, *Andromeda rosmarinifolia*, etc. etc.

Saturday, September 1, the rain kept us from traveling, and one of the Indians was sick. The cause seemed to be blocked perspiration. He had been wet with rain the preceding day, and he had slept in his soaked blanket. In the afternoon the weather lightened up a trifle, and we went on despite the rain. All night long there was rain, thunder, and lightning. We traveled about 15 miles through a lake and very narrow streams, only a canoe's width.

Sunday, September 2, it was very overcast in the morning, but it developed into melting sleet. The cold was less intense, but we had a portage of about one and three-quarter miles across *Sphagnum* savannah where we waded to mid-calf, and despite the hailstorm that went on all day, we continued our journey as the Indians and I wanted to get to Mistassini as soon as possible for fear that the snows and the cold would become worse. We had to cross three lakes and traveled about 25 miles.

September 3, the frost was approximately a line [ca. 3/32 inches or 2.25 mm] thick. By midnight I saw white frost on the shrubs and plants around the hearth where we were camping. The weather appeared better at least during the day, but around seven o'clock it became cloudy, and we had rain, alternating with hail and snow with intervals of radiant sunshine. We saw a reindeer in a prairie, but the Indians could not reach him because of the wind's direction. At eleven o'clock we entered a large river that flows north. We were able to travel 30–35 miles because of a favorable current. The soil seemed to be better.

September 4, we made three portages because of the rapid currents over the rocks. At 10:15 we entered Lake Mistassini.[40] Near the lake we saw *Bartsia pallida*, *Gentiana*?,[41] *Narthecium* calyx with equal parts, *Lycopodium*?, etc. etc. We traveled on the lake about 25–30 miles and camped on the northwest bank [of the outlet river], 15 miles from the lake.

September 5, we traveled 20–25 miles and had dinner on the Goelands River, 40 miles from the lake.[42] We killed a ring-necked goose.[43] We took five fish that were 1.5–2 feet long. At night, we camped.

*Abies nigra*, *Larix*, *Betula pumila*,[44] *B. alba*, *Sorbus aucuparia*,[45] *Myrica gale*, *Cornus* of Canada (*Cornus* red osier), *Ribes . . .*, *R. . . .*, *R. . . .*,[46] *Pinguicula alpina*?[47] *Vaccinium niveum* 8 stamens, *Vaccinium atoca*, *V. uliginosum*?, *Epigea*, *Avena nuda*, *Arundo* glumes 2-flowered, *Andromeda rosmarinifolia*, *Kalmia angustifolia*, *K. glauca*, *Sarracenia purpurea*, *Vaccinium vitis idaea*, *Pteris aquilina*, *Osmunda regalis*, *Hieracium paludosum*?,[48] *Linnea borealis*, *Vaccinium corymbosum minus*. [We traveled] 55 miles total. We camped near the Atchouke River (River des Loups-Marins).[49]

September 6, we returned to [Lake] Mistassini: 65 miles.[50] The soil on the hills surrounding Lake Mistassini is sandy on the surface and clayey beneath.

The stones and rocks are blended quartz, mixed with clay and sometimes with humus. The stones on the banks worn by the movement of the waves exhibit extraordinary shapes because the layers of clay and softer materials are more eroded. There are also stones of quartz with mica and tourmaline, but very little pure quartz. I did not see any calcareous rocks, no sign of volcanic rocks. The terrain is low in this area; the hills are far in the distance. The flow of water from this lake is toward the north and northwest by several rivers that flow into Hudson Bay. We could get there [Hudson Bay] in four days, but it would take ten days to return. Our trek was about 70 miles since the natives were very eager to return. The trees of lower Canada are not found at the higher elevations, although these trees and plants are at their most vigorous in the lower parts of Canada.

SEPTEMBER 7, we left [Lake] Mistassini and our journey was approximately 42–45 miles because of the swift river currents. The head Indian who was leading me killed an otter that was swimming across the river and [that] lifted its head above the water from time to time. We had been on our way from 6 o'clock in the morning until about 6 at night, despite the fog and the cold.

SEPTEMBER 8, frost and ice [accumulated] in a tin container. Nice weather all day. At about 60 miles from [Lake] Mistassini, near a prairie, I collected samples of *Juniperus communis*, and although I saw 40 specimens in this location, I did not have the satisfaction of seeing any with fructification. My Indian guide killed a muskrat (*Castor zibaticus* Linn.).[51] That night, joined by his comrades, he roasted and ate it, but he did not want to eat the otter that he had killed the previous day. We traveled approximately 50 miles with the current over several rapids, while we had to portage while going upstream.

SUNDAY, SEPTEMBER 9, we went by Lac des Cygnes and slept at the mountain, Monte-à-Peine. Our journey was about 50 miles. *Andromeda calyculata*, *Kalmia angustifolia*, *Ledum palustre* cover the surface of hills and mountains where trees have been burned. The area that was burned two years ago at most is covered by *Vaccinium corymbosum minus*, *Pinus abies nigra*, *P. larix*, and *P.* needles in twos, short, form the dominant trees. There are large areas of swamp covered by *Sphagnum* in which one sinks knee deep. There the only plants that grow are *Andromeda rosmarinifolia*, *Kalmia glauca*, *Betula pumila*, *Vaccinium atoca*, *Sarracenia purpurea*. These swamps are never dry, and the wettest ones produce only *Andromeda rosmarinifolia* and *Kalmia glauca*. Our course was about 75 miles.

SEPTEMBER [date blotted out] 1792, I had moss gathered to pack plants collected around the lake. In the morning, I went to gather plants, and upon returning, I saw four large fish taken in nets that the Indians stretched the night before. After breakfast, I continued my plant collecting around the peninsula where we camped, and I visited some places west, north, and east, east-southeast, west-northwest. I recognized *Pinus abies nigra, P. larix, P.* leaves in 2s, *Betula alba, B. pumila, Sorbus aucuparia americana, Mespilus canadensis arborea, Rubus occidentalis, Rubus arcticus,*[52] *Potentilla fruticosa, Myrica gale.*

Above there is an error in the date. [This is Michaux's note, perhaps referring to the date blotted out following September 9.]

The Indians killed several scavenger birds for me. I saw a small woodpecker; its upper body was black mixed with white spots and more gray on the sides and the tip of the wings, the underside whitish, some of the tail feathers white at the tip.[53]

Two birds of the same genus as the magpie, the top of the head white, toward the front, white, the back and wings ash brown, breast and throat whitish as well as the area below the eyes, the eyes black, ears large, tip of tail with white border . . .[54]

Some of Michaux's text here is omitted because it repeats his text for September 6 describing the soil and rocks around Lake Mistassini.

On . . . [date blotted out in the original] we took five fish that were 1.5–2 feet in length. The quadrupeds that I had occasion to see from Lac Saint-Jean and Lake Mistassini are: *renne* [reindeer], called *caribou* by the Canadians, *Atták ko* by the Indians; *Castor* [beaver], called *Amish-Ko* by the Indians; *Loutre* [otter] *Netchako; martes* [marten], *marmottes* [marmots], called *sifleux* [whistlers] by the Canadians; there are also *linx* [lynxes], *renards* [foxes], *ours* [bears], and a very sly animal that the Canadians call *carcajou* [wolverine] and the Indians *kouikouatchou,* which does not run fast but knows how to determine a reindeer's path, climbs a tree, and jumps on top of it. The Indians told me that more often it walks noiselessly to surprise the reindeer, and when it jumps on it, there is no way to get rid of it unless the reindeer finds a river, then the *carcajou* will let go of his prey.

I propose to take up again the trees and plants that I noticed in this, the most northern country through which I traveled in America, and I will take care to mention the site where each species begins. Hills surrounding Lake Mistassini:

*Pinus abies nigra, P. larix, P. balsamifera, P.* needles in 2s, *Betula pumila, B. alba, Sorbus aucuparia americana, Cerasus corymbosus, Juniperus sabina?, Myrica gale, Cornus canadensis, C.* (red osier dogwood of the Canadians), *Rubus occidentalis, R. arcticus, Ribes . . . , R. . . . , R., Potentilla fruticosa, Vaccinium corymbosum* 10 stamens, *V.* (dwarf with 10 stamens),[55] *V.* streamside, 8 stamens, *V. atoca* 8 stamens, *V. niveum* 8 stamens, *Andromeda calyculata, A. rosmarinifolia, Kalmia angustifolia, K. glauca, Linnea borealis, Sarracenia purpurea, Bartsia pallida, Euphrasia odontites,[56] Rhinanthus crista-galli,[57] Pinguicula . . . , Cacalia hastata, C. incana,[58] Vaccinium vitis idaea* 8 stamens, *Hieracium paludosum, Pteris aquilina, Osmunda regalis, Osmunda filiculifolia.*

The short plant list and text below is set off from the remainder of Michaux's extensive journal notes included under September 9, possibly because it is a list of plants that he anticipated finding on this leg of his journey but did not.

*Pinus strobus, Thuya occidentalis, Populus balsamifera, Betula nigra, Gaultheria procumbens, Rubus odoratus, Adiantum pedatum* are not found in the upper parts of Canada although . . .

September 10, we had a heavy white frost at the top of the mountain, and in the small streams, the branches of small shrubs that were splashed with water were covered with icicles. On arriving on the other side, the southern part, frost had its effect, but *Convallaria* and other tender plants showed little damage.[59] *Lonicera diervilla* begins here and continues abundantly as far as Albany. *Achillea millefolium* begins here and is found in Canada and even on Lake Champlain. The Indians and I killed nine perdrix, *Tetrao lagopus,* called partridge by the Canadians. These birds keep together and fly a short distance into trees, and they let themselves be killed to the last one. They eat fruits of *Vaccinium, Carex,* and larch buds as I observed.[60]

We encountered two families of Indians; one made me a present of a blueberry cake (*Vaccinium corymbosum*), cooked in resin and dried. I gave him in exchange flour and salt pork from my provisions, and he gave me a second cake. We traveled about 55 miles as we had an unfavorable wind that kept us from getting the full benefit of the river's currents. At night one of the Indians that we had met brought a bear that he had just taken from one of his traps. I had them give him some supper in hope of receiving some fresh meat from his hunt.

SEPTEMBER 11, at daybreak I saw the wife of the hunter who was beginning to dress the bear, and I had a large kettle put on the fire that was made nearby for that purpose. In effect, he brought me the head and a large piece of filet. There were at least eight to nine pounds of meat, about six pounds not counting the bones. I gave him two bushels of flour and a piece of salt pork. We ate with good appetites, and only the bones remained. The interpreter said that I had eaten almost as much as the Indians. Myself I ate three times as much in Canada as I did while living in the Carolinas. Despite the fatigue of this trip, the suffering occasioned by midges, mosquitoes, and very small bees that fill the air, and the "no see em," little flies that one cannot see without magnification, my health had been restored completely. At about nine o'clock we embarked; we went down several rapids without making portages, and after about 40 miles we arrived at the Great Rapids. Here *Potentilla tridentata* begins. Eight miles below the Great Rapids, we begin to see ash and elms. It must be mentioned that none are seen from the Great Rapids to Mistassini. Ten miles below the Great Rapids, I saw the first *Pinus strobus*; I had not seen a single one from here to Mistassini going or coming. The country is mountainous from Lac des Cygnes to the Great Rapids, and after that until Lac Saint-Jean, the terrain is flat with no mountains.

It is clear that the country between Lac des Cygnes and Lake Mistassini has the highest [elevation], and Lake Mistassini empties into Hudson Bay by way of the Rupert River,[61] which runs northwest. Lac des Cygnes flows into the St. Lawrence via the Mistassini River, by Lac Saint-Jean, by the Chicoutimi River,[62] and finally by the Saguenay River until Tadoussac, where it meets the St. Lawrence. I have trouble with the naming of the Mistassini River, the river that runs from Lac des Cygnes until Lac Saint-Jean. I made this observation to Canadians who go into this country to trade with the Indians. They said that it was believed earlier that you could go up this river to reach Lake Mistassini and that it was named thus by the Jesuit missionaries.

We traveled about 35 miles and camped near the first Weymouth pine (*Pinus strobus*) that one encounters coming down from Mistassini.

SEPTEMBER 12, strong wind and cold rain prevailed. I noticed while coming down that the low country is uniformly flat; one sees no mountain on either the left or right side of the river, which is 3.75–5 miles wide and about 37 miles long, before it empties into the lake; it is intersected by banks of sand and is quite shallow. We arrived at about seven o'clock in the evening at the Post of Lac Saint-Jean and had traveled about 37 miles.

SEPTEMBER 13, I botanized in the area surrounding the lake. I collected several species of seeds. I skinned several species of birds and quadrupeds and prepared to continue my journey. *Circea canadensis*, *Mitella aphylla*; saw the large raven (*Corvus corax*).

September 14, a strong wind from the southwest made it impossible to set out with the canoes, and I kept the Indians busy collecting seeds the whole day.

The post established on Lac Saint-Jean for trading with the Indians is located on the northwest [side] of the lake. The soil is generally sandy, but there are extensive banks of calcareous stone. The calcareous stones are flattened bands and are sometimes composed of schist. We can see petrifactions of marine shells and ammonites that are barely in the shape of a horn, but about the same size throughout and about the thickness of a finger. There are also some rocks made of quartz toward the north.

Note: I forgot to mention that from Monte-à-Peine, the mountains are of calcareous rocks, but there are also large stretches where the hills are pure sand or a mixture of sand and rounded stones, called tumbleland by the Canadians.

September 15, we left the Post on Lac Saint-Jean. Three and a half miles farther there is a little river that empties into the lake, River Choua-mouchouan; it empties into the lake exactly in the west as seen by the September sun; the Mistassini enters the lake on the west-southwest.[63] I saw two rivers that enter the lake; finally at night we reached . . . River [Belle Rivière], which was supposed to lead us to Chicoutimi, and we camped nearby. In fact skirting the lake from east to south and west there are five large rivers that empty into the lake. The largest discharge is into the Saguenay to the northeast. I don't know if there are others.

Sunday, September 16, we left the lake [Lac Saint-Jean] and camped at the southern end of Lake Kenogami.[64] The lake is not much more than a mile wide at its widest. It is 17 miles long. At the beginning of the lake, on the north, I noted *Acer rubrum*, *Medeola virginica*, *Cypripedium calceolaria* red flower, but this latter grows also on the hills that surround Lac des Cygnes, thus it cannot be considered that it is starting [its distribution] here. The rocky mountains that surround the lake are steep although of an average height, and the woods are thick and full of large trees as if growing on fertile soil.

Monday, September 17, we reached Chicoutimi. Plants noticed again: *Polygonum aviculare*, *P. hydropiper*, *Lamium* . . . , *Lappa* . . .[65]

The distance from Lac Saint-Jean to Chicoutimi is about 100 miles.[66]

SEPTEMBER 18, left Chicoutimi; wind was favorable, and we had the advantage of the ebbing tide.

SEPTEMBER 19, we arrived in Tadoussac.

SEPTEMBER 20, I gathered Labrador tea and other types of seeds.

SEPTEMBER 21 . . . [the manuscript page is left blank and there are no further entries until September 27 on the top of the facing page]

SEPTEMBER 27, we left Tadoussac.

SEPTEMBER 28, we reached Malbaie.

SEPTEMBER 29, I botanized and . . .

SUNDAY, SEPTEMBER 30, I recognized *Salicornia* . . . , *Salsola* . . . , *Lappa* . . . , *Ranunculus* . . . ,[67] *Trifolium* . . . ,[68] *Lithospermum* . . .

OCTOBER 1, I left on a ship for Quebec.

There are no entries for October 2–16, the period of time that Michaux spent in Quebec.

WEDNESDAY, OCTOBER 17, left [the city of Quebec] and spent the night at Pointe aux Trembles.[69]

OCTOBER 18, I went by Pointe aux Trembles [and] Jacques-Cartier River and spent the night at Sainte-Anne[70] at [the home of] Mr. . . . .

OCTOBER 19, I went by Batiscan [and] Trois Rivières and slept in Yamachiche: *Juglans hiccory*,[71] *Celastrus scandens* in Trois Rivières, *Populus (fastigatus?)* also at Trois Rivières, as well as *Triosteum, Ulmus, Carpinus, Quercus alba*,[72] *Pinus canadensis* . . . , *Spiraea tomentosa* and *Spiraea opulifolia, Adiantum pedatum, Fagus sylvatica americana* at Trois Rivières but more certainly at Berthier.[73] *Cephalanthus occidentalis*, known from River de L'Assomption. *Ledum palustre* ends near River de L'Assomption, as does *Kalmia glauca*, which I saw at Batiscan.

OCTOBER 20, I spent the night near River de L'Assomption.[74]

SUNDAY, OCTOBER 21, I arrived in Montreal.

OCTOBER 22, in the vicinity of Montreal, I saw *Crataegus coccinea* [and] *C. lutea, Cephalanthus occidentalis, Prinos verticillatus*.

OCTOBER 24, I lunched at Mr. . . . Henry's.

OCTOBER 27, I lunched at Mr. Frobicher's.

OCTOBER 28, I lunched at Mr. John Dease's.[75]

OCTOBER 30, I lunched at Mr. Selby's.[76]

NOVEMBER 7, 1792, I left Montreal, and the fog was so thick that the drivers lost their way. The boat ran aground on rocks where we spent the night. The boat took on water. My books and a portion of my baggage became wet.

NOVEMBER 8, we passed Longueil and arrived at La Prairie. I had breakfast with Mr. La Croix, Esq., the next day.[77]

NOVEMBER 9, I paid 2 piasters to transport my baggage to Saint Jean. We paid communally 1.5 piasters to have a carriage take us from Saint Jean to La Prairie. This is a distance of 15 miles.

NOVEMBER 10, I visited Col. Gordon and lunched with the officers of the garrison.[78]

SUNDAY, NOVEMBER 11, I had breakfast at Col. Gordon's. All day was occupied drying my books and my belongings.

NOVEMBER 12, I lunched with Col. Gordon.

NOVEMBER 13, I left and spent the night across from Isle aux Noix [Quebec]: 15 miles.

NOVEMBER 14, we traveled 10 miles.

NOVEMBER 15, after 5 miles on the road we passed by the line that separates Canada from the United States. This line is found 9 miles south of the Isle aux Noix. Then [we] passed in front of La Pointe au Fer [New York]; although it is in the territory of the United States, it is occupied by the 26th Regiment of English soldiers, whose commandant is Capt. Hope.[79] Pointe au Fer is 15 miles from Isle aux Noix, and we spent the night at Cumberland Head [New York], 26 miles from La Pointe au Fer and 56 miles from Saint Jean [on the Richelieu River in Quebec].

NOVEMBER 16, a storm with snow forced us to remain here.

NOVEMBER 17, we left Cumberland Head and arrived in the state of Vermont at a place called Shelburne:[80] *Platanus occidentalis, Ceanothus americanus.* We slept in Vermont opposite Split Rock,[81] 39 miles from Cumberland Head.

SUNDAY, NOVEMBER 18, the wind coming from the south was very violent and unfavorable and obliged us to remain in place. I saw *Ceanothus americanus*, *Hyppophae canadensis*, *Acorus*.

NOVEMBER 19, we had lunch at Basin Harbor,[82] 6 miles away. We went by Crown Point,[83] 12 miles, and spent the night at Ticonderoga, which is 35 miles from Split Rock or Rocher Fendu: *Pinus bifolia*, *Hyppophae*, *Juniperus communis*.

NOVEMBER 20, we stopped in two places in the territory of Vermont, and we spent the night at Skeensborough, [now] known as Whitehall.

NOVEMBER 21 and 22, we stayed in Whitehall to dry my seeds that were damaged in the boat on Lake Champlain.

The distance from Montreal to Whitehall, southernmost point of Lake Champlain:

| | |
|---|---|
| From Montreal to La Prairie | 6 miles |
| From La Prairie to Saint Jean | 18 |
| From Saint Jean to Isle aux Noix | 15 |
| (From Isle aux Noix to boundary line between United States and Canada—9 miles) | |
| From Isle aux Noix to Pointe au Fer | 15 |
| From Pointe au Fer to Cumberland Head | 24 |
| From Cumberland Head to Split Rock | 39 |
| From Split Rock to Harbor Basin | 6 |
| From Harbor Basin to Crown Point | 12 |
| From Crown Point to Ticonderoga | 15 |
| From Ticonderoga to Whitehall | 28 |
| Total from Montreal to Whitehall | 178 |

NOVEMBER 23, I left Whitehall and arrived at Fort Ann for breakfast, 12 miles away. I noted the following plants: *Pinus strobus, P. canadensis, Acer sacharinum, Alnus glauca, Liquidambar peregrinum, Acorus,* etc. etc.

I spent the night at Fort Edward, 24 miles from Whitehall. Between Fort Ann and Fort Edward,[84] I saw: *Laurus benzoin, Liquidambar peregrinum, Pinus* needles in 3s, *P. strobus, P. canadensis, Andromeda racemosa, Hamamelis virginiana.*

NOVEMBER 25 [24], there was much snow so I remained at Fort Edward at the home of Capt. Baldwin.[85]

SUNDAY, NOVEMBER 25, I left and slept at Saratoga, 20 miles away. *Fagus castanea americana* begins near Saratoga.[86]

NOVEMBER 26, I continued my journey on the side opposite Saratoga. I had breakfast at Easton.[87] *Cornus florida, Laurus sassafras, Liriodendron tulipifera* begin in the neighborhood of Easton, 10 miles from Saratoga. I spent the night in Albany, 36 miles from Saratoga. Total distance from Whitehall [then Skeensborough] to Albany is 80 miles.

NOVEMBER 27, I embarked on a sloop in Albany for New York on the Hudson River.

NOVEMBER 28–29, adverse winds prevailed.

NOVEMBER 30, the wind was from the northwest; we stopped on the bank opposite Poughkeepsie as the wind had ripped the sail. I went to botanize on the hillside, and I recognized *Azalea viscosa, Kalmia latifolia,* which begins close to this area, *Liriodendron tulipifera; Juniperus virginiana* begins here, and *J. communis* ends here; *Thuya occidentalis* ends here in low areas, but in the mountains it continues in several places in New Jersey. *Nyssa aquatica,* or rather *N. montana* leaves with villous petioles, begins near Albany. *Quercus* . . . chestnut oak begins at Albany.

I noted also on the rocks 10 miles south on the opposite banks from Poughkeepsie: *Arbutus (acadiensis?)* leaves entire. *Liquidambar styraciflua* begins in the higher elevations in the Catskills.

SATURDAY, DECEMBER 1, 1792, I went by Crown Point [West Point is meant].

TARRYTOWN IS a small village 32 miles from New York.[88] There are mountains and a lake on the opposite side.

SUNDAY, DECEMBER 2, I arrived in New York.

| | |
|---|---|
| From Montreal to Whitehall | 178 miles |
| From Whitehall [then Skeensborough] to Albany | 80 |
| From Albany to New York | 164 |
| Total | 422 |

DECEMBER 6, I left New York.

DECEMBER 8, I arrived in Philadelphia.

Soon after he arrived in Philadelphia, André Michaux paid another visit to Dr. Benjamin Rush, who recorded the visit in his diary.

### Commonplace book (diaries) of Dr. Benjamin Rush (excerpt)[89]

December 15 [1792] Mr Mecheaux [sic], a French botanist, drank tea with me. He had just returned from a journey of 650 leagues beyond and 150 leagues to the north of Quebec in search of plants. He found there many of plants of similar latitudes in Europe, among others the Labrador Tea. He says that the scattered French people whom he found in that cold country had coarse skins from scorbutics complaints. They lived chiefly on salt meat and seals; their blood when effused had a color blacker than natural. The Indians who eat wild fruits plentifully escape the scurvy.

Later in December, André Michaux wrote two letters to his son with essentially the same content summarizing his journey to Canada and then sent these letters to France by different ships.

### AM to François André Michaux, December 30, 1792, from Philadelphia[90]

I wrote to you, my dear son, a few days ago by way of Charleston, and I announced my return two weeks ago from Canada. I was well received by the government officers who are English. They helped me in all manners in terms of the voyages to the interior that is not inhabited. The tradesmen who work from the mouth of the Saguenay River until the territory around Hudson Bay furnished me with two canoes serviced by three natives, with whom I went up the Sagney or Saguenay River. I went by Tadoussac, Chicoutimi, Lac Saint-Jean, and then I arrived to the height of the land, that is to say that at this elevation, the waters go toward Hudson Bay but also toward the south. The whole surface of the soil is continually underwater; in these sterile and cold sites there are only mountains, lakes, or rivers. Between the Mistassini Lakes and the lake above, there are hundreds of unknown lakes, which lack names. From Chicoutimi River to Lake Mistassini, I went through 32 lakes and 82 portages. The three natives, who were my sole companions during nine weeks, were well behaved and were even attentive and obliging. The other natives that I met obtained fresh meat, beaver, bear, lynx, geese, ducks, rabbit, etc. My natives, while generally lazy, were alert enough to kill birds and to catch insects. As to plants, I did not find an abundance of new species, as in more temperate latitudes, but nevertheless it would be interesting to know the botany of this extremity of America. Aside from a rather large number of herbarium specimens from these areas, I also brought back a rather nice collection of live plants from Lake Mistassini that were in good shape upon my arrival in Philadelphia, and I will send them at the first occasion to the Jardin des Plantes. I am very happy to have visited these distant regions, but one makes this kind of voyage only once. In all respects I am more pleased than I had expected. I spoke to Canadians who have seen the source of the Mississippi. But that of the Missouri is not known yet, even though several had gone beyond the brilliant mountains; they went through these chains of mountains at high northern latitude. They met natives who had never seen white man or Europeans, and other natives who brought them gold bells taken from the neck of cows from the areas established by the Spanish; they brought back to Montreal this last summer two skins of a type of mountain lamb, which they say belongs to the genus of the *Lamas*, etc.

# CHAPTER NINE

# Philadelphia, Western Expeditions
# Considered, 1793

SUMMARY: AFTER RETURNING FROM HIS Canadian journey, André Michaux went from New York to Philadelphia and subsequently made only a few sporadic entries in his journal. There are only seventeen entries from December 10, 1792, until early June 1793. Letters and other documents written either by Michaux or by others, concerning his activities, provide a more complete record of this six-month period.

Michaux exchanged letters with the new French consul in Charleston, who had recently been appointed by Louis XVI during the short-lived constitutional monarchy, but who held political views that soon made him an active functionary of the Girondist regime of the French Republic. Michaux also wrote letters to his son, to his close friend and colleague in France, Louis Bosc, and to other correspondents. He shipped some of the natural history collections that he had made on the Canadian journey to France. A political awareness on the botanist's part is indicated by the fact that he joined a group of Frenchmen who espoused and promoted the ideals of the new French Republic. He became a member of La société française des amis de la liberté et de l'égalité de Philadelphie on February 17, 1793, and was said to have been a diligent member throughout that spring. In May he proposed for membership in the organization the young Canadian Henri Mézière, who was trying to foment a revolution in Canada.[1]

The outstanding event of this six-month period, however, was Michaux's proposed exploration of the territory west of the Mississippi River for the American Philosophical Society (APS). This territory was claimed by neither the United States nor France but by Spain. Nonetheless, prominent Americans led by Thomas Jefferson wanted reports about the vast essentially unknown land that lay between the Mississippi and the Pacific Ocean. Negotiations between Michaux and the APS as well as fundraising and planning for this proposed expedition are well documented in extant records. The proposal brought Michaux into personal contact with some of the most prominent scientific and political figures in the United States. Michaux's failure to carry out this planned western expedition and his return to the service of France in a political role was the consequence of events planned by politicians many months earlier on the far side of the Atlantic. The Canadian expedition had only whetted Michaux's appetite for exploring unknown regions. Indications are that Michaux would continue to dream of exploring the American West for many years to come. Jefferson, although disappointed in 1793, boldly seized the opportunity for the United States to purchase the Louisiana Territory a decade later while he was US president. He then launched the celebrated expedition of Lewis and Clark. Jefferson's instructions for Lewis and Clark echo his earlier instructions for Michaux.

### Notebook 7 continues

DECEMBER 6, I left New York.

DECEMBER 8, I arrived in Philadelphia.

DECEMBER 10, I proposed to several members of the Philosophical Society the advantages for the United States to have geographical knowledge of the country west of the Mississippi and asked that they subsidize my journey with the sum of 3,600 pounds. With the help of this sum, I am ready to travel to the sources of the Missouri and even explore the rivers that flow into the Pacific Ocean.

My proposition having been accepted, I gave to Mr. Jefferson, secretary of state, the conditions under which I am prepared to undertake this journey. According to these conditions, I do not intend to accept the 5,000 piasters, the sum of the subscription made by the members of the Philosophical Society, but only an advance of 3,600 pounds mentioned above from which reimbursement will be made based on the salary that is due to me. I offer to communicate all the discoveries and geographical knowledge to the Philosophical Society, and I reserve for my own use all the knowledge acquired on natural history that I will obtain on this trip.[2]

All, or perhaps only the second paragraph, of the journal entry for December 10 must have been written later because it includes information about Michaux's proposed trip west of the Mississippi River that he would not have known on December 10.

Although there are no entries in Michaux's journal dated between December 10, 1792, and January 20, 1793, his proposal was under consideration during this period. The members of the APS who initially met with Michaux are not identified, but soon after the initial meeting, Thomas Jefferson, vice president of the APS and secretary of state of the United States, took the lead in dealing with Michaux on the society's behalf. Jefferson wrote to Benjamin Smith Barton, asking him to meet with Michaux and discuss the proposal, and Barton complied immediately. Jefferson's request to Barton and the latter's report to Jefferson are reproduced below.

### Thomas Jefferson to Benjamin Smith Barton, labeled December 2, 1792 (but most likely January 2, 1793, is meant) (excerpt)[3]

Th. Jefferson presents his compliments to Dr. Barton and has the pleasure to inform him that the Indian is now in Philadelphia.[4] He [Jefferson] is now under inoculation;[5] but whenever well enough he will ask the favor of Dr. Barton and Mr. Michaux to meet him here and have a conference on the expedition. He thinks the return of these Indians will afford Mr. Michaux an excellent opportunity of being conveyed to Kaskaskia in perfect safety and without expence,[6] and that such a lift as this should by no means be neglected. He mentioned it to Michaux, who seemed to have some idea of a previous trip to S. Carolina which it would be well to dissuade him from if possible.

### Benjamin Smith Barton to Thomas Jefferson, January 4, 1793 (excerpt)[7]

In consequence of your note, I have visited M. Michaux. He assures me that he will relinquish all thoughts of his journey to SC, and that he will engage in his scheme; as soon as you think proper. He seems much pleased with the prospect of having so valuable a guide to Kaskaskia as the one you have pointed out, and I will be happy to have the opportunity of conversing with the Indian whenever you shall appoint a time for the purpose.

I have ventured this morning to be very explicit with my friend on the pecuniary head. He seems content to undertake the arduous task (for such, it undoubtedly is) with a very modest assistance in the off-set. This assistance he does not even ask for until his arrival at Kaskaskia, where he thinks, it would be [tear at margin / word missing] that he should have the "power" of drawing for the sum on one hundred guineas. Upon his return, he supposes (provided he shall make discoveries of interesting importance) he shall be entitled to something handsome.

Benjamin Smith Barton's visit also prompted Michaux to write to Jefferson, and the next entry in his journal recorded this communication. The text of this letter to Jefferson follows Michaux's journal entry.

### Notebook 7 continues

JANUARY 20, I communicated to Mr. Jefferson the conditions under which I am willing to undertake the trip to the west of the Mississippi.

### AM in Philadelphia, observations on journey west of the Mississippi, sent to Thomas Jefferson in Philadelphia, January 20, 1793[8]

1. In order to have more freedom to act and to take the course that will be imposed on me by circumstances, I would prefer to undertake this expedition at my expense. My scrupulousness does not allow me to accept the proposed sum for a venture that perhaps will not be carried out completely.

2. In order to not delay this expedition and to remedy the difficulties that I sustain actually to withdraw [funds] from the administration, I would consider it a great favor to accept my drafts up to the amount of 3,600 pounds tournois based on [collateral] of 17,520 pounds owed to me for my salary and funds by the administration. If Mr. de la Forest does not wish to endorse my draft for the sum of 3,600 even if authorized by the Minister of Marine and if I am not assured of the sum before departing Philadelphia, I will not undertake this expedition.

3. I will receive all the letters of recommendation necessary for merchants in Illinois, Indian chiefs, etc.

4. All the knowledge, observations, and geographical information will be communicated to the Philosophical Society.

5. The other discoveries in natural history will be for my immediate benefit and then destined for the general usage.

6. I will undertake this expedition only after having settled my affairs in Carolina, and I only ask for three days after the reception of letters that I await in Charleston.

Upon receipt of this document, Jefferson acted promptly, drawing up a contract for Michaux that also served as a fundraising document and presenting it to George Washington about January 22. Washington then signed the document as a private individual and made the first pledge of one hundred dollars (an enormous sum by the standards of the day) toward the expedition. By March 2, Jefferson had obtained at least twenty-nine additional pledges and begun the fourth column on the verso of the document. He collected signatures in a hierarchical order, regardless of whether the signatories were members of the APS. The first column begins with the president and vice president and continues with seven US senators. The second column begins with three cabinet officers, continues with seven more senators, and ends with the governor of Pennsylvania. The third column has the signatures of thirteen US representatives, including James Madison.[9]

### American Philosophical Society's subscription agreement for André Michaux's western expedition, circa January 22, 1793[10]

Whereas Andrew Michaux, a native of France, and an inhabitant of the United States has undertaken to explore the interior country of North America from the Missisipi along the Missouri, and Westwardly to the Pacific ocean, or in such other direction as shall be advised by the American Philosophical society, and on his return to communicate to the said society the information he shall have acquired of the geography of the said country it's inhabitants, soil, climate, animals, vegetables, minerals and other circumstances of note: WE THE SUBSCRIBERS, desirous of obtaining for ourselves relative to the land we live on, and of communicating to the world, information so interesting to curiosity, to science, and to the future prospects of mankind, promise for ourselves, our heirs executors and administrators, that we will pay the said Andrew Michaux, or his assigns, the sums herein affixed to our names respectively, one fourth part thereof on demand, the remaining three fourths whenever, after his return, the said Philosophical society shall declare themselves satisfied that he has performed the said journey & that he has communicated to them freely all the information which he shall have acquired & they demanded of him: or if the said Andrew Michaux shall not proceed to the Pacific ocean, and shall reach the sources of the waters running into it, then we will pay him such part only of the remaining three fourths, as the said Philosophical society shall deem duly proportioned to the extent of unknown country explored by him, in the direction prescribed, as compared with that omitted to be so explored. And we consent that the bills of exchange of the said Andrew Michaux, for monies said to be due to him in France, shall be received to the amount of two hundred Louis, & shall be negociated by the said Philosophical society, and the proceeds thereof retained in their hands, to be delivered to the said Andrew Michaux, on his return, after having performed the journey to their satisfaction, or, if not if not to their satisfaction, then to be applied towards reimbursing the subscribers the fourth of their subscription advanced to the said Andrew Michaux. We consent also that the said Andrew Michaux shall take to himself all benefit arising from the publication of the discoveries he shall make in the three departments of Natural history, Animal, Vegetable and Mineral, he concerting with the said Philosophical society such measures for securing to himself the said benefit, as shall be consistent with the due publication of the said discoveries. In witness thereof we have hereto subscribed our names and affixed the sums we engage respectively to contribute.

| | | |
|---|---|---|
| Go: Washington $100 | H. Knox $50 | Jona: Trumbull $20 |
| John Adams $20 | Th: Jefferson $50 | James Madison Jr. $20 |
| Benjamin Hawkins $20 | Alexander Hamilton $50 | J: Parker $20 |
| Ra. Izard $20 | Rufus King $20 | Alexr White $20 |
| Sam Johnston $20 | John Langdon $20 | John Page $20 |
| Robt Morris $80 | John Edwards $16 | John B. Ashe $20 |
| Jno Henry $10 | John Brown $20 | Wm Smith $20 |
| G Cabot $10 | Tho Mifflin $20 | Jere Wadsworth $30 |
| John Rutherfurd $20 | | Richard Bland Lee $10 |
| | | Thos. FitzSimons $20 |
| | | Saml. Griffin $10 |
| | | Wm B. Giles $10 |
| | | Jno W: Kittera $10 |

The fourth column, on the back of the document, begins with another US representative and continues with seven additional signers. APS secretary Nicholas Collin, whose name appears in the fourth column, made a copy of the document and continued to solicit funds for the expedition, which ultimately included seventy-six pledges.[11]

John F. Mercer $12

Sam Magaw $16

Nicholas Collin $16

Jona Williams $20

John Bleakley $10

Jos Parker Norris $10

Wm: White $10

John Ross $20

While he negotiated with the APS, Michaux also remained in contact with French officials and colleagues. In Charleston, French consul Jean Baptiste Petry had been replaced in September 1792 by Michel-Ange-Bernard Mangourit. Appointed by Louis XVI with the support of the Brittany delegation in the assembly, Mangourit nonetheless possessed excellent revolutionary credentials. He had participated in the storming of the Bastille and had founded a revolutionary journal. In January 1793, Michaux wrote to the new consul to explain his present and past circumstances.

### AM in Philadelphia to Michel-Ange-Bernard Mangourit (presumably in Charleston), January 15, 1793 (excerpt)[12]

Having been sent by the administration to N[orth] A[merica] about 7 years ago, I had established my main residence in the Carolinas in order to fulfill my mission better, which was as botanist to research trees and plants from this country to send to France. About 2 years ago I received a notice from the director of buildings to consider my mission as completed and to sell the two establishments that I had made, one near Charleston and the other near New York. These establishments were set up for depositing my collection of trees and plants brought from far away, to assure their survival and multiply them in order to ship them later to France; it was very difficult for me to abandon and to lose the fruits of many expenses, fatigues, and dangers. I was thwarted on the one hand by my desire to keep my botanical acquisitions and on the other [by my desire] to submit to the order of the director general who had decided more positively on the sale of the Carolina establishment. Finally, in order not to be reproached, I used all means to bring out the value of the property, to announce the sale in three gazettes, and to proceed to other formalities. Then I became the owner by retrocession of Mr. Himely, to whom the propriety had been adjudged.

Last April, having finished things that concerned the service of the

FIGURE 17. Michel-Ange-Bernard Mangourit (Courtesy of Bibliothèque nationale de France, Paris)

department, I decided to make a trip to Canada, which was a continuation of my research. Not having previous resources, I proposed to the Soc. of Ag. of Carolina [Agricultural Society of Charleston] to assist in the livelihood of the gardener and of negroes that remained for culture, under the condition that trees and plants obtained from the botanical garden would benefit this society and that the plants of my discoveries in America would be for my use, that is, to be shipped to France in the future at a convenient period. These propositions, which were reciprocal, were accepted by Mr. Peter Smith, Col. Vanderhorst, and Gen. Pinckney.[13] I arrived in Phila. last May and continued on my way toward Canada in such a way that I arrived in the territory of Hudson Bay and on the borders of Lake Mistassini in August. I left this extremity of Canada that is uninhabited in September, and I returned to Phila. last December. I had planned to go to Charleston, and it would have been satisfying, sir, to congratulate myself with you on the victories won by our country against the tyrants.

The letters that I found here in May announced that the national operation by which the establishments that I made in America had to be considered as national properties. Upon my return I saw by letters of August that
this business [was] sent back to a committee, [that it] had not been decided,
and that it may be abandoned. Having visited the southern portions as well
as the mountains of N[orth] A[merica] from Cape Canaveral in the south to
Lake Mistassini in Canada, it remains to complete the task that I had imposed on myself and to complete my research to know the portions west of
the Mississippi. I do not know if the administration can be concerned now
with this objective and subsidize these voyages. Not having been able to draw
on my appointment, not even on advances that I had made on my own funds,
I found myself without funds and exposed to wasting my time. To remedy
these difficulties, I had recourse to the Philosophical Society of Philadelphia
[APS] and communicated with some of its members the advantages that one
could hope from a trip to the source of the Missouri River and even beyond.
Several of these men have agreed to help me for the necessary sum. I would
not want to undertake this voyage before spring, but there is one occasion by
which I would have to leave in order to profit by it in six weeks and perhaps
even before. Not having heard from my son, nor from our friends in several
months, I would like to have the satisfaction of seeing the letters that were
sent to me in Charleston before starting this trip, which would last perhaps
more than 2 years. I do not know whether M. Petry would have communicated to you the affairs that I wished to finish and whether he had given you
funds that I had left him for this purpose. If you allow it I would entrust
under your vigilance the garden that I established in the Carolinas, not
wanting to give you more work than to have the gardener report to you from
time to time on the state of culture, its care, improvements, etc. If there were
something in this garden that you liked, I would be delighted if you could
use it. Other than the letters that would have gone to M. Petry, I would ask
you to find others that might have been addressed to several others, such as
M. Himely, Mr. Robinest, M. May, Mme Bars, M. Chion, etc.[14] If you had
the goodness to reimburse the expenses for me made by any of these people,
I would give this amount to Mr. Homassell or to some other individual that
you might want to indicate.[15] Please excuse, dear sir, the unguarded manner
in which I place you in terms of my affairs. Honesty and patriotism lead me
to act with confidence. I offer you the converse, and I hope to be useful to you
for something here.

If by chance there were a boat destined for Philadelphia directly, have the goodness to profit to send me my letters, especially since I would like to profit from the opportunity of friendly Indians who must go to Illinois.

Although Michaux had not directly asked for Mangourit's aid in this letter, Mangourit wrote to Minister LeBrun on February 12, 1793. He enclosed a copy of Michaux's letter saying that Michaux had been "abandoned by despotism" and asked that the botanist be helped "to continue his voyages of discovery."

### Mangourit, consul in Charleston, to LeBrun, minister of foreign affairs, in Paris (excerpt)[16]

You will find enclosed a copy of the letter of French C[itizen] M[ichaux], abandoned by despotism. The minister of liberty will give him a gracious hand. You will not allow that his useful establishments pass into foreign hands. The National Convention should hasten against dangers to national establishments. You would agonize with me to see this celebrated botanist who in his way is as necessary as Abbé Friard is in his,[17] without support from his country. This fellow citizen was obliged to put his hand out so to speak to individuals in Philadelphia to continue his voyages of discovery.

During the same period, some of Michaux's colleagues in France were also advocating that the government continue to support his mission in America, and on May 26, 1793, LeBrun officially directed the new French minister in America, Genet, to give financial aid to Michaux.[18]

With the next entries in his journal, Michaux announced shipments of collections from his Canadian journey. We are reminded that Michaux is not only a botanist but also a collector of all kinds of natural history objects. Although he no longer had the resources to make the large shipments that he had made (principally to Rambouillet) before 1791, he continued to make collections during his travels and to send the seeds and other natural history objects to correspondents in France.[19]

### Notebook 7 continues

JANUARY 29, I made a shipment of seeds from Canada. By this same shipment of January 29, I sent birds, quadrupeds, etc., insects and plants.

FEBRUARY 10, I sent a shipment of live plants from Canada by La Rochelle.

FEBRUARY 18, I announced the contract for 1,200 pounds in my letter to my son via the vessel *Le Suffrein*.

FEBRUARY 29, I wrote to Dr. Afzelius via Baron de Nolken in London.[20]

### AM to Dr. Pehr von Afzelius, professor of medicine, University of Uppsala, Sweden, from Philadelphia, February 20, 1793 (excerpt)[21]

I am sending you a collection of seeds from Canada obtained last year while on a trip made up to the 51° latitude.[22] I have sent two shipments from Carolina [to Sweden] in the last three years, one by Mr. Risberg, and the other by the Rev. Minister Collin of Philadelphia. Now I anticipate going to visit the borders of the Missouri River, and I will always be happy to contribute to the progress of science by sending you objects of natural history when it is possible. If there is no inconvenience to send you shipments through your ambassador in London, you can let me know through letters addressed: to the care of Rev. Collin, minister of the Swedish Church, Philadelphia.

Michaux's next entry in his journal, on March 2, mentions a letter that he wrote to his son. Writing to their colleague and mutual friend Auguste Broussonet in June 1796, François André Michaux quoted extensively from this letter, whose content was apparently similar to his father's earlier letters describing his Canadian journey.[23]

### Notebook 7 continues

MARCH 2, I wrote to my son.

Although not mentioned in his journal, Michaux also wrote an urgent letter to Consul Mangourit in Charleston on March 5, 1793. The botanist was seeking an important mail packet that he believed had been sent from France to him in Charleston.

### AM in Philadelphia to Mangourit, French consul in Charleston, March 5, 1793 (excerpt)[24]

The information that I received of a packet sent from London to Charleston last November leads me to trouble you for the last time. According to the message that I received from France, it contains my commission as well as a letter from the minister of the interior to continue collecting all natural history objects for the natural history collection of the republic and the Jardin des Plantes. I am about to depart for the great voyage to the west of the Mississippi, and I will do everything to wait for your response. Nevertheless I would like to benefit from the escort of friendly Indians who have to leave in a few weeks from Philadelphia, in order to travel safely through the territories of the Ohio.

I may be away two years on this trip, and I am not without worry in seeing the increase in enemies against the republic. Each person must give himself to the most pressing needs of his country, even the weakest according to his capacities, and I have to admit that I hesitated about returning to France. But I refrained because of the advice that I received, and for the reason that if I did not execute this voyage that would complete my research in natural history, I despair that in waiting longer, I would never be able to undertake it. Those are the sentiments of

Your cocitizen, A. Michaux

Since the Delaware River is not being blocked by ice at this time, the most direct route for my address is: Michaux at Mr. Homassel, Philadelphia.

It's Citizen Genest [*sic*] who replaces M. Ternan, who is recalled.

The documents that Michaux sought in his March 5, 1793, letter to Consul Mangourit have not come to light. Nonetheless, it appears, from the contents of a letter that he sent less than a month later to his colleague and close friend Louis Bosc in France, that during March he had either received this packet or learned specific details of its contents. This reassured Michaux that he could continue his work in North America with the support of the republic.

### Notebook 7 continues

APRIL 1, I wrote from New York to Louis Bosc and to my son. I sent birds, squirrels, insects, seeds, samples of plants, etc.

The letter from Michaux to Louis Bosc offers an insight into Michaux's thinking on a variety of subjects during spring 1793, but perhaps most importantly for our consideration of Michaux's proposed exploration of the lands west of the Mississippi, this letter signals a positive change in the botanist's financial circumstances and confirms the resumption of his employment as a French botanist in America.

### AM in New York to Louis Bosc, administrator of the post office, in Paris, April 1, 1793 (excerpt)[25]

I am sending to your address a case of animals, quadrupeds, birds, insects, seeds, and plants. You will share with my son; as to the birds and the box of insects, they are for you. You will find a turtle with claws; I noted several

FIGURE 18. Louis Bosc (Courtesy of Muséum
nationale d'historie naturelle, Paris)

species of this genus. My son will relate to you the remarks on the type of
pigeon that I am sending, two species, the only two that exist in North
America. The box of insects is not that rich in variety, but there are many for
the actual season that we are in, where the snow has just left the ground. It
is true that this winter is one of the mildest that we have had in America in
man's memory. It is very extraordinary to see passenger pigeons here in the
winter, but we have attributed this to the mild winter and the snow [that]
came only in intervals and the large quantity of red oak acorn[s] that are
still found in abundance on the forest floor. The bird that I call turtle dove is
probably only a pigeon, and it has the voice of the pigeon with long tail that I
sent, only less strong.

I send you my thanks for the trouble that you took to allow me to continue my research in natural history. Probably I will obtain more ample collection than ever, not having specific work to take up my time, like cultivation that I had to do under the Old Regime. I am very satisfied with the sum that I have been given even though I will have to economize to do something with this sum. Nevertheless I will do the best and will make many acquisitions in natural history objects because I know how to travel with more economy than anyone else, and I can live with one *écu* longer than anyone [else] with four *écus*. However, if I could draw on my stipend that is due and amounts to approximately one thousand pounds yearly, this would help in certain circumstances in which I might be troubled. I would not use it badly, and I only bring this up if the prospects of our republic have the expectation of continued success. I could not say yet where I would go; I intended for a long time to go and visit the far-reaching areas to the west of North Louisiana, but since I have no other objective than to acquire the largest number possible of objects in natural history, I will decide depending on the circumstances for this goal. When I will be in Illinois, if I do not anticipate obstacles from the Spanish government from Lower Louisiana, I will go there easily.

We have just learned of the execution of Louis Capet and the French war declaration against the European powers that have not recognized the republic. This last tiding is not confirmed; however I am not without worry over these events.[26]

Michaux's journal for April 1793 has only two additional entries.

### Notebook 7 continues

APRIL 24, I sent through Le Havre a box of insects, sample of *Panax*, etc.

APRIL 30, I communicated to the Philosophical Society the incentives that prepare me to undertake the trip west of the Mississippi.

### American Philosophical Society's instructions to André Michaux, circa April 30, 1793[27]

Sundry persons having subscribed certain sums of money for your encouragement to explore the country along the Missouri, & thence Westwardly to the Pacific Ocean, having submitted the plan of the enterprise to the direction of the American Philosophical society, & the Society having accepted of the trust, they proceed to give you the following instructions.

They observe to you that the chief objects of your journey are to find the shortest & most convenient route of communication between the US. and the Pacific ocean, within the temperate latitudes, & to learn such particulars as can be obtained of the country through which it passes, its productions, inhabitants & other interesting circumstances.

As a channel of communication between these states and the Pacific Ocean, the Missouri, so far as it extends, presents itself under the circumstances of unquestioned preference. It has therefore been declared as a fundamental object of the subscription, (not to be dispensed with) that this river shall be considered & explored as a part of the communication sought for. To the neighborhood of this river therefore, that is to say to the town of Kaskaskia, the society will procure you a conveyance in company with the Indians of that town now in Philadelphia.

From thence you will cross the Missisipi & pass by land to the nearest part of the Missouri above the Spanish settlements, that you may avoid the risk of being stopped.

You will then pursue such of the largest streams of that river, as shall lead by the shortest way, & lowest latitudes to the Pacific ocean.

When, pursuing these streams, you shall find yourself at the point from whence you may get by the shortest & most convenient route to some principal river of the Pacific Ocean, you are to proceed to such river, & pursue its course to the ocean. It would seem by the latest maps as if a river called Oregon interlocked with the Missouri for a considerable distance, & entered the Pacific ocean, not far Southward of Notka sound. But the Society is aware that these maps are not to be trusted so far as to be the ground of any positive instruction to you. They therefore only mention the fact, leaving to yourself to verify it, or to follow such other as you shall find to be the real truth.

You will, in the course of your journey, take notice of the country you pass through, its general face, soil, rivers, mountains, its productions animal, vegetable, and mineral so far as they may be new to us & may also be useful or very curious; the latitude of places or materials for calculating it by such simple methods as your situation may admit you to practice, the names, numbers, & dwellings of the inhabitants, and such particularities as you can learn of their history, connections with each other, languages, manners, state of society & of the arts and commerce among them.

Under the head of Animal history, that of the Mammoth is particularly recommended to your enquiries, as it is also to learn whether the Lama, or Paca of Peru is found in those parts of the continent, or how far north they come.

The method of preserving your observations is left to yourself, according to the means which shall be in your power. It is only suggested that the noting them on the skin might be best for such as are most important, & that further details may be committed to the bark of the paper birch, a substance which may not excite suspicions among the Indians, & little liable to injury from wet, or other common accidents. By the means of the same substance you may perhaps find opportunities, from time to time, of communicating to the society information of your progress, & of the particulars you shall have noted.

When you shall have reached the Pacific Ocean, if you find yourself within convenient distance of any settlement of Europeans, go to them, commit to writing a narrative of your journey & observations & take the best measures you can for conveying it by duplicates or triplicates thence to the society by sea.

Return by the same, or such other route, as you shall think likely to fulfill with most satisfaction and certainty the objects of your mission; furnishing yourself with the best proofs the nature of the case will admit of the reality & extent of your progress. Whether this shall be by certificates from Europeans settled on the Western coast of America, or by what other means, must depend on circumstances.

Ignorance of the country thro' which you are to pass and confidence in your judgment, zeal, & discretion, prevent the society from attempting more minute instructions, and even from exacting rigorous observance of those already given, except indeed what is the first of all objects, that you seek for and pursue that route which shall form the shortest & most convenient communication between the higher parts of the Missouri & the Pacific ocean.

It is strongly recommended to you to expose yourself in no case to unnecessary dangers, whether such as might affect your health or your personal safety: and to consider this not merely as your personal concern, but as the injunction of Science in general which expects its enlargement from your enquiries, and of the inhabitants of the US in particular, to whom your Report will open new fields & subjects of Commerce, Intercourse, & Observation.

If you reach the Pacific ocean & return, the Society assign to you all the benefits of the subscription before mentioned. If you reach the waters only which run into that ocean, the society reserve to the apportionment of the reward according to the conditions expressed in the subscription.

They will expect you to return to the city of Philadelphia to give in to them a full narrative of your journey & observations, and to answer the

enquiries they shall make of you, still reserving to yourself the benefits arising from the publication of them.

Michaux's situation changed in the months between his original offer in January and his receipt in April of the detailed instructions prepared by Jefferson.
In the intervening months, the botanist became aware that France would continue to employ him as a natural history collector in America, and his primary
allegiance was to France. His restatement of his willingness to undertake the
journey for the APS now contained reservations. His original offer had listed
six conditions that he planned to observe in carrying out the exploration, but
his response to the APS now included only three. First, he restated his com-
mitment to make the journey at his own expense because he was concerned
that he might not be able to fully carry it out. Second, he reaffirmed his promise that he would communicate all the knowledge and geographical observations that he made on the journey directly to the APS. His third point, however, was most important to Michaux. He insisted that he must retain personal
control of any natural history discoveries he might make on the expedition.[28]

Jefferson's instructions emphasized that Michaux was being employed by
the APS and was responsible for reporting his findings to that organization.
Michaux, however, wanted to retain part of his collections and to be able to
make use of his discoveries in natural history before making them available to
the APS.[29] Nonetheless, it is possible
that these differences might have been resolved with further negotiation had
Edmund Charles Genet, the new minister representing the Girondist government of the French Republic, not arrived in Philadelphia after an overland journey from Charleston. Genet did not reach Philadelphia until May 16;
however, news of his impending arrival preceded him.[30] As we have seen in
his March 5 letter to Mangourit, Michaux had been aware of Genet's appointment for several weeks. Now he waited to meet the new minister and offer his
services to his own country rather than continue his discussions with the APS
or immediately begin the journey.

Only four days after Genet's arrival in Philadelphia, Michaux submitted a
memorandum to the minister that described his situation.

### AM to Edmund Charles Genet, Philadelphia, May 20, 1793 (excerpt)[31]

After my return from Canada I was informed by letters dated at Paris last
August [1792] that the report made to the legislative assembly in the preceding January had not been put on the agenda at that time by reason of the

workload of the assembly. Uncertain of being continued in the service of the Republic and not being able to get drafts on my back salary accepted, I found myself more than ever destitute of means and threatened with wasting my time. To remedy these difficulties, I have communicated to several members of the Philosophical Society of Philadelphia my observations on the usefulness of a journey to the sources of the Missouri for geographical knowledge. These gentlemen have subscribed five thousand dollars of which the sum of one thousand was to be advanced to me for the first expenses of this undertaking and the rest after having furnished proof that I should have reached the rivers which flow into the Pacific Ocean.

But not wanting to depart from the Principles which have always directed me to accord to my country the profit and honor of my enterprises and of my discoveries, I have resolved not to accept the advances supplied by the Philosophical Society except as a deed of reimbursable loan on my salary and I have submitted to the Secretary of the United States the following conditions:

Towards the end of March it seemed to me by the instructions given concerning this journey that I was considered as completely foreign to my own country and employed exclusively for the advantage of the Philosophical Society. I submitted to the president of this Society the following memoir to make known the incentives which inspire me, while working for my country, to be useful (secondarily) to America.

Michaux inserted the list of conditions that he had given to the APS on January 20 with his memorandum. He followed this memorandum with a memoir, submitted to Genet a few days later, that described his North American travels and the plan to explore the West for the APS. This memoir closed with the following statement.[32]

Dedicated to my mission in the service of the Republic, I wish to be employed in the way most useful to my country, and I will consider myself fortunate to respond to this mark of confidence by my zeal and my success.
Philadelphia, the 21st [of] May, 1793, the 2nd year of the French Republic.
A. Michaux

Genet thus became aware immediately after his arrival in Philadelphia that in Michaux he had a patriotic, English-speaking Frenchman who was experienced in frontier journeys and who wanted to make a journey west of the

Appalachians. He learned also that Michaux was so well regarded by the Americans that a group of the most prominent citizens of the United States were on the verge of selecting him to lead a voyage of exploration to the Pacific Ocean.

Michaux's journal entries in May and June continued to be sparse, sporadic, and somewhat jumbled chronologically, suggesting that they may have been written after events transpired. Michaux no doubt became involved in a whirlwind of activity in the first weeks after Genet's arrival.[33]

**Notebook 7 continues**

(MAY 10, I sent Bosc some insects but the shipment did not leave until June 9.)

MAY 29, I wrote to Madame Desaint.[34]

JUNE 9, I sent some insects to Louis Bosc (these two shipments are really only one).

MAY . . . , Citizen Genet, minister plenipotentiary of the French Republic, arrived in Philadelphia.

MAY 18, I communicated to Citizen Genet a memorandum of observations on French colonies in North America, in Louisiana, [in] Illinois, and in Canada.

MAY 22, I gave an abbreviated memorandum of my journeys in North America.

JUNE . . . , 1793, gave an account of the sums received and my expenses since my departure from France for North America.

By June Michaux was involved in the political schemes of Genet, and there is no further mention of the expedition for the APS. Nonetheless, this proposed scientific and geographical expedition to the West for Jefferson and the APS ultimately provided such an excellent cover story for Michaux's political journey to Kentucky that it still causes confusion among historians. Some writers have used a passage in a memoir of Meriwether Lewis written by Jefferson in 1813 to suggest that in the beginning, Michaux acted on the APS's instructions or combined the two missions,[35] but the overwhelming evidence suggests that he was acting only on Genet's instructions from July until December 1793.[36]

There is no evidence in Michaux's journal or letters that, after reaching Kentucky, he continued west to Illinois, as some have suggested,[37] nor can evidence be found that he left Kentucky to meet with any Indian tribes in the region, as he was authorized to do by Genet. He did not explore the lands beyond the Mississippi but rather faithfully followed the instructions of Minister Genet to assist Gen. George Rogers Clark in Kentucky with his plot to use force to expel the Spanish from the Mississippi Valley and New Orleans.[38]

# CHAPTER TEN

# Kentucky Journey for Genet, 1793

SUMMARY: MICHAUX'S JOURNEY TO KENTUCKY in 1793, described in notebook 7, was not a botanical journey but rather a political mission for France. His return to Philadelphia is recounted in notebook 8. The monarchy had been overthrown, the king executed, and a legislative body—the National Convention—ruled republican France. The National Convention was dominated by the Girondins, a party that wished to spread revolution and faced opposition both within France and without, from European monarchies who viewed republican France as a threat to the stability of their kingdoms.

In these perilous times the new government's first diplomatic representative to the United States was Edmond Charles Genet, a bright, energetic young man who spoke English fluently. Part of Genet's instructions were to secure a new treaty and to improve relations with the United States, France's only ally, while another part directed him to liberate areas in North America that were under the control of Spain and Britain, France's enemies. This complex and delicate mission called for exceptional diplomatic skills, but events soon revealed that Genet lacked the capacity to deal successfully with the American government.

The aim of the Girondins to spread the revolution meshed with a long-cherished French ambition to recover the territory in North America that France had lost at the end of the French and Indian War. After siding with

FIGURE 19. Edmond Charles Genet (Courtesy of the New York Public Library)

the American revolutionaries and opening the Mississippi waterway to them during the revolution, Spain had closed the Port of New Orleans to American commerce in 1784. This restriction caused hardship in American settlements west of the Appalachians, whose natural commercial outlet led down the Ohio and Mississippi Rivers through the Port of New Orleans. Resentment of Spanish restrictions ran high in Kentucky, and in early 1793 American Revolutionary War hero Gen. George Rogers Clark sent the French an offer of his services to raise an army in the western settlements and lead it down the Mississippi to forcibly liberate New Orleans from Spain and open the river to free navigation.[1]

Arriving in the capital, Philadelphia, in mid-May, Genet received Gen. Clark's offer and met with Michaux several times, received his written reports, and learned that the botanist was ready to serve France. Genet then assigned him a central role in this plan to "liberate" Louisiana from Spanish rule, a plan that became known as the Genet Affair. The botanist went to Kentucky as the minister's personal emissary to Gen. Clark, who needed French support and financing to carry out his plans. Though Genet sent Michaux to assist the general and to assure him of French support, he would provide little money for Clark's expedition.[2]

FIGURE 20. Gen. George Rogers Clark, painted by James Barton Longacre, ca. 1830, watercolor on paper (Courtesy of the National Portrait Gallery, Smithsonian Institution)

In June Genet continued to enjoy the sympathetic ear of Thomas Jefferson, the secretary of state. Before sending Michaux off to Kentucky as his courier, the French minister first attempted to persuade the secretary to accept him as French consul in Kentucky. In this Genet was not successful, but he soon obtained a letter of introduction for Michaux from Jefferson to the Kentucky governor, Isaac Shelby. About the same time, Genet also informed Jefferson directly of his instructions to Michaux by reading them aloud to "Mr. Jefferson," speaking to the latter as a private individual and not in his official capacity as US secretary of state.[3] The instructions were absolutely clear that the botanist would be acting as Genet's courier to Gen. Clark in Kentucky for the purpose of raising an army of American frontier fighters to drive the Spanish from the Mississippi Valley and New Orleans.[4]

Although it is unlikely that Michaux needed additional incentive to undertake this mission for France, it must have pleased him that he would now be paid by the new government for this political work as well as for his work as a botanist. The demise of the Old Regime had ended his funding, with his salary and expenses considerably in arrears. In financial distress, he had sought support from the Americans to explore the territory west of the Mississippi, but what he desired most was to be able to continue working for his own country. The mission to Kentucky must have seemed especially fortuitous to the botanist as he had wanted to visit Kentucky at least since his aborted trip in 1789. Genet now gave him the opportunity to be the first botanist to visit the area, while promising him both an annual government stipend and an even larger amount for expenses.[5]

Accompanied by two traveling companions who were later identified as French artillerymen, Michaux began his journey from Philadelphia on July 16, 1793. He first traveled across Pennsylvania by road through Lancaster, Carlisle, and Bedford to Pittsburgh.

Michaux's route to Kentucky was probably chosen because he traveled with heavy baggage;[6] he also wanted to visit Hugh Henry Brackenridge.[7] It was not the quickest, safest, or best way to Kentucky from Philadelphia.[8] When Michaux arrived in Pittsburgh, the water in the Ohio River was low, so he encountered a delay of more than two weeks before he could begin the second leg of his journey, down the river to Kentucky. Two more weeks were spent traveling downriver by boat and visiting the settlements scattered along the river. He left the river at Maysville (then known as Limestone), Kentucky, while his French traveling companions continued down the river. From Maysville, Michaux traveled overland on horseback to Lexington and then on to Danville, the capital of the state. Along his route he visited a number of individuals for whom he carried letters of introduction, including the governor of Kentucky, Isaac Shelby, who lived near Danville. Leaving Danville he continued overland to deliver Genet's messages to Gen. Clark in Louisville. He then promptly returned to Danville, where he had rented lodgings to await Clark's response. A short round-trip from Danville to Lexington followed, while Michaux and Clark exchanged letters, and while Michaux attempted to raise money for the expedition from Lexington merchants. After Clark accepted the French commission that Michaux offered him, the botanist then returned to Philadelphia in order to seek additional funds for the project from Genet.

In his next notebook, he recorded this return journey and continued to make entries that reflected his continued work as Genet's political agent even after he returned to Philadelphia. In February 1794, with Genet's approval, he returned his commission as French agent in Kentucky and began the overland journey to the garden near Charleston. He made this journey carrying Genet's messages to Consul Mangourit regarding a proposed attack on Spanish Florida with some expectations that he would be sent to Kentucky again, but his political duties were at an end. While Michaux was en route to South Carolina, the French diplomatic delegation led by Joseph Fauchet, representing the new Jacobin regime of the French Republic, arrived in Philadelphia to replace Genet. Fauchet promptly revoked all of Genet's commissions and ended French support for the projected attacks on Louisiana and Florida. Michaux was again free to make journeys of botanical discovery.

Michaux began his entries concerning the forthcoming journey to Kentucky with remarks on his preparations and listed many people to whom he was carrying letters. He also noted the name of a French trader with whom he was asked to make contact in the Illinois country. Not all the people listed were to be informed of the plans of Genet and Clark; some of them were only concerned with the botanist's safe passage west.

When Michaux made this journey down the Ohio River, travelers between Pittsburgh and Kentucky remained subject to hostile attack from Native Americans, as did the settlements in the region. It would be more than a year before Gen. Anthony Wayne's troops would decisively defeat the hostile tribes at the battle of Fallen Timbers, near present-day Toledo, Ohio, and make the region relatively safe for white settlers and other travelers on the river.

### Notebook 7 continues

[JUNE] . . . consulted and conferred with Citizen Genet on my mission to Kentucky.

JUNE 22, 23, 24, 25, and 26, prepared for the journey to Kentucky.
[I have] letters of recommendation for H. H. Brackenridge, Esq., from Pittsburgh; [for] Maj. Isaac Craig, from Maj. Sgt. Stagg; for Capt. John Pratt, commanding troops on their march to the western frontier [also from Stagg];[9] for Brig. Gen. Geo. Rogers Clark; for Isaac Shelby, Esq., governor of the state of Kentucky;[10] for Alexander D. Orr, Esq., near Limestone;[11] [for] Dr. Adam Rankin, Danville; [and for] James Brown, Esq., Lexington.[12]

JULY 1, I packed my belongings.
Letters of recommendation for Thomas Craighead, Springhill; James Brown, Lexington; Dr. Adam Rankin, Danville; Col. Alexander Orr, near

Limestone; Maj. Gen. Benjamin Logan, Lincoln County;[13] James Speed Jr., Danville;[14] Gen. Clark, Louisville; Joseph Simpson, Lexington; Governor Shelby, Esq.; and Brig. Gen. James Wilkinson.[15]

Mr. Robert recommended that I visit M. Tardibeau at Kaskaskia for him.[16]

JULY 15, I took my leave of Citizen Genet, French minister to the United States, and I left Philadelphia the same day at 10 o'clock at night to avoid the heat and to travel by moonlight.[17]

JULY 16, I was in the company of . . . Humeau and . . . LeBlanc; we traveled 40 miles.[18]

JULY 17, we went through Lancaster and traveled 35 miles.[19]

JULY 18, we went through Carlisle . . . miles and spent the night at Shippensburg.[20]

JULY 19, we slept at Upper Strasburg . . . miles.

SUNDAY, JULY 20, we left Upper Strasburg, small town at the foot of the mountains.[21] We only traveled 21 miles as one of our horses was sick. I observed *Magnolia acuminata, Azalea octandra, Kalmia latifolia, Fagus castanea, F. pumila, Pinus* 2 leaves, [*P.*] 3 leaves, [*P.*] *strobus,*[22] *Abies canadensis, Quercus castaneafolio*, and *Juglans nigra*.

JULY 21, we left Wells Tavern, went by Juniata River . . . ,[23] and observed *Rhododendron maximum, Hydrangea frutescens, Trillium erectum*. We slept at Bedford: 21 miles.[24]

JULY 22, we left Bedford and breakfasted four miles away, where the road for Pittsburgh divides in two. We took the one on the right, and the rain obliged us to stop and spend the night only 12 miles from Bedford.

JULY 23, we traveled 24 miles and went beyond the summit of the Alleghenies.

JULY 24, we covered 25 miles.

JULY 25, we went by Greensburg and made 31 miles.[25]

JULY 26, because of the rain, we only went . . . miles.

JULY 27, we traveled 19 miles and arrived in Pittsburgh. Total of 32 [320] miles from Philadelphia.

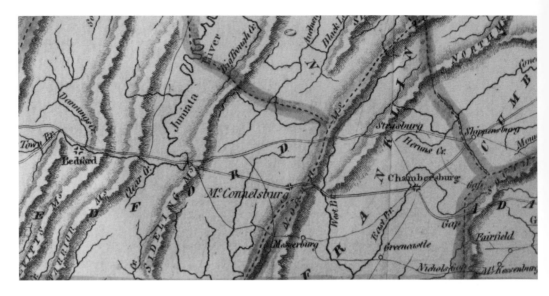

MAP 12. Pennsylvania map excerpt, Matthew Carey, Philadelphia, 1814 (Courtesy of the Philadelphia Print Shop, Philadelphia, PA)

JULY 28, I visited Mr. H. Brackenridge, Esq.

JULY 29, I botanized and recognized on the banks of the Monongahela *Dracocephalum virginianum, Bignonia radicans, Crotalaria alba*?[26] These plants are submerged when the water is high.

JULY 30, I recognized a plant of the genus *Ziziphora . . .* , *Cunila pulegioides* flowers 4-parted, *Teucrium canadense, Eupatorium aromaticum, Sigesbeckia . . .* , *Verbenae* several species.

AUGUST 1, I botanized and recognized: *Cassia marylandica, Monarda didyma, Sanicula marylandica, Triosteum perfoliatum, Sicyos angulata, Acer rubrum,* [*A.*] *saccharum, Campanula . . .* , *Cercis canadensis, Menispermum canadense, Actea spicata, Tilia americana, Urtica divaricata, Arum triphyllum, Celtis occidentalis, Panax quinquefolium, Staphylea trifoliata, Asarum canadense, Rhus typhina,* [*R.*] *glabra,* [*R.*] *vernix,* [*R.*] *copallinum,* [*R.*] *radicans,* [*R.*] *toxicodendron, Clinopodium vulgare,* [*C.*] *incanum.*

AUGUST 2, I recognized *Aristolochia sipho seu* [or] *macrophylla, Panax quinquefolium, Lobelia siphilitica, Convallaria* several species, *Veronica . . .* , *Oxalis stricta.*

AUGUST 3 and 4, I botanized and saw: *Cacalia* 2 species, *Phryma leptostachia, Leontice thalictroides, Lobelia siphilitica,* [*L.*] *inflata,* [*L.*] *cardinalis, Eupatorium perfoliatum,* [*E.*] *maculatum,* [*E.*] *odoratum,* [*E.*] *celestinum,*[27] *Actea spicata, Podophyllum peltatum, Asarum, Hydrophyllum canadense, Trillium cernuum, Panax quinquefolium, Aristolochia sipho seu, Menispermum* . . . , *Sambucus canadensis* black fruit, *Sambucus* . . . red fruit, tomentose leaves, *Tilia americana, Laurus sassafras,* [*L.*] *benzoin, Robinia pseudoacacia, Juglans oblonga, Juglans hiccory, Platanus occidentalis, Acer rubrum,* [*A.*] *saccharum, Ulmus* . . . , *Hamamelis* . . . , *Cynoglossum* 3 species, *Vitis vulpina, Dioscorea* fruit inferior, *Teucrium canadense, Scrophularia marylandica, Dracocephalum virginiana, Dianthera* . . . , *Sophora* leaflets in 3s, stipulate, broadly lanceolate, flowers blue standard, petal shorter than corolla, *Mimulus ringens, Bignonia radicans, Cercis canadensis, Fagus sylvatica americana, Circaea canadensis, Urtica inermis, Erigeron canadense, Cornus florida, Rubus odorata, Rubus occidentalis, Penthorum sedoides, Cephalanthus occidentalis, Polygonum aviculare,* [*P.*] *hydropiper,* [*P.*] *amphybium,* [*P.*] *scandens, Sanguinaria canadensis.*

AUGUST 6, on the banks of the Monongahela opposite from Pittsburgh I saw a coal mine whose entrance appears to be 15 feet thick of pure mineral without any mixture; sometimes one can distinguish between some of the layers a tinge of iron. In several areas we can find soft rocks that appear to be good for sharpening large instruments, their makeup appears to be a mixture of sand, clay, and iron particles, very rarely with mica fragments.

The soil around Pittsburgh is generally clayey, and the stones are calcareous, of a brown color, and containing much silty clay. The soil between the two rivers on which Pittsburgh is built is of alluvium; one can find on land set aside for well building, at a depth of 30 feet, rounded stones, worn by the action of the stream.

AUGUST 9, I prepared to leave, when the pilot of the boat on which I had put my baggage came to tell me that he was waiting for the boats that were to carry troops, as this boat appeared to be overloaded as happens when the water is low. It looked as if it might rain.

AUGUST 10, the river appeared to get lower.

AUGUST 11, 12 and 13, we stayed there while waiting departure.

AUGUST 13, three boats arrived from Illinois belonging to Mr. Vigo. They were guided by about 30 French Canadian or Illinois oarsmen.[28]

A Frenchman who has lived in America for 14 years, who was charged to send provisions of flour to New Orleans, told me that he would give me letters for Illinois addressed to the leader of the Post at Saint Louis. He is now established in Pittsburgh, and his name is Audrain. This man Audrain, they say, is an associate of a man named Louisière or Delouisère expatriated from France for being known to have been involved in the plot to deliver Le Havre to the united fleets of England and Spain.[29] This Delouisère is not present in Pittsburgh. There is another Frenchman, an inhabitant of Pittsburgh, M. Lucas de Pentareau, an excellent democrat, actually also absent. He is said to be an educated individual who is knowledgeable about laws.[30]

Pittsburgh is situated at the confluence of the two rivers, the Monongahela and the Allegheny. These two rivers when joined form the Ohio, or La Belle Rivière. There are many more houses on the Monongahela than on the Allegheny. There are about 250 houses, and every year the number grows considerably. We can still see the ditches that acted as an entrenchment for Fort Duquesne, built by the French. Since then, the English built another fort, almost next to it, at the angle formed by the junction of the two rivers. It was built of bricks, and the Americans had it demolished to use the bricks in the construction of houses that are being built daily at Fort Pitt.

The Americans have a stockade built behind the city on the banks of the Allegheny River; it serves to harbor the troops that are being sent out against the natives and as a depot for ammunition that is being sent from Philadelphia.[31]

WEDNESDAY, AUGUST 14, I left Pittsburgh and only stayed the night at two miles from the point of a small island on which I recognized *Acer negundo*, [*A.*] *rubrum*, [*A.*] *saccharum*, *Evonimus* capsules glabrous.

AUGUST 15, I recognized 20 miles from Pittsburgh *Pavia lutea*, *Panax quinquefolium*, a *Bryonia* plant monoecious calyx 5-parted, corolla 5-parted, male flowers spicate, axillary, female flowers each axillary, fruits with harmless spines.[32] Our journey was 28 miles.

AUGUST 16, we passed the state line between Pennsylvania and Virginia. This line is marked by trees cut at about . . . feet high to the right and the left of the Ohio or Belle Rivière; this site is at 45 miles from Pittsburgh. The same day we arrived at night at Buffalo Creek, 79 miles from Pittsburgh.[33]

AUGUST 17, we went by Wheeling, 92 miles from Pittsburgh; this site as well as Buffalo Creek are settled by about 12 families each. Because of contrary winds, we traveled only 30 miles.

SUNDAY, AUGUST 18, I saw several flocks of wild turkey; wind is unfavorable.

August 19, we traveled 50 miles. There is no settlement between Wheeling [then Virginia, now West Virginia] and Marietta, little town found at the mouth of the Muskingum River [in Ohio]. We slept at a place called Fort Harmar, on the right bank of the Muskingum River across this river from Marietta.[34] *Dianthera americana.*

August 20, we spent the day there.

August 21, we passed by the Little Kanawha River, Belpré, and Belleville: 34 miles.[35]

August 22, we saw no settlements. I recognized *Polymnia canadensis,*[36] *Acer rubrum* leaves glaucous on lower surface, *A. negundo, A. saccharum, Acer* leaves rugose with somewhat wooly veins, *Annona triloba, Pavia lutea, Platanus occidentalis.*

August 23, we went by the Great Kanawha [River], 4 miles before arriving at Gallipolis, on the opposite bank. We reached the settlement of Gallipolis on the left bank of the Ohio River. The houses are all built of squared timbers with grooves only at their ends instead of mortise (log house).

August 24, I stayed over and paid a visit to Dr. Petit. He inspired in me great respect because of his knowledge, character, and quality. It would seem that the only motive that kept him in this unfortunate colony was a humanitarian one. Of the 600 people that came to settle, only 150 are left.[37]

Sunday, August 25, we left Gallipolis; at 35 miles I recognized *Iresine celosioides* on the banks of the Ohio River, submerged by large floods;[38] we passed a river named Guyandotte, with no settlement: 40 miles.

August 26, we did not see any houses; we passed Scioto River, . . . miles.[39]

August 27, we saw a settlement of several houses at a site called Three Islands,[40] ten miles before arriving at Limestone; these settlements are thought to be the first belonging to Kentucky. Toward nightfall we arrived in Limestone.

Limestone is known as the Port of Kentucky (landing place).[41] There they unload merchandise sent from Philadelphia for Danville, Lexington, etc. Four miles away on the road to Lexington is a little town established ten years ago, named Washington, [which] is already flourishing since it is situated in a very fertile area.

August 28, I visited Col. Alexander D. Orr.[42]

August 29, I parted from the two companions that I had had since Philadelphia. They continued on toward Louisville, while I was going toward interior settlements. Col. Orr offered me his company to go to Lexington, where he planned to go in a few days.

August 30 and 31, I botanized while waiting for horses for the trip to Lexington. *Guilandina dioica, Fraxinus (quadrangularis),*[43] *Gleditsia triacanthos, Serratula praealta, Eupatorium aromaticum, Crepis sibirica?,* etc.[44]

Sunday, September 1, I had dinner with Col. Lee.[45]

September 2, I dined at . . . Fox and made arrangements for my baggage in anticipation of departure.

September 3, the journey has been put off for the next day; the soil around Washington is clayey, blackish and very rich;[46] the stones are calcareous, dull blue, full of petrifactions of marine shells. The bones of monstrous animals that are thought to be elephants are found nearby. It must be presumed that these bones belonged to marine animals because of the great abundance of marine life remains found in these parts.[47]

September 4, we left Washington and went by an area where the soil is full of saline substances and where the buffalo come in large numbers to lick the particles of salt that collect at the surface of the soil. In this area there are springs whose water is acrid, putrid, blackish, and which release a terrible smell that is given off by bubbles released from the soil as one approaches the spring. The inhabitants in the neighborhood have set up furnaces and caldrons to obtain salt by boiling off the water. We traveled 33 miles.

September 5, we covered 27 miles, and we arrived early in Lexington, principal city of the state of Kentucky.[48] We went by a small settlement, named Paris, capital of Bourbon County.[49] There are about 28 houses and country houses along the road, and travelers can go without danger from Limestone to Lexington, a 66-mile journey.

September 6, I visited two persons living in Lexington for whom I had letters of recommendation.[50]

September 7, botanized

Sunday, September 8, I was obliged to stay as I did not find a horse for hire.

SEPTEMBER 9, I left Lexington and went through woods, interrupted by very few plantations. I passed the Kentucky River, whose banks are drawn close together, and when the water level is low there is more than 100 feet from this river to the land through which it flows. People tell me that during times of flood, the water can rise 40 feet in one day. When you arrive there, you think that you are between two steep mountains, but in effect it is only a river or stream whose bed has been deeply hollowed out. The rocks on the banks are calcareous. Several shrubs and plants from the Carolinas are found on the southern exposure, thus shielded from the cold by this favorable situation of the great depth of the river.

SEPTEMBER 10, I arrived in Danville and visited several individuals for whom I had letters;[51] Col. Barbee, etc.,[52] Peter Tardivau, captain, man of character, etc. etc.[53]

SEPTEMBER 11, I visited Gen. Benjamin Logan, whose residence is twelve miles from Danville. He showed confidence in the mission with which I had been charged and indicated that he would have been pleased to take part in this enterprise, but that he had received a few days before a letter from J. Brown,[54] which informed him that negotiations had been started between the United States and the Spaniards and the Creek Indians concerning navigation on the Mississippi. That a messenger had been sent to Madrid, and if before his return on December 1 those in the United States were to begin to act with hostility against the Spaniards, they would be blamed by the federal government. He had to leave the next day for his property at Bullskin Creek, and after I had conferred with Gen. Clark, he hoped that he could confer with us all together, etc. etc.

SEPTEMBER 12, I returned to Danville.

SEPTEMBER 13, I visited (His Excellency) the governor of the state of Kentucky, Isaac Shelby; I visited the hills known as Knob Licks; I saw several plants, particularly in the salty areas enclosed in the interior of Kentucky, *Andromeda arborea* . . .

Michaux presented letters of introduction from both Thomas Jefferson and Kentucky senator John Brown to Governor Isaac Shelby.[55] The letters employed similar wording, and on the surface each was simply an innocuous introduction for the naturalist, but they subtly informed Shelby that Michaux was more than a naturalist; he was the new French minister's representative.

**Secretary of State Thomas Jefferson to Isaac Shelby,
governor of Kentucky, July 5, 1793[56]**

The bearer hereof Mr. Michaux is a citizen of the French republic who has resided several years in the US as the Conductor of a botanical establishment belonging to the French nation. He is a man of science and merit, & goes beyond Kentuckey in pursuit of objects of Natural history & botany, to augment the literary acquirements of the two republics. Mr. Genet the Minister of France here, having expressed to me his esteem for Mr. Michaud and good opinion of him, and his wish that he should be made known to you, I take the liberty of recommending him to your notice, your counsels, & good offices. His character here persuades me they will be worthily bestowed on him, and that your further knowledge of him will justify the liberty I take of making him known to you.

**Notebook 7 continues**

SEPTEMBER 14, I left Danville for Louisville, spent the night at Cumberland, 19 miles from Danville.

SUNDAY, SEPTEMBER 15, 1793, at 22 miles from Danville, I found a type of *Tragia*, a monoecious plant with fruits like *Euphorbias*.[57] A little before arriving in Bardstown, I recognized rocks and stones of calcareous nature that had the form of fossil coral. The top of these hills through which one travels 3–4 miles before reaching Bardstown are made up entirely of these coralline fossils. I found many plants that are not found elsewhere. [I did also find] *Fagara* from the state of New York, *Rhamnus* (Carolinian), and *Rhamnus* . . . , etc. etc. The surroundings [of Bardstown] are very interesting to visit for a botanist. I had dinner in Bardstown and spent the night 6 miles farther, 31 miles [in all].[58] From Bardstown to Louisville, the countryside is of no interest to a botanist.

SEPTEMBER 16, I arrived in Louisville, having traveled by a new route: 29 miles. A total of 79 miles from Danville.

SEPTEMBER 17, I visited Gen. Clarke. I gave him the letters of the minister and told him the object of my mission. He answered that the enterprise in question was dear to his heart, but since he had written so long ago about it without any answer, he had thought that the project had been abandoned. I told him that his letter had fallen into foreign hands and that the minister had only received it indirectly once arrived in Philadelphia. He told me that a new circumstance appeared to put an obstacle in the way.

September 18, I stayed in Louisville and botanized.

September 19, I returned to visit Gen. Clarke.

September 20, I left Louisville and went by Gen. Clarke and spent the night near Salt River.[59]

September 21, I went by Bardstown; *Evonimus* branches 4-angled, capsules with spines.

Sunday, September 22, I arrived again in Danville at 5 o'clock in the evening. I wrote to Minister Genet on that day and sent it by the Philadelphia post.

September 23, I rested.

September 24, I left for Lexington and slept at the crossing of the Kentucky River.

September 25, I noted that my horse had been lost since I stayed in an inn where there was no stable; the horse had jumped over the enclosure, and I spent the whole day looking for him.

During this time I noted on the sandy shores: *Iresine celosioides, Mollugo verticillata.*

On the rocks: *Heuchera americana, Asplenium rhizophyllum, Pteris* new species, *Parietaria . . . , Hydrangea arborescens.*

On the calcareous hills: *Serratula,* two unknown species,[60] *Cuphea viscosa,* an undetermined gymnosperm in a new genus with two pairs of stamens of different lengths, an undetermined new angiosperm with two pairs of stamens of different length.

On the side of the Dix River: *Dirca palustris, Sophora* with blue flowers.

In the shaded forests: *Acer* with leaves silvery and red, *Acer saccharum, Fraxinus* leaflets subentire, *Fraxinus* leaflets serrate, branches 4-angled,[61] *Gleditsia triacanthos, Guilandina dioica, Robinia pseudoacacia, Evonimus* branches almost round, capsules smooth.

September 26, it rained all day. I slept one mile from the Kentucky River at . . . Hogan,[62] who was kind enough to lend me a horse without charge to find mine.

September 27, I arrived in Lexington, only 20 miles from the crossing of the Kentucky River called Hickman Junction.

October 5, I left Lexington.

FIGURE 21. US post office, Danville, Kentucky, 1792. Mail service to Kentucky began in 1792, and, in 1793, Michaux used the mail to communicate with Genet in Philadelphia, visiting this building to post and receive his letters. (Photo by Charlie Williams)

OCTOBER 6, I arrived in Danville. This same day wrote to Citizen Minister Genet.

OCTOBER 7, I lodged at Puvit and obtained my baggage.[63]

OCTOBER 10, I sent a messenger to Louisville.

SUNDAY, OCTOBER 13, I returned to Lexington and came back to Danville.

SUNDAY, OCTOBER 20, I could not profit from the post to write to the minister in Philadelphia since I had no answer from Gen. Clark.

OCTOBER 21, I received an answer from Gen. Clark.[64]

Notebook 7 ends here. Michaux then began to make arrangements to return to Philadelphia on the Wilderness Road through Cumberland Gap, and his next journal entry records the beginning of this journey in a new notebook.

## Notebook 8 begins

NOVEMBER 10, 1793, year 2 of the French Republic; I left Danville for Philadelphia after visiting Col. George Nicholas, Esq., near Danville.[65] He insisted on the plan that he had proposed the preceding day concerning navigation on the Mississippi, namely that the naval forces of the republic, on taking possession of the mouth of the Mississippi, would declare that this territory belonged to them by right of conquest and invite the Americans from the West to profit from this freedom to navigate. Thus if the Spaniards, located farther upriver, attacked the loads of provisions transported by the Americans, these would be in their right to fight force with force. In this way the Spanish government would have no reason to complain to the United States against breaking the pact since the French Republic was said to be in possession of this part of the country. I spent the night at Crab Orchard, 22 miles from Danville.[66]

NOVEMBER 11, 1793, I left Crab Orchard in the company of 12 persons who were assembled in this place to cross the uninhabited woods, which were frequented by Indians. The distance between Crab Orchard and the Holston settlement is 130 miles and is referred to as Les Wilderness. I slept at Langford Station, 10 miles away.

NOVEMBER 12, I spent the night at Modrel Station: 28 miles.[67]

NOVEMBER 13, I slept at Middleton Station: 28 miles.[68]

NOVEMBER 14, we went through low and marshy areas where the water was brown and stagnant. At 6 miles from the Middleton Post and 18 miles before arriving at the top of Cumberland Gap, I saw a climbing fern that covered six acres near the road.[69] At this season, where frost had produced a layer of ice, ¼ to ⁵⁄₁₆ inches [7–10 mm] thick, this plant was not damaged in the least. In this territory there are two areas with names, one called Flat Lick, and the other Stinking Creek.[70] I saw a *corbeau* (raven) around a decomposing deer. Davis's Station, 2 miles from . . . Cumberland Gap: 26 miles.[71]

NOVEMBER 15, we traveled in the high mountains; [we] crossed the Clinch River and spent the night at Holston Station, at . . . 27 miles.[72]

NOVEMBER 16, followed the Holston River and slept at the home of . . . Amis, Esq., at 3 miles beyond Hawkins Courthouse: 26 miles.[73]

SUNDAY, NOVEMBER 17, the rain obliged me to stay in a little cabin near the North Fork of the Holston: 25 miles.[74]

November 18, my horse was so tired from the speed and the bad roads in Les Wilderness that I had to stop after only 11 miles.

November 19, we left at dawn. At the foot of the house where I stayed the Kentucky road divides:[75] the right one leads to Burke Courthouse, North Carolina, passing by the mouth of the Watauga River; the other leads to Abingdon Courthouse, first town in Virginia. My horse [was] still tired; I traveled only 20 miles.

November 20, I traveled 15 miles and arrived in Abingdon.

November 21, I stayed the night 22 miles from Abingdon near Seven Mile Ford,[76] the middle branch of the Holston.

November 22, crossed Seven Mile Ford; the Holston River is made up of three main branches, which are North Fork, Seven Miles Fork, and South Fork.

Six miles after passing this little river, I observed on the north-facing hillsides bordering several small rivers: *Pinus abies canadensis, Thuya occidentalis, Rhododendron maximum,* and also *Magnolia acuminata,* where the soil was very rich; *Fagus chinquapin,* soil clayey, rocks of quartz with iron, slate rare calcareous stones with veins of white quartz. Gray squirrel. (I forgot to mention that in Abingdon I saw a turtle, 8 inches in diameter, that was petrified in black limestone like the rocks that are abundant in this territory.) Our journey was 23 miles.

November 23, I spent the night at a German's house. During the night my horses strayed. I saw *Abies canadensis* and *Thuya occidentalis* in the mountains between Abingdon and Wythe Courthouse.[77]

Sunday, November 24, passed Wythe Courthouse, and 18 miles away in the steep mountains, I saw: *Pinus strobus, Pinus* leaves three, *Pinus* leaves two, *P. abies canadensis, Rhododendron maximum, Kalmia latifolia, Gaultheria procumbens, Epigea repens.*

In drier areas: *Fagus chinquapin, F. castanea americana, F. sylvatica americana, Andromeda arborea, Hypericum kalm.*[78] In the moist rocks or watered by the streams: rocks of silex and even somewhat transparent agate.

From Seven Mile Ford to Wythe Courthouse: 36 miles.

November 25, I crossed Peppers Ferry on New River, then crossed over to the west side on the eastern side of the Alleghenies. Spent the night on a branch of the James River named Catawba Creek, which runs to the east, while the New River runs west of the mountains.[79]

NOVEMBER 26, I continued my way toward Botetourt Courthouse: 30 miles.[80]

NOVEMBER 27, I passed by Botetourt Courthouse and the southern branch of the James River, 12 miles from Botetourt.

NOVEMBER 28, I passed by Lexington, 40 miles from Botetourt, and by the northern branch of the James River, a mile from Lexington:[81] *Thuya occidentalis, Pinus strobus.*

NOVEMBER 29, stayed over at the home of MacDowall; [my] horse had a swollen leg so that he could not walk.[82]

NOVEMBER 30, I walked 27 miles.

SUNDAY, DECEMBER 1, went by Staunton, rather flourishing little town, 120 miles from Richmond and 75 miles from Botetourt.[83]

DECEMBER 2, I went through Rockyham or Rockytown [Harrisonburg], 20 miles from Staunton.[84]

DECEMBER 3, I passed by Woodstock,[85] another little town, 37 miles from Rockytown. Between Staunton and Woodstock, the terrain is mountainous, the soil rather fertile with clay and calcareous stones called blue limestone. *Quercus rubra*, [*Q.*] *alba, Fagus chinquapin*, and *Pinus* with two leaves, cones with rigid and prickly scales.[86] Three miles before arriving in this town, I saw on the north side of a hill near the road: *Thuya occidentalis, Pinus* with two leaves, *Juniperus virginiana* L.

DECEMBER 4, I left Woodstock and went through Newtown.[87]

DECEMBER 5, I went by Winchester,[88] 35 miles from Woodstock, formerly Millerstown.

DECEMBER 6, I passed by Charles Town,[89] 22 miles from Winchester. [I] went by Harpers Ferry on the Potomac River, 8 miles from Charles Town, and entered Maryland.

DECEMBER 7, I went by Frederick, 20 miles from the (Potomac River) Harpers Ferry, and 50 miles from Winchester.[90]

December 8, I went through Woodberry and Littlestown, 35 miles from Frederick.[91]

December 9, I passed by Hanover, formerly called Macallister Town, 42 miles from Frederick, and by York, 18 miles from Macallister Town now Hanover.[92]

December 10, I crossed the Susquehanna River and entered Pennsylvania,[93] eleven miles from York. I passed Lancaster, 12 miles from Harris Ferry [Wright's Ferry] on the Susquehanna River, and 24 miles from York.[94]

December 11, I traveled 30 miles.

Thursday, December 12, I arrived in Philadelphia, 66 miles from Lancaster.

December 13, I visited Citizen Genet, minister plenipotentiary of the French Republic.

December 14, I visited Mr. Jefferson, Mr. Rittenhouse, and . . .[95]

Sunday, December 15, recapitulation of the road taken:

| | |
|---|---|
| From Danville to Lincoln[96] | 12 miles |
| From Lincoln to Crab Orchard | 10 |
| From Crab Orchard to Langford Station | 10 |
| From Langford to Modrel Station | 28 |
| | 60 |

*Continuation*

| | |
|---|---|
| Modrel to Middleton Station | 28 |
| Middleton to Cumberland Gap | 24 |
| Cumberland to Davis Station | 2 |
| Davis to Holston | 27 |
| Holston to Hawkins Courthouse | 22 |
| Hawkins to . . . Amis | 3 |
| Amis to North Fork of Holston | 25 |

| | |
|---|---|
| North Fork to the Carolina Fork | 31 |
| Carolina Fork to Abingdon before Washington Courthouse, Virginia | 15 |
| Abingdon to Seven Mile Ford, then Seven Mile Ford to Wythe Courthouse | 60 |
| Wythe to Peppers Ferry | 33 |
| Peppers Ferry to Botetourt Courthouse [Fincastle] | 50 |
| Botetourt to James River South Fork | 12 |
| James River South Fork to Lexington | 28 |
| Lexington to Staunton | 35 |
| Staunton to Rocktown [Harrisonburg] | 20 |
| Rocktown to Woodstock | 37 |
| Woodstock to Winchester | 35 |
| Winchester to Charles Town [WV] | 22 |
| Charles Town to Harpers Ferry on Potomac | 8 |
| Potomac to Frederick [MD] | 20 |
| Frederick to Littlestown [PA] | 35 |
| Littlestown to Hanover, formerly known as MacAlister | 7 |
| Hanover to York | 18 |
| York to Susquehanna, Harris Ferry [Wright's Ferry] | 11 |
| Susquehanna to Lancaster | 12 |
| Lancaster to Philadelphia | 66 |
| | |
| Total | 746 miles |

Danville to Lexington—33 miles

Danville to Louisville—84 miles

Sunday, December 16, 1793, I dined at [the home of] M. Genet.

December 17, I sent my horses to the Bartrams.

December 18, I visited Dr. Colin, minister of the Swedish Church.[97]

December 19, I visited Mr. Peale, keeper of the museum.[98]

December 20, I skinned several squirrels.

December 21, I changed lodging.

December 22, I worked on my accounts.

December 23, I saw Minister Genet and Citizen Bournonville.[99]

December 24, I went to look at my seeds; I divided them to send to France in two different shipments.

December 25, I worked on getting my Kentucky collection in order.

December 26, I visited Mr. Rittenhouse, president of the Philosophical Society.

December 27, I wrote and was busy with different things.[100]

December 28, I visited Mr. Jefferson, Minister Genet, etc.

December 29, I hunted birds.

December 30, I stuffed the birds killed the preceding day.

December 31, I was busy all day writing.

1794

Wednesday, January 1, I hunted birds, killed two crossbills; I prepared and stuffed them.[101]

January 2, I made several visits, and I learned of the arrival in Baltimore of a ship coming from Le Havre bringing favorable news regarding the French Republic.

January 3, I was told to prepare myself for a trip to the Carolinas,[102] and I went to the botanist Bartram to let him know so that he could give me a list of plants that he wanted.

January 4, 1794, I visited Dr. Barton, and he loaned me *Systema Naturae* of Linnaeus.[103]

Sunday, January 5, I copied and abstracted from the history of Mammalia, Quadrupeds, and Birds.

January 6, I brought to Citizen Bournonville my account for expenses during my trip to Kentucky, and because he was very busy, he asked me to return the day after tomorrow.

January 7, I continued to make an abstract of *Systema Naturae*.

January 8 and 9, I continued the same task.

January 10, Citizen Bournonville still didn't have time to verify my accounts.

I returned to the minister the open-ended document that he had entrusted me to show to Gen. Clark,[104] as well as a memorandum on the state of the harvest with regard to supply of wheat for France. He told me that the trip to the Carolinas was no longer as urgent as he had previously supposed. I told him that I wanted to employ my time as best as possible on research in natural history; however, if the minister had another mission in mind for the service of the republic, I would devote myself to it; if not, then I wanted to go to the Carolinas to retrieve and put my collections in order. He agreed to my proposition and told me that upon my return he would give me a commission for Kentucky. He recommended that in the meantime that I visit with the deputies from the state of Kentucky in Congress.

January 11, 1794, I was occupied all day with writing.

Sunday, January 12, I visited Mr. Brown and Col. Orr, members of Congress, deputies from Kentucky. I conferred with them regarding the inclinations of the federal government and the execution of the plan of Gen. Clark.[105]

January 14, I wrote to Gen. Clark to let him know of the intentions of the minister and to send him $400.

January 16, I drew the $400 mentioned above and . . .

January 17 and 18, I wrote several letters to different persons in Kentucky and . . .

January 18, I wrote a memorandum for a motion to be made to the Society of Friends of Liberty and Equality in Philadelphia in order to mitigate the fate of French prisoners at the hand of the English.[106]

Sunday, January 19, I skinned and stuffed several birds.

There are no entries in Michaux's journal between January 19 and February 9, 1794. Probably some of this time was spent with the Bartrams because he left from there on his way back to Charleston. During the time that he spent in Philadelphia, Michaux sent a report to Clark in Kentucky via the US mail, directed to the attention of Col. Barbee, the postmaster in Danville.

### AM in Philadelphia to Clark in Louisville, December 27, 1793 (excerpt)[107]

Having arrived two weeks ago in this City I have seen several times Mr. Genet, Minister Plenipotentiary of the Republic. He charge[s] me to inform you that he persist[s] always in the execution of the Plan you have proposed to him, he has been much pleased with the particulars which I have related to him of your operations. The difficulty or rather the impossibility to create a diversion with the navy, forces the Minister to delay those operation[s] until next Spring. The Squadron which was here at his dispositions has been sent to an other destination by the unhappy events in the French West Indies Islands. If we except this part of the French Empire where aristocracy has displayed all its fury, the affairs of our Republic are every where in the most successful situation. The execution is only delayed to execute it at a better opportunity. The Minister thinks to send you French volunteers as you have desired me to mention to him. He will soon write you more particularly not-withstanding the business he is throng [?] with[.]

I send you a Bill of Exchange of four hundred Dollars on Monsieur Morrison, merchant at Lexington[108] which remain to me of the necessary purchases that I have made. You will please to take seventy Dollrs of that sum every three mounth to give to the two artillery men Humeau and Le Blanc for the Minister has recommended me not let those men wait for their salaries.

As for the concern you take to our success, I sent [send] the most certain account we have from Europe . . .

The letter continues with a recital of reports Michaux had heard about French military successes against the European powers that were attempting to bring down the French Republic and restore the monarchy.

## CHAPTER ELEVEN

# North Carolina Mountains, 1794

SUMMARY: IN FEBRUARY 1794, MICHAUX left Philadelphia for Charleston, South Carolina, traveling overland along the same route through Wilmington, Delaware; Baltimore, Maryland; Alexandria and Richmond, Virginia; and Wilmington, North Carolina, that he had followed in 1789. In 1789 he had made this journey in twenty-nine days accompanied by his son and an enslaved black assistant. Although he does not mention it, this time he may have had as a companion a young black man purchased in Philadelphia.[1]

While in Philadelphia Michaux had continued his study of natural history. The journal includes several observations of birds and the northern limits of a number of woody species encountered on this winter journey as well as his usual remarks about the qualities of the land, distances traveled, directions, and names of people.

A few days after Michaux's departure from Philadelphia, Genet's mission came to an end. His successor arrived in the capital and canceled all Genet's projects, including the attacks on Louisiana and Florida. Although Consul Mangourit in Charleston had become aware of Genet's recall before Michaux's arrival in mid-March, he did not cease his efforts to launch the expedition against Spanish Florida until after his own successor arrived in early April.

With his political duties at an end, Michaux resumed full-time botanical work. After spending only a few weeks at the Charleston garden, he embarked on a new, more thorough botanical exploration of the southern mountains. Leaving Charleston on July 14, he traveled familiar roads through Camden into North Carolina near Charlotte and headed northwest along the route that he had followed to these high mountains in 1789. Arriving at Turkey Cove, near modern Marion in McDowell County, on July 30, he began to explore the surrounding area using Turkey Cove as his base. As he had in 1789, he made a series of ascents in the Black Mountains, home of five of the ten highest peaks in eastern North America, but evidence is lacking that he reached any of the summits on this long high ridge.

In mid-August, Martin Davenport, who lived along the North Toe River, became his new mountain guide. With his base at Davenport's homestead, and this guide's extensive knowledge of the country, Michaux continued his mountain explorations with climbs to the summit of Roan in the Unaka Mountains and Grandfather, the highest peak in the Blue Ridge Mountains. Reaching the splendid isolation of the peak of Grandfather on a clear day, he looked out over a spectacular panorama of mountains in all directions. Mistakenly believing that he had climbed the highest mountain in North America, he burst into a rendition of "La Marseillaise." Continuing to explore mountain summits, Michaux then ascended the less lofty, but remote and difficult mountaintops of Table Rock and Hawksbill, overlooking Linville Gorge, before returning to Morganton.

From Morganton the botanist traveled east across North Carolina through Lincolnton and Salisbury, crossing the landscapes of the ancient Uwharrie Mountains and the Sandhills to reach Fayetteville on the Cape Fear River. Reentering South Carolina from Fayetteville, he proceeded to Long Bluff (now Society Hill) and on through South Carolina's inner coastal plain counties of Marlborough, Darlington, Florence, Williamsburg, and Berkeley before returning to the garden near Charleston on October 2. Two weeks later a serious attack of fever prevented him from working at his usual pace for the remainder of October. He made no additional entries in his journal until the following April.

### Notebook 8 continues

SUNDAY, FEBRUARY 9, 1794, I left the Bartrams.[2] Snow fell all day and obliged me to spend the night 7 miles from Philadelphia.

FEBRUARY 10, I slept in Wilmington, 28 miles from Philadelphia.

FEBRUARY 11, I lodged 24 miles farther.

February 12, it snowed most of the day.

February 13, I observed several *mésanges* (titmice),[3] very closely related to the blue titmouse, *Parus coeruleus*. I arrived in Baltimore.

February 14, I stayed here because I had to buy a horse and sell mine.

February 15, I left Baltimore and saw several birds. . . . The male has red at the ends of the lower wing feathers, like lacquer or sealing wax, the tip of the tail is yellow, the body is ash-colored, the head is crested, and there is a velvety dark black circle around the eye;[4] he is feeding on *Diospyros* at this season. I saw several birds . . . called blue birds by the Americans. The soil is sandy with yellowish clay and iron ore. There are several iron mines that are in operation in this part of Maryland.[5] Black oaks are common in these parts.

Sunday, February 16, between Bladensburg and Alexandria,[6] the soil is sandy sometimes with very red clay, iron mines. The birds are *Parus americanus*; the male has the upper part of its body blackish, and the lower part gray; the female is gray. This bird appears to feed only on seeds of herbaceous plants like *Sarothra gentianoides*, etc. It inhabits forests but is abundant along hedges and fences and is associated with a small sparrow (*friquet d'Amérique*) during the winter, etc. . . . *Parus* . . . bird, which has a great affinity with the blue titmouse of France, does not seem to feed on grains but flutters and goes from branch to branch with great animation and speed, particular to this bird. . . . Carolina cardinal, this bird spends the winter in sandy areas in the Carolinas as well as [in] Virginia and even in the low maritime sandy areas of Maryland. I saw it 15 miles before reaching the Potomac, which separates Maryland from Virginia. I spent the night in Alexandria, first town in Virginia on the southern side of the Potomac River.[7]

February 17, the soil is alternately clayey and sandy; I saw the American sparrow, the cardinal, the *moquer*, the two species of titmice mentioned previously; I saw a pine with three leaves near Dumfries,[8] pine with two leaves with scales that do not recurve after the seeds have fallen out but are only spread apart and concave, leaves long and straight, a large tree. It is the same pine that is abundant in certain parts of the Carolinas. I also saw in cold, arid, and mountainous areas a pine with 2 leaves whose scales are sharp and much rougher than the preceding species, scales recurved and needles shorter and slightly twisted.[9] This species is found on the hills along the Schuylkill River in Pennsylvania. I stayed overnight in Dumfries, 28 miles from Alexandria.

FEBRUARY 18, I went through Fredericksburg.

FEBRUARY 19, I went by Bowling Green and Hanover.[10] From Fredericksburg to Hanover Courthouse, the soil is sandy with many pines with two and three needles intermingled on the same branch. The cones are smaller than those of the pine with 3 needles from southern Virginia, whose scales are soft with barely a spine. Near Bowling Green, 22 miles from Fredericksburg, the pine with 3 needles begins. The pine has rough scales with rather long leaves. It is a smaller version of the pine with long leaves named *P. palustris*, and I refer to it as pine with 3 needles of southern Virginia and the Carolinas.[11]

FEBRUARY 20, from Hanover to Richmond is 22 miles.

FEBRUARY 21, I left Richmond; one and a half miles on the road to Petersburg, I saw an American elm with spongy bark. This spongy bark does not go around the branches but forms two wings or flat membranes that intersect where the buds appear to come out.[12] It is the same elm that I saw in large numbers in Kentucky between Louisville and Bardstown. Nine miles farther near a stream or small river, I noticed . . .

At 12 miles I saw *Smilax laurifolia* and *Smilax* with red berries in the same kind of site that we find these species in the Carolinas; at 20 miles I saw *Ilex aestivalis*. I slept at Petersburg: 25 miles.

FEBRUARY 22, 18 miles farther, I saw *Bignonia crucigera*, *Vaccinium arboreum*; at 30 miles I saw *Laurus aestivalis* and frequent *V. arboreum* and *Ilex aestivalis*. Along the rivers I noted several times the *Ulmus* with spongy bark. *Cunila* . . . ends between Petersburg and Halifax, 38 miles from Petersburg at Tompkin Shop, where I slept.

SUNDAY, FEBRUARY 23, 1794, rain kept me from starting before 11 o'clock; I went through Emporia [then Hick's Ford], small village 28 miles from Halifax, which is the first town in North Carolina. The line on this road that separates Virginia from North Carolina is 12 miles from Emporia and 16 miles from Halifax in North Carolina. At 10 miles from Emporia, and 2 miles before leaving Virginia, I saw *Bignonia sempervirens* near the creek called Fountain's Creek. I also noted *Hopea tinctoria* one mile before entering North Carolina. A mile from the border between Virginia and North Carolina, and in North Carolina, I saw *Cyrilla racemiflora* in a large swamp, three miles before arriving at Paterson's Tavern where I spent the night,[13] 16 miles from Emporia and 12 miles from Halifax; 23 miles [in total].

FEBRUARY 24, 10 miles from Halifax and six miles from the Virginia-Carolina line, *Pinus palustris* long leaves, large cones, begins. *Quercus palustris* with deltoid leaves begins also in this area. Pine with 3 needles but with mid-size cones, which begins in Bowling Green, is found here among pines with 2 or 3 needles,[14] *Bignonia crucigera* and *B. sempervirens*, *Hopea tinctoria* are seen in abundance south of Halifax, as well as *Nyssa dentata* and *Cyrilla racemiflora* in the swamps. I spent the night at Endfield Courthouse at the home of Col. Brandt: 25 miles.[15]

FEBRUARY 25, I had lunch at [the home of] Col. Phillips,[16] 16 miles, and passed by Tar River, 4 miles from Tetts Bridge.[17] I saw a *Sophora* known as yellow lupine, whose stems were all dried up. I collected the seeds that were still in their fruits, which are assembled in a cluster.[18] Twelve miles farther I went through Town Creek Bridge and slept 3 miles farther, 35 miles [in total].[19]

FEBRUARY 26, soil still sandy, covered by pines called *Pinus palustris*; these trees are notched, and their bark removed on one side of the tree, one foot long by two feet wide. The base of the notch, which is deeper, is for collecting resin, which is called turpentine. They take away the turpentine when the basin formed by the cut is full.[20] Twelve miles before arriving at Peacock Bridge,[21] *Laurus borbonia* appears, and three miles before Peacock Bridge begins *Andromeda wilmingtonia*.[22] *Stewartia malacodendron* is also found near this Peacock Bridge. It is approximately 21 miles from Town Creek Bridge to Peacock Bridge. The three species of Carolina *Myrica* begin in this district,[23] as does the large Carolina *Rhexia*.

FEBRUARY 27, I traveled toward Neuse River and at the site called Whitfield Ferry, passing by the home of . . . It is about 24 miles from Peacock Bridge to Whitfield Ferry.[24]

FEBRUARY 28, I traveled from Whitfield Ferry to Kenansville [then Duplin Courthouse or Dixon]: 31 miles.[25] Fifteen miles before one's arrival at Kenansville, *Andromeda axillaris* begins; that is 65 miles north of Wilmington. I also saw an abundance of *Vaccinium* with evergreen leaves, with prostrate stems, black fruits, as well as [A.] *wilmingtonia*, [A.] *paniculata*, [A.] *racemosa*, etc. *Bignonia crucigera*, [B.] *sempervirens*, [B.] *radicans*, and *Catalpa*.

SATURDAY, MARCH 1, I saw *Andromeda nitida* or *lucida* of the swamps of the Carolinas; it begins at 40 miles north of Wilmington. I saw large amounts of [A.] *wilmingtonia*, [A.] *axillaris*, [A.] *racemosa*, and [A.] *nitida*.[26]

I went through [South] Washington, 18 miles from Kenansville;[27] *Gordonia* begins 3 miles north of [South] Washington, about 38 miles north of Wilmington; *Ilex angustifolia* . . . begins 26 miles north of Wilmington. It is about 35 miles from Washington Courthouse to Wilmington.[28]

SUNDAY, MARCH 2, I saw in the dry sands *Lupinus perennis* and *L. pilosus*, *Atraphaxia*? shrub with thin branches, thick leaves, persisting during winter;[29] *Vaccinium sempervirens*, etc., [which] Bartram saw on the road to Warmspring;[30] *Chamaerops acaulis*, which begins 15 miles north of Wilmington. *Olea americana* is found near Wilmington and begins in that area. *Stillingia herbacea* begins 30 miles north of Wilmington. Rain forced me to stay overnight 8 miles from Wilmington.

MARCH 3, I arrived in Wilmington; I had to rest several days as my horse was extremely tired.[31] I saw M. Verrier, Frenchman from the Islands, a true Republican,[32] as well as Dr. LaRoque, established in Wilmington.[33] Mr. Josselin, keeper of the large tavern in Wilmington, is a great friend of the French Republic.[34]

MARCH 4, I went to dig up an *Andromeda* that I had noticed four years ago as well as an *Ixia* of the Carolinas, and I made a box with these plants to send them by sea with Capt. Mitchell's ship, the sloop . . . , to Charleston.

MARCH 5, I packed my collections and put them on board the ship.

MARCH 6, rain forced me to delay, and in the surrounding area of Wilmington I saw *Dionoea muscipula*, *Olea americana*, *Andromeda mariana*, [*A.*] *paniculata*, [*A.*] *racemosa*, [*A.*] *axillaris*, [*A.*] *nitida*, [*A.*] *wilmingtonia*, *Vaccinium arboreum*, [*V.*] *repens* black fruit, etc., *Bignonia sempervirens*, [*B.*] *crucigera*.

MARCH 7, I left Wilmington; [I] went through Town Creek, 12 miles away, by Lockwood Folly, 15 miles from T[own] Creek (by Shallotte Bridge, 8 miles).[35]

MARCH 8, I went by Charlotte Bridge and the home of W. Gauss, Esq. (wooden leg), 13 miles from the Ross Tavern or Lockwood Folly.[36]

SUNDAY, MARCH 9, I left the Foster residence, a violent aristocrat.[37] Along the sea I saw *Pisonia inermis*, a shrub with berries, [and with] branches and leaves opposite. It begins in North Carolina and is found in South Carolina, Georgia, and Florida etc . . .[38]

I also saw *Magnolia grandiflora*, 6 miles north of the line that separates the two Carolinas. At eleven thirty I arrived in South Carolina; at noon [I] went through a little hamlet of 4–5 houses on the banks of the Little River, inhabited by two French Democrats to whom I was pleased to impart the latest favorable news regarding the French Republic. One of them, named Jouvenceau, while drinking with an American Tory who spoke with contempt about the French Revolution, hit him twice with his fists, and the American retaliated by shooting him in the stomach. This Jouvenceau was an old soldier, and he was sick in bed. The surgeon hoped that the patient would recover despite the danger that he was in. The Foster mentioned earlier does not have a tavern, and it is 15 miles from William Gauss, Esq., to Green (it is important for travelers to have provisions of half a gallon of corn or unthrashed rice as . . .). I slept at Wren, 9 miles from Green.[39]

MARCH 10, I went through Long Bay,[40] about 9 miles from Wreen, had breakfast at the [home of the] widow . . . ; as the inhabitants along this coast do not maintain a tavern but receive travelers, one cannot expect nourishment for one's horse, and I had to be happy with a very courteous reception, but my horse had to do without food. The same day I spent the night at [the home of] Mr. Macgill, who married a daughter of the French family Balouin, a Frenchman who was a refugee earlier because of his religion.[41] I was very well received in this house. I was obliged to buy rice from the enslaved blacks to feed my horse.

MARCH 11, 12 miles farther, I had breakfast at the home of Dr. Mazie,[42] and fortunately the rice provision that I had taken could feed my horse, which was worn out by fatigue in the sterile sands through which we traveled for several days.[43] I finally arrived at Pittcock Ferry,[44] 22 miles from Macgill. My horse could go no farther. This ferry is a little south of Georgetown, and it is 1.5 miles to cross the river and 4 miles to reach Georgetown. I slept at the house of the ferry, a poor tavern, but my horse was well taken care of.

MARCH 12, I went across the river very early in the morning, and I had breakfast at Cooke, otherwise known as Cook's Ferry, on the Santee River, 12 miles from the Waccamaw River.[45]

I dined at the [home of the] Widow Morell (very good inn for horses). I slept there, 10 miles from Cook's Ferry, in all 22 miles without counting the crossings of rivers, very time consuming and often dangerous.

MARCH 13, I left the Widow Morell's place; after 7 miles I shifted to the right to go to the Manigault plantation. From the Manigault plantation,[46] I went to the Wiggfall plantation.[47] I saw a plant *Justicia*? a little before entering the cultivated field; near the middle to the left the road leads to Clement's Ferry, 5 miles from Wiggfall. I saw *Andromeda wilmingtonia*. At night I arrived at Clement's Ferry by a sandy road without an inn,[48] and [it was] the most disagreeable and inhospitable (site) that one could find from Philadelphia to Charleston. It is about 32 miles from Morell's Tavern to Clement's Ferry.

March 14, I arrived in Charleston, 5 miles from Clement's Ferry. In general in all of North and South Carolina and Georgia the roads are sandy, dangerous during the rainy season, which washes away the bridges; the inns are very poor, and often there are no settlements; one sometimes finds breakfast or dinner given without charge, but it would be considered discourteous to ask for food for one's horse; the best way is to carry a supply when it is possible to buy some, either corn or rice, called rough rice. When I could buy it from enslaved blacks, I was never without some provisions; that is why one always needs to have some small change. I visited Citizen Mangourit, consul of the French Republic.[49]

March 15, I visited the botanical garden that I had entrusted to the gardener before my trip to Canada.[50]

Sunday, March 16, dined at the home of Citizen Consul Mangourit.

March 17, I went back to my house and settled matters dealing with several agricultural tasks.

March 18, I received the plant collection that I had sent from Wilmington, and I planted it.

March 19, I had many trees transplanted.

March 20–21, the same work.

March 22, I conferred with M. Mangourit on the projected expedition of Minister Genet toward the conquest of East and West Florida.[51]

Sunday, March 23, I looked for plants.

March 24, I searched for plants and worked in the garden. I trimmed and pruned trees from the nursery.

March 25, I trimmed and pruned and settled with the gardener as to what needed to be done in the coming week.

March 26, I went to Charleston.

Michaux made no further entries in his journal until July 14, but a surviving letter of this period provides insights into his thinking and activities.

### AM in Charleston to Rev. Nicholas Collin in Philadelphia, April 20, 1794 (excerpt)[52]

I am sending you, as promised, seeds of South Carolina. I found my collections in the best order possible upon my arrival here. I am preparing to go botanize to the high mountains to the west of the Carolinas. It is possible that I will not return to Philadelphia this year but will return to Charleston, where I will be delighted to receive news of you . . .

The next entry in Michaux's journal, twelve weeks later, marks the beginning of his new journey.

### Notebook 8 continues

JULY 14, I left the house and slept at Moncks Corner,[53] noted near the bridge at Goose Creek *Eryngium* with lanceolate leaves.[54]

JULY 15, two and a half miles from Moncks Corner, I found *Menispermum* . . . , *Smilax laurifolia* in flower. I went by Eutaw Springs, and then taking the road to Manigault Ferry,[55] I slept 5 miles away. I noted frequently *Serratula fistulosa*,[56] *Heliotropium* . . . ,[57] *Sida* . . . ,[58] *Rhexia* . . . , [unknown woody plant] the base with spongy bark.

JULY 16, I crossed the Manigault Ferry because of the overflow of water that kept us from going by way of Neilson Ferry. The rain came down all day, and we spent the night at the entrance of the territory called Santee High Hills.

JULY 17, crossed the Santee High Hills, noticed *Phlox* . . . , *Coreopsis verticillata* leaves ovate, *Carduus virginicus*. We spent the night at Stateburg. Soil partly clay and better. Red oak with long petioles, the acorns short, sessile and thick; it is not the same as that from Pennsylvania and Canada; it is the true scarlet oak of Wangenh.[59]

JULY 18, I went through Camden. On leaving Camden to go to North Carolina, one finds, two miles farther, sands that are called Pine Barrens.

Four or five miles farther there is a stream or creek (swamps) full of sphagnum, *Azalea*, *Eriophorum*, and other aquatic plants; among them, along the road is a *Kalmia* that has not been described and probably has never been seen before.[60] A plant of the 9th class,[61] *Sophora* with yellow flowers, *Carduus virginicus, Lupinus pilosus.* We slept a mile beyond this swamp and six miles from Camden.

JULY 19, I passed by Johnston House and spent the night at [the home of] Wm. Graim: 35 miles.[62]

SUNDAY, JULY 20, had breakfast 3 miles before arriving at [the home of] John Cry,[63] and [I] lodged for the night 7 miles farther at the miserable and detestable Huston's Tavern.[64]

JULY 21, left early in the morning; the rain made us stop several times. Slept at the home of John Spring, horse merchant, a rich man, very honest, and whose home is very decent and agreeable.[65] I noticed *Rhus glabrum, Rhus* winged between the leaflets, and individuals with male and perfect or rather female flowers growing on different plants, *Rhus . . . ,*[66] *Delphinium.*[67]

JULY 22, I passed through Charlotte in Mecklenburg; soil is red clay, stones with quartz, clear water in contrast with what was seen before, the water with the color of dead leaves or dry tobacco. The vegetation is made up of oaks, red, black, white, etc. etc. *Actea spicata,*

Here in the journal, Michaux placed a line of six widely spaced periods, perhaps intending to add additional plant names at a later date.

Slept six miles from Tuckasege Ford.

JULY 23, I passed by Ben Smith's, 20 miles from Charlotte.[68] [At] two and three miles before arriving there, I saw *Magnolia* tomentose-glaucous, with long cordate leaves, a new *Stewartia*?[69] I spent the night six miles from B. Smith.

JULY 24, I went through Lincolnton and lunched at Reinhart's.[70] I saw *Calamus aromaticus*; I stayed at the old shoemaker's.

JULY 25, I went by Henry Watner, now Robertson.[71]

JULY 26, I arrived at Morganton, previously known as Burke Courthouse, 30 miles from Robertson's. Shrub like a *Calycanthus.*

SUNDAY, JULY 27, I remained because of the rain and the water in the creeks that could only be crossed by swimming.

JULY 28, I remained there.

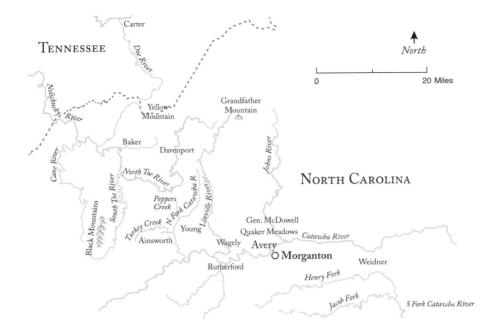

MAP 13. Northwestern North Carolina (Map by Brad Sanders)

JULY 29, I left and stayed overnight at John Ratherford's, near whose house I went across Muddy Creek on a bridge.[72]

JULY 30, I returned on the usual road that leads to Turkey Cove and arrived at the home of a man named Ainswort.[73]

JULY 31, I botanized on the high mountains of Linville, to the southeast of Ainswort's house and on rocks and mountains without trees.[74] I dug up a little shrub, *Clethra buxifolia*?

FRIDAY, AUGUST 1, I botanized on the mountains to the northeast, where the soil is very rich: *Veratrum viride album*? *Convallaria majalis, C.? umbellata*; I measured a tulip tree that was 23 French feet in circumference.[75]

SATURDAY, AUGUST 2, botanized in the northern mountains; *Convallaria umbellata* leaves with entire margins and [word illegible], softly hairy, flowers in umbels, and blue berries,[76] *C. racemosa, C. multiflora, C. majalis* leaves entire, glabrous, flowers in a simple raceme with flowers on one side, and blue berries.[77]

SUNDAY, AUGUST 3, I botanized among sedges and other aquatic plants.

AUGUST 4, I prepared for the trip to Black Mountain.

AUGUST 5, I delayed because of lack of provisions.

AUGUST 6, I left and arrived at a site called Crab Tree.[78] I noticed the following plants: *Azalea lutea* style very long, *Veratum viride album*.

AUGUST 7, I botanized on the mountains around Crabtree: *Clethra montana, Cassine . . . , Rhododendron maximum, Kalmia latifolia, Convallaria bifolia, Trillium cernuum* erect with red fruits, *Magnolia auriculata,* [M.] *acuminata* flowers glaucous, a shrub like an *Azalea,*[79] *Vaccinium* leaves with ciliate margins, reticulate on the surface, axillary peduncles with single flower, corolla revolute, 4-parted, stamens 8, fruit inferior, a pear-shaped berry, red with 4 locules,[80] *Cypripedium calceolaria* two species,[81] *Veratrum viride album* on the hillsides, *Melanthium . . . , Veratrum luteum* in the brooks, *Spiraea (paniculata) trifoliata, Robinia pseudoacacia,* [R.] *viscosa,* [R.] *hispida, Monarda coccinea* in the streams, [M.] *fistulosa, Quercus prinus-glauca.*

AUGUST 8, I botanized: *Hamamelis . . . , Nyssa . . . , Halesia tetraptera, Convallaria majalis?* with yellow berries, *C. umbellata* with blue berries.

AUGUST 9, I continued my botanical search: *Abies canadensis, A. nigra?* leaves scattered on all sides,[82] *Spiraea . . . , Spiraea . . . ,*[83] *Pinus strobus.*

SUNDAY, AUGUST 10, I arrived at the base of Black Mountain; *Podophyllum* flowers, blue berries,[84] *Vaccinium coccineum, Fagus castanea americana,* etc.

AUGUST 11, I arrived on the . . . side of Black Mountain: *Abies nigra, Diervilla,*[85] *Acer pensylvanicum, Sedum* with lower leaves dentate and upper leaves entire,[86] *Sorbus aucuparia, Rubus odoratus, Rhododendron maximum, Kalmia latifolia, Vaccinium stamineum,* [V.] *resinosum, Andromeda arborea,* [A.] *axillaris,* [A.] *racemosa,*[87] *Clethra montana,* shrub similar to *Azalea, Vitis* with lower surface of leaves tomentose, large berries, fox grapes, good to eat.

AUGUST 12, I returned from the mountain.

AUGUST 13, I arrived at the home of Mr. Ainsworth.

AUGUST 14, the thick fog made it difficult to travel in the high mountains; I botanized in the valleys.

AUGUST 15, rain.

August 16, I traveled toward Yellow Mountain and Roan Mountain; I arrived at the North Toe River and the Bright settlement. The main inhabitants of this establishment are Davinport and Wiseman . . .[88]

I botanized: *Azalea coccinea lutea*, [*A.*] *flava*,[89] [*A.*] *alba* and *rosea*;[90] all these varieties of *Azalea nudiflora* occur in this area;[91] *Vaccinium* cranberry related to *Oxicoccus*, *Pinus strobus*, *Abies canadensis*, etc. etc. *Gaultheria procumbens*, *Epigea repens*.

Sunday, August 17, I made arrangements with a hunter to go to the mountains.[92]

August 18, I botanized and wrote descriptions of several plants of *Syngenesie frustanée*,[93] *Helianthus atrorubens*, *Rudbeckia*, etc etc.

August 19, left to go to the high mountains.

August 20, I botanized in the mountains: *Acer pensylvanicum*, [*A.*] *canadense*, etc.

August 21, I arrived at the summit of Roan Mountain and saw an abundance of a little shrub with leaves of boxwood that I had called before *Ledum buxifolium*, but whose capsule had three locules and dehisced at the tip: flowers many, pedunculate, terminal, flowering in June, calyx deeply 5-parted, the lobes narrow, horizontal after blooming, close together, petals 5 ovate or subcordate, with obtuse apices, inserted below the receptacle, white, entirely deciduous, stamens 10, filaments the length of the corolla, erect-spreading, white, anthers sub-rounded, in pairs, versatile, pale red, ovary ovate, the style filiform the length of the stamens, stigma obtuse, capsule 3-loculed . . . fruit, shrub resembling boxwood, evergreen. *Potentilla tridentata*, *Sorbus aucuparia*, *Pinus abies balsamifera*.

August 22, we reached the top of Yellow Mountain.[94]

August 23, we returned to Davinport's home.

Sunday, August 24, I worked at putting my collection in order.

August 25, rain.

August 26, we left for Grandfather Mountain, the highest of those that make up the Alleghany and the Appalachian.[95]

August 27, we arrived at the base of the highest mountain.

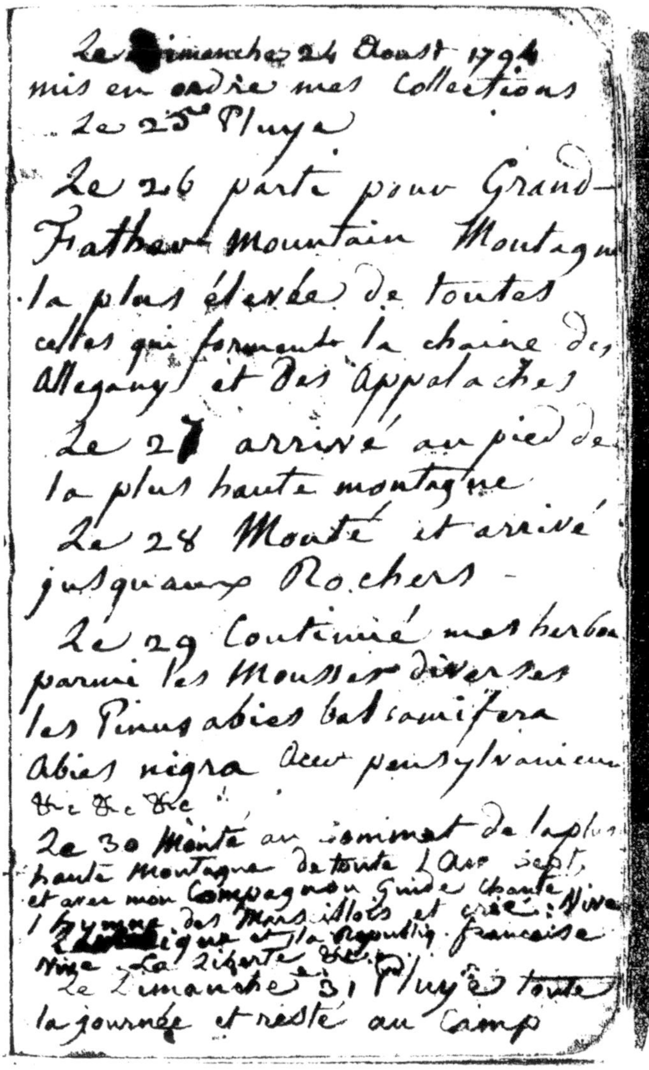

Le Dimanche 24 Aoust 1794
mis en ordre mes Collections
Le 25 Pluye
Le 26 parti pour Grand-
Father mountain Montagne
la plus élevée de toutes
celles qui forment la chaine des
Allegany et des Appalaches
Le 27 arrivé au pied de
la plus haute montagne
Le 28 Monté et arrivé
jusqu'aux Rochers -
Le 29 Continué mes herbori
parmi les Mousses diverses
les Pinus abies balsamifera
Abies nigra Acer pensylvanicum
&c &c &c
Le 30 Monté au sommet de la plus
haute Montagne de toute l'Am. sept.
et avec mon Compagnon Guide chanté
l'hymne des Marseillois et crié: Vive
l'Amérique et la Republiq. Française
Vive la Liberté &c.
Le Dimanche 31 Pluye toute
la journée et resté au Camp

FIGURE 22. Michaux's journal account of the climb of Grandfather Mountain, North Carolina (Courtesy of the American Philosophical Society)

AUGUST 28, we climbed as far as the boulders.

AUGUST 29, I continued my botanizing; among the various mosses,[96] there were: *Pinus abies balsamifera*, *Abies nigra*,[97] *Acer pensylvanicum*, etc. etc.

AUGUST 30, we climbed to the summit of the highest mountain of all North America, and with my companion and guide, [we] sang the hymn "La Marseillaise" and yelled "Long life to America and the French Republic, Long life to Liberty," etc. etc.[98]

Sunday, August 31, we stayed in camp as it rained all day.

Monday, September 1, we returned to the home of my guide, Davinport.[99]

September 2, rain, and [I] botanized.

September 3, I made a list of my collections.

September 4, continued the same task.

September 5, I left for Table Rock Mountain.[100]

September 6, I visited the rocks on Hawksbill Mountain and Table Rock; these rocks are very barren, and the only rare plant is the new shrub *Ledum? buxifolium*. It is there in abundance.[101] I stayed overnight 6 miles away at . . . Park's.[102]

September 7, I left for Burke Courthouse or Morganton; I spent the night at [the home of] Gen. Mac Douwal;[103] I saw *Spiraea tomentosa* in abundance near his house.

The original manuscript has a horizontal line drawn here to separate the following note about distances to familiar stops on the route west from the preceding note about finding *Spiraea* at Gen. McDowell's plantation. The printer's layout of this passage in Dugger's frequently reprinted translation of this journey lacks this fine point and has misled some readers. Michaux did not visit these people between stopping at McDowell's on September 7 and reaching Morganton on September 8, but earlier on his journey.[104]

From Burke to John Wagely, it is approximately 12 miles; from John Wagely to Th. Young . . . ; from Thomas Young to Davinport, 8 miles.

September 8, reached Burke Courthouse, or Morganton, I visited Col. Avery and slept there.[105]

September 9, I left Morganton in the evening and lodged three miles away. I met a resident of Statesville, Mr. Atkinson, who invited me to stay with him.

September 10, I arrived at the home of Robertson, 30 miles from Morganton.

September 11, I lodged at Reinhart's, Lincoln Courthouse, 15 miles from Robertson.

September 12, I left for Yadkin River and Salisbury; I spent the night at Catawba Springs, 18 miles from Lincoln.[106]

September 13, I went through Beatties Ford on the Catawba River, 20 miles from Lincoln.[107] Annual plant, with erect opposite four-angled branches, leaves ovate, 3-nerved, subsessile, the peduncles axillary with one flower, calyx 5-parted, epicalyx of two scales, lobes of calyx ovate, acuminate, suberect, corolla tubular, the cylindrical tube the length of the calyx, the lobes irregularly 5-parted, ovate, laciniate, the upper two erect, stamens, 4 didynamous, filaments as long as the corolla, filiform, anthers almost round, ovary 4-angled, style filiform, the length of the stamens; stigma divided, lobes equal; "seeds" 4, at the base of the calyx, ovate, rugose. This annual plant flowers in July and August; flowers, filaments, and pistil blue, the anthers the color of hyacinth (sea green). It grows in scattered areas of Virginia and North Carolina in rocky places.[108]

I slept at a farm 8 miles before arriving at Salisbury, which is the junction and meeting point of the three roads to Philadelphia, to Charleston, and to Kentucky.[109]

Sunday, September 14, I went through Salisbury, a town whose appearance is less miserable than others in North Carolina referred to as Courthouse. It is 50 miles from Lincoln to Salisbury. I continued on my way to Fayetteville, went by Yadkin River, and slept at 14 miles from Salisbury.

September 15, I went by several creeks and low but very rocky mountains.[110]

September 16, I left the very rocky road. I saw *Magnolia acuminata* yellow flowers,[111] *Collinsonia tuberosa*, then encountered sandy soils. I spent the night at the home of the storekeeper Martin.[112]

September 17, I continued across the sandy hills.

September 18, I arrived at 6 miles from Fayetteville and lost my two horses.

September 19 and 20, these two days were occupied in searching for my horses.

Sunday, September 21, I found one of them and . . .

September 22, I was once again in Fayetteville before Cross Creek. The Cape Fear River passes near this town.[113] I saw during botanizing in the swamps that surround this town: *Cupressus disticha*, [*C.*] *thyoides* often to-gether, *Andromeda wilmingtonia*, and *Nymphaea hastata*.

TUESDAY, SEPTEMBER 23, I left Fayetteville after having had the satisfaction of reading the news that had come from Philadelphia the day before concerning the glorious success of the republic.[114] I stayed at the [home of the] old man MacCay, 15 miles from Fayetteville on the road to Salisbury.[115]

SEPTEMBER 24, I took the road on the left for Charleston and crossed Drowning Creek at McLauchland Bridge.[116] However the most direct road from Fayetteville to Charleston is to go to Widow Campbell's bridge, 40? miles from Fayetteville. It is 10 miles from Campbell Bridge to Gum Swamp, which is the point that separates North and South Carolina.

SEPTEMBER 25, I went through Gum Swamp and lodged at 8 miles beyond. [The distance is unknown from] Fayetteville. I saw *Cupressus thyoides* and *C. disticha* in many swamps. I saw *Andromeda wilmingtonia* in large numbers in all the swamps. *Liquidambar peregrinum*, etc.[117] Two miles from Gum Swamp, one enters South Carolina.

An ink stain on the page of the original journal renders parts of Michaux's entry for September 25 particularly difficult to decipher and interpret: *Liquidambar peregrinum* is on a separate line after his comment about the abundance of *Andromeda wilmingtonia* in the swamps, while "Two miles from Gum Swamp" is written in a very tiny script at the bottom of the page and appears to have been inserted later.

SEPTEMBER 26, I passed through Society Hill [then Long Bluff], a small hamlet found two miles south of the Great Pee Dee River, 74 miles from Fayetteville.[118]

SEPTEMBER 27, I went through Black Swamp, 22 miles from Society Hill. Col. Benton,[119] 12 miles from Society Hill; Black Creek, 10 miles from Society Hill; Jeffries Creek, 10 miles from Society Hill.[120]

SUNDAY, SEPTEMBER 28, I went through Lynch's Creek, 40 miles from Society Hill.[121]

SEPTEMBER 29, I crossed Black River, 30 miles from Lynch's Creek.[122] A man named Lorry is in charge of the ferry on the Black River.

SEPTEMBER 30, I arrived at Murray's Ferry on the Santee River, 15 miles from Black River and 20 miles from Moncks Corner.[123] Crossing by ferry was dangerous, and I had to go to Lenud's Ferry. It is 25 miles from Murray's Ferry to Lenud's Ferry.

OCTOBER 1, left Lenud's Ferry and crossed Strawberry Ferry 25 miles from Lenud's Ferry and 28 miles from Charleston.[124] I arrived at the house near Ten Mile House.[125]

OCTOBER 2, I left for Charleston.

I was busy until the end of November with collecting fall plants. Toward October 10, I became sick with the fever of the climate. I had it for about 12 days, and it took me more than six weeks to recover from it. I worked as much at repairing the garden as in putting my plant collection in order until the end of December.

A letter to Rev. Collin provides information not found in the journal.

### AM in Charleston to Rev. Nicholas Collin in Philadelphia, November 14, 1794 (excerpt)[126]

It is only six weeks ago that I returned from my botanical trip in the Carolina mountains. This trip lasted three months, and I had the satisfaction in using this time to complete my objective. Upon my return, I was attacked by fevers of the climate, and it is only in the last few days that I am able to return to work. Just before I left I had the pleasure of receiving your letter. As my residence is several miles from Charleston, I was unable to reply. As you mentioned seeds for your friend in Sweden, I will send some at the first opportunity. Among the seeds that I will send you, there are several species that are especially interesting for Sweden as they are more likely to succeed in the coldest countries. Since I have the custom of sending to Mr. Bartram everything that one can find here, I presume it is not necessary to give him [some] from those that I sent you, as there are several that can endure the heat of the summer.

The letter, written in French up until this point, continues in English.

If you receive the work *Systema Naturae* etc, please to call at Mr. John Bartram for the return of the expenses or rather at Mr. Petry consul at Philadelphia as I have the opportunity to send him [funds for payment].

My compliments to Mr. Peale I hope that I can send him some birds in the course of this winter . . .

My best compliments to Mr. Lebours.[127] Excuse my heedlessness for writing part of this in another language than at the beginning.

When you have received the botanical book please send it to the following address

Michaux, botanist, to the care of John James Himely, watch maker Charleston.

# CHAPTER TWELVE

# Journey West to the Mississippi River, 1795

SUMMARY: ANDRÉ MICHAUX ENDED 1794 and began 1795 working in the garden near Charleston. During winter 1794–95, he made plans and arranged the financing for the long journey to the western frontier that he began in spring 1795.

While the botanist made no entries in his journal until he began his next journey the following April, we have other evidence of his activities. Although the war between France and the European powers had made shipment of plants and seeds to France more difficult, in January 1795, Michaux sent a shipment of seeds to France on a ship traveling under a flag of truce. Other information about Michaux's activities that winter comes from his neighbor Charles Drayton's (1743–1820) diary, kept from 1784 until 1820; he recorded a visit that Michaux made to his plantation, Drayton Hall, on November 10, 1794.[1] Based on seed lists that Drayton kept, the two men must have visited fairly frequently. He also specifically notes an earlier visit to Michaux's garden on February 17, 1793. Although the botanist was away in Philadelphia, Drayton obtained several plants, including a "green tea plant." During the November 1794 visit, Michaux's horse died, and his host loaned him another one. Upon returning to his garden, Michaux sent the horse back with a gift of plants and his thanks. Three months later, on February 23, 1795, Drayton recorded another visit to the Michaux garden in his diary; he again returned with plants and shrubs.[2] While it is likely that Michaux visited and received visits from other local citizens as well, specific evidence has not come to light.

With the demise of the Old Regime, Michaux had been transformed into a botanist in the service of the French Republic. Genet had authorized a salary and expenses for the botanist just prior to his departure on the political mission to Kentucky in 1793, but in the suddenness of Genet's recall, it is not clear whether or how this funding was continued.[3] Michaux's relations with Consul Mangourit in Charleston had been very warm, but his relationship with Mangourit's successor, Antoine-Louis Fontpertius, is not known. The botanist had received news about the fate of his recent shipments to France and was aware of the important developments in the realm of natural history in Paris. Between 1790 and 1793, funding for the elaborate state support of science under the Old Regime had come under attack in the ever more radical National Convention. However, through the adroit political work of some of its scientists and their allies, especially the young deputy Joseph Lakanal, the Jardin du Roi was transformed into the Muséum national d'histoire naturelle on June 10, 1793.[4] Thus the stigma of the Old Regime was removed, and the Jardin des Plantes, now part of the Muséum, came to be viewed as a national asset belonging to the people of France.

Michaux announced his intention to leave on a new journey in a 1795 letter to Rev. Collin.

### AM in Charleston to the Rev. Nicholas Collin in Philadelphia, March 28, 1795 (excerpt)[5]

Since I am about to leave for the West (Western territories), I am sending you a small collection of seeds for your Swedish friend. Some are from the Carolinas, but there is a small package separated from the rest, which comes from the coldest parts of America. I recommend that you send these [seeds] particularly as there is no other country where these seeds are more likely to naturalize, and thus [they] should be considered as a precious acquisition.

About three month[s] ago I had the pleasure of writing to you to ask you to address M. Petry, consul of France in Phila[delphia], to reimburse [you] for the book *Systema naturae*, etc. I hope that you received my letter and that if you receive this volume that you will get reimbursement from Petry or Bartram to whom I can send funds and to ask you to keep the book until my return.

If in the next ten months you find it appropriate to write to me please send to Michaux, botanist, to the care of Gen. Th. Barbee, Postmaster, Danville, Kentucky . . .

Permit me to send my sincere compliments to Mr. Peale.

Within the same week that he sent this letter to Collin, Michaux sent another letter and a collection of seeds to his colleague André Thouin.[6] He sent two copies of this important letter by different means, and both letters were received, but not all of the seeds arrived. Michaux sent one copy with the diplomatic correspondence through Citizen Droict-Bussy, chancellor of the consulate in New York; the other letter was carried directly to François André by Citizen Theric, a resident of Charleston traveling to France. When he received the seeds and his father's letter in June 1795, François André promptly added a cover letter and sent the material on to Thouin, whose notes on one copy of the elder Michaux's letters indicate that he did not process this correspondence until October 1, 1795.

### François André Michaux to Citizen Thouin, director of the Jardin des Plantes, Paris, June 18, 1795 (18 messidor year 3) (excerpt)[7]

Citizen Theric, who arrives from the Carolinas, has given me a collection of seeds that he received from my father and [that is] destined for the Jardin des Plantes; I am sending you this in all haste with a sealed letter that was included with the one addressed to me. Citizen, if you would like to write to my father, I have a favorable occasion [to send letters] direct to Charleston.

### AM to André Thouin, from Charleston, April 1, 1795; annotated in Paris by Thouin, October 1, 1795 (excerpt)[8]

I have seen, Citizen, the translation of the report of the convention by one of the members of the committee on public instruction in which it is cited with justice that the garden of the republic in the Carolinas is in a flourishing state. This is very true, and the idea that one has of it is undoubtedly inferior to its actual state, as, with the exception of Minister Genet,[9] who has some knowledge of botany, this garden has not been visited by those worthy of judging it. In view of the interest that is being placed on useful establishments, from now on my shipments must be addressed directly to the Jardin des Plantes. You, Citizen, have even more the right to claim them since I am in the service of the republic, and it is to the Jardin des Plantes that my acquisitions will be sent, instead of [fulfilling] the different requirements of the Old Regime.[10] I would like to receive new instructions on the quantity and the nature of the objects wanted in general and [on] those of more urgent interest and also especially on new measures to be taken to assure the security of my shipments. I am very mortified to see that our correspondence is almost nil, either because of the disloyalty of the people in French ports who are scheduled to receive my shipments, or because of [some] other reason of which I know nothing. As for me, I can only take the precautions that my shipments are sent from here to the port in France, where the ship is destined to arrive. In effect, none of the numerous shipments sent during the Old Regime failed to arrive. I never trusted anyone else; I delivered them myself [to the ships] and was there during the placement on board not only so that my conditions would be met but also [so that I could] obtain the certificate or letter of the enterprise; sometimes I even asked that they be placed in cabins, and that was rarely refused.

Since the war, occasions [for shipments] are less frequent and less secure; not only are there dangers of English pirates, but I no longer know to whom I can recommend my shipments in the French ports. Last year I made two shipments of seeds from Kentucky; one was given to Minister Genet, and the other to the marine agent also in Philadelphia. I could not ship these cases because there was no immediate opportunity. It was in the month of February, and I was leaving for the Carolinas, and the cases could not be shipped [until] the large convoy for France left in April. The first one did arrive. The distance is another circumstance that makes the occasions [for shipping] less frequent and more difficult.

I am actually obliged because of my research to be at three or four hundred leagues from any port in the United States. You know that I have visited North America from the point of Florida to the territories around Hudson Bay, beyond the Mistassini Lakes. In many places I traveled throughout the eastern portions of the Allegany and Appalachian Mountains between the sea and these mountains. To fulfill the task that results from my mission and to know North America better, it remains for me to visit the western regions (Western territories), areas beyond the Appalachians and the Mississippi. If the objects that come from these regions are more difficult to obtain because of the dangers and obstacles to be surmounted, such as traversing through Native American territories always at war with the US inhabitants, crossing rivers by swimming, etc.,[11] they are no less precious. They deserve more care, more precautions and more important recommendations than for those shipments that can [be] made more often, such as the plants from South Carolina and Pennsylvania.

It is with the greatest zeal that I undertake the first obstacles until the shipments are placed on board. I still regret the shipment of plants obtained around the Mistassini Lakes, for which I crossed while returning [by way of] 22 lakes and 84 portages. These plants arrived in New York. I divided them while giving a portion to the garden in New York, [and] one portion to Bartram, and most were shipped to France. I received notice that they arrived, but they never reached Paris. It would seem that trouble in Brittany was the cause. One day, one will be able to judge their importance by [studying] my herbarium collection.

In view of this revelation, here is what I propose unless better instructions are given in this matter:

1) That my shipment be made directly to the Jardin des Plantes.

2) That our letters be registered at the minister of the republic in Philadelphia under his security, and the interior envelope unsealed.

3) That the shipments from America for the Jardin des Plantes be sent and received by the mayor and municipal officers from the port in France where the ships will arrive.

4) That the shipments for the Jardin des Plantes will not be examined by the custom officers. In this regard it would suffice to place them in the customs office warehouse, until, as was the old practice, the authorities have received the message to let them go without examination.

Finally, that the municipal officers be authorized to be reimbursed for the cost for sea and land transport.

The 17 nivose (January 6) I made a shipment of seeds by a truce ship that left from here. By this actual occasion I am sending you, my dear cocitizen, a package of seeds from South Carolina and those from the highest mountains.

In a year's time, my address will be Michaux, botanist, Western territories, Danville, Kentucky.

I have, Citizen, many interesting observations to communicate regarding methods of acquiring North American plants in larger numbers than has been done until now.

Would it not it be suitable that after visiting the country of the West (Western territories) for two years that I return to France?

This letter concludes with an extensive seed list. Less than three weeks after sending this letter, Michaux began the journey west and resumed making entries in his eighth notebook when he left Charleston.

On this new journey west, the botanist followed his previous route to Moncks Corner, again crossed the Santee at Nelson's Ferry, and continued north through Camden along the same route that he had taken in 1789 and 1794. North of Camden at Hanging Rock (near modern Heath Springs), he chose the road to the northwest instead of the road northeast through the Lancaster District as he had done on previous journeys. Then, after crossing the Catawba River at Land's Ford, northwest of the modern town of Lancaster, he entered North Carolina along the west bank of this river, thus bypassing Charlotte. After crossing the South Fork of the Catawba River at Armstrong Ford (modern Belmont), he rejoined his previous route to the mountains through Lincolnton and Morganton and reached North Cove near the Turkey Cove settlement where he had begun his previous mountain explorations. After ascending the Blue Ridge for a week of springtime explorations in the mountains with Martin Davenport, his guide from 1794, he crossed the highest mountains into Tennessee, where he visited Col. John Tipton (in modern Johnson City). From Tipton's, he proceeded southwest through Jonesboro and Greeneville to Knoxville, where he remained for several days before continuing west to Fort Southwest Point on the Clinch River. There a group of travelers assembled to cross the dangerous wilderness between the Clinch and Cumberland Rivers under militia protection. Crossing the Cumberland near modern Gainesboro, he safely emerged from the wilderness at Bledsoe's Station (modern Castilian Springs) and continued west to Nashville. From Nashville he headed directly north into Kentucky. He crossed the Big Barren River at modern Bowling Green before swinging northeast through the Kentucky Barrens to Danville for a sojourn of almost two weeks. From Danville he retraced his route of 1793 through Bardstown to Louisville, where he ended his eighth notebook.

Michaux began his ninth notebook in Louisville. Traveling overland across southern Indiana, he quickly reached Vincennes on the Wabash River, then he continued across southern Illinois to Kaskaskia on the Mississippi River. He remained in this area for over three months, visiting the old French settlements in Illinois, collecting botanical specimens, and exploring the region. He traveled overland to Fort Massac, located on the Ohio River near its confluence with the Mississippi, and explored the region around Fort Massac, including brief trips upstream on both the Tennessee and Cumberland Rivers, which enter the Ohio near the fort. There is no evidence Michaux made any exploration west of the Mississippi River in Missouri, but he did contact and make friendly overtures to an important French nobleman in exile who lived in upper Louisiana (modern Missouri) and who served the Spanish government.

Michaux's efforts in this regard are evidence that he wanted Spanish approval for a planned future exploration west of the Mississippi. In mid-December 1795, he began the first leg of his return journey to Charleston, traveling by water down the Mississippi and then up the Ohio to its confluence with the Cumberland, where he continued up the Cumberland toward Nashville.

## Notebook 8 continues

SUNDAY, APRIL 19, 1795 (30 germinal of year 3 of the French Republic, one and indivisible), I left to go botanize in the high mountains of the Carolinas and then to visit the West (Western Territories). Plants seen before arriving at Moncks Corner: *Heuchera . . . ,*[12] *Vicia, Smilax herbacea erecta, Melanpodium? . . . , Polyg. necess., Silene virginica, Phlox lanceolata* in flower, *Valeriana.* I slept at 45 Mile House.[13]

APRIL 20 (1 florial), in the neighborhood of Forty Five Mile House, *Valeriana*; 3 miles before Nelson's Ferry, *Gnaphalium dioicum, Uvularia? . . . ,* a new tree from the Santee River with elm-like leaves, with rough capsule with one subovate seed; its seeds were almost ripe.[14] *Celtis occidentalis* flowers . . . and male flowers lower. I spent the night at 77 miles from Charleston.

APRIL 21, I noted on the Santee High Hills: *Phlox* with white flowers and *Phlox* with pink flowers, two different species, a very small *Phlox* with lanceolate leaves.[15] I saw in the vicinity of Moncks Corner *Lupinus hirsutus* in flower, and [I] dined at Dr. . . . and slept at Stateburg.[16]

APRIL 22, I went through Camden; five miles farther I saw a new *Kalmia*; it was not yet in bloom.[17] I slept 10 miles beyond Camden.

APRIL 23, I passed Flat Rock and Hanging Rock Creek,[18] and [I] spent the night at Cane Creek, Lancaster County, at the home of Mr. May.[19] During the night, my horse escaped; by following his tracks we could see that he had gone to . . . Lee, Esq.

APRIL 24, I was obliged to search for him all day. Mr. Lee also sent his son and his black man to look for him. He obtained a horse for my search, and afterward he invited me to stay with him; he overwhelmed me with courtesies.[20]

APRIL 25, the horse came of its own accord to the home of Mr. Lee; the plants that I saw near the creek were: *Dodecatheon meadia, Asarum canadense, Claytonia virginica, Erythronium dens-leonis.*

Sunday, April 26, I left Cane Creek; I went through Land's Ford on the Catawba River.[21] But the best route is to ask for [directions at?] the home or plantation of Col. Crawford on Waxhaw Creek, then pass by McClenegans Ferry on the Catawba;[22] from there straight to the ironworks known as Hill's Ironworks, operated by Col. Hill.[23] Thus from Cane Creek to Waxhaw [Creek] . . . miles; from Waxhaw [Creek] to ironworks, York County . . .

April 27, I went through the ironworks, about 32 miles from Cane Creek.

April 28, I passed by Armstrong Ford on the southern branch of the Catawba, 12 miles from the ironworks.[24] On that day I went by the home of Bennet Smith, where there is a . . . *Magnolia*, 12 miles from Armstrong Ford.[25]

April 29, I went through Lincoln, 12 miles from Bennet Smith and 36 miles from the ironworks.

Thursday, April 30, I went by the home of the good man Wilson,[26] 9 miles from Lincoln and 6 miles from Robertson; 15 miles from Lincoln to Robertson; I arrived at Morganton, 30 miles from Robertson.

May 1, I spent the day botanizing in the neighborhood.

May 2, I spent the day at Col. Avery's, 4 miles from Morganton.[27]

Sunday, May 3, I left for the mountains; at 14 miles from Burke, one finds the Wagely house. The mountains of Linville, at the foot of which this house is situated, have abundant *Magnolia auriculata*.[28] They were in flower. It is 8 miles from Wagely to Capt. Young's.[29]

May 4, I left Young's home. Ainsworth is two miles away, but by taking the right-hand road one arrives at the base of a very high mountain, 3 miles from Young's; the summit is 5 miles from Young's home.

From the summit of the mountain to the home of Y. Bright, known as Bright settlement, it is 3 miles; and from Bright to Davinport, 2 miles; in all 10 miles from Young to Davinport.[30]

May 5, I botanized in the area of the homes of Davinport and Wiseman.

May 6, I left for the mountains, that is, Roan Mountain and Yellow Mountain; the Toe River runs between these mountains.[31] All the *Convallaria* were in flower,[32] as were *Podophyllum diphyllum* and *P. umbellatum*.[33]

Sunday, May 10, I returned from the mountains to the home of Davinport.

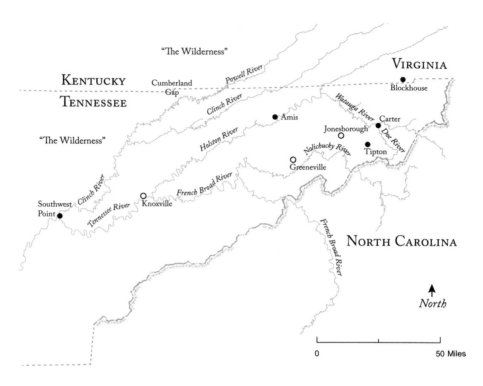

MAP 14. Eastern Tennessee, 1790s (Map by Brad Sanders)

MAY 11, I botanized on the mountains opposite the house. It is about 3 miles to get to the summit of the Blue Ridge on the portion called Rompback [Humpback];[34] on the first mountains we can see a great abundance of *Azalea* leaves with glandular tip, *Azalea lutea*. There is no other Azalea on the hills that surround the homes of Davinport and Wiseman apart from this one with yellow flowers. The ones that are found along rivers usually is that with brilliant red flowers and that with white flowers.[35]

MAY 12, I climbed to the summit of the Blue Ridge and saw *Rhododendron minus* in flower and *Cypripedium luteum*.

MAY 13, I left to continue my travels. I arrived at noon at the base of Yellow Mountain: 10 miles. At night I lodged at the home of John Miller,[36] 12 miles from the mountains. Thus, it is 22 miles from Davinport's to Miller's; one and a half mile farther is the beginning of the crossing of the Doe River.

MAY 14, I followed and crossed the Doe River 27 times. It is a dangerous river

FIGURE 23. Col. John Tipton's home, ca. 1795–96, when André
Michaux was a guest there (Drawing by Hugh Pruitt, courtesy
of the Tipton-Haynes Tennessee State Historic Site)

when the waters are high.[37] I slept at Col. Tipton's,[38] 20 miles from Miller's.

MAY 15, I went by Jonesborough,[39] 10 miles from Col. Tipton's home and
84 miles from Morganton, North Carolina [then often called Burke Court-
house]. I slept at Anthony Moore's, near the Nolichucky River.[40] During the
night my horse escaped.

MAY 16, Sunday 17, and 18, I spent time searching for my horse.

WEDNESDAY, MAY 19, I bought another horse for $50 from an inhabitant of
Nolichucky River named . . . Earnest, neighbor of Andrew Fox.[41] There is an
abundance of *Magnolia tripetala* on the banks of the Nolichucky.

MAY 20, I passed through Greeneville,[42] 27 miles from Jonesborough, and
took the road to Kentucky by taking the road to the right and going by way of
. . . Ferry on the Holston River. If one continues straight, it will lead to Knoxville, while if one takes the left a little before Greeneville, that road will lead
to the French Broad River.[43] It is 27 miles from Jonesborough to Greeneville.

MAY 21, I went through Bull's Gap, 18 miles from Greeneville.[44]

MAY 22, I passed through Iron Works, 30 miles from Bull's Gap.[45] It is only four miles to the Holston River. At two miles from Iron Works, there is a mineral rock whose pieces, if ground into a powder, give a red color to cotton; one can boil this mineral, etc.[46]

MAY 23, as my horse was injured, I had to remain at one mile from Iron Works on Mossy Creek at [the home of] a man named Newman.[47] Near his house (½ mile), one finds the mineral that I suppose is antimony.[48]

SUNDAY, MAY 24, I arrived at [the home of] Col. King,[49] on the Holston River at the site known as McBee's Ferry,[50] 15 miles from Iron Works.

MAY 25, I crossed on the ferry and arrived in Knoxville, 15 miles from McBee's Ferry, residence of the governor of the Western territories,[51] 110 miles from Jonesboro. Plants and trees of the Knoxville area and surroundings:[52] *Quercus prinus saxosa, Q. prinus humilis, Q. rubra, Q. praemorsa, Q. tomentosa, Q. pinnatifida,*[53] *Q. alba, Ulmus viscosa, U. fungosa, Fraxinus, Diospyros virginiana, Liquidambar styraciflua, Juglans nigra, [J.] alba* or *oblonga,* pignut hickory, *Platanus occidentalis, Nyssa aquatica,*[54] *Fagus castanea americana, F. pumila, F. sylvatica americana, Magnolia acuminata, Betula alnus americanus, Cercis canadensis, Cornus florida, Evonimus latifolius, E. americanus, Podophyllum peltatum, Jeffersonia, Sanguinaria canadensis, Trillium sessile.*

I stayed all week in Knoxville and botanized in the surrounding areas while waiting for a caravan large enough to go through the Wilderness.[55]

SUNDAY, MAY 31, I received word that 25 armed travelers were about to arrive in Knoxville.

MONDAY, JUNE 1, old style, the journey was again postponed.

THURSDAY, JUNE 4, left Knoxville and spent the night 15 miles away at Capt. Campbell's at a site called Campbell's Station.[56]

FRIDAY, JUNE 5, I stayed overnight at a site named West Point on the Clinch River, a military post to guard the frontier of the territory, 25 miles from Campbell's Station.[57]

JUNE 6, I left and crossed the river in a ferry belonging to West Point station. We walked more than 10 miles. Fifteen armed men and more than 30 women and children made up the travelers.

SUNDAY, JUNE 7, we crossed the Cumberland Mountains: 22 miles.[58]

JUNE 8, we continued our passage into the mountains: 23 miles. *Magnolia* with purple at base of the petals.[59]

TUESDAY, JUNE 9, we alternated climbing and descending the mountains: 25 miles. In the distance I saw many *Magnolia tripetala*.

JUNE 10, after 10 miles we reached the Cumberland River, and [we] slept 20 miles beyond.

JUNE 11, we arrived at Blodsoe [Bledsoe] Lick at Blodsoe Station: 20 miles.[60] In total, 120 miles of Wilderness. I slept here as one can find provisions for men and horses.

FRIDAY, JUNE 12, I went one mile to [the home of] Col. Winchester, where I stayed two nights to give myself and my horse a rest.[61]

SUNDAY JUNE 14, I botanized.

JUNE 15, I arrived at the home of a settler, a Mr. Jackson, on fertile ground near the river.[62] Oaks: *Quercus prinus*, Q. *rubra*, Q. with large acorns, fruit enclosed, called overcup white oak,[63] *Q. tomentosa*, *Q. praemorsa*: 25 miles.

JUNE 16, I reached Nashville after 12 miles. A total of 197 miles from Knoxville to Nashville, capital of the Cumberland settlements, located on the Cumberland River.[64]

JUNE 17, I visited several people: Daniel Smith,[65] Col. Robertson,[66] Capt. Gordon,[67] . . . Deaderick,[68] Dr. White,[69] Th. Craighead,[70] etc. etc.

The following days, I looked for trees of the Nashville territory: *Quercus prinus*, *Q. phellos latifolia*, *Q. pinnatifida*, *Q.* leaves lyrate, underneath tomentose, cups large with laciniate margins, generally enclosing an acorn, named overcup white oak, *Q. rubra*, *Q. tomentosa*, *Acer saccharum*, *A. negundo*, *A. rubrum*, *Juglans nigra*, [*J.*] *oblonga*, [*J.*] *hiccory*,[71] *Platanus occidentalis*, *Liquidambar styraciflua*, *Ulmus viscosa fungosa*, *Carpinus ostrya americana*, *Rhamnus alaternus latifolius*,[72] *R. frangula*? shrub with plums, *Juniperus virginiana*. [Along the] banks of the Cumberland River, *Philadelphus inodorus*, *Aristolochia sipho-tom*, *Mimosa erecta-herbacea*, *Mirabilis clandestina* or *umbellata* or *parviflora*, *Hypericum kalmianum grandiflorum*.

The soil of Nashville is stony, rich in clay. The rocks are calcareous like those of Kentucky; the rocks are arranged horizontally, rarely with veins of quartz, with many marine petrifactions.

SUNDAY, JUNE 21, I killed and skinned several birds. Birds: robin, cardinal, grouse, the rare kingbird, a quantity of flycatchers of the genus *Muscicapa*,[73] a few woodpeckers from the genus *Picus*,[74] [and] wild turkeys. Quadrupeds: muskrat, beaver, elk, dwarf deer, bear, buffalos, wolves, small gray squirrels.

Minerals: clay soil, calcareous rocks always in horizontal layers, impure slate, fossils of land and freshwater shells.

MONDAY, JUNE 22 (old style), 4 messidor of year 3 of the republic, I left Nashville for Kentucky; I went by Mansko's Lick [Mansker's Station, now Goodlettsville], 12 miles from Nashville; [I] spent the night at the place of Maj. Sharp, 29 miles from Nashville.[75]

JUNE 23, I crossed the barren of oaks and slept on . . . Creek. There is no house in this area. The land produces only black oaks: 30 miles.

JUNE 24, I crossed Big Barren River. The man who runs the ferry is well supplied with provisions.[76] There are 3 miles from the creek . . . I crossed the barrens and slept on the ground, without a fire and without letting my horse graze at a distance, for fear of hostile natives.

JUNE 25, I crossed Little Barren River; the first house is 43 miles from the Big Barren River. Then I crossed the Green River, 6 miles from the Little Barren River.[77]

JUNE 26, I passed Rolling Fork, the head of Salt River, 30 miles from Green River.[78]

JUNE 27, I reached Danville, 35 miles from Rolling Old Fork. It is 117 miles from Nashville to Danville, the oldest city in Kentucky.[79]

SUNDAY, JUNE 28, I rested.

JUNE 29, I skinned 3 striped squirrels (*Sciurus striatus*).[80]

JUNE 30, I botanized.

WEDNESDAY, JULY 1, I visited several residents.[81]

JULY 2, it rained continually.

JULY 3, I put my old collection in order.

JULY 4, [entry blank]

SUNDAY, JULY 5, [entry blank]

JULY 12, I dined with the governor of the state of Kentucky, Isaac Shelby.

JULY 16, I left Danville.

JULY 17, I passed through Bardstown, 40 miles from Danville.

JULY 18, I reached Standford's,[82] near Mann's Lick.

SUNDAY, JULY 19, I remained there to wait for my baggage.[83]

JULY 20, I stayed on, and since I was obliged to wait, I observed the salt production operation. The pits for drawing saltwater are dug to a depth of about
. . . feet. One finds muddy clay to the depth of . . . feet. Then . . . feet of shale
rock. When the rock has been pierced, there is saltwater for more than . . .
feet. This shale burns in the fire as if it were impregnated with bituminous coal
or composed entirely of this substance. They have found the bones of large
marine animals, which are rather frequent on the banks of the Ohio River, in
the impure clay that they had to dig through to get to the slate rock.[84]

JULY 21, I arrived in Louisville, 40 miles from Bardstown.

JULY 22 and 23, I remained there and searched for plants.

JULY 24, I returned to Mann's Lick, 16 miles from Louisville.

JULY 25, I returned to Louisville.

SUNDAY, JULY 26, I botanized. Plants of the Louisville area: *Quercus cerroides*,[85] *Q. rubra*, *Q. alba*, *Q. prinus*, *Liriodendron*, *Fagus castanea*, *Fagus sylvatica*, *Rhus* with winged leaves, dioecious, *Hibiscus* with hastate leaves, outer
calyx subulate and laciniate, flowers pale pink.[86]
Observations on American grapes:[87]

[1] Lincoln, North Carolina. *Vitis* with tomentose leaves, berries very
large, fruiting at the beginning of August, is called fox grape.

2) *Vitis* with tomentose leaves, smaller berries, fruiting around the
10th of September, is called summer grape; [it] is the best for eating
and very good if one allows it to ripen completely.

3) *Vitis* with glabrous leaves, berries very large, is also known to be
good-tasting and to make wine; it is known as muscadine grape by the
local inhabitants, fruiting around September 20.

4) *Vitis* with reticulate glabrous leaves, with small berries, grows
along streams and rivers.

FIGURE 24. *Quercus macrocarpa*, acorn with fringed cap
(Photo by Charlie Williams)

The fifth item in this list appears on a loose sheet that was inserted at the end of notebook 8.

> [Supplement 5] *Vitis* with crenate, acuminate glabrous leaves, stems prostrate, *Vitis repens* or *Vitis riparia*.

The note below appears to have been inserted at another time.

> (Time of sunset in Charleston
> [in] July at approx. 7 o'clock
> August 5:15 . . . 6:45 to 6.30
> September 5:45 . . . 6:15)

Notebook 8 ends. The last, brief, note in the eighth notebook is a short plant description in Latin, but the condition of the page renders the text undecipherable.

Michaux began his ninth notebook with a calculation that the coming year, 1796, would be a leap year. He listed the months followed by the number of days in each month.

### Notebook 9 begins

SATURDAY, AUGUST 1, I prepared to leave for Wabash and Illinois.

SUNDAY, AUGUST 2, I was invited for dinner at the home of a Frenchman named Lacassagne, a resident of Louisville for more than 15 years.[88]

Trees, shrubs, and plants of the Louisville territory: *Liriodendron tulipifera*,

*Platanus occidentalis, Acer rubrum* leaves silvery on the lower surface,[89] *Fagus sylvatica americana, Quercus rubra, Q. alba, Q. praemorsa, Q. prinus, Q. cerroides, Tilia americana, Juglans nigra, Juglans alba,*[90] *Juglans hiccory, (Juglans pecan* rare), *Gleditsia tricanthos, Guilandinia dioica.*

SUNDAY, AUGUST 9, I left Louisville and spent the night at Clarksville, two miles from Louisville on the opposite side of the Ohio.[91]

AUGUST 10, we left on our journey and arrived at Post Vincennes on the Wabash River on Thursday night, August 13. The distance is approximately 125 miles. The day of our arrival, we crossed a river about 20 miles before arriving at Post Vincennes, and although the water was very low, we were able to make a raft, as the countryside is not inhabited on this road. Of all the voyages that I have made in America in the last ten years, this was one of the most painful because of the vast number of trees uprooted by storms, because of the thick underbrush through which we had to cross, and because of the vast quantity of ticks that devoured us, etc.[92]

AUGUST 14, 15, and Sunday 16, I was obliged to rest as I was almost sick when I arrived. While jumping to get across a trunk of a large fallen tree, my horse threw me quite a distance, and for several days I was in pain due to a wound at the base of the chest on the left side caused by the lock of my gun, which struck me there.

AUGUST 17, I spent part of the day botanizing along the Wabash River. I continued the following days.

AUGUST 18, 1795. List of Plants noted on the Wabash:

No. 1. *Verbena urticifolia* stems erect, panicles spreading, bracts shorter than the flowers, flowers white.

No. 2. *Verbena* . . . , stem erect, panicle with clustered branches, bracts and calyx pilose, flowers purplish blue.

No. 3. *Verbena* stems erect, panicles erect, leaves ovate, tomentose, twice-serrate.[93]

No. 4. *Verbena* . . . ,

No. 5. Verbena creeping stems, leaves pinnatifid, very long bracts.[94]

*Silphium perfoliatum, S. connatum,*[95] *S. laciniatum, S. grandifolium, S. trifoliatum, S. pinnatifidum. Andropogon muticum, Holcus?* . . . ,[96] *Poa* . . . , *Quercus cerroides* curly oak [or] overcup white oak, *Quercus latifolia* shingle oak [or] ram's oak, *Quercus . . . ,*[97] *Polygonum aviculare* with 5 stamens and 3 styles, *Polygonum aviculare* larger, with 5 stamens and 3 styles, *Trifolium? pentandrum* larger, *Trifolium? pentandrum* flowers purple,[98] *Sanicula marylandica* called root of Becquel by the French of Illinois and Sakintépouah by the Piankeshaw tribe. A decoction of the root is an excellent remedy for several diseases and for chronic venereal diseases.[99]

SUNDAY, AUGUST 23, I left Post Vincennes on the Wabash for the land of the Illinois on the Mississippi. We went 6 miles and camped on the bank of a small river [Embarras River].[100] I had no other companion but a Native American and his wife. I had hired the native for ten piasters, and [I] promised him two more if his horse would carry all my baggage.

AUGUST 24, we traveled about 25 miles, but as the native was sick, we had to stop three hours before sunset.

AUGUST 25, we crossed several prairies. I noted a new species of *Gerardia* stems usually unbranched, leaves sessile, opposite, ovate, flowers axillary, purplish.[101]

AUGUST 26, since the meat provision had been eaten, the native stopped very early as he saw a favorable place for hunting; besides, it started to rain rather hard around three o'clock. An hour after we set up camp, the native returned burdened by a young bear and two rumps of a much older one. We boiled the pot twice, and we had enough to satiate us. We roasted the leftover.

AUGUST 27, the Native American killed two deer; we stopped very early to dry the skins and to eat because the native and his wife ate five meals per day. In addition they feasted on the marrow in the bones, which they ate raw. As they could not carry the meat, they contented themselves with a piece from the kidneys of the animal.

AUGUST 28, as much as I wanted to see game the first two days, I was afraid of seeing more because of the loss of time. I wanted to go on, especially since it was raining every day. I had already been obliged to dry by the fire my baggage, which had been completely soaked, in particular four books of botany and mineralogy that I had with me because I was unwilling to expose them to the hazards of the river by which I had sent two trunks with drying paper, powder, lead, alum, boxes for holding insects, and all other objects necessary for collecting plants, animals, insects, and minerals.

Sunday, August 30, I reached the village of Kaskaskia, two miles from the Mississippi and half a mile from the Kaskaskia River.[102] It is inhabited by former Frenchmen now under the American government. There are about 45 families. Its location is pleasing, but the number of inhabitants has decreased, and one can see only ruined and abandoned houses because the French people of Illinois, having been brought up and accustomed to trade in skins with the natives, have become the laziest and most ignorant of mankind. They live and dress for the most part in the manner of the Native Americans. They do not wear pants but rather have a piece of cloth, less than half a meter long, between their thighs, which is held in front and in back, above the kidneys, with a belt.[103]

August 31, I botanized.

Tuesday, September 1, continued botanizing; also on September 2, 3, and 4.

September 5, I left for the village called Prairie du Rocher, 15 miles from Kaskaskia. I went by the village of St. Philippe, abandoned by the French people and now occupied by three American families.[104] This village is 9 miles from Prairie du Rocher.[105]

September 6, I arrived at Cahokia,[106] near the Mississippi, at . . . miles from Prairie du Rocher.

September 7, I botanized and visited the vicinity of Cahokia.

September 8, left for Kaskaskia and arrived on September 9.

September 10, I continued botanizing around Kaskaskia until September 13.

Sunday, September 13, I went across to the southern bank of the Kaskaskia River with an Native American guide and continued to botanize in this area until September 18.

September 18 and 19, it rained continuously. I put my collections in order and rested my horse.

Sunday, September 20, . . . Kaskaskia, 45 families; Prairie du Rocher, 22 or 24 families; St. Philippe, 3 American families; Fort de Chartres, in ruin; Cahokia, 120 families; Americans at Elkhorn and Bellefontaine, 35 families.[107] St. Louis, Florissant,[108] Petite Coll.,[109] Vide Poche,[110] Ste. Geneviève,[111] Rivière des Arcs,[112] Natchez, Mobile, Naquitoches, Wachitas, Apalusas, Acatopas, New Orleans.[113]

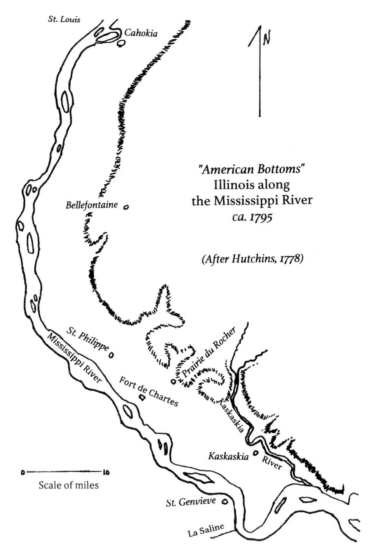

MAP 15. "American Bottom," along the Mississippi River
(Drawing by Charlie Williams, after Hutchins 1778)

There are no journal entries dated between September 20 and October 2, 1795.

Friday, October 2, I left to go overland to where the Ohio empties into the Mississippi, because of the difficulty in crossing the Kaskaskia River, we walked only 12 miles.

October 3 and Sunday, [October] 4, it rained, and we crossed several prairies; we walked approximately 27 miles.

OCTOBER 5, we went across more prairies interrupted by edges of woodlands. My guide killed an elk, called by Canadian and Frenchmen Illinois deer. This animal is much bigger (twice as big) as the dwarf deer of the United States, which is also abundant in Illinois and which the French people refer to as roe deer. Its antlers are twice as big as that of the European deer. There is under each eye a cavity that remains closed, but on separating the two lids, one can put an inch of one's finger in the cavity. This cavity seems to be made for the secretion of certain substances. In fact, when I opened this cavity, I found a substance of the shape and consistency of a rabbit pellet, but the size of an acorn. This animal has canine teeth above and below like those of a horse, called hooks. Hunters say that this animal is always very fat. This one was excessively so. We traveled about 32 miles.

OCTOBER 6, we went through several forests and rivers. Traveled . . . miles.

OCTOBER 7, my guide killed a buffalo that he judged to be about 4 years old. It seemed to weigh more than 900 pounds. As it was not very fat, my guide told me that it was not unusual for an animal this age to be 1,200 pounds. He appeared to be larger than any oxen from France, in both length and weight.

THURSDAY, OCTOBER 8, I saw another buffalo at about 200 feet from our path. We stopped to observe him. He was walking very slowly, but after two minutes, he stopped, and being aware of our presence, he ran with extraordinary speed. We arrived the same day at Fort Cherokee, called by the Americans Fort Massac; [in total] 125 miles.[114]

OCTOBER 9, I botanized along the Mississippi River [Ohio River]:[115] *Platanus*, *Liquidambar*. Bonducs, pecans, and hickories, known as hard nuts by the French, also pungent nut, American nut, bitternut, and nut tree with round nut. White oak, *Quercus rubra* much branched, *Quercus cerroides* known by the French as curly oak, by the Americans [as] overcup white oak, *Q. prinus*, *Quercus integrifolia* young as well as mature leaves always entire with undulate margins and a bristle at tip. This species is abundant in the Illinois country. It loses its leaves later than other oaks. The French inhabitants call it the lath oak. In South Carolina, it is rather rare, but there it keeps its leaves until February or March. It appears to be closely related to the live oak but differs in the shape of its acorn.[116] *Nyssa montana* rather rare, *Gleditsia tricanthos*, *Robinia pseudoacacia* (referred to by the French people as *fevier*); *Gleditsia tricanthos* goes under the name *spiny fevier*, and *Guilandina dioica* [under that of] *big fevier*, and the seeds are like broad beans. Note: there is along the Illinois

FIGURE 25. Fort Massac reconstruction, Fort Massac State Park, Illinois. The fort would have appeared much like this when André Michaux visited it in 1795. (Photo by Charlie Williams)

River a species or variety of *Guilandina dioica* whose seeds are twice as large as those from the banks of the Mississippi, from the Cumberland, etc. climbing vine *Rajanioides*,[117] *Anonymos ligustroides*,[118] *Vitis monosperma*. This [last] species is found along rivers but never in the woods; I saw it on the Kaskaskia River, along the Mississippi in the region of Fort Massac, [and] along the Tennessee River, but it covers entirely the banks of the Cumberland River for 45 miles beyond its mouth.

SUNDAY, OCTOBER 11, I left with a guide to go by canoe up the Cumberland River, the Shawnee, but rain forced our return.

TUESDAY, OCTOBER 13, I hired two men at a piaster per day for each to go up the river of the Cherokee territory. We left from the Cherokee fort, known as Fort Massac. It is six miles to the mouth of the Tennessee River, as it is called by the French, [or] the Illinois River, [as it is called] by the Cherokees. This river is very large and wide. After going up for about 6 miles, we saw bear tracks on the banks; we stopped, and as we entered the woods, a female bear and three cubs presented themselves. The dog pursued the mother, and the young ones having climbed a tree, I killed one, and the guides killed the other two. We spent the night at this site.

OCTOBER 14, the fog was very thick; we only walked 4 miles. It started to rain at about noon.

OCTOBER 15, we rowed or paddled approximately ten miles as a strong wind, which had come up before a storm the night before, continued [for] part of the day. We camped across from an island or a chain of rocks, which extends almost across the whole river. Despite this there is a rather deep channel on the south side of the river, sufficient for the passage of large boats. List of the plants I saw on the banks of the Tennessee River: *Platanus, Juglans pacane,* [*J.*] *hiccori,* [*J.*] *pignut, Liquidambar, Quercus rubra, Q. prinus, Anonymos carpinoides, Anonymos ligustroides. Betula australis,* [a] birch with gray bark that is found all over America from Virginia to the Floridas; it differs from *B. papyrifera.*[119] *Bignonia catalpa, Ulmus, Fraxinus, Vitis rubra* or *monosperma, Gleditsia tricanthos, Diospyros, Smilax pseudochina,*[120] *Bignonia crucigera,* [*B.*] *radicans, Rajania . . .* vine, dioecious with 8 stamens, *Populus caroliniana* known by the French Creoles as *liard* and by the Americans [as] cotton tree. (Note: the poplar of Canada is called tremble by the Canadians and quaking aspen by the English of Canada.) *Acer rubrum, A. saccharinum, A. negundo, Anonymos ligustroides, Anonymos ulmoides.*

The note below concerning French officials in the United States appears at the bottom of the page in smaller letters. It is in Michaux's handwriting, but it appears to have been written with a different (finer) pen than the entries immediately preceding or following. Note also that Michaux made no entries dated either October 10 or October 12, but he dated two entries October 15.[121]

On June 22, 1795, according to the *Gazette,* the following agents of the French Republic were recognized by President Washington:[122]

Philip Joseph Latombe, Consul General
Theod. Charl. Mozard, Consul at Boston
Jean Anth. Bern Rozier, Consul in New York
Leon Delaunay, Pennsylania
Louis Etienne Duhait, Maryland

OCTOBER 15, I botanized.

October 16, I went down the river and camped at the mouth of the Shavanon River, called Cumberland by the Americans, 18 miles from Fort Massac. I killed a Canada goose, called by French of Canada and Illinois *outarde*. I killed two moor hens, an American fisher martin, an American pelican.

October 17, we went back up the river ten miles; the banks were full of wild turkeys. The oarsmen and I killed five while passing in our canoe, without landing.

October 18, we continued our course up the river.

October 19, we turned back downriver.

Tuesday, October 20, we returned to Fort Cherokee, known as Fort Massac.

Trees and plants near the banks of the Ohio:

*Platanus occidentalis*, called sycamore by the Americans and *cotonnier* by the French of Illinois; *Populus*, known as cotton tree by the Americans and *liard* by the French of Illinois; *Celtis occidentalis*, called hackberry by the Americans and by the French [as] unknown wood; *Liquidambar styraciflua*, known by the French of Louisiana as *copalm* and by the Americans. . . .[123] A Frenchman who traded with the Cherokee cured himself of scabies [a skin disease caused by mites] by drinking a decoction from shavings of this tree for ten days that he called copalm and that is the true *Liquidambar*. *Gleditsia tricanthos*, known as *fevier* by the French and sweet locust by the Americans; *Guilandina dioica*.

Sunday, October 25, *Spiraea trifoliata* is a purgative used by the Native Americans and the French of Illinois. They call it *papiconah*. In the vicinity of Fort Massac one can find also a *Geranium* called Becquet's herb, or rather root, which is given for several weeks for chronic illnesses. Often *Veronica virginica* is added; it goes under the name by the French of the four-leaved herb.[124]

Sunday, November 1, I had to delay my departure as my horse could not be found.[125]

Friday, November 6, my horse was brought back to the fort, and I prepared to leave immediately for the Illinois country. Left that day and walked about 18 miles.

November 7, the rain started in the morning and lasted all day. I stayed in camp under a cliff where I had stopped with my guide the night before.

SUNDAY, NOVEMBER 8, we traveled through the woods and the hills.

NOVEMBER 9, I did the same thing.

NOVEMBER 10, we reached the prairies toward nightfall.

NOVEMBER 11, went across the prairies.

NOVEMBER 12, toward evening, we entered the forest once again and slept 7 miles from the Kaskaskia River.

NOVEMBER 13, we arrived at Kaskaskia before the morning meal, approximately 130 miles from Fort Massac. I rested.

SUNDAY, NOVEMBER 14, I went hunting for Canada geese.

NOVEMBER 15, put my seed collection in order.

NOVEMBER 16, I continued the same.

NOVEMBER 17, I went hunting.

THURSDAY, NOVEMBER 18, left to go to Prairie du Rocher.

NOVEMBER 19, went hunting for ducks.

NOVEMBER 20, went hunting for geese.

SUNDAY, NOVEMBER 22, I visited several people.

NOVEMBER 23–28, I visited the Rock Mountains, which are around the inhabited countryside.[126] I saw opossums, raccoons, aquatic birds, etc.

SUNDAY, NOVEMBER 29, I went to St. Philippe's Village, called the Petit Village.

NOVEMBER 30, I visited Fort de Chartres.[127]

TUESDAY, DECEMBER 1, left for Kaskasia and remained there.

DECEMBER 2 and 3, I made arrangements with Richard to go by water to Cumberland.[128]

DECEMBER 4, I returned to Prairie du Rocher.

DECEMBER 5, I prepared to leave. I stuffed a wild goose with a white head.

DECEMBER 6, I left again for Kaskaskia.

DECEMBER 7, it was confirmed again that the secondary bark of *Celtis occidentalis*, called by the people of Illinois *bois connu* (known wood), and toward New Orleans *bois inconnu* (unknown wood), is an excellent remedy to cure jaundice. They add to it a handful of the roots or leaves of *Smilax sarsaparilla*. It is used for about a week as a decoction.

DECEMBER 8, the French Creoles call the species of *Smilax* that is found in Illinois *squine*. It is the only species that grows that is spiny; it loses its leaves in autumn. The other species is herbaceous and climbing.

DECEMBER 9, the root of *Fagara* as a decoction is a powerful remedy to cure illnesses of the spleen; there is no doubt that one can use those of *Zanthoxylum clava herculis* for obstructions of the liver and spleen.

DECEMBER 10, *Bignonia catalpa*, known by the French Creoles as Shawnee wood; *Cercis canadensis*, black wood; *Liriodendron tulipifera*, yellow wood; *Nyssa*, olive wood. The workmen use wood of the following trees for making wheels of carriages: *Padus virginiana* for the rims, elm for the hubs, and white oak for the spokes.

DECEMBER 11, confirmed again that a decoction of the root of *Veronica virginica*, commonly known as herb with four leaves, used for a month is useful in curing venereal diseases. One has to boil four or five of these roots. As this tea is purgative, one has to increase or decrease the strength of this tea by putting more or less of root material or boil it more or less according to the effect that it produces; it is sufficient that the bowels are relaxed and looser the first few days. It is not unusual on the first day to have three or four bowel movements.

Michaux separated the entry above from the one below with a horizontal line across the page, and the handwriting is smaller below.

I have been told by the [people in] Illinois that the Scotsman Mackey, and Even, a Welshman, left at the end of July 1795 from Saint Louis on a barge with four oars to go up the Missouri. They are aided by a company of which ClaMorgan, a Creole from the islands, is the manager.[129]
December 1795.

SUNDAY, DECEMBER 13, made the last preparations for the journey to Cumberland.

The following day, Michaux began the first leg of his return journey to Charleston traveling down the Mississippi by boat. Michaux made no note in his journal to indicate that before leaving Illinois, he had written a letter to an influential Frenchman who lived in Spanish territory across the Mississippi, Pierre Charles Dehault Delassus de Luzières, who was a wealthy aristocrat who fled France with his family because of the revolution and remained a royalist all his life, and whom Michaux had mentioned when he visited Pittsburgh (see Michaux's entry for August 13, 1793). After de Luzières arrived in Spanish territory, Baron de Carondelet, governor and intendant of Louisiana, had appointed him to establish a colony of French royalists at New Bourbon, near Ste. Geneviève, Missouri.

De Luzières had become familiar with Michaux's name in 1793. In a letter to Carondelet dated December 26, 1793, remarking on Genet's plans with George Rogers Clark, de Luzières had written, "the army in question [i.e., Clark's] is composed of American vagabonds and some French renegades, Messieurs Despot [Charles de Pauw], la Chaise and [André] Michaux are the principal officers."[130]

Although he was preparing to leave the territory, Michaux wanted to be on good terms with the Spaniards and the French, even the royalists, at least partly because he wished to explore west of the Mississippi in the future. His letter to de Luzières quoted below indicates that the botanist understood that he was viewed with suspicion. With his letter, Michaux attempted to cast himself in a different light by warning de Luzières that, with the breakdown in peace talks between France and Spain, there might be trouble ahead in the territory. He also attempted to win favor with these suspicious men by offering both de Luzières and Manuel Gayoso de Lemos (whom he was soon to meet) some rare seeds.

### AM to M. Louisiere de la Suze (de Luzières), at Ste. Geneviève, December 2, 1795[131]

I am very upset that I did not have the honor of seeing you the last time that you came here. This annoyance was caused by the people of this place, who had planned to keep me from knowing the circumstances of the arrival of Governor Gayoso, thinking that my company would not please him. Upon my questioning regarding the people who would accompany him, I was told simply that he would come with an officer.

A few days ago, upon learning that the negotiations for peace between France and Spain had broken, I had wanted to inform you that probably the expedition against Louisiana would be renewed. The sound from the public appears to support this conjecture. As it is very possible that I will be informed of the circumstances, be persuaded that I will do all that is possible to contribute to the peace of the people of the territory in which you live, to calm their fears, and you will have, dear Sir, proof of my attachment to you.

Knowing, dear Sir, that you are interested in the culture of gardens, I had promised to your son a collection of very rare seeds.

If I could meet someone connected to you Sunday or Monday, I would give it to them, as I will be leaving next Tuesday for Cumberland [Nashville] and Knoxville. I will find there news of my botanical establishment in the Carolinas and even a collection of seeds sent to me to distribute to those who are interested in the rarest objects that botany can offer. As I can furnish in this regard objects of usefulness as well as of ornament, if I could learn the things that might please you more, I would reserve those for you.

Because I also have seeds of rare trees that are likely to do well in the climate of Natchez, if you know that Governor Gayoso is curious concerning this type of object and could make good us[e] of them, I would make a special collection that I would give you so that you could transmit them to him.

I lent to your son a brochure and a map from Kentucky. This map has the description of this country, and the work would be defaced if they were separated; I would be obliged to him to send [the map] back to me.

# CHAPTER THIRTEEN

# Kaskaskia to Charleston, December 1795–April 1796

Summary: Michaux left Kaskaskia on December 14, 1795, headed down the Mississippi River on a river barge (*bercha*).[1] He risked arrest while traveling down the Mississippi or when the boat docked for the evening in Spanish territory because Spanish authorities regarded the entire Mississippi River as their territory and patrolled the waterway with a fleet of gunboats propelled by oarsmen. At the junction of the Ohio and the Mississippi, the boat carrying Michaux as a passenger turned upstream on the Ohio. It was pursued by gunboats and soon visited by curious Spanish officials, even though it had landed in US territory. After a surprisingly friendly meeting with the Spanish governor of Upper Louisiana, Michaux continued on his journey.

Leaving the Ohio at the mouth of the Cumberland River, the botanist continued upstream on the flooding Cumberland in winter weather until the boat reached Clarksville, Tennessee. There he purchased a horse and continued overland to Nashville. After a brief visit there, where he knew a number of leading citizens from his earlier visit, Michaux made a journey on horseback to Louisville, Kentucky. The cold, snowy weather that he encountered on this route made this part of his journey especially difficult and dangerous.

In Louisville he gathered the natural history collections that he had left there for safekeeping and met once again with Gen. Clark. He promptly returned to Nashville, where he purchased an additional horse to carry his collections and began his journey east along the same route that he had followed west the previous year. On the upper Cumberland River near modern Gainesboro (Fort Blount), he stopped once again to botanize. He collected seeds of the yellow-wood tree and wrote a letter about this discovery to the governor. His letter was published in the Knoxville newspaper. Continuing on his journey he visited the governor in Knoxville, and he retraced his 1795 route through East Tennessee to Col. Tipton's plantation in what is now Johnson City. Michaux then crossed into North Carolina on a less-frequented trail over Iron Mountain rather than taking the Yellow Mountain route that he had taken in 1795. After arriving in North Carolina, he detoured slightly to revisit his mountain guide, Martin Davenport, for a week of botanical collecting. When he resumed his journey eastward, Michaux retraced his route through Morganton and Lincolnton, soon crossed the Catawba River near Charlotte, and returned to South Carolina, completing this long journey to his garden north of Charleston. It was to be Michaux's last journey in North America.

### Notebook 9 continues

December 14, left for Cumberland, went by the salt springs on Spanish territory;[2] noticed *Tagetoides*. Learned of the news of peace between France and Spain.[3] Slept six miles from the salt springs. Observed on the banks of the Mississippi *Equisetum*, which the French Creoles call *prêle*;[4] here this plant is one inch in diameter and 4 feet high.

December 15, went by Cape Cinque Hommes, at the foot of which the Mississippi makes an angle.[5] Fish are caught here in abundance; it is 18 miles from Kaskaskia. Camped at Cape Girardeau, 40 miles from Kaskaskia.[6]

December 16, continued for six hours; there were rocks and hills on the banks of the river followed by low grounds. We camped at the mouth of the Belle River [Ohio River] on the Mississippi. On the opposite bank, Governor Don Gayoso, governor of Natchez and Upper Louisiana, was encamped. He sent over a boat to find out who we were, and learning that I was a traveler, he came to see me. He told me of the news of the peace between France and Spain. He offered me his services.[7] It is 54 miles from Cape Girardeau to the mouth of the Belle River and in total 105 miles from the Illinois.

December 17, camped about 21 miles away.

DECEMBER 18, arrived near Fort Massac: 21 miles.

DECEMBER 19, camped across the junction of the Cherokee River or Tennessee River.

SUNDAY, DECEMBER 20, went through the Pacanière; it is a large marsh situated on the northwest side [of the Ohio River], bordered by pecan trees, found across or more precisely a little above the mouth of the Cumberland River. That day [we] entered the Shavanon [Shawnee] River, known [to the Americans] as the Cumberland River, which begins about 18 miles from Fort Massac. [We] spent the night six miles beyond the mouth.

DECEMBER 21, we rowed about 24 miles.

DECEMBER 22, we rowed about 21 miles, and we spent the night at the large eddy, which is estimated to be 45 miles from the mouth.[8]

DECEMBER 23, we camped above Isle of the Willows, a distance of about 12 miles or 4 leagues.

DECEMBER 24, we stayed in camp; it rained all day. The river, which had been easy to navigate up till today, widened appreciably and spread into the forest.

DECEMBER 25, the rain continued and was mixed with hail; we stayed in camp.

DECEMBER 26, we remained in camp as the river rose, and the current was very fast.

SUNDAY, DECEMBER 27, we rowed only approximately 4 miles because of the difficulty in rowing against the current. We camped at the mouth of the Little River.[9]

DECEMBER 28, we crossed to the opposite bank; the current, which was as fast as in preceding days, forced us to camp. There was white frost.

DECEMBER 29, again there was a heavy rain; we stayed in camp.

DECEMBER 30, the river having overflowed its banks and submerged all the woods, we had to leave camp and returned to the Little River. We ascended until we found a hill high enough for us not to fear the flood. Rain continued.

DECEMBER 31, the weather cleared, [and] the wind passed to the north, but the river continued to overflow. Most of the group went to hunt wild turkeys.

FRIDAY, JANUARY 1, 1796, the wind is from the north; it is freezing. The river rose one inch during the night. The surrounding countryside of the Little River is sprinkled with hills; the soil is clayey with rich vegetation, the rocks of flint have little iron, and the calcareous rocks are bluish.

Animals: raccoons, dwarf deer, opossums, buffalo, bears, gray squirrels, beavers, otters, [and] muskrats (the last three are very rare).

Birds: crow, large species of owl, cardinals, blue jays, a small species of green parrot with yellowish head,[10] woodpecker with red throat and head.

Trees and plants: *Liriodendron, Liquidambar*, chestnut oak, red oak, *Annona*, hornbeam.

JANUARY 2, we stayed in camp at the same spot. The weather is overcast; the river has gone down only two inches.

SUNDAY, JANUARY 3, strong wind. *Nyssa montana* the black gum tree is called by the French wild olive tree and [by] the Americans from Kentucky black gum tree, and people from Pennsylvania call it tupelo. Since I had nothing to do, I made ink from the gallnuts that I collected from oak trees near our camp; it was made in less than five minutes and will serve me as a sample. Around Little River, I saw *Liriodendron, Liquidambar, Carpinus ostrya, Ulmus fungosa, Padus virginiana minor, Laurus benzoin*, etc.

JANUARY 4, we rowed 4–5 miles and camped near a rather high hill; the soil was unstable with worn pebbles. *Carpinus ostrya, Ulmus fungosa, Padus virginiana minor, Philadelphus inodorus, Nyssa montana* known as black gum by the Americans, *Acer rubrum, Viscum* parasite, *Fagus americana*, and *Orobanche virginiana* parasite on the roots of *Fagus* in America, and *Betula spuria* known to the French as bastard birch.

TUESDAY, JANUARY 5, we traveled 7 miles and camped opposite Diev Island, 12 miles from the Little River.

JANUARY 6, the snow that fell during the night made it cold. There were sharp limestone cliffs that continued for a mile on the eastern side; we rowed about 8 miles.

JANUARY 7, the river had come down 19 inches during the night; the freeze that lowered the water made us hope for easier rowing against the current of this river, which is naturally hemmed in between hills. We rowed about 8 miles.

JANUARY 8, the river had gone down 19 inches during the night; we went by the island on a line drawn between Cumberland and Kentucky.[11] Plants along the banks: *Platanus occidentalis, Betula australis* or *spuria, Acer rubrum, Ulmus americana, Fraxinus, Salix* on the low islands, *Anonymos ligustroides.*[12] We rowed about 10 miles.

JANUARY 9, the river had gone down about 5 feet during the night. We rowed about 10 miles.

SUNDAY, JANUARY 10, the river had gone down 4 feet during the night; continual rain and snow. We passed by the Yellow River (Yellow Creek), 16 miles before arriving in Clarksville. We passed by Blooming Grove Creek? 13 miles before Clarksville;[13] rocks and hills. We went by Dixon Island? 10 miles before Clarksville, and actually the most remote area in the Cumberland territory. This establishment is made up of 15 families who came here three months ago. The principal town is known as Blountsborough or Blountsville [now Palmyra].

JANUARY 11, rain fell throughout the preceding night and part of the day. We passed by a chain of hills and a rock by the name of Red Painted Rock [Red Paint Hill] on the right side of the river; that is, on the northern side of the river, 2 miles from Clarksville. Then we went by the Red River, whose mouth is also on the northern side about a quarter mile from Clarksville. Finally we arrived in Clarksville.[14]

JANUARY 12, 1796, stayed in Clarksville because of the rising river.

JANUARY 13, Dr. Brown from the Carolinas,[15] [who] came to establish the new town of Blountsborough, 10 miles from Clarksville.

JANUARY 15, I bought a horse for one hundred dollars.

JANUARY 16, I left; my horse got away, but I caught up with him, 6 miles from Clarksville, 10 miles from the mill.

SUNDAY, JANUARY 17, I dined 10 miles from Nashville, one quarter of a mile from the mill at [the home of] Ebenston; [he is] an old man from Pennsylvania, a well-educated man who was up to date on foreign news.[16] I lodged at Crokes, 18 miles from Ebenston. The Widow Martin lives nearby, and her house is preferable for travelers.

JANUARY 18, I crossed the Ridges, 15 miles without seeing any houses until White Creek. The old man Stump lives 5 miles from White Creek.[17]

January 19, I left Stump's home and arrived in Nashville, 5 miles farther. Total mileage from Clarksville to Nashville, 54 miles by land and 70 by water.

| | |
|---|---|
| From Saint Louis to Kaskaskia | 94 miles |
| From Kaskaskia to the mouth of the Ohio to the Mississippi | 95 |
| From there to Fort Massac | 45 |
| From there to the mouth of the Cumberland River | 18 |
| From there to Clarksville on the Red River | 120 |
| From there to Nashville | 60 |
| Total | 432 miles |

The list of thirteen people below was omitted from the 1889 transcription of Michaux's journals and thus from all subsequent reports about Michaux in Tennessee.

People known in Nashville: Mr. De Montbrun;[18] John and Mr. Geo. Deaderick, merchants;[19] Tetts, merchant;[20] Seth Lewis, attorney;[21] Overtun, attorney;[22] Mr. Gordon, merchant;[23] Dr. Sapinton;[24] Dr. Holland;[25] Maxville, tavern keeper; Betts, tavern keeper;[26] Col. Hay, 7 miles from Nashville;[27] Mr. Jackson, attorney.[28]

Prices: dinner, 2 shillings; lunch or supper, 1 shilling 4 pence; ½ pint of whiskey, 1 shilling; hay and corn for horse, 2 shillings. Total is 6 shillings, equivalent to one dollar.

January 20–22, stayed in Nashville.

January 23, I left Nashville and traveled 29¾ miles and stayed overnight at Maj. Sharp's [inn].[29]

Sunday, January 24, arrived at a creek, 29 miles, near where a man named Chapman had lodgings;[30] 3.5 miles [farther] MacFaddin has a ferry and lodging on the Big Barren River. Total, 32.5 miles.[31]

January 25, rain and snow.

January 26, I left for Green River; the ground was covered with snow, the roads were rough, and my horse became lame; I was obliged to go on foot. I

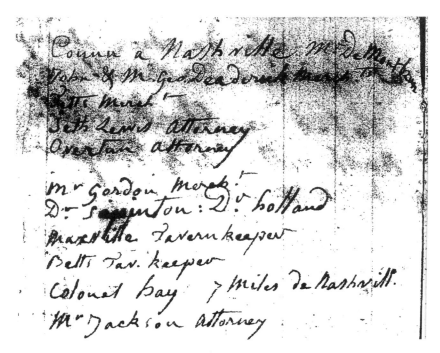

FIGURE 26. Michaux's journal page with his list of people known in Nashville
(Courtesy of the American Philosophical Society, Philadelphia, PA)

traveled 12 miles. It was impossible for me to make a fire; the trees and the woods were covered with icicles. I spent the night almost frozen. Around two o'clock, the moon having risen, I decided to return to MacFaddin. I arrived there at 10 o'clock in the morning.

JANUARY 27, I was overcome by cold and weariness, having had to walk on foot, not having had anything to eat since the morning of the previous day, and not having slept the night. I found that the toes of my right foot were inflamed. I bathed my feet in cold water several times during the night, and no sores developed, but for several days, the toes of my feet were numb and almost without sensation.

JANUARY 28, I had to go seven miles to have my horse shod, and I came to the home of Mr. Maddisson, who had his plantation nearby, and where I spent the night.[32]

JANUARY 29, I left early in the morning since I had to travel 38 miles without finding an inn or another house. I had been received with all the courtesies that one can hope for from a man who had a much better education than those of people from the country. This Mr. Maddisson was from Virginia and related to the famous Maddison, member of Congress.[33] He was a true Republican in his beliefs and principles, and I had spent a very interesting and pleasant evening at his house. His wife did all she could to favor me with her hospitality; it is very rare except among people with an education that is superior to that of the common inhabitant. This lady suggested that I use booties of coarse wool to cover my shoes. She herself cut out a pair, and I was so surprised by the benefits that I obtained the next few days that I resolved never to travel in times of snow or ice without such a pair in my wardrobe. I arrived at night, 3 miles from the Green River, and spent the night at the home of a man named Walter; I slept on the floor, and my horse slept under the stars, but I was accustomed to this.

January 30, in the morning, I crossed Green River with the ferry.[34] The cold was extreme, such as had not been experienced in several years. Nine miles farther, I went by Bacon Creek and the cabin of a man newly settled, who had nothing, not even corn to provide for his household. Twenty-two miles from the Green River, one finds the house of a man named Ragar, and I hurried to get to some better homes before nightfall. Twenty-six miles from the Green River, I saw a house about 400 meters [1,200 feet] from the road, along a creek. The house was inhabited by a German who had settled here about a year ago. He had a good stable with wheat straw and corn fodder for my horse, and I ate wheat bread for the first time since I left Illinois. My supper was bread and milk, and I was very well treated. My host's name was George Cloes of German origin;[35] his house is on the South Fork of the Nolin River.

Sunday, January 31, went by Huggins's Mill on the Nolin River (good lodging),[36] a quarter of a mile farther, the road to the right goes to Bardstown. Two and a half miles farther, the new cut road is straight. Nine miles farther, I passed by Rolling Fork; and four miles beyond, I spent the night at [the home of] Mr. Scoth on Beech Fork.[37]

Monday, February 1, 1796, I passed Dr. Smith's house, 8 miles from Beech Fork, and McKinsey's, 9 miles from Beech Fork. From Mac Kinsy to Long Lick Creek, it is 6.5 miles. From Long Lick Creek to Shepherdsville on Salt River, it is 4 miles. From Shepherdsville to Standeford's (a good inn) is 9 miles.[38] From Standeford's to Prince Old Station is 8 miles.[39] From Prince to Louisville is 6 miles.

February 2, I left Prince's home and arrived in Louisville. Three and a half miles before arriving, I measured a *Liriodendron tulipifera* on the left side of the road whose circumference was 22 feet, which makes for a diameter of more than 7 feet.

(Mr. Serpé is Mr. La Cassagne and St. James Bauvais's representative in New Orleans. In Philadelphia, La Cassagne's representative is Gequir and Holmes, merchants).[40]

Prices: Dinner, supper, and breakfast, each 1 shilling 6 pence; lodging, 9 shillings; ½ pint of brandy, 2 shillings 3 pence; hay and corn for horse daily, 3 shillings 9 pence.

February 3, 4, [and] 5, I remained in Louisville, busy, getting together my collection, which I had left at the home of the previously mentioned, La Cassagne.

February 6, I saw Gen. Clark, and he told me of the visit of Col. Fulton, who had come from France a few months ago.

Sunday, February 7, I breakfasted at the home of Gen. Clark's father, which is 3 miles from Louisville.[41] I wanted to get more detailed information regarding Lt. Col. Fulton. They told me that he was to go to Philadelphia immediately after he had been to Georgia and that he would leave for France and hoped to be back in America at the end of the summer of this year, 1796.[42]

The same day, I left to return to Nashville. I slept at Standeford's, 14 miles from Louisville. Supper, 1 shilling; lodging, 6 shillings; hay for the horse, 1 shilling; 8 quarts of corn, 1 shilling 4 pence.

Monday, February 8, breakfast, 1 shilling. I went by Shepherdsville, 9 miles from Standeford. Three quarts of corn for the horse, 9 pence, Virginia money, as in all parts of Kentucky and Cumberland. I went by Long Lick Creek, where they make salt as well as in Shepherdsville;[43] I slept 4 miles from Shepherdsville at the home of Mackinsy, 7 miles from Long Lick Creek.

In the marshy areas around Long Lick Creek: *Quercus alba, Q. cerroides, Fraxinus . . . , Nyssa, Laurus benzoin, Sassafras, Mitchella repens, Fagus sylvatica americana*. On the hillsides: *Pinus* with two leaves, smallish oblong cone, scales acute, recurved. I saw planks of this tree at a man's house. The wood appeared as heavy as that of the three-needle Carolina pine. They also make tar from it in this part of Kentucky.[44]

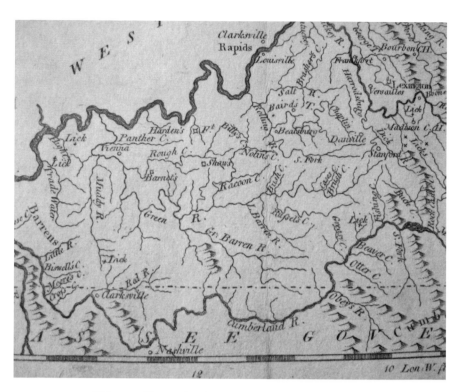

MAP 16. Western Kentucky, from Carey's *American Pocket Atlas*, 1796
(Courtesy of the Philadelphia Print Shop, Philadelphia, PA)

FEBRUARY 9, I left very early from Mackinsy's. I had been very well received; that is to say, he had provided me with a supper of boiled pork, and the same for breakfast; my horse was well provided for as to fodder, corn, and a stable that was not dirty, unlike those in America when one lodges with Americans or Irishmen. I paid 3 shillings, which was 1 shilling 6 pence for my horse, and the same amount for me. I had paid 5 shillings for lodging the night before, and I had not been as satisfied. As the daughter of this household gave more efficient service than I had seen anywhere in America, I gave her a quarter of a piaster, and the old man offered me a savory tongue, but I thanked him as I was not a lover of salt meat.

The rain began an hour after I left, but I had the good fortune to go by Beech Fork and Rolling Fork, 13 miles from Mac Kinsy's.

I had to stop at the home of a settler a mile and a half from the crossing, and the rain kept me there overnight.

In this neighborhood, there is *Liriodendron* with yellow wood, and in some districts, *Liriodendron* with white wood. The inhabitants like the one with yellow wood better.[45]

Wednesday, February 10, I had supped the night before on tea made from the shrub called spice wood. One boils a handful of young wood or branches, and after boiling for at least a quarter of an hour, one adds sugar as one would for regular tea. There was no milk, but I was told that milk makes it much more pleasant. This drink is supposed to revive one's strength, and it produced this effect, as I arrived very tired. This shrub is *Laurus benzoin* Linn. The French from Illinois call it pepper plant, and hunters season their meat with several pieces of the wood.

There is a plant in this area of the orchid family whose leaf persists all winter; rarely there are two; its form is oval, entire, with parallel veins. The roots bear 2–3 very viscous bulbs, which are used to glue broken crockery. It is called Adam and Eve.[46] This plant is more common in the rich lowlands west of the Alleghenies. I have seen it in South Carolina, but it is very rare. It is not rare in Illinois.

Rain continued all day, and I had to spend the night in a house near Nolin Creek because the water had overflowed.

February 11, I arrived at [the home of] Huggins, 12 miles from Rolling Fork.[47]

February 12, I crossed a countryside with grasses and oaks that had been burned yearly and thus no longer formed forests. They call this land barren, although it is not really sterile; the grasses dominate. *Salix pumila*, *Quercus nigra*,[48] and *Q. alba*, known as mountain white oak. *Gnaphalium dioicum* also grows there in abundance; it is called by the Americans white plantain. The same day I went by Bacon Creek, a site newly established, 19 miles from Huggins Mill; I arrived at Green River, 9 miles from Bacon Creek. I spent the night three miles farther at the home of a man named Walter.

February 13, I traveled 37 miles without finding any houses in the area called Barren Lands. *Salix pumila* is common around here and is the same species found in the prairies of Illinois from Post Vincennes to Kaskaskia. I spent the night beyond Big Barren River.

Sunday, February 14, I traveled about 30 miles. In all the houses, the children were sick with whooping cough. This illness develops from a simple cold, but the poor diet of smoked or salted meat that they cooked in a frying pan, producing acrid smells, made it more difficult to expectorate.

February 15, I traveled 27 miles and arrived in Nashville. Supper, lodging, breakfast, 2 shillings.

FEBRUARY 16, I left to visit Col. Hays,[49] a rich resident, to whom I had been recommended the year before by Governor Blount, governor of the area known as the Western territories, southwest of the Ohio. The country, estimated to have a population of sixty thousand because of the numerous annual immigrants and the increasing population, had just been established as a state, governed by its own representatives, with the new name of the state of Tennessee, from the name of a large river that flows through the land of Holston, the Cumberland, the country of the Cherokees, and other adjacent areas. This great river discharges into the Ohio, 9 miles above Fort Massac. It was known by the French, who were the first to explore the interior of North America, by the name of Cherokee River and is thus designated on French maps. I saw at Col. Hays's home several inhabitants of the neighborhood who came to discuss current affairs and the election of military and civil officers.

FEBRUARY 17 and 18, I stayed at [the home of] Col. Hays because of the bad weather.

FEBRUARY 19, I finished bargaining for the purchase of a horse to carry the plant and bird collection and other objects that I had to bring back from Illinois and more recently from Kentucky. I returned the same day to spend the night in Nashville.

FEBRUARY 20, I was occupied all day assembling and packing my collections. I saw French travelers who spend their whole lives trading with the natives and asked them for advice to obtain a guide to go up the Missouri River. One of them named . . . said that he would be willing to enlist voluntarily for a year for the price of 500 piasters in furs, or the equivalent of 1,000 piasters in silver; another asked for 2,000 piasters in silver.[50]

FEBRUARY 21, I prepared for my trip.

FEBRUARY 22, I had my two horses shod.

FEBRUARY 23, I left and traveled 2 miles but was obliged to turn back because . . .

FEBRUARY 25, I left to return to the Carolinas and slept 10 miles away at the home of Col. Mansko, a declared enemy of the French people because they had killed their king, so he said.[51] Although I had not had dinner, I did not want to accept his supper, believing that a Republican should not have any obligations toward a fanatic partisan of royalty. I was mortified that nightfall, and rain forced me to stay at his home. But I slept on my deerskin and paid for the corn that he provided me for the journey through the Wilderness.

FEBRUARY 26 . . .

SUNDAY, FEBRUARY 28, I stayed 10 miles from the river because of the rain and the overflowing creeks.

FEBRUARY 29, at nightfall, I went across the creeks and slept in the woods near the road in a place full of reeds or canes. This species of grass, which is abundant in areas that have not been settled, will be destroyed if domesticated animals browse on it thoroughly; the pigs kill it while rooting in the soil and breaking the roots. The size of the stem is sometimes that of the shaft of a goose feather, but in the rich soils that are found between the rivers and the mountains, stems are as much as 2–3 inches; their height may be up to 25–30 feet. This grass may branch but rarely forms fruits in the territory of Kentucky, Tennessee, or the Carolinas. This grass is found beginning in the southern coastal region of Virginia. As one goes south into the Carolinas, Floridas, and southern Louisiana, we find this reed in abundance. It snowed all night, and the next morning my two horses that had been hitched had swollen legs because of the cold and the muddy roads by which I had traveled the last few days.

MARCH 1, I arrived at Fort Blount, found on the Cumberland River.[52] Snow continued to fall part of the day.

MARCH 2, I stayed in the area to dig up young plants of a new *Sophora* that I had noticed near Flynn's Creek at about 12 miles from the fort.[53] Snow covered the ground, and I could not dig up young plants, but Capt. William Jr.,[54] who resided at the fort, cut some trees, and I found several good seeds. I also dug up some roots of this tree so that I could replant them in my garden in Carolina.

The same day, I had the opportunity to write to Governor Blount.[55]

MARCH 3, I continued my travel, went across Flynn's Creek several times. I saw again the little bulbous umbellifer that I had noticed several days before.[56] Toward evening, the road became less muddy.

MARCH 4, I reached the mountains known as Cumberland.

MARCH 5, I went by several creeks and rivers on which grows in abundance a climbing fern belonging to the genus . . .

The land crossed by these streams is less fertile than that around Nashville, known as Cumberland settlement, and pines with two needles are abundant here.

SUNDAY, MARCH 6, I arrived at West Point on the Clinch River.[57]

MARCH 7, I slept 15 miles farther, near the junction of the Holston River and the Tennessee River.

MARCH 8, I reached Knoxville.

MARCH 9, I dined at Gov. Blount's house.[58]

MARCH 10, I took lodging at Capt. Looney's,[59] near the Cumberland River [Holston].[60]

MARCH 11, I botanized on the opposite bank, bordered with steep rocky cliffs, covered by saxifrage,[61] bulbous umbellifer.

MARCH 12, I continued botanizing.

SUNDAY, MARCH 13, I visited Capt. Rickard, commandant of the garrison.[62]

MARCH 14, I botanized, saw in flower *Anemone hepatica*, *Claytonia virginica*, *Sanguinaria*. I saw a new genus of plant designated by Linnaeus, *Podophyllum diphyllum*, and [which I] discovered several years ago in Virginia, while passing by Fort Chiswell. This plant is less rare in the fertile areas of Kentucky and Cumberland. It can be found around Knoxville. Dr. Barton gave it the name of *Jeffersonia* in a description of this plant, after having seen the flower of the plants that I had brought back to the botanist Bartram in Philadelphia. The time of flowering around Knoxville is around March 10.[63]

MARCH 15, I received a letter from Governor Blount in response to the one I had sent him on the discovery of the new *Sophora* near Fort Blount.[64] I left the same day and spent the night 7 miles farther. I paid 2 shillings 3 pence for supper and corn and forage for the horses, and 2 pence for a bundle of fodder.

MARCH 16, I spent the night a mile from Jefferson City [Iron Works] at the home of Mr. Rice,[65] a lawyer, 30 miles from Knoxville. I noticed flowers on *Ulmus viscoa*, *Acer rubrum* male flowers on one individual and female flowers on another tree.

MARCH 17, I slept near Bull's Gap, 30 miles from Iron Works.

MARCH 18, I went by Lick Creek and Greeneville, 18 miles from Bull's Gap.[66]

FIGURE 27. André Michaux imagined at the Tipton Farm. Artist Jenny Noseworthy drew this image of Michaux exploring on Colonel Tipton's farm in present-day Johnson City, Tennessee. Notice both her depiction of a medium-sized man with powerful calf muscles and her over-the-shoulder view that leaves his facial features to the imagination. (Courtesy of the Tipton-Haynes Tennessee State Historic Site)

MARCH 19, I passed Jonesborough, 25 miles from Greeneville. There are several merchants established in Jonesborough who obtain their merchandise overland from Philadelphia.

SUNDAY, MARCH 20, I left Jonesborough. I saw in passing Mr. Overton from Kentucky; Maj. Carter from Watauga, at whose home I had stayed several years before with my son; and Col. Avery.[67]

I noticed *Corylus americana* in flower, the female flowers have purplish styles or stigmas, *Ulmus* viscosa twin golden flowers with 4–5 or 6 stamens and purple stigmas. I slept at [the home of] Col. Tipton, 10 miles from Jonesborough.[68]

MARCH 21, I saw in several areas mountains covered with:

*Sanguinaria*, *Claytonia*, and *Erythronium* with spotted leaves. These plants were in bloom. *Magnolia acuminata* and [*M.*] *auriculata*, *Rhododendron*,

*Kalmia*, *Pinus abies canadensis*, *P. strobus*, *Azalea*, etc. etc. are in abundance at the foot of the mountains. I reached Limestone Cove and spent the night at the home of Ch. Collier,[69] 18 miles from Col. Tipton.

MARCH 22, I crossed Iron Mountains and arrived at night at the home of David Becker, 23 miles without seeing any houses.[70]

MARCH 23, left the Beckers' home on Cane Creek; to Rider, 6 miles; from Rider to the Widow Nigh, 7 miles; from Nigh to Sam Ramsey,[71] 2 miles; from Ramsey to David Cox on Pepper Creek, 4 miles; from Cox to Young, one mile; from Sam Ramsey to Davinport, 8 miles. Total 23 miles. I lodged at Davinport's.[72] I saw *Salix capreoides* in flower on the banks of the creeks.

MARCH 24, I visited the high mountains opposite Davinport's home. I dug up several hundred plants: *Azalea lutea*, *A. fulva*, *Anonymos azaleoides*,[73] *Rhododendron minus*, etc. etc.

MARCH 25, I saw *Corylus cornuta* in flower, male catkins in pairs or as often solitary, the scales ciliate at apex, anthers ciliate at apex, female flowers with reddish styles. This species blooms about two weeks later than *C. americana*, which is found in all the climates of North America, even in South Carolina near Charleston. *Corylus cornuta* is found only on the highest mountains and in Canada. *Corylus americana* male catkins solitary, outer surface of scales tomentose, margin glabrous, female flowers with reddish styles.

MARCH 26, I botanized and pulled cuttings of shrubs and young plants for transport to the garden of the republic in Carolina.

SUNDAY, MARCH 27, . . .

MARCH 28, I prepared and packaged my collections of fresh plants from the mountains.

MARCH 29, I left Davinport and spent the night at Young's.[74] Violet with reniform, dentate leaves, hairy petiole, flowers yellow, in full bloom on the banks of the creeks and cool areas.

MARCH 30, I continued on my way, but by mistake I took the road to the right, which leads to Wilkes.[75] Another *Viola* [violet] yellow, scapose, leaves hastate, in flower in cool but less moist areas than the preceding one and also flowering a little later.

MARCH 31, I arrived at [the home of] Col. Avery and spent the night in Morganton, called Burke Courthouse.[76]

FIGURE 28. *Magnolia macrophylla* in flower. At a site near Stanley in Gaston
County, North Carolina, Michaux collected *Magnolia macrophylla* seedlings
that were later planted in Empress Josephine's garden at Malmaison. The
collection site is now part of the Jack C. Moore Preserve protected by the
Catawba Lands Conservancy. (Drawing courtesy of Frank J. McLaurin)

FRIDAY, APRIL 1, I left Morganton and spent the night at Robertson, formerly the home of Henry Waggner, 30 miles from Morganton.[77]

APRIL 2, *Epigea repens* in full bloom as on preceding days. Several individual
plants had female flowers without vestigial stamens, and in other plants, all
the flowers were hermaphroditic. I arrived at noon at the home of Christian
Reinhart near Lincoln.[78] I stayed all day to pull cuttings of *Spiraea tomentosa*,
which grows in swampy regions.

SUNDAY, APRIL 3, I arrived at Bennet Smith's place, 12 miles from Lincoln;
stayed there all day to dig up young plants of a new *Magnolia* with very large
leaves auriculate, oblong, glaucous, silky, especially the young leaves and
the buds, flowers with white petals purple at the base, stamens yellow, etc.[79]
Along the creek on the banks of which the *Magnolia* grew, I also saw *Kalmia
latifolia*, *Viola* [violet] yellow with hastate leaves, *Ulmus viscosa* in fruit, *Halesia*, *Stewartia pentagyna*.

APRIL 4, I left and crossed Tuckasegee Ford on the Catawba River, 10 miles
from Bennet Smith. I took the left road instead of going through Charlotte
and slept 11 miles from the Catawba River.

   Note: Before crossing the ford, I had breakfasted at . . . Alexander, a very
respectable gentleman who received me with much courtesy.[80]

April 5, 1796, 12 miles farther; I rejoined the road that goes from Camden to Charlotte.

Note: If one doesn't want to go through Charlotte when going to Lincoln, it is necessary, 12 or 15 miles before arriving, to inquire about the road that goes to the left, to go across the Tuckasegee Ford.

I took young plants of *Calamus aromaticus*, which are found in damp areas near Charlotte and Lincoln, *Rhus pumilum*.[81]

I slept close to Waxhaw Creek in South Carolina, about 35 miles from Tuckasege Ford.

April 6, at the home of Col. Crawford,[82] near Waxhaw Creek, I saw a plant without a name, leaves 4 at a node, perfoliate, entire, and glabrous. This same plant is found in settlements in Cumberland and Kentucky, *Frasera foetida*.

I went by Hanging Rock; it is 22 miles from Waxhaw to Hanging Rock. To go to Morganton, known as Burke Courthouse, one should not go through Charlotte but take the road on the left 3.5 miles from Hanging Rock: about 200 feet from the junction of the two roads, one of which leads to Charlotte, one finds an unnamed shrub with red root that has the appearance of *Calycanthus*. This shrub is the one that I saw near Morganton. I slept near Hanging Rock.

Thursday, April 7, 1796, I arrived in Camden. Five or six miles before arriving, I pulled up young plants from a new *Kalmia* that I had seen several years before. It is 26 miles between Hanging Rock and Camden.

April 8, left Camden and went through Stateburg, 22 miles from Camden; and [I] slept in Manchester, 30 miles from Camden.[83]

April 9, my horses strayed during the night after breaking down the fence that enclosed them.

In the streams, there is *Callitriche americana*: simple fruit, axillary, sessile, calyx with two segments, stamen 1, long filament, pistil laterally flattened, 2 parted?, styles two, as long as stamens, stigmas acute.

*Silene* . . . calyx 5-parted, cylindrical, corolla with 5 petals, divided to the base into 5 parts, narrow claws, flat, laciniate, lobes obtuse, stamens 10, inserted at base of corolla, pistil oblong, styles 3, stigmas acute, capsule unilocular, very many seeds, flowers pinkish.

I left in the afternoon and slept 15 miles away, having gone through 10 miles of sands known as the High Santee Hills, in whose area I noted *Phlox*, *Silene* . . . , *Dianthus* in bloom, *Lupinus perennis*, and [*L.*] *pilosus* in flower.

Sunday, April 10, I arrived at a place on the Santee River known as Manigault Ferry; I noted before arriving *Verbena* (*Aubletia*?) and, on the banks of the Santee, an unnamed tree whose muricate fruits, which were almost mature, were covered by soft spines. The Manigault Ferry is 28 miles from Manchester.[84]

Two miles farther, one takes the road to the right, called Gaillard Road, shorter than the ordinary road but muddy during the winter.[85] I slept at the Widow Stuard's, 18 miles from the Manigault Ferry, a dirty tavern without forage for the horses.

April 11, I left early in the morning, 5 miles farther I noted *Lupinus perennis* and *L. pilosus* in flower; distance from Charleston is 40–43 miles. I reached the garden of the republic, 37 miles from the Widow Stuard; that is, 47 miles from Charleston.

Summary of the road from Illinois to Charleston:

| | |
|---|---|
| From Saint Louis of the Illinois to Cahokia | 4 miles |
| To the village of St. Philippe | 45 |
| To Prairie du Rocher | 9 |
| To Kaskaskia | 45 |
| To the junction of the Mississippi and Belle [Ohio] Rivers | 95 |
| To Fort Massac | 45 |
| To the junction of Cumberland on the Ohio River [from Fort Massac] | 18 |
| To Clarksville on the Red River | 120 |
| To Nashville | 60 |
| To Bledsoe's Lick | 30 |
| To Fort Blount on the Cumberland River | 40 |
| To West Point on the Clinch River | 90 |
| To Knoxville on the Holston River | 40 |
| From Knoxville to Iron Work [Jefferson City] | 30 |

| | |
|---|---|
| To Bull's Gap | 30 |
| To Greeneville | 25 |
| To Jonesborough | 25 |
| To Col. Tipton [Johnson City] | 10 |
| To Limestone Cove | 18 |
| From David Baker beyond Iron Mountain | 23 |
| From Baker to Young | 20 |
| To Morganton, known as Burke | 22 |
| To Robinson | 30 |
| To Lincoln | 16 |
| To Tuckasege | 22 |
| To Waxhaw Creek | 35 |
| To Hanging Rock | 22 |
| To Camden | 26 |
| To Manchester | 30 |
| To Manigault Ferry | 28 |
| To Charleston | 70 |
| Total | 1123 miles |
| Leagues | 374⅓ |

# CHAPTER FOURTEEN

# Charleston, Spring and Summer 1796

SUMMARY: AFTER REACHING CHARLESTON IN April, Michaux did not resume his journal until he set sail for France in August. With the exception of two letters to André Thouin, our only information about this period when Michaux made his decision to leave the United States comes from letters and documents written by others.

Michaux, who had not been able to receive any letters from France during his journey, immediately responded to a letter from his colleague Thouin at the Jardin des Plantes. Michaux's response includes a list of plants gathered on his recent journey west that he did not mention in his journal.

### AM in Charleston to André Thouin in Paris, April 13, 1796 (excerpt)[1]

I arrived, dear Citizen, yesterday from a voyage that lasted almost one year, as I left last year on April 19. During the course of this trip, I visited the highest mountains of the Alleghenies in the spring. . . . I had traveled through them during other seasons in earlier years. . . . I visited the new establishments of Cumberland and of . . . ,[2] emancipated since last March 1 under the name of [the] state of Tennessee, which adds a sixteenth state to the United States. Several weeks later I went to Kentucky, and I arrived in July at Vincennes [Indiana]. In early August I botanized on the banks of the Mississippi in the country of the Illinois. I went up to the Illinois River [and] to the French establishments and the junction of the Missouri.[3] Returning to [Kaskaskia] Illinois, I went by land to [Fort Massac on the Ohio] then to the junction of the Tennessee (French maps: Cheroquis [Cherokee] River) and then on to the Cumberland River (French maps: Shavanon [Shawnee] River). I went up this one for 180 miles. Returning to Illinois, after gathering [of] seeds at the end of autumn and some collecting of birds: I left December 14 of last year and finally arrived here yesterday, April 12.

The new consul, Citizen Dupont,[4] gave me your interesting letter dated 14 fructidor year 3 of the republic (August 31, 1795). I experienced much satisfaction as, since the revolution, I wanted a plan of instruction. It was not that I was doubtful about the type of work or the sites that I visited, but I congratulated myself that my ideas corresponded with the intentions of the administration, and it is an inexplicable satisfaction while working ardently for one's country, while escaping dangers and obtaining success, that the other citizens be aware of this. I am . . . an old soldier whose greatest reward is to have the approval of fellow citizens. During the course of this last voyage, I met with several Indian Nations; there is no more danger than visiting more civilized people, but it depends much on the attitude and the manner in which one behaves with them. I could give political information on the Nations of North America from the high points of the Missouri and the Mississippi, on the French people from these areas. For several years, I have been occupied in collecting all the information, not only on plants but also on quadrupeds, on the native country of migrating birds, on their migratory transits, on lake fishes, [and] on minerals and all the fossils of these distant lands while waiting until I can go there myself.

I come back to your letter as I have no time to lose; I must be in Charleston tomorrow at noon to give it to Capt. . . . who leaves for Bordeaux. I must also make haste to make a small box of seeds from my last trip. I will make a duplicate for you of this shipment, which will be considerable if I have the time to prepare it. I begin to answer all the points of your last letter.

In the shipment of 1792 there was a grass *Uniola montana*; it is now abundant in the garden and reseeds every year. There is also another one that I name [*U.*] *maritima*; it is the wild oat of Catesby.

In the shipment of March 3, 1792, you are disappointed with . . . *Vaccinium arboreum*, [*V.*] *resinosum*, *Styrax tomentosum*, *Stillingia herbacea*, *Aletris aurea*. If you did not want these plants, I would not have dared to send them, as Abbé Nolin received them by the thousands, practically all the plants and shrubs from South Carolina. At the moment, I cannot satisfy [your request]

for plants from South Carolina and Virginia because my voyage took me beyond the mountains to the west of the United States and on the banks of the
rivers of the interior of this country. You will also have the indulgence to be
persuaded that when one travels more than 400 leagues [1,200 miles] from
Charleston or Philadelphia without being able to send or receive one single
letter, correspondence must be often interrupted. There is no post office in the
country where I traveled; settlements are spaced out by four, six, or even eight
days' travel through forests, sometimes interrupted by the overflow of waters,
obliged to make rafts to cross rivers and make the horses swim across; all
these circumstances make communication difficult and scarce. The spaces between settlements is called "Wilderness," for example those between Holston
and Cumberland, between Cumberland and Kentucky, between Kentucky
and Vincennes, between this station and the Illinois, and [between] Illinois
and Prairie du Chien,[5] where there are only two families and half-breed
wood hunters, much worse than the true Indians. It may be that there is mut-
tering that I am not sending to some individuals some seeds of dried plants,
some birds, some insects, but when my herbarium is deposited and closed
up for one year or more, I cannot upon my return open the cases and upset
everything to send about fifty plants to pick out among ten thousand; my first
care is rather to renew the interior with other fragrant resins and to paint the
outer surfaces of the cases in order to conserve my earlier precious harvest.
When I am three or six months at home, apart from the work of the nursery,
the work on my herbarium to classify new acquisitions, the writing of the
descriptions that must accompany them, the shipment if there are some to
be made, etc. etc., all this takes up my time night and day. [I do] as much as
possible. In addition, the dangers that one runs from English pirates do not
tempt me to send objects that are not available in duplicate.

To return to your letters, the shipment of 4 messidor [June 16, 1794?] was
in duplicate; it seems to me that you only received one. I replace in this shipment several seeds that I sent [to you] earlier and that you asked for again.

The summary of instructions that you gave me regarding objects to enrich
the Muséum again gives me great satisfaction because I can do much more
than many others with regard to living animals. When the opportunity
presents itself, I will send you a list of quadrupeds of North America, some
unknown.

I stop to work on the shipment; it will be small but you will be more
pleased by it than [by] others that were larger; in any case time does not allow
me to send you all that I brought back.

Michaux added a two-paragraph postscript to this letter.

> Dombey, I believe died in Montserrat.[6] Rouelle lives someplace in Virginia with his wife and children.[7] Bauvois is in Charleston for several weeks, but I have not seen him yet.[8] Nectoux came last year to the United States.[9]
>
> I cannot write this time to my son. Tell him please that I have not forgotten his account of saltpeter and that in . . . I will compare the methods used by the . . . to refine saltpeter, and I will study the different types of wood used for making charcoal, etc. etc. There is no ship for Le Havre, and I am sorry not to be able to send him some rice.

Michaux ended his letter to Thouin with an annotated list of thirty-nine species of seeds that he was sending with the letter. Many of these species are not listed in his travel narrative.

> List of seeds from Kentucky with several remarks:
> *Acer saccharinum*; this species differs from the sugar maple that is used in the central part of Canada, near Montreal and Quebec. It is not found in the United States east of the Alleghenies, but it is abundant in the new areas called the (Western territories) country of the West; perhaps it has never been seen in Europe.
> *Fraxinus*, branches quadrangular with a membrane found in the angles.[10]
> *Guilandina dioica*; the pod of this tree contains a sugary pulp but with bitter aftertaste; I did not see the flower.[11]
> *Panax*, 5 leaflets, [from] shady and fertile sites in Kentucky.[12]
> *Panax trifoliatum*, sandy sites, low, moist and shaded, Pennsylvania.[13]
> *Hypericum*, new species of subshrub, calcareous and dry sites.[14]
> *Silphium*, leaves very large, a new species?[15]
> *Phryma leptostachia*, cool and damp sites.[16]
> *Tetradynamia* [cabbage family], perennial on dry rocks near the river, Kentucky.
> *Dracocephalum virginiana*, wet areas.[17]
> *Podophyllum peltatum*, rich shaded soils.
> *Cinna arundinacea*, perennial grass from Canada to the Carolinas.
> Grass . . . monoecious, bulbous, buffalo grass, Kentucky.[18]
> *Sophora*, blue, cool sites, submerged during winter, perennial, springs and banks of the Ohio.[19]
> *Campanula*, many flowered, not campanulate.[20]

*Serratula*, plumose pappus.[21]

*Serratula?* new genus, phyllaries imbricate, outer ones largest, plant showy.

*Lobelia siphilitica.*

*Cuphea viscososa*, annual plant.

*Tragia volubilis.*[22]

*Ziziphora*, new species.

*Leontice thalictroides*, humid fertile sites.[23]

*Illecebrum divaricatum*, plant annual, erect.[24]

*Rhinanthus*, 4-foot-tall plant similar to . . .[25]

*Scrophularia*, new, 3 leaves, opposite, ternate, pinnatifid, moist areas.

*Viburnum*, leaves dentate, fruit compressed, new species.[26]

*Iresine celesioides*, dioecious plant; this plant must be grown in the orangerie. It grows only in cold climates, where the sites are underwater during the winter.

*Polymnia uvedalia.*[27]

*Polymnia canadensis*, ray florets very small, plant rare in America, shaded and fertile sites.

*Annona triloba.*

*Nepeta altissima.*[28]

*Rhus toxicodendron*, may be the *trifolium* of l'Héritier. I have not seen the flower.

*Aster grandiflorus.*[29]

*Urtica divaricata*, smooth; and other species.[30]

*Evonimus* (*latifolius?*), flowers 4-merous, capsules glabrous, leaves petiolate.

*Evonimus americanus*, flowers 5-merous, capsules warty, leaves subsessile.

*Sarothra gentianoides*, plant annual.[31]

*Mimosa mississipi*, perennial, erect, spineless, given by Bartram.[32]

*Crepis* . . . new species.

On March 31, 1796, two Frenchmen arrived in Charleston from Philadelphia by ship. One was François-Alexandre-Frédéric, duc de la Rochefoucauld-Liancourt, a refugee of the French Revolution, who was to write an eight-volume travelogue of North America. The other was the botanist Ambroise Marie François Joseph Palisot de Beauvois,[33] whom Michaux had met in Philadelphia in May 1792, and who had come to inspect the "French Garden" on behalf of Consul General Létombe of Philadelphia.[34] Palisot de Beauvois subsequently wrote a report that was sent to France. Michaux said that he did not meet with Palisot de Beauvois at that time; in fact, he wrote that Palisot de

Beauvois avoided him, and he learned of this report only after he returned to France. There is little doubt that la Rochefoucauld-Liancourt met Michaux at the garden;[35] however, given Michaux's later statement, there is some doubt whether Palisot de Beauvois met face to face with Michaux either at the garden or in Charleston, although Palisot de Beauvois indicated in a letter to Caspar Wistar in Philadelphia that he had met Michaux in Charleston.[36]

### Report on the garden of the republic found two miles from Charleston and under the direction of Citizen Michaux, botanist, by Palisot de Beauvois (undated, but likely 1796)[37]

To answer perfectly the wishes of Citizen Létombe, consul general, I think I must divide the report into two parts.

In the first part I will deal with the usefulness of such a garden and to what extent such an establishment is even necessary; in the second [part] I will discuss this garden specifically.

Natural history encompasses the study and knowledge of all the productions found on the globe. It is the source without interruption from which man can obtain abundantly all his needs. It thus should be protected and encouraged by the government after the incontestable principle recognized by France, whose branch of administration has always been the object of its care; it is unnecessary to expand on the usefulness [of a place] where are assembled, grown, and cultivated all the trees, shrubs, and plants of this country especially those that can be naturalized in France and that can be useful to science and to the arts.

The garden established for the benefit of the republic in the neighborhood of Charleston fulfills a portion of this purpose, and I do not doubt that soon the territory of France will be enriched because of the care, zeal, and hard work of C[itizen] Michaux with a large number of plants that, joined to the ones it already possesses naturally, can only contribute to the splendor of the state, to the well-being and the general advantage of its inhabitants.

But this establishment as it exists today is very imperfect. One cannot cultivate the native plants from the cold climate of which many are useful. I am convinced that the trees that are common both in the north and in the south of America, like among others, the large number of species of oak, those that come from the south would not be as successful in France as those from the north. When C. Michaux planted items from the north in the garden of which he is in charge, despite careful daily tending, he was not able to do it; the seeds do not grow, or [they] perish soon after germination. The young plants are weak. If we compare the Lord Weymouth pine, a beautiful tree up north, [it] is here only 3–4 feet high and is already 40 or 50 years old, and it is already crowned as maritime pine that is planted in the wrong place.

It is evident that this establishment in the south, despite zeal and unquestionable usefulness, is insufficient; [an] insufficiency that can easily remedied with few expenses in forming in the neighborhood of Philadelphia, under the eyes of the minister, a second establishment, not only for growing northern plants, but for trying to grow plants from the south, which when successful could be used in the temperate regions of France. Under this scenario France would not take long in naturalizing southern plants and trees that would enrich the southern areas as well as the northern ones of temperate climate.

One could give to these establishments a greater utility not only by making them a depot for plants but [by making them a depot] for all the other branches, no less useful, of natural history.

Botany is without a doubt an essential part of natural history. It furnishes man his nourishment and clothing; it is in the plant kingdom that we get shelter from bad weather; it's with the aid of plants that we are able to move from pole to pole, even in the most isolated areas . . . but the vegetable production is not the only [factor] to enhance the life of man with usefulness and pleasure. The animal kingdom is no less worthy of the government's attention, if [it] furnishes quadrupeds so useful for their furs, birds equally important for their plumage, and both of which along with fishes are equally precious for their flesh, and the latter also yields oil; there remain insects among which the industrious bee, the timid and rich silkworm, the brilliant and colorful cochineal. Likewise in the mineral kingdom, [of] whose importance one does not have to give details. This area is even more precious in this country as it is unknown and sought after by French naturalists, especially C[itizen] Daubenton.[38]

One could thus easily, without much additional expense, which would consist only in the appointment of two directors and the maintenance of the garden, make [the garden] a depot of all natural history objects, which would be shipped to France where they would soon become naturalized. Thus soon we would see the opossum, curious by its nature, useful because of its fur, which is used for making hats, and its flesh, which is very good eating; the raccoon (*Castor* Lin),[39] equally precious because of its fur; the groundchuck [woodchuck?]; the polecat [skunk or related member of that family]; all the different species of squirrels and foxes as well as a large number of domesticated birds and others.

It might be objected that most of these animals are destructive; the destruction that they can cause is small compared to the real and consequent usefulness of acquiring these animals.

I don't think it is necessary to say more to prove 1) the great usefulness of the establishment entrusted to C. Michaux, [and] 2) the necessity of establishing a similar one in the north, supporting each other, as an extension of the garden of which I spoke.[40]

As to the garden under the guidance of C. Michaux, it is according to its report about 100 acres. The soil is, as is most of that in S[outh] Carolina and G[eorgia] close to the sea, pure sand; nevertheless, it has pieces of shells and plant material mixed with it. The largest part is cleared and well kept up. C. Michaux gathered there a large quantity of plants that appeared in good health and of interest to be sent to France. All those that he brought back from his different trips are grown with care; some come from nearby areas; others are young plants that he brought and others that he dug up a long distance away, which were difficult to bring in good health, [and] which he [is] able to multiply by skillful manipulations such as grafting on other plants that are related. I noted, among others, a pretty grouping of *Andromeda*, *Vaccinium*, *Magnolia*, etc., and many other plants and trees that are new and interesting.

I expected to find there numerous nurseries of each species and principally different species of oak, beech, ash, [and] wild cherry, whose wood is useful for different arts, but C. Michaux told me that he had shipped them all to France.[41] However, I think that in order to fill the demand of the republic it is not a question of only sending several individual [specimens], which for one only become objects of pure curiosity, but to the contrary [sending] numerous shipments of each object, in order that they might spread out and multiply in various parts of France, [and so that] they would increase the number of useful productions.

In addition the garden appeared to be in good order; it could not be in better hands than those of the zealous and hardworking C. Michaux.

C. Michaux was obliged to be absent for his research; it seems to me that it would be appropriate if the consul had an exact number 1) of each type of plant that is cultivated in the garden, with notes of the site and the soil quality where it can be found in the garden, 2) of the different shipments made to France, [and] 3) of the plants that are there and generally an exact statement of the situation in the garden, where double the number would be sent every year with increases and changes given to the consul general, so that in case of absence of C. Michaux, the minister and the consul would know what has been sent [and] that which is likely to be shipped and make a judgment of the daily improvement of such a useful establishment. Beauvois

Two years later Michaux wrote a revealing letter about this episode to his close friend Bosc who was then in Charleston.

### AM in Paris to Bosc in Charleston, August 22, 1798 (5 fructidor year 6) (excerpt)[42]

A little before leaving Charleston for France, I found out that de Bauvois had made a visit to the garden with Consul DuPont during my absence. As people were reluctant to tell me the reason for this visit, I did not refuse to see him, but he avoided me, and I left without guessing the cause [of this avoidance]. On my arrival in Paris, the minister of the interior sent me a report that had been addressed to him by Bauvois,[43] who had made a trip to Charleston to answer the wishes of Citizen Létombe [consul general of France]. And finally while praising my work and my care, etc., he said that the objectives were wanting if the establishment remained in the Carolinas. That it should be transferred to Philadelphia and put under his jurisdiction to be expanded by the addition of other objects of natural history in which living animals would thrive. This first demand was not granted. A second petition signed by Rouelle in the names of Rouelle and Bauvois arrived; they asked that the establishment be transferred to Rouelle's manor in Virginia, with the sum of 12,000 pounds to be shared by Rouelle and Bauvois. Rouelle offered his property to the government. He said that the property in Carolina consumed 12,000 pounds from the government. I answered that the establishment only costs 3,000 pounds. I cited my accounts submitted six months before to the minister of the interior, and this sum is taken from my own resources since the revolution. That between the interval when my mission stopped, I sustained it with my own funds, etc. etc.; the Minister was very satisfied. But a third memorandum to the professors at the Muséum was strongly supported by de Jussieu in favor of Bauvois, his protégé, whom he had just eliminated from the list of refugees.[44] Rouelle is protected by the Thouin family.[45] Finally apprehension that the establishment would fall into bad hands led me to write a report to Citizen la Révellière.[46] Justice is rendered to me. It was given to me by most of the members of the institute; by the minister of the interior; by Lamarck, one of the professors of the Muséum whom I judge to be a true Republican. While his character is a little forbidding, he is nevertheless worthy; those that love their country here are so rare.[47] Jussieu, who had the most knowledge of my work, whom I honored by giving him all my plants from North America,[48] reproached me for not having sent any or [for having sent only a] few plants to the Muséum [but having] sent some [specimens and plants] to others such as Bosc, but I answered completely regarding these objections. Now these people appear to be my great friends, and I am at peace in this regard. I presume that La Révellière thinks well of me; I gave him several useful reports. He arranged for me to receive 4,000 F [francs], which I asked for on my advances and old stipends, but this sum is used up because I had only that to live on . . . a sum of 72,000 pounds that I had when I began in the sciences was spent partly on voyages, partly on paper money that was reimbursed in silver that I had invested, and the remainder is due to me. Such as it is, if I have enough to publish my work, I will be satisfied. I will ask for another installment, and after that, if no justice is rendered, I will be able to cope by my activity and the experience that I have obtained.

Michaux shipped a collection of plants and seeds to Thouin in June, and the letter accompanying this shipment contains valuable insights about his future plans.

### AM in Charleston to André Thouin in Paris, June 8, 1796 (excerpt)[49]

There is an important subject on which I would like to explain my viewpoint and to have your opinion. It is concerning my return: you must be aware that the size of my collection must be considerable. Despite the fact that there are descriptions that accompany the new plants, there is an infinite number of interesting facts to add, and it would be high time to put [them] in my collection in exact order, to edit [them], and to have them published. I see that several plants that I knew about before [other authors did] have been published in English texts.[50] The garden in the Carolinas contains many new genera and species. During my absence someone steals that which I want to communicate to my country before all others. The thing that is most disagreeable is that it is this vain nation, proud and treacherous, that is able to harm me the most in this regard because of the ease of communication between Charleston and England.

All the educated Americans who live on the borders of the Wilderness and who are more aware of its dangers are surprised and reproach me for wanting to begin new travels before publishing what I have already acquired. After all

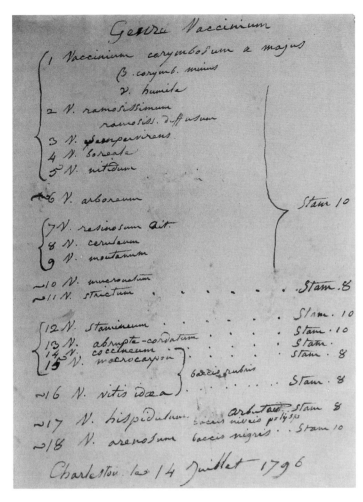

FIGURE 29. *Vaccinium* study found on loose sheet from Michaux's journal
(Courtesy of the American Philosophical Society, Philadelphia, PA)

these considerations, I propose to interrupt for some time the voyages that I
planned to undertake west of the Mississippi. If I can get my collection safely
to France, I could in 8 or 10 months put my collections in order for publication and be back at the Mississippi before this space of time. I must return
because I left many objects and things necessary for the great journeys in the
interior of America. I have, by the way, much information to tell you on all
the subjects of your letter concerning collections of animals and minerals. I
must finish this letter because I fear missing the boat's departure. In a few
weeks, I will go to Philadelphia to discuss with the minister the way to send
my collections to France and to go myself, or I will go down the Ohio again
to continue my travels to the most western regions.

We can be confident that Michaux continued his botanical studies because of a one-page document on the genus *Vaccinium* dated July 14, 1796, found among his papers. The microfiche of his herbarium contains forty-five specimens that he had placed in this genus, and this document reveals that he made a study of these specimens in summer 1796 and attempted to classify the *Vaccinium* specimens according to several morphological traits.

During summer 1796 Michaux made the decision to return to France with his collections of plants and seeds and with his herbarium. The consul in Charleston, Victor DuPont, wrote to Pierre Auguste Adet, the French minister in Philadelphia, on July 30, 1796, that Michaux had rejected his advice to travel to Philadelphia and find passage to France from there, but was leaving immediately for France by way of Amsterdam. "He [Michaux] has decided on this voyage to France as promptly as he prepared for those to Hudson Bay or to Tennessee, and he perseveres with the same zeal and eagerness."[51]

Because mail traveled so slowly, Michaux did not know of the renewed interest of officials in France toward his American mission. On July 12, the minister of the interior, Pierre Bénézech, had written a letter commissioning Michaux to send American seeds and plants for the national nurseries.[52] One week before Bénézech wrote directly to Michaux, the minister of external affairs had written a letter to Adet,[53] inquiring of that official what he knew concerning Michaux's fate. His letter was answered promptly.

**Pierre Auguste Adet, minister plenipotentiary of France in the United States, Philadelphia, to Charles Delacroix, minister of external affairs, in Paris, 7 brumaire year 5 (October 28, 1796) (excerpt)[54]**

I received yesterday the letter that you did the honor of writing me the 17 messidor year 4 (July 5, 1796) regarding Citizen Michaux, naturalist sent by the government to the United States.

As soon as I arrived in the United States, I attempted to obtain news of the fate of this man, precious to science. I learned that he was deep in the territory west of Virginia and that he would finish his voyage by coming back through the far end of the Carolinas. I charged the consul in Charleston to ascertain his fate and to give him help in the pursuit of his research.

Having learned that the republic had a botanical garden in Charleston established due to the care of Citizen Michaux, I ordered this same consul to have it taken care of and cultivated and to assemble there the plants that would be sent there by said naturalist or by another naturalist of which I will speak further.

I was waiting with some anxiety [for] news of Citizen Michaux when, the previous 12 thermidor (July 30, 1796), the consul wrote that he returned with a large collection of plants and animals, fruits of 12 years of hard work and research.

Virtually overnight he decided to leave for France via Amsterdam with this rich harvest after he had given the consul the title to the property of the botanic garden.

Unfortunately Citizen Michaux, like all other artists, is firm in his ideas and is more conscious of glory than of money and did not take the time to clear up his monetary demands, and I don't know if he had enough for his trip, but unfortunately he did not leave behind any duplicates that he might have had and which would be of inestimable value, should his transit not be a happy one.[55]

The other naturalist of whom I want to speak is Citizen Beauvois, associate of the Institut National who has already by the order of the government [made] a trip along the African coast. He has just gone toward the vast areas beyond Georgia, and I received a letter from him recently that he had gone deeper in Indian land than Michaux had, his course being more to the south; he promises results that are no less rich and more varied. He began this journey with the mediocre sum of $600,[56] under the condition that he would keep accounts in case the government did not approve of this disbursement.

# CHAPTER FIFTEEN

## Return Voyage to France and Shipwreck, August 1796–January 1797

SUMMARY: NOTEBOOK 10, MICHAUX'S LAST, covers only about four months of Michaux's travels and adventures and takes him from Charleston to Paris. He returned to a France far, far different from the country that he had left in 1785. Cataclysmic political, social, and economic upheaval had not only swept away the monarchy, it had removed from power and influence the group who had directed and supported his American mission. Some had lost their lives, some had lost their their fortunes, and others had fled the country. Although he had attempted to remain in contact with his son and some of his colleagues, Michaux may not have formed a realistic understanding of how the changes in his homeland would affect him financially. We can be certain that he hoped for a warm welcome in France for his years of dangerous and difficult work in America and that he expected support to resume his botanical explorations, including a journey beyond the Mississippi.

The loss of patronage combined with an especially dramatic decline in US–French relations would halt Michaux's plans to return to America. Instead, he would remain in Paris in reduced financial circumstances, attempting to collect his unpaid salary, organizing his collections, and preparing the manuscripts of his books. When he did finally leave France on another expedition, it would be not to America but to the South Seas.

His return journey to France was marked by disaster; the ship wrecked on the Dutch coast in a storm. Although the shipwreck was traumatic and life-threatening, Michaux was not seriously injured, and he regained his focus and energy very quickly. His personal possessions were lost, but through his amazing resilience and hard work, Michaux was able to save most of his precious cargo.

The Batavian Republic (now the Netherlands), where he landed, had been proclaimed in January 1795 and was founded with the armed assistance of the French Republic; it was considered France's "Sister Republic." At this time, the Belgian states had been annexed by France. This situation undoubtedly explains the ease with which Michaux was able to make his way through these countries and ship his large collection back to Paris.

Although he does not mention it, Michaux undoubtedly brought back to France an eleven- or twelve-year-old black boy who would eventually accompany him to the South Seas. The child was not technically enslaved (slavery had been abolished in France); Michaux appears to have acted more in the role of a surrogate parent.

François André Michaux, who lived in Paris, did not travel to the Netherlands to assist his father, whom he had not seen in six years, but he must have been there to greet him and offer him a place to live when he arrived in the French capital. André Michaux wrote to Interior Minister Bénézech on December 25, 1796, only two days after his arrival in Paris, and gave his son's address as his own.

André Michaux had wanted to return to America very soon to continue his exploration in lands west of the Mississippi. Unfortunately, the deteriorating relations between the two countries prevented his return. US negotiation and ratification of the treaty known as "Jay's Treaty" with Great Britain was viewed as a betrayal in France. The atmosphere was no doubt made more poisonous by the American minister to France, James Monroe, who misread events at home and assured the French that the treaty would be rejected. Smarting from what it considered a double betrayal, the French government angrily refused to receive Monroe's successor, Charles Cotesworth Pinckney (a friend of Michaux's) and ordered him to leave France. Relations between the two countries quickly deteriorated into a "quasi war." The French navy and French privateers began to seize American ships, and the reaction in America to French attacks on American commerce led to the rebirth of the US Navy. Naval fighting between the two countries continued, especially in the Caribbean area, until 1800, when another French government, led by Napoleon Bonaparte, then first consul of the French Republic, began peaceful overtures and curbed the activities of the privateers and the navy. Hostilities formally ended in September 1800, but by then Michaux was at sea on the new expedition to the South Seas.

Michaux used two styles of dating in this notebook. The new style used the French Revolutionary calendar calculated from September 22, 1792, the day that the republic was first proclaimed. Weeks became ten days long, and months were thirty days long and given new names including vendemaire, thermidor, and fructidor. When Michaux used this calendar in his journal, he often also inserted the old-style dates in parentheses; however, he sometimes made dating errors that we have corrected by placing the correct dates in brackets. Any dates in brackets found in this chapter are editorial insertions to assist readers.

### Notebook 10 begins

AUGUST 13, 1796, in the old style, 27 [26] thermidor of year 4 [5] of the French Republic, One and Indivisible, we embarked from the harbor of Charleston, South Carolina, on board the ship *Ophir* with Capt. Johnston, destination Amsterdam. The 14th and 15th, [we] stayed at anchor.

AUGUST 16 (30 [29] thermidor), we lifted anchor and set sail.

AUGUST 18, lost sight of land.

SEPTEMBER 15 (30 [29] fructidor), there was a storm that lasted until the 16th at night.

OCTOBER 5, we passed in the middle of an English squadron commanded by Admiral Roger Curtis, made up of 14 warships: 8 ships with 2 decks, 2 with 3 decks, and 4 frigates. One of the frigates, the *Melpomene*, came toward us, and having sent an officer aboard our ship, he looked at the papers and bill of lading of the captain. After having verified that all the information given by the captain was true, he was satisfied and wished him a good trip. In conversation, he said that the war with France was getting wearisome for the sailors, that they were not capturing any ships, but [he] hoped that the war with Spain would be more profitable and that the first expedition would be against Manila. This squadron was at the entrance of the English Channel, closer to the Isles of Scilly than Ushant.[1]

OCTOBER 9, 1796 (18 vendemiaire year 5), the wind was favorable and good, but at 5 p.m., there arose a storm that became extremely severe in less than two hours; it continued throughout the night doubling its furor, and the wind that came from the east forced us toward shore. At midnight the captain had prepared the axes to cut the masts. Finally daylight arrived without the boat landing, but on 19 vendemiaire (October 10), around 8 o'clock, the captain, seeing that the sounding line did not show an adequate depth of water, decided to run the ship aground; after 4 or 5 violent jolts, it stopped; then the waves moved with such rage and violence that everything on deck was swept overboard. The sails were in shreds in less than 15 minutes. A mast was broken; the ship was half overturned and was shaken severely for one-half hour. Then the waves became more violent, and we were completely drenched so that all the crew and I were losing our strength. Several trunks that had been brought on deck were thrown into the sea, and the inhabitants of a town named Egmond, approximately one league away, were removing everything that came near shore.[2] There were 200 people, of whom 25 men belonged to a troop with an officer, sent to try to rescue us if possible. Finally several seamen, having lost hope, attached themselves to pieces of wood that had been thrown overboard, and [they] reached the shore. As for myself, I had attached myself to some rigging and had my legs under a spar that had become detached during the night and became attached to the deck. Having been beaten by the waves for three hours, I felt my strength ebbing, and I went down to the lower deck and waited for the end of my suffering and [for] death.[3] I lost consciousness immediately, as I remember nothing of the circumstances that passed until the moment when I was transported to the village, was undressed, and had my clothes changed. They made me drink two little glasses of wine, and they brought me near a big fire, where I regained consciousness a half-hour later, but I was shaking during the whole day. I only know what was told to me, since I had lost consciousness; that around eleven, the captain, having seen the ship's boat fall beside the ship, got three men who had stayed behind to transport me in this boat as well as another man who was in the same situation. Then they put me in a cart to transport me to the village, and toward one o'clock in the afternoon I became conscious. I was in a room near a big fire with new clothes and about 40 or 50 inhabitants of the area. I thought immediately of my boxes and trunks that contained my collections, several of which I had seen thrown in the sea three hours before. The people said that everything that fell off the ship or had been thrown overboard was arriving on shore, and that the detachment of troops was on the lookout to keep the country people from removing any items. The captain, who was the last to stay behind on the ship, jumped into the water and swam to shore around two o'clock, and [he] arrived at the village in a cart extremely exhausted, as were all the sailors.

The inhabitants of the area furnished us all the help possible, shirts, suits, bread, meat, brandy, etc., and toward evening all the people who had been shipwrecked were relieved and recovered. My collection in total was made up of sixteen cases and four trunks, of which only five or six they said had arrived on shore. The wind continued blowing with the same fury, and the general consensus in everybody's mind was that by the next morning we would no longer see any sign of the ship. The wind, they said, was a little less violent during the night, and the next day the ship . . .

A man who did not know how to swim had stayed on board until nightfall and would have perished without the humanitarian spirit of a man from a neighboring village. He had attached a bar in a cross configuration at the end of a little mast and had sat on it, armed with rope that anchored him against the violent waves; while seven or eight men were advancing the mast by the opposite end, they were able to bring it to the ship; then this man who was on the piece of wood threw the rope to the one still on board the ship. The latter took the rope and put it around his body, and having knotted it, he let himself fall in the sea, and thus he was able to be brought to shore. A man named . . . who had been captain of a ship in the Dutch Navy, having learned of this humanitarian act, came to get this man. He hosted him at his own home for several days and presented him with a silver snuffbox on which was engraved the date of this act. In addition, the captain took it on himself to obtain from the municipality a testimonial honoring the man. The rescuer was asked to go to Amsterdam, where he was presented a public prize consisting of a silver box full of money [and] engraved with the details of his bravery, etc.

Sunday, October 9, the day before the storm, there had come aboard the ship two little birds, male and female, which I recognized to be Pinson d'Ardenne.[4]

The day after the storm, we found an aquatic marine bird on shore, named by the English as Gannet.

Although he made no additional entries in his journal until he left Egmond and traveled to Amsterdam on November 25, Michaux started corresponding with friends and colleagues almost immediately after the shipwreck. While his colleague André Thouin was not in Paris to receive the letter reproduced next, it provides another detailed firsthand account written only five days after the shipwreck.

### AM in Amsterdam to André Thouin (thought to be in Paris), 23 vendemiaire year 5 (October 14, 1796)[5]

I arrived here today, and I take advantage of an extraordinary courier who will leave tomorrow morning to inform you that the ship on which I arrived from Charleston was wrecked last Monday at 8 a.m., near a village by the name of Egmond. Almost all my collections were saved; that is, my herbarium specimens (with the exception of two notebooks or bundle of plants, unordered and duplicates of those in the general collection). I also saved a strong box with birds and quadrupeds, all the descriptions and the memoranda concerning plants and animals; thus, despite this tragedy, the main principle [task] will be accomplished. I do regret one case with birds that was too heavy and did not arrive on shore [and] a little suitcase with papers, in which there were additional memoranda, which I can recall from memory with regard to the most important observations. I lost all my personal belongings and my clothing except for what I was wearing, but this is of little importance in view of the great difficulty by which I saved my life. The boat was broken by the waves, and I was transported on land unconscious along with two other people; others had saved themselves by swimming. Despite this, I was only sick for two days. Citizen Fousenberte, consul of the republic, is of great assistance because of the steps that he is taking toward the government so that I can be in possession of my collections, because according to the laws of the land there are multiple formalities after a shipwreck. It is also to the consul that I am indebted for my most pressing needs, buying clothing [and] 15–18 reams of gray paper, because it is necessary that the whole herbarium collection be gone through and changed; this will take me 12–15 days at least. I would like that 25–30 louis be loaned to me to buy the books necessary to my work, such as *Systema Vegetabilum*, etc. etc., as I have lost most of mine, and probably here is the best place to replace this loss. Please employ your good graces for this advance with the minister of the interior, while observing that this advance will be reimbursed on my old or new stipends, whose authenticity and title have been verified by the minister and consul of the republic in the United States.

I wanted to let you know that I was able to save a precious collection of seeds of the interior of America, which were not damaged because they were wrapped in oilcloth. I was able to save three living ducks belonging to the species "Beau Canard of Louisiana" Buffon, but I lost three different species of another genus of birds.

Inform Citizens l'Héritier, Jussieu, Cels, etc. etc., of my arrival, and tell them that despite my misfortune and difficulties, they will be satisfied. Please inform my son of my arrival here as I do not have his address.

News of his shipwreck traveled quickly, and only three weeks later Michaux was answering a letter from the eminent botanist l'Héritier de Brutelle.

### AM in Egmond to Charles Louis L'Héritier de Brutelle in Paris, 17 brumaire year 5 (November 7, 1796) (excerpt)[6]

I received a few days ago with the most heartfelt gratitude the letter in which you express interest in my misfortune. Since the shipwreck I have been busy here repairing the damage, having had eight cases and three trunks completely submerged. I was compelled to go through all the plants and to change paper several times. The first pass took 16 days while working from 4 a.m. until 8 p.m., and I used up 17 reams of paper. I finished today the work on plants, and I am very happy that nothing will be lost as to the usefulness that one will be able to obtain from my trips and my collections. I did not lose one single plant; only the colors have changed. There are six cases that did not suffer any damage.

### Notebook 10 continues

5 FRIMAIRE (November 25, 1796), I left Egmond and arrived in Amsterdam.[7]

6 FRIMAIRE [November 26], I wrapped and marked cases and trunks.

7 FRIMAIRE 7 [November 27], I dined at Citizen Fousenbarte's house.

8 FRIMAIRE [November 28], I put my cases on a covered boat to Brussels, addressed to Citizen l' Endormi. This boat was to pass through Antwerp.

9 FRIMAIRE 9 [November 29], I obtained passports from the admiralty for transit of my collections without having to be searched by the Batavian inspectors.

10 FRIMAIRE (November 30), I wrote to Citizens Bosc, Chion, Bussy, Rev. Nicholas Collin, and Gen. Charles Cotesworth Pinckney,[8] by way of New York. I left Amsterdam for Leiden, slept in Haarlem.

Perhaps the most revealing letter about Michaux's plans for future travel in the United States is the letter excerpted below that he wrote to his friend Bosc. Ten years earlier, in winter 1786, Michaux had mentioned in a letter to Thouin that he wished that Bosc could join him so that they could explore North America together. Now, in a twist of fate, their ships had each traveled across the Atlantic and had passed one another in mid-ocean. Bosc was now in Charleston, but Michaux was in Europe. In order to ensure that Bosc received his letter, Michaux addressed it to him in the care of his friend P. G. Chion, who had an office at 39 Bay Street in Charleston.

### AM in Amsterdam to Bosc in Charleston, November 25, 1796 (5 frimaire year 5) (excerpt)[9]

I learned, dear citizen and friend, that you left for Charleston without my having the pleasure of seeing you before my departure. Take care . . . to keep your health, and you will see me in Charleston before the end of the year. Then if it is possible for us to get together and to acquire a good knowledge of the natural history of North America, I have no doubts that we will have the greatest success together. The great amount of practice that I have had traveling and the experience that I've had have taught me to travel for less money, to conserve my health, and to keep out of life-threatening dangers, [lessons] that are all important in using one's time wisely and not being slowed by obstacles, and to profit in all favorable circumstances to visit the most interesting sites in a suitable period. I will not speak of the advantages that are personal to you and superior to mine, but since we both desire passionately the same outcome, there is chance that we will succeed.

If it is convenient for you, I would be pleased if you lived at the house. The black man will be able to give you some information on the trees of the garden; even though he doesn't know their names, he knows which were of greatest interest for me. As I united [grafted] the two individuals of those that are dioecious, the black man will be able to point those out while waiting for the flowering season to show them up. If you find it appropriate, ask the people of the house which room was mine;[10] you can sleep there yourself and use my books while having it repaired, and [make use of] everything that would suit you that I placed in a case, but it will be essential that you do not leave the key to the room in the door when you are not there.

Write me as soon as you can, and do not spare me in telling me what you need.

11 FRIMAIRE (December 1, 1796, old style), I arrived in Leiden . . . distance from Haarlem and Amsterdam . . .

I VISITED natural history professor Brugmans to whom I gave several seeds from America. I bought several natural history books.[11]

13 FRIMAIRE (December 2 [3]), I left for The Hague and dined the same day at the house of the French minister of the Batavian Republic.

14 FRIMAIRE (December 3 [4]), dined with the minister.

15 FRIMAIRE [December 5], I left for Rotterdam, 5 leagues from The Hague.

16 FRIMAIRE [December 6], I visited the Gevers brothers, whose bird collection is one of the rarest and best prepared of the ones that I had ever seen.[12] I visited Dr. Van Noorden, Consul Le Roux la Ville . . .

17 FRIMAIRE [December 7], I left Rotterdam, passed by Dordrecht, and arrived at Mordick, mouth of many rivers and very dangerous to cross. I slept near Breda, a very fortified town, 9 leagues from Rotterdam.

18 FRIMAIRE [December 8], I arrived in Antwerp, 10 leagues from Breda.

19 FRIMAIRE [December 9], I went for information to the Office of Customs concerning my cases and trunks sent from Amsterdam to Brussels.

20 FRIMAIRE [December 10], the office was closed. I could not transact any business.

21 FRIMAIRE [December 11], I visited Citizen Bruslé, comissioner of the Executive Directory, and Citizen Petit-Mongin [also spelled Mangin], director of customs. I was very pleased with the patriotism and national spirit of Citizen P.-Mongin as well as his honesty and strong intellect. I finished my affair concerning the safety and transport of my cases.

22 FRIMAIRE [December 12], I left for Brussels.

23 FRIMAIRE [December 13], I settled with Citizen J. B. Champon Jr. for the transit of my cases and trunks.

24 FRIMAIRE [December 14], I visited Baron de Reynegom and bought from him ducks from the Mississippi to replace those that were lost on the wreck of 19 vendemiaire.

25 FRIMAIRE (November 15 [December 15], old style), left Brussels for Ghent, arrived the next morning.

26 FRIMAIRE [December 16], I visited M. Van Aken.

27 FRIMAIRE [December 17], I left for Lille.

28 FRIMAIRE [December 18] . . .

29 FRIMAIRE [December 19] . . .

30 FRIMAIRE [December 20], I left Lille.

1 NIVOSE (Wednesday, December 21, old style), passed by Douai [and] Cambrai.[13]

3 NIVOSE [December 23], I arrived in Paris.

4 NIVOSE [December 24], I sent to the Muséum national four ducks (*Anas sponsa*) from the Mississippi and two ducks (*A. galericulata*) from China. Visited Citizens Thouin, Daubenton, Richard, Desfontaines.[14]

5 NIVOSE [December 25], I visited Citizens Cels, Tessier, and Andrieux, all three attached to the 4th division of the Department of the Ministry of Interior Agriculture.[15] Visited l'Héritier, curator of the botanical directory, etc.[16]

### AM to Bénézech, minister of the interior, December 25, 1796 (5 nivose year 5)[17]

I arrived yesterday in Paris; my first obligation is to express my gratitude for the help that you had the kindness to provide me. The damage that my natural history collections suffered from the shipwreck has been so well repaired that nothing will be lost of general utility.

As soon as they arrive in Paris, I will have the honor to take your orders as to the means to accelerate their publication and to make useful to the republic the results of several years' travels in North America.

Michaux was to have considerable correspondence with Minister Bénézech, and two months later Michaux sent him a résumé of his work in Persia and North America.

### Notebook 10 continues

6 NIVOSE 6 [December 26], I visited Mangourit;[18] Citizen de la Croix [Delacroix], minister of external affairs; Col. Fulton, etc.[19] I assisted at the session

of the Institut National de France. Visited Citizens Lamarque [Lamarck], Jussieu, etc.

The following letter of Mangourit (former French consul in Charleston) to the minister of external affairs is included here to show how well the former understood his friend Michaux and how he wanted to do everything he could so that Michaux would stay in France until he had finished organizing his collections and written his books. Mangourit's flowery style is in sharp contrast with Michaux's plain writing.

### Mangourit, chargé d'affaires of the French Republic for the United States, to Citizen Charles de la Croix (Delacroix), minister of external affairs, Paris, October 24, 1796 (3 brunaire year 5)[20]

One of the men who will do most honor for science in France, Citizen Michaux, botanist of the republic, has just arrived in Holland from the United States with a precious collection that has escaped the vandalism of the sea. He is getting ready to bring it to Paris and to return to the field of nature, where his insatiable curiosity calls him. If you yield to this inclination, Citizen Michaux's discoveries will be buried in the catacombs of the Jardin des Plantes; they will be lost to science, lacking any proof. One who collects with so much fearlessness and constancy a large number of plants and seeds from unknown properties will have only served his passion if you do not stop him and make him describe and tell us in what soil of the republic these new organisms can be naturalized.

Would that Michaux, one of the most extraordinary men, known for this type of research that he chose, cease his recruitment in the plant world to increase the number here and enrich our soil. Stop him, and place at his side an editor already initiated in botany and an illustrator who will be able to reveal the detailed appearance with minute accuracy that plants require. Here is the account that I demand of Michaux: the editing of his notes, the telling of his memoirs; this is a treasure of which he is the depository that resides in him and that the eternal republic has the right to demand from a mortal Republican.

If you agree with my desire, Citizen Minister, you are more interested in the prosperity of agriculture and the improvement of forests, not to classify the work of Michaux in terms of useful discoveries and in new ornamental introductions. In the past, botany, as did other sciences, strove only for the pleasures of the monarchs and the parks and gardens of the aristocracy; all that would tend to enrich public land was forgotten, while many portions of France that were stupidly said to be sterile await still the foreign grasses that will make them fertile, or the American tree that, Citizen Minister, relies on you to summon for [planting] in the sandy or swampy areas.

Citizen Michaux has made several discoveries of a useful nature, but his imagination roams too much for him to have left them to posterity: it's a swallow only who knows the composition of its nest. Let us stop this bird of passage so that he can give us the secret of his putty: he knows the smallest flower of the robe of Latone [mother of Apollo and Diana] in North America. His research, or rather his conquests, stretched from Florida through Lake Mistassini. Charles XII often slept with his boots. The conquering Michaux often traveled shoeless. The ferocious and venomous beasts, the heat of the equator, the ice of the pole, the escarpments of mountains, and the depth of precipices never interfered with his progress nor disturbed his sleep. Michaux is one of the purest, cleverest Republicans that I have known. At the time of the war with Spain, he was in charge of sounding out the inclinations of the people from the Louisiana area and of the Indian nations. He acquitted himself perfectly well.

Michaux, who has obtained contributions from so many countries, should have his turn. I swear to you, Citizen Minister, that of all the contributors to the republic, he is perhaps the only one who is poor and deems it an honor to be poor.

This letter led to the following exchange between Minister Bénézech and Minister Delacroix:

### Minister of the interior to minister of exterior affairs, 16 frimaire year 5 (December 6, 1796) (excerpt)[21]

You are in accord, my dear colleague, with the whole of scientific France in praising the work and discoveries of Citizen Michaux. Few naturalists have had as much courage as well as success. The letter of Citizen Mangourit that you sent me [on] the 7 of this month appears to have the objective of making him stay in Paris, to give him the means and the necessary rest to gather and publish his observations. But on the other hand wouldn't it be as valuable for him to repair his losses made in the vegetable kingdom? Powerful motives support both objectives. I think that Citizen Michaux must be consulted on both, but I believe in his zeal for the progress of science and his love for the splendor of the republic. Only these sentiments will dictate his decision. I will write to him concerning this and communicate the ideas of our chargé d'affaires for the United States.

### Minister of interior affairs to AM, 16 frimaire year 5 (December 6, 1796) (excerpt)[22]

The minister of external affairs sent me a letter that he received from Citizen Mangourit, chargé d'affaires of the French Republic for the United States. This envoy, while giving the greatest compliments regarding your courage, your zeal, the breadth of your knowledge, without saying too much, because it is impossible to exaggerate while talking of your talents, this envoy seems to want you to renounce new travels so that you will remain in France to put together your observations, write your notes, and compose your memoirs on your arduous travels and your rich discoveries; he is afraid that if you continue your research, you will leave to France only the material of your discoveries and that [France] will be deprived of a much more valuable gift, your thoughts.

If, according to this account, we would wish your continued presence on the soil of the republic, your presence in the United States will perhaps be necessary for the conservation of the living plants that you left behind or for making up for losses from the plant kingdom that you have sustained. France is eager to benefit from the riches that you have obtained for her, but she desires fervently to see you enlarge your conquests; no one deserves more rest than yourself.

Despite all this apparent concern for Michaux's success, he was able to get little support from the French government.

### Notebook 10 continues

7 NIVOSE [December 27], I wrote to the minister of interior, to Mangourit, to Champon in Bruxelles [Brussels]. I went to Versailles and slept in Satory.[23]

8 NIVOSE [December 28], I slept and dined at Satory.

9 NIVOSE [December 29], I visited Lemonnier and dined with him.[24]

10 NIVOSE [December 30], I visited l'Héritier at his home with G[en.] Pinckney; dined at Cels.

11 NIVOSE [December 31], I visited Jean Thouin; Mme Gilbert; Mme Le Clerc; Mme Trouvé, previously known as Gorelli, wife of the editor of the *Moniteur*.[25]

12 NIVOSE [January 1, 1797], I visited Gen. Pinckney, the botanic garden, and collections of natural history. Dined with M. Goy and visited M. Barquet.[26]

13 NIVOSE [January 2], I looked for lodging.

14 NIVOSE [January 3], I again visited l'Héritier [and] M. Dupont, and [I] dined at the home of a member of the Directory, la Révellière-Lépeau.[27]

15 NIVOSE [January 4], public meeting of the Institut National des Sciences et Arts.

16 NIVOSE [January 5], I visited Richard, Thouin; meeting of the Institut.

17 NIVOSE [January 6], wrote to Citizen Petit-Mangin, inspector of customs in Anvers [Antwerp]; and Citizen Champon in Bruxelles [Brussels]. Dined at Remi Claye opposite the Pont au Change.[28]

18 NIVOSE [January 7], I worked on moving.

19 NIVOSE [January 8], I dined at [the home of] Citizens Redouté, painters at the Louvre.[29]

20 NIVOSE [January 9], I dined at [the home of] Cels.

21 NIVOSE [January 10], I went to the Institut; dissertation of Ventenat on the Phallus from Cayenne.[30]

22 NIVOSE [January 11], I visited the Panthéon.[31]

23 NIVOSE [January 12], I bought several pieces for my household. I visited M. Dubois and Minister Bénézech.[32]

24 NIVOSE [January 13], I visited Thouin, Delaunay,[33] and Desfontaines. I dined with Mme Barquet.

25 NIVOSE [January 14], had the carpenter do some work. Wrote to Brussels.

26 NIVOSE [January 15], I visited Mangourit; meeting of the Institut, dissertation on the Rhinoceros with one and two horns; received a letter of Citizen Petit Mangin; he said that my collections had not yet arrived in Antwerp on 22 [23] nivose [January 12].

27 NIVOSE [January 16], I wrote several letters.

28 NIVOSE [January 17], I went to Thouin's; I met member of the Directory, La Révellière-Lépeau.

29 NIVOSE [January 18], I visited Citizen Louvet.[34]

30 NIVOSE [January 19], I went to Citizen Cels's home.

1 PLUVIOSE [January 20], I went to the Institut National; gave Cels a letter from the minister of the interior to be sent to the consul in Charleston.

2 PLUVIOSE [January 21], I wrote to Citizen du Pont and sent the letter of the minister of the interior.

3 PLUVIOSE [January 22], I went to the office of the minister of the navy and to Gen. Pinckney [and] Bernard de Ste Afrique.[35]

4 PLUVIOSE [January 23], I wrote several letters to: Bosc in duplicate, Capt. Baas, Mme Duverney, M. Duverney, Dupont at Charleston, Bussy in New York, Chion, Saulnier.[36]

5 PLUVIOSE [January 24], I dined at [the home of] Gen. Pinckney.

6 PLUVIOSE [January 25], I wrote to the minister of the navy and sent papers concerning Spillard, Institut National des Sciences.

7 PLUVIOSE [January 26], I wrote to Himely in Switzerland and to Mrs. Himely in Charleston.[37]

8, 9, 10 PLUVIOSE [January 27–29], I worked on putting in order the seed collection from Illinois; dined at Cels's home and gave him a collection of these seeds.

End of Journal

# Epilogue

*Michaux's Last Years, 1797–1802*

SUMMARY: LITTLE ATTENTION HAS BEEN directed to Michaux's last years in Paris and his voyage to the South Seas with Capt. Baudin. Two recent studies, "André Michaux in Madagascar" by Smithsonian Institution botanist Laurence J. Dorr, presented at the 2002 André Michaux International Symposium and revised in 2004, and "Un portrait du botaniste André Michaux" by Michaux descendant Régis Pluchet, presented at the 2009 symposium in Brussels on the Baudin Expedition and published in 2010,[1] shed new light on this last period of Michaux's life.

## FRANCE

Upon his return to Paris, Michaux was very busy reestablishing old contacts and forming new ones. He continued his correspondence; we have letters from Michaux to Samuel du Pont, giving him advice on the roads in the eastern United States;[2] to Per Afzelius, Swedish scientist whom he had met in Paris before his departure for the United States;[3] to Carl Thunberg, Swedish botanist to whom he sent plants and birds;[4] and to Rev. Nicholas Collin, pastor of the Swedish Church and secretary of APS;[5] as well as a letter from Alexander von Humboldt asking Michaux about the growth of trees in the United States.[6] He gave approximately five hundred of his specimens from America to Antoine-Laurent de Jussieu, specimens that are now housed in the Jussieu collection in the Muséum national d'histoire naturelle in Paris. He attended many meetings of the Institut de France and, with his son's assistance, continued working on his two proposed publications, *Histoire des chênes de l'Amérique* and *Histoire naturelle de l'Amérique septentrionale*. Michaux also continued the training of Merlot, said to be from Saint-Domingue (Haiti), a young black man whom he had brought from the United States.[7]

In July 1798, when an expedition toward Australia to be led by Capt. Nicholas Baudin (1756–1803) was proposed, Alexander von Humboldt and Aimé Bonpland were asked to join the voyage. Delays in the expedition's funding eventually caused the scientists to turn their attention to South America.[8] In spring 1800, Napoléon decided to finance the project, and Michaux was selected as senior botanist by the commission appointed by the Institut de France with the strong backing of de Jussieu. Michaux had been trying for several years to resolve his financial situation. Despite promises of several ministers, he had only received meager subsidies. He negotiated firmly until his departure for Australia,[9] and he was finally given 1,200 francs for preparations for his voyage and a salary of 4,200 francs per year for his participation in the Baudin expedition.[10]

Michaux placed conditions on his participation in the expedition. After leaving Australia, he wanted to be deposited either in Indonesia or in the Philippines, then he planned to travel to Chile and Peru, and from there he would return to the United States.[11] He asked to bring an apprentice gardener of his choice. This is the only known document written by Michaux where he hints at Merlot, the young black man mentioned earlier.[12]

His conditions were met, and the ministers wrote for the necessary passports for the countries that he wanted to visit. De Jussieu wrote to Jan Hendrik van Swinden, a director of the Dutch Republic who responded with the papers necessary for Michaux to travel in the Dutch colonies.[13] Michaux corresponded with the noted Spanish botanist Antonio José Cavanilles to tell him that the minister of exterior affairs had communicated with the Spanish government to request a visa, should he want to land on Spanish territory.[14] Michaux asked Cavanilles to help him and offered him seeds.

In September 1800, Michaux exhibited an archeological specimen in the Cabinet of Antiquities that he had obtained near Baghdad on his earlier voyage to the Middle East. This black serpentine stone with inscriptions and markings, a kudurru, was to become an important key in deciphering cuneiform writing. It was the first example of Babylonian writing to arrive in Europe, and it dated back to the eleventh century BCE. Lucien Bonaparte, minister of the interior, bought it from Michaux for the French government in exchange for three thousand francs used toward the expenses of printing Michaux's books. This stone, referred to as Caillou Michaux, is on permanent

FIGURE 30. Caillou Michaux, the kudurru that Michaux had obtained near Baghdad in a voyage to the Middle East and later exhibited as an archaeological specimen in the Cabinet of Antiquities. It became an important key in deciphering cuneiform writing. (Photo by Eliane M. Norman)

exhibit at the Cabinet des médailles of the Bibliothèque nationale de France in Paris.[15]

After giving his plant collection to the Muséum in Paris[16] and announcing his departure at the Institut,[17] Michaux visited a notary and made his son executor of his estate.[18] He left for Le Havre, where he wrote to de Jussieu[19] that there was good rapport among the scientists.[20] Michaux sailed on October 19 aboard the *Naturaliste*, captained by Emmanuel Hamelin. He was put in charge of the library. Capt. Baudin, the expedition leader, sailed on the *Géographe*, the other ship.

## Voyage to the South Seas

In April 1801, Michaux was among the ten scientists and artists who left the expedition at Mauritius (Isle de France). Unlike the others, Michaux's departure does not seem to have come from hostility toward Capt. Baudin; his reason appears to have been more subtle. Unlike the others, who left under the pretext of "indisposition," he announced his departure at the very last moment and only after much hesitation.[21] There was a decree, apparently only revealed after the expedition departed from France, that all the collections made on the voyage were destined for the Muséum, except for a secret collection intended specifically for Mme Bonaparte.[22] Baudin wrote: "Of all the scientists who disembarked, Michaux is the only one that I will miss. I spoke to him in a friendly fashion on the inconveniences that might result. Everything was for naught, he could not get used to the idea that he would have to give to the government all the objects that he would be able to collect during the time with us. This is the only reason that he gave for abandoning us."[23] This idea that he wanted to have control over his collections was not new. In 1793, in Philadelphia, while negotiating with APS on the proposed voyage west of the Mississippi, Michaux stated that he wanted the discoveries in natural history to be for his immediate use and only later for the public's benefit.

On Mauritius, Michaux established himself a few miles south of Port Louis, where he was received by Paul Martin de Moncamp (1758–1827), a physician with whom he had traveled on his journey in the Middle East. The latter had a sugarcane plantation of over 1,250 acres, and he obtained lodging, a servant, and land to make a garden for Michaux.[24] Michaux botanized with Merlot and worked with several eminent botanists: Jean Nicolas Céré (1737–1810), head of the Pamplemousses and Mon Plaisir gardens; Joseph Charpentier de Cossigny (1736–1809); Louis Marie Aubert du Petit-Thouars (1758–1831); and Jean-Frédéric Stadtmann (1762–1807).[25] Du Petit-Thouars remarked on Michaux's frugal lifestyle.[26]

Michaux wrote to his son in October 1801, asking about the publication of his books,[27] not knowing that *Histoire des chênes de l'Amérique* had already appeared. He sent a collection of seeds to de Jussieu, to his nurseryman-friend Cels, and to the Spanish botanist Cavanilles. In September 1802, Ventenat,

Michaux's correspondent at the Institut, announced that Michaux had sent him seeds of *Ayapana triplinervis*, among a collection of very interesting seeds of the island.[28] He also attempted to naturalize some plants there; he planted oaks from Europe and America, as well as walnuts in the Plain of Moka near Port Louis. He encouraged planters to increase the quality of their fruits by grafting, demonstrating the technique himself.[29] At that time Mauritius was a very important center for the movement of useful plants throughout the tropics.[30]

Because the botany of Mauritius had already been well studied, Michaux's main objective was the "Grande Ile," Madagascar. He probably left Mauritius with Merlot in early June 1802 to document the completely unknown flora at the center of Madagascar. First, however, he wanted to establish a small garden on the east coast to facilitate transport of plants to Port Louis and subsequently to France. He was assisted by André Thouin's young gardener Louis-Armand Chapelier, who had moved to Madagascar in 1799. Chapelier lived near Tamatave on the border of the Ivondro River and helped Michaux find property for the garden. As was his custom, Michaux put all his energy into the task. Four to five months after his arrival, he contracted a tropical fever, and he died on October 11, 1802, at the age of fifty-six.[31]

In June 1803, Jean Nicolas Céré, director of the Pamplemousse Garden on Mauritius, wrote to Denis Decrès, minister of marine, news of the deaths of Michaux on Madagascar and Anselme Riedlé, a gardener on Timor, and of the declining health of Chapelier, who was suffering from tropical fever.[32] The *Naturaliste*, the first vessel of the Baudin expedition to return to Mauritius from the South Seas, then brought the letter back to France.

Following Michaux's death, Chapelier occupied the botanist's lodging and reported "I am cultivating about forty exotic trees such as mango, guava, avocado, litchi, Japanese plum, rougey de la Réunion,[33] tea, and coffee that were brought by the worthy and unfortunate Michaux."[34]

Geoffroy Saint-Hilaire, professor of zoology at the Jardin des Plantes, reported to his colleagues on 21 germinal year 11 FR (April 11, 1804) on the fate of Merlot, Michaux's young black gardener. The professor had gone to Lorient when the second vessel, the *Géographe*, arrived home, carrying collections for the Jardin, and he had the opportunity to speak to the survivors:

> Merlot is now in the neighborhood of Soullepointe [Foulpointe], Madagascar. He wanted to go back to Isle de France [after Michaux's death], but King Sacavola [name probably misinterpreted], while recognizing the

justice of this demand and the merit of the petition, pointed out the laws of his kingdom: no black man, who was not sold, could leave the island. He thus proposed to Merlot to fulfill this formality by paying [for him] two piasters, but Merlot refused, as he did not believe in the honesty of the European who would serve as go-between in this affair and didn't want to live as a slave on Isle de France. Instead Merlot established himself as a gardener and came to be employed by King Sacavola and other nobles; he is comfortable, but since he has the habit of making collections, it would be worthwhile if he were to help us in this way, through the intermediary of the government of Isle de France. Our colleague Jussieu should transmit this information to the state counselor, Dupuis.[35]

King Sacavola, a cruel tyrant, however, was soon killed, and it appears that Merlot was sold as a slave despite the complaints made by Michaux's son.[36] However, he may have regained his independence. M. Leschenault, a former member of Baudin's expedition who later visited Java and gathered information on arrow poisons, included the following information:[37]

A young black man, named Bognam-nonen-derega, of the Macpas people of western Africa, attached to the service of the celebrated explorer Michaux, the elder, told me that in his country, they poisoned arrows by dipping them in the juice of a plant, to which was added the venom of an animal, which I judged to be a species of *Scolopendra* [centipede].

In 1801, François André was sent on an official mission to the United States. He returned to France in March 1803 at the time that his father's second book, *Flora Boreali-Americana*, came out. He probably learned of his father's death upon the return of the *Naturaliste* in June. He requested that Paul Martin de Moncamp, his father's host in Mauritius, gather his father's belongings and take care of his accounts.[38] Michaux's papers were brought back to Mauritius, but most of them were lost "because of negligence of the marine office of this colony."[39] Nevertheless, Geoffroy Saint-Hilaire indicated that "one case containing dried plants collected by Michaux from Mauritius and Madagascar was found, opened, and resealed as it was in good condition."[40]

In 1940, Edmond François, president of the Malagasy Academy, found the tombs of Michaux and Chapelier.[41] Lucille Allorge, botanist at the Muséum national d'histoire naturelle in Paris, rediscovered the tombs in 2003 and noted their good condition. The same year, Madeleine Ly-Tio Fane, historian of Mauritius, located and published Michaux's death certificate in Port Louis.[42]

## New Jersey Garden

Neither of the gardens that Michaux established in North America survived for more than a few decades. Pierre-Paul Saunier, the gardener of the New Jersey property, made no shipment of plants to France between 1792 and 1802. He received no salary, but he paid taxes and made essential repairs.[43] Economist Pierre Samuel du Pont de Nemours (1739–1817), who was acquainted with Michaux, arrived in the United States with his family in 1800 and lived for two years at Bergen Point, about ten miles from the New Jersey garden. He wrote to the Institut de France urging that both of the "French gardens" be preserved. The Institut forwarded the appeal to the minister of the interior and the Bureau of Agriculture. This led to François André Michaux's voyage to America in 1801, in part to settle these properties. Initially his instructions were to sell both gardens. However appeals from Saunier and the du Ponts—both the father and the younger son, Eleuthère Irénée (E. I.)—led to a more favorable situation for Saunier. It was agreed in 1802 by François André Michaux that the French government would pay for back taxes and repairs, and it offered Saunier one thousand francs for yearly shipments of seeds and plants. It is not clear whether or when the property was officially transferred to Saunier. The official records of the 1840s reveal that the property was in the name of Michel Michaux Saunier, eldest son of Pierre-Paul. Shipments of plant material continued into the 1830s, long after Pierre-Paul's death.[44] Some of these shipments were subsidized by E. I. du Pont.[45] Some years afterward, a substantial portion of the property became the Hoboken Cemetery, which it remains today. No marker at the site or in the area indicates that it was the location of Michaux's New Jersey garden, and Pierre-Paul Saunier is buried in the cemetery of the English Neighborhood Reformed Church of Ridgefield Park, New Jersey.[46]

## Charleston Garden

The 111-acre property outside Charleston that Michaux had purchased for the French government in November 1786, and where he had cultivated thousands of plants for nearly ten years, was left in the hands of his friend J. J. Himely in August 1796. When François André arrived in 1801, he made a short list of the plants in his father's garden, a list that Thouin published soon after.[47] In 1802, the property was sold to Himely, who in turn resold it shortly afterward to the Agricultural Society of South Carolina,[48] which held it until 1820. In 1803, horticulturist Philippe Noisette, a refugee from Saint-Domingue (Haiti), became manager of the garden through the backing of Gen. Pinckney and François André Michaux.[49]

FIGURE 31. Dedication of South Carolina Historical Marker at French Botanic Garden site, North Charleston, South Carolina, March 7, 2008. *Left to right*: Marie Arnaud, Alliance Française de Charleston; Dr. George W. Williams, Michaux Committee of Charleston; Mayor Keith Summey of North Charleston, South Carolina; Colonel Frank Jones, US Air Force, Charleston Air Force Base; and Charlie Williams, André Michaux International Society. (Courtesy of Lydia Williams)

In Charleston, affectionately known as "a city set in a garden" and appreciated for its fine gardens since colonial days,[50] the memory of Michaux persisted. He had come to know the leading families there and had generously shared the treasures in his garden with them. Charleston's positive memory of Michaux led to action by its citizens to honor him. In the early decades of the twentieth century, there were unsuccessful efforts led by the Charleston Museum to preserve the garden site, which was located near the city's airport. Instead it came under the jurisdiction of the US Army Air Corps during World War II. After the war, the Garden Club of Charleston took up the cause, and the garden property was transferred to the City of Charleston and later turned over to the SCANA Corporation. Inspired by the 1986 bicentennial of Michaux's arrival in Charleston, the Charleston Natural History Society continued the effort and arranged for the first preliminary archaeological investigations of the site by Dee Dee Joyce and her students at the College of Charleston.[51] The property later returned to the control of the US Air Force, but the small section believed to include the Michaux house site remains free of construction.

In 2008, through the efforts of George Walton Williams with the Michaux Committee of Charleston, a new local group specifically dedicated to honoring Michaux, a South Carolina Historical Highway Marker was installed on Aviation Avenue in the city of North Charleston at the garden site. The Garden Club of Charleston had marked the site with a large stone marker in 1954, but this stone had been moved and obscured by decades of airport expansion. More recently, the Michaux Committee sponsored noted artist Karl Beckwith Smith's large mural depicting Michaux's Charleston years for the visitor terminal of the Charleston International Airport.[52] Happily, the Garden Club of Charleston's large stone marker has been moved once again, and it now occupies a conspicuous location beneath this striking new mural.

In the early part of the twentieth century, when William Chambers Coker visited and described the garden site, there were still plants from Michaux's period, including *Gingko biloba* and a majestic *Magnolia grandiflora*.[53] As late as the early 1940s, some other plants that Michaux had cultivated still survived, notably *Pinckneya bracteata*,[54] but today it is principally of interest as an archaeological site.

Even though Michaux made strenuous efforts to fulfill his chief mission of sending trees and shrubs to reforest his native country, this ultimately proved unsuccessful because of many impediments. However, because of his perseverance and intelligence, his self-imposed mission to acquaint the world with the flora of North America became a lasting accomplishment resulting in the publication of two very important books, *Histoire des chênes de l'Amérique* (1801) and *Flora Boreali-Americana* (1803). The persistence of André Michaux's name in botany was assured with the hundreds of plants that he described and named for the first time in these two books.

## The Michaux Legacy

André Michaux's legacy was further enhanced by the work of his son, François André, who twice returned to America, continued his father's study of North American trees, worked at naturalizing American trees in France,

FIGURE 32. Portrait of François André Michaux (1770–1855), only son of André Michaux and author of the *North American Sylva*, painted by Rembrandt Peale, ca. 1808 (Courtesy of the American Philosophical Society, Philadelphia, PA)

and published his own landmark book, *Histoire des arbes forestiers de l'Amérique septentrionale* (Paris: 1810–13). The easily understood text in this monumental work was aimed not at specialists but at all persons interested in trees. Published in English (1817–19) as the *North American Sylva*, it was enhanced by 156 color illustrations prepared by the celebrated botanical artists Pierre J. Redouté, Pierre's brother Henri, and Pancrae Bessa. The *Sylva* continued to be the standard reference work on North American trees throughout the nineteenth century, being updated and reprinted several times with the last printing appearing in 1871.[55]

François André Michaux gathered honors on both sides of the Atlantic, including being named a member of the Agricultural Society of Charleston and the American Philosophical Society of Philadelphia. He maintained close ties

FIGURE 33. Portrait of an older François André Michaux (1770–1855), unknown
   artist, ca. 1819 (Courtesy of the Royal Botanic Gardens, Kew, England)

with the APS throughout his long life, and for many years he sent shipments
of the latest European scientific journals to the APS library. He also made two
very important personal gifts to the APS. In 1824 he donated his father's nine
original journal notebooks to the APS, ensuring that this important documen-
tary record of his father's career could be examined in the country where his father worked. Then, at the end of his life, François André Michaux left the APS
a substantial monetary bequest in his will as well as a smaller amount to the
Massachusetts Society for Promoting Agriculture. The APS used its bequest
to create a Michaux Fund, the income from which was soon used to promote
early efforts at forest conservation in Pennsylvania and in subsequent decades
has funded many other projects. The first forest reserve in Pennsylvania was
ultimately named the Michaux State Forest as a tribute to both Michauxs.[56]

Over Cup White Oak.

*Quercus macrocarpa.*

FIGURE 34.  Bur oak, by Pierre J. Redouté, from the *North American Sylva*, 1819, by François André Michaux (Photo by Charlie Williams)

FIGURE 35. Southern red oak, by Pancrae Bessa, from the *North American Sylva*, 1819, by François André Michaux (Photo by Charlie Williams)

FIGURE 36.  Fever-tree, by Pierre J. Redouté, from the *North American Sylva*, 1819, by François André Michaux (Photo by Charlie Williams)

FIGURE 37.  Yellow-wood, by Henri J. Redouté, from the *North American Sylva*, 1819, by François André Michaux (Photo by Charlie Williams)

FIGURE 38. Michaux State Forest sign, Chambersburg, Pennsylvania.
As André Michaux rode west between Carlisle and Shippensburg
through an immense forest on his way to Pittsburgh, he could not have
imagined that 140 years later the ridgetop on his left would become
part of the Michaux State Forest. The name of Pennsylvania's first state
forest honors both him and his son. (Photo by Charlie Williams)

# Appendix

## *Plants and Animals Michaux Encountered*

THE APPENDIX CONTAINS PHOTOGRAPHS OF a selection of North American plants André Michaux encountered (starting on this page) and tables listing the plants and animals described in Michaux's North American journals and letters (starting on pages 373 and 415, respectively). Visit the web page for this book at www.uapress.ua.edu to download searchable PDFs of the tables.

### PHOTOGRAPHS OF PLANTS

Photographs of several dozen plants André Michaux is known to have encountered in North America follow.

FIGURE A1. Bristly locust, *Robinia hispida* (Photo by Charlie Williams)

FIGURE A2. Bottlebrush buckeye, *Aesculus parviflora* (Photo by Charlie Williams)

FIGURE A3. Painted trillium, *Trillium undulatum* (Photo courtesy of Jacques Cayouette, Agriculture and Agri-food, Canada)

FIGURE A4. River birch, *Betula nigra* (Photo courtesy of Lenny L. Lampel, James F. Matthews Center for Biodiversity Studies, Charlotte, NC)

FIGURE A5. Common pawpaw, *Asimina triloba* (Photo courtesy of Lenny L. Lampel, James F. Matthews Center for Biodiversity Studies, Charlotte, NC)

FIGURE A6. Bigleaf magnolia, *Magnolia macrophylla* (Photo by Charlie Williams)

FIGURE A7. Lily-of-the-valley, *Convallaria majalis* (Photo by Charlie Williams)

FIGURE A8. Cross-vine, *Bignonia capreolata* (Photo by Charlie Williams)

FIGURE A9. Sweet-shrub, *Calycanthus floridus* (Photo by Charlie Williams)

FIGURE A10. Mountain doghobble, *Leucothoe fontanesiana* (Photo by Charlie Williams)

FIGURE A11. Common partridge-pea, *Chamaecrista fasciculata* (Photo by Charlie Williams)

FIGURE A12. Fringe-tree, *Chionanthus virginicus* (Photo courtesy of Rob D. McHenry, James F. Matthews Center for Biodiversity Studies, Charlotte, NC)

FIGURE A13. Mountain sweet-pepperbush, *Clethra acuminata* (Photo by Charlie Williams)

FIGURE A14. Bunchberry, *Cornus canadensis* (Photo by Charlie Williams)

FIGURE A15. Hearts-a-bustin', *Euonymus americanus* (Photo courtesy of Lenny L. Lampel, James F. Matthews Center for Biodiversity Studies, Charlotte, NC)

FIGURE A16. Stemless arctic bramble, *Rubus arcticus* var. *acaulis* (Photo courtesy of Jacques Cayouette, Agriculture and Agri-food, Canada)

FIGURE A17. *A*, Carolina buckthorn, *Frangula caroliniana*. *B*, Carolina buckthorn, *Frangula caroliniana* (flower insert). (Photos courtesy of Catherine M. Luckenbaugh, James F. Matthews Center for Biodiversity Studies, Charlotte, NC)

FIGURE A18. Franklinia, *Franklinia alatamaha* (Photo by Charlie Williams)

FIGURE A19. *A*, Wintergreen *Gaultheria procumbens* (flower). *B*, Wintergreen *Gaultheria procumbens* (fruit) with *Galax* leaf. (Photos by Charlie Williams)

FIGURE A20. Mountain Indian-physic, *Gillenia trifoliata* (Photo by Charlie Williams)

FIGURE A21. Pipevine, *Isotrema macrophyllum* (Photo by Charlie Williams)

FIGURE A22. Mountain laurel, *Kalmia latifolia* (Photo by Charlie Williams)

FIGURE A23. White wicky, *Kalmia cuneata* (Photo by Charlie Williams)

FIGURE A24. *A*, Scaly blazing-star, *Liatris squarrosa*. *B*, Scaly blazing-star, *Liatris squarrosa* (flower insert). (Photos courtesy of Lenny L. Lampel [*A*] and Catherine M. Luckenbaugh [*B*], James F. Matthews Center for Biodiversity Studies, Charlotte, NC)

FIGURE A25. Red elderberry, *Sambucus racemosa* var. *pubens* (Photo courtesy of Jacques Cayouette, Agriculture and Agri-food, Canada)

FIGURE A26. Rattlesnake-root, *Prenanthes racemosa* (Photo courtesy of Jacques Cayouette, Agriculture and Agri-food, Canada)

FIGURE A27. Umbrella-tree, *Magnolia tripetala* (Photo by Charlie Williams)

FIGURE A28. Northern sweet-bay, *Magnolia virginiana* (Photo by Charlie Williams)

FIGURE A29. Eastern sensitive-brier, *Mimosa microphylla* (Photo by Charlie Williams)

FIGURE A30. Common pyxie-moss, *Pyxidanthera barbulata* (with *Campsomeris* wasp) (Photo by Charlie Williams)

FIGURE A31.  Bee-balm, *Monarda didyma* (Photo by Charlie Williams)

FIGURE A32.  Nestronia, *Nestronia umbellula* (Photo courtesy of Lenny L. Lampel, James F. Matthews Center for Biodiversity Studies, Charlotte, NC)

FIGURE A33.  Allegheny-spurge, *Pachysandra procumbens* (Photo by Charlie Williams)

FIGURE A34.  Buffalo-nut, *Pyrularia pubera* (Photo by Charlie Williams)

FIGURE A35.  Tough bumelia, *Sideroxylon tenax* (Photo by Charlie Williams)

FIGURE A36.  Carolina lily, *Lilium michauxii* (Photo by Charlie Williams)

FIGURE A37. Flame azalea, *Rhododendron calendulaceum* (Photo by Charlie Williams)

FIGURE A38. Catawba rhododendron, *Rhododendron catawbiense* (Photo by Charlie Williams)

FIGURE A39. Oconee azalea, *Rhododendron flammeum* (Photo by Charlie Williams)

FIGURE A40. Pinxterflower, *Rhododendron periclymenoides* (Photo courtesy of Kat Sweeny, James F. Matthews Center for Biodiversity Studies, Charlotte, NC)

FIGURE A41. Bloodroot, *Sanguinaria canadensis* (Photo courtesy of Catherine M. Luckenbaugh, James F. Matthews Center for Biodiversity Studies, Charlotte, NC)

FIGURE A42. Yellow pitcherplant, *Sarracenia flava* (Photo by Charlie Williams)

FIGURE A43. Saw palmetto, *Serenoa repens* (Photo courtesy of Lenny L. Lampel, James F. Matthews Center for Biodiversity Studies, Charlotte, NC)

FIGURE A44. Red-osier dogwood, *Cornus stolonifera* (Photo courtesy of Jacques Cayouette, Agriculture and Agri-food, Canada)

FIGURE A45. Oconee bells, *Shortia galacifolia* (Photo by Charlie Williams)

FIGURE A46. Fire-pink, *Silene virginica* (Photo by Charlie Williams)

FIGURE A47. Prairie-dock, *Silphium terebinthinaceum* (Photo courtesy of Lenny L. Lampel, James F. Matthews Center for Biodiversity Studies, Charlotte, NC)

FIGURE A48. Pinkroot, *Spigelia marilandica* (Photo by Charlie Williams)

FIGURE A49. *A*, Carolina hemlock, *Tsuga caroliniana* (compare needle pattern with *T. canadensis*). *B*, Eastern hemlock, *Tsuga canadensis*. (*A*, photo by Charlie Williams; *B*, photo courtesy of Lenny L. Lampel, James F. Matthews Center for Biodiversity Studies, Charlotte, NC)

FIGURE A50. Mountain-ash, *Sorbus americana* (Photo by Charlie Williams)

FIGURE A51. Perfoliate bellwort, *Uvularia perfoliata* (Photo courtesy of Lenny L. Lampel, James F. Matthews Center for Biodiversity Studies, Charlotte, NC)

FIGURE A52. Deerberry, *Vaccinium stamineum* (Photo courtesy of Lenny L. Lampel, James F. Matthews Center for Biodiversity Studies, Charlotte, NC)

FIGURE A53. Halberd-leaf violet, *Viola hastata* (Photo by Charlie Williams)

FIGURE A54. Zenobia, *Zenobia pulverulenta* (Photo by Charlie Williams)

FIGURE A55. Yellow-wood, *Cladrastis kentukea* (Courtesy of Ronny and Donnietta West, Jackson County Historical Society, Gainesboro, Tennessee)

FIGURE A56. Canada germander, *Teucrium canadense* (Photo by Walter Kingsley Taylor)

FIGURE A57. Swamp rosemallow, *Hibiscus grandiflorus* (Photo by Walter Kingsley Taylor)

FIGURE A58. Azure sage, *Salvia azurea* (Photo by Walter Kingsley Taylor)

FIGURE A59. Southern dewberry, *Rubus trivalis* (Photo by Walter Kingsley Taylor)

FIGURE A60. Pipe-plant, *Agarista populi-folia* (Photo by Walter Kingsley Taylor)

FIGURE A61. Pond apple, *Annona glabra* (Photo by Walter Kingsley Taylor)

FIGURE A62. Devil's-walking-stick, *Aralia spinosa* (Photo by Walter Kingsley Taylor)

FIGURE A63. Big flower pawpaw, *Asimina obovata* (Photo by Walter Kingsley Taylor)

FIGURE A64. Tarflower, *Bejaria racemosa* (Photo by Walter Kingsley Taylor)

FIGURE A65. *A*, Small-flowered pawpaw, *Asimina parviflora*. *B*, Small-flowered paw-paw, *Asimina parviflora* (flower) (Photos by Walter Kingsley Taylor)

FIGURE A66. Gumbo limbo, *Bursera si-maruba* (Photo by Walter Kingsley Taylor)

FIGURE A67. Pepper cinnamon, *Canella winterana* (Photo by Walter Kingsley Taylor)

FIGURE A68. Black ti-ti, *Cliftonia mono-phylla* (Photo by Walter Kingsley Taylor)

FIGURE A69. Coastal fetterbush, *Eubotrys racemosus* (Photo by Walter Kingsley Taylor)

FIGURE A70. Carolina Jessamine, *Gelsemium sempervirens* (Photo by Walter Kingsley Taylor)

FIGURE A71. Seven-year apple, *Casasia clusiifolia* (Photo by Walter Kingsley. Taylor)

FIGURE A72. Two-wing silverbell, *Halesia diptera* (Photo by Walter Kingsley Taylor)

FIGURE A73. Florida star-anise, *Illicium floridanum* (Photo by Walter Kingsley Taylor)

FIGURE A74. Hairy wicky, *Kalmia hirsuta* (Photo by Walter Kingsley Taylor)

FIGURE A75. White mangrove, *Laguncularia racemosa* (Photo by Walter Kingsley Taylor)

FIGURE A76. Blue sandhill lupine, *Lupinus diffusus* (Photo by Walter Kingsley Taylor)

FIGURE A77. Sundial lupine, *Lupinus perennis* (Photo by Walter Kingsley Taylor)

FIGURE A78. Staggerbush, *Lyonia fruticosa* (Photo by Walter Kingsley Taylor)

FIGURE A79. Parrot pitcherplant, *Sarracenia psittacina* (Photo by Walter Kingsley Taylor)

FIGURE A80. Lakeshore sedge, *Carex lenticularis* (Photo courtesy of Marie-Ève Garon-Labrecque, Carleton University, Canada)

FIGURE A81. Lingonberry, *Vaccinium vitis-idaea* (Photo courtesy of Marie-Ève Garon-Labrecque, Carleton University, Canada)

FIGURE A82. Caudate wormwood, *Artemisia campestris* (Photo courtesy of Marie-Ève Garon-Labrecque, Carleton University, Canada)

FIGURE A83. Bluejoint, *Calamagrostis canadensis* (Photo courtesy of Marie-Ève Garon-Labrecque, Carleton University, Canada)

FIGURE A84. Russet buffalo-berry, *Shepherdia canadensis* (Photo courtesy of Marie-Ève Garon-Labrecque, Carleton University, Canada)

FIGURE A85. Scotch false asphodel, *Tofieldia pusilla* (Photo courtesy of Marie-Ève Garon-Labrecque, Carleton University, Canada)

FIGURE A86. Northern bog asphodel, *Triantha glutinosa* (Photo courtesy of Marie-Ève Garon-Labrecque, Carleton University, Canada)

FIGURE A87. Rough cinquefoil, *Potentilla norvegica* (Photo courtesy of Marie-Ève Garon-Labrecque, Carleton University, Canada)

FIGURE A88. Limber honeysuckle, *Lonicera dioica* (Photo courtesy of Marie-Ève Garon-Labrecque, Carleton University, Canada)

FIGURE A89. Purle marshlocks, *Comarum palustre* (Photo courtesy of Marie-Ève Garon-Labrecque, Carleton University, Canada)

FIGURE A90. Mountain fly-honeysuckle, *Lonicera villosa* (Photo courtesy of Marie-Ève Garon-Labrecque, Carleton University, Canada)

FIGURE A91. Seven-angled pipewort, *Eriocaulon aquaticum*, and sundew, *Drosera* sp. (Photo courtesy of Marie-Ève Garon-Labrecque, Carleton University, Canada)

FIGURE A92. Tufted hairgrass, *Deschampsia cespitosa* (Photo courtesy of Marie-Ève Garon-Labrecque, Carleton University, Canada)

FIGURE A93. Northern sheepkill, *Kalmia angustifolia* (Photo courtesy of Marie-Ève Garon-Labrecque, Carleton University, Canada)

FIGURE A94. Common juniper, *Juniperus communis* (Photo courtesy of Marie-Ève Garon-Labrecque, Carleton University, Canada)

FIGURE A95. Sourwood, *Oxydendrum arboreum* (Photo by Charlie Williams)

FIGURE A96. Punctatum, *Rhododendron minus* (Photo by Eliane M. Norman)

FIGURE A97. Coastal witch-alder, *Fothergilla gardenii* (Photo by Charlie Williams)

FIGURE A98. Northern bush-honeysuckle, *Diervilla lonicera* (Photo by Charlie Williams)

FIG. A99. Fever-tree, *Pinckneya bracteata* (Photo by Walter Kingsley Taylor)

Michaux collected many plants that, while not described in his journals or letters, were nonetheless important discoveries. Images of nine of these plants are included among the preceding photographs: *Pyxidanthera barbulata* Michx., Common pyxie-moss (p. 360); *Pachysandra procumbens* Michx., Allegheny-spurge (p. 361); *Lilium michauxii* Poir., Carolina lily (p. 361); *Rhododendron catawbiense* Michx., Catawba rhododendron (p. 362); *Shortia galacifolia* Torr. & A. Gray, Oconee bells (p. 363); *Hibiscus grandiflorus* Michx., swamp rosemallow (p. 365); *Salvia azurea* Michx., ex Lam., azure sage (p. 365); *Rubus trivialis* Michx., southern dewberry (p. 365); *Carex lenticularis* Michx., lakeshore sedge (p. 369).

# Table of Described Plants

The following table lists the plants André Michaux described in his North American journals and letters. It is organized according to Michaux's names for the species, and modern scientific and common names are provided for each species. Note: an asterisk (*) indicates that the name applied by Michaux is recognized for a different species, and page numbers in italics refer to photographs.

| Michaux's names for a particular species | Modern binomial (scientific name) | Common name(s) | Page(s) in this volume |
|---|---|---|---|
| Abies / Abies balsamea / Pinus balsamea / Pinus balsamifera / Pinus, needles notched at the tip | *Abies balsamea* (L.) Mill. | balsam fir | 178, 179, 184, 186, 189, 193 |
| Abies canadensis / Abies spruce / Pine or Sapinette / Pinus abies canadensis / Pinus canadensis / Thuya canadensis / Hemlock pines | *Tsuga canadensis* (L.) Carrière | eastern hemlock | 43, 73, 111, 113, 174, 175, 179, 196, 227, 238, 256, 257, 306, *364* |
| Abies nigra / Pinus abies nigra / Pinus abies rubra / Pinus fol. denticulatis | *Picea mariana* (Mill.) BSP. or *Picea rubens* Sarg. | black spruce / red spruce | 186, 188, 189, 190, 191, 192, 193, 256 |
| Abies nigra? / Abies leaves scattered on all sides [NC] | *Tsuga caroliniana* Engelm. | Carolina hemlock | 256, 258, *364* |
| Abies, with sparse leaves on all sides / Pinus, with few leaves on all sides / Pinus abies | *Picea* sp. | spruce | 175, 178, 179, 185 |
| Acacia | *Acacia* sp. | undetermined | 152 |
| Acacia de cayenne / Mimosa gum Arabic | *Acacia nilotica* (L.) Willd. ex Delile | acacia de cayenne / gum Arabic tree | 151 |
| Acacia from India | *Senegalia catechu* (L. f.) P. J. H. Hurter & Mabb. | khair | 151 |
| Acer | *Acer* sp. | maple | 109 |
| Acer canadense | *Acer spicatum* Lam. | mountain maple | 257 |
| Acer, leaves rugose with somewhat wooly veins | *Acer nigrum* Michx. f. | black maple | 231 |
| Acer negundo | *Acer negundo* L. | box elder | 111, 230, 231, 275, 285 |
| Acer pensylvanicum | *Acer pensylvanicum* L. | striped maple | 174, 176, 179, 184, 256, 257, 258 |
| Acer rubrum, leaves glaucous on lower surface / Acer rubrum, leaves silvery on the lower surface / Acer saccharinum / Acer, with leaves silvery and red | *Acer saccharinum* L. | silver maple | 199, 231, 235, 279, 314 |

*Table of Described Plants (continued)*

| Michaux's names for a particular species | Modern binomial (scientific name) | Common name(s) | Page(s) in this volume |
|---|---|---|---|
| Acer rubrum / red maple | *Acer rubrum* L. | red maple | 89, 93, 95, 157, 176, 195, 228, 229, 231, 279, 294, 295, 304 |
| Acer sacchariferum canadense / Acer saccharum | *Acer saccharum* Marshall | sugar maple | 176, 228, 229, 230, 235, 275 |
| Achillea millefolium | *Achillea millefolium* L. | yarrow | 193 |
| Aconitum uncinatum, known as Ti-savoyanne / Helleborus trifolius | *Coptis trifolia* (L.) Salisb. or *Coptis trifolia* (L.) var. *groenlandica* (Oeder) Fassett | goldthread / goldenroot | 177 |
| Acorus / Calamus aromaticus | *Acorus calamus* L. or *Acorus americanus* (Raf.) Raf. | American calamus / European calamus / sweetflag | 198, 199, 254, 308 |
| Acrostichum aureum | *Acrostichum danaeifolium* Langsd. & Fisch. | giant leather fern | 93 |
| Actea, fruit red | *Actaea rubra* (Aiton) Willd. subsp. *rubra* | red baneberry | 178 |
| Actea spicata / Actea spicata alba / Actea spicata, fruit white | *Actaea racemosa* L. or *Actaea pachypoda* Elliott or *Actaea rubra* (Aiton.) Willd. forma *neglecta* (Gillman) B. L. Rob. | common black cohosh / white baneberry / doll's-eyes | 176, 178, 228, 229, 254 |
| Actea spicata alba / Actea spicata, fruit white | *Actaea rubra* (Aiton) Willd. forma *neglecta* (Gillman) B. L. Rob. | white baneberry / doll's-eyes | 175, 178, 186 |
| Actea spicata | *Actaea* sp. | baneberry | 176 |
| Adiantum pedatum | *Adiantum pedatum* L. | northern maidenhair | 193, 196 |
| Aeschynomene grandiflora, Agathy | *Sesbania grandiflora* (L.) Poir. | agati / hummingbird tree / tiger tongue | 151 |
| Aeschynomene sesban | *Sesbania sesban* (L.) Merr. | Egyptian riverhemp | 149, 151 |
| Alaterne from Carolina / Rhamnus (Carolinian) / Rhamnus alaternus latifolius / Shrub with plums / Rhamnus frangula? | *Frangula caroliniana* (Walter) A. Gray | Carolina buckthorn | 105, 234, 275, *358* |
| Aletris aurea | *Aletris aurea* Walter | golden colic-root | 312 |
| Alisma . . . | *Alisma triviale* Pursh | northern water-plantain | 177 |
| Alisma subulata | *Sagittaria graminea* Michx. | grassy arrowhead | 184 |
| Almond tree | *Terminalia catappa* L. | tropical almond | 150 |

| Michaux's names for a particular species | Modern binomial (scientific name) | Common name(s) | Page(s) in this volume |
|---|---|---|---|
| Alnus / Betula alnus / Betula alnus americanus | *Alnus serrulata* (Aiton) Willd. | tag alder / smooth alder | 118, 157, 274 |
| Alnus glauca / Alnus glauca, stipules lanceolate | *Alnus incana* (L.) Moench subsp. *rugosa* (DuRoi) R. T. Clausen | speckled alder | 184, 186, 199 |
| American elm, spongy bark / Ulmus fungosa / Ulmus, spongy bark | *Ulmus alata* Michx. | winged elm | 248, 274, 294 |
| Amorpha minor | *Amorpha herbacea* Walter var. *herbacea* | dwarf indigo-bush | 154 |
| Amyris elemifera | *Amyris elemifera* L. | sea torchwood | 120 |
| Andromeda arborea | *Oxydendrum arboreum* (L.) DC. | sourwood | 62, 64, 68, 69, 89, 96, 110, 138, 145, 157, 233, 238, 256, *371* |
| Andromeda axillaris [coastal] | *Leucothoe axillaris* (Lam.) D. Don | coastal doghobble | 162, 249, 250 |
| Andromeda axillaris [mountain] | *Leucothoe fontanesiana* (Steud.) Sleumer | mountain doghobble | 256, *357* |
| Andromeda calyculata | *Chamaedaphne calyculata* (L.) Moench | leatherleaf / cassandra | 157, 177, 178, 182, 184, 186, 187, 188, 189, 191, 193 |
| Andromeda coriacea | *Lyonia fruticosa* (Michx.) G. S. Torr. | staggerbush / poor-grub | 157, *369* |
| Andromeda ferruginea | *Lyonia ferruginea* (Walter) Nutt. | crookedwood / staggerbush | 63, 86, 163, 435n6 |
| Andromeda formosissima | *Agarista populifolia* (Lam.) Judd | Agarista / pipe-plant | 89, 97, 157, *366* |
| Andromeda glauca / Andromeda wilmingtonia / Andromeda, glaucous leaves | *Zenobia pulverulenta* (W. Bartram ex Willd.) Pollard | Zenobia / honey-cups | 148, 157, 249, 250, 252, 261, *365* |
| Andromeda mariana | *Lyonia mariana* (L.) D. Don | staggerbush | 157, 250 |
| Andromeda nitida or lucida | *Lyonia lucida* (Lam.) K. Koch | shining fetterbush | 108, 157, 249, 250 |
| Andromeda paniculata [NC] | *Lyonia ligustrina* (L.) DC. var. *foliosiflora* (Michx.) Fernald | southern maleberry / male blueberry | 157, 249, 250 |
| Andromeda paniculata [VT] | *Lyonia ligustrina* (L.) DC. var. *ligustrina* | northern maleberry | 175 |
| Andromeda polifolia / Andromeda rosmarinifolia | *Andromeda polifolia* L. | bog rosemary | 157, 177, 184, 187, 188, 189, 190, 191, 193 |
| Andromeda racemosa [coastal] | *Eubotrys racemosus* (L.) Nutt. | coastal fetterbush / swamp doghobble | 199, 249, 250, *367* |

| Michaux's names for a particular species | Modern binomial (scientific name) | Common name(s) | Page(s) in this volume |
|---|---|---|---|
| Andromeda racemosa [mountain] | *Eubotrys recurvus* (Buckley) Britton | mountain fetterbush | 256 |
| Andropogon muticum | *Andropogon virginicus* L. | old-field broomstraw / broomsedge | 279 |
| Anemone dichotoma | *Anemone canadensis* L. | Canadian anemone | 176 |
| Anemone hepatica | *Anemone americana* (DC.) H. Hara | round-lobed hepatica | 176 |
| Annona | *Asimina* sp. | pawpaw | 61, 65, 66, 74, 89, 93, 96, 109, 115, 120, 129, 152, 294 |
| Annona glabra | *Asimina parviflora* (Michx.) Dunal | small-flowered pawpaw | 95, *366* |
| Annona glabra | *Annona glabra* L. | pond apple | 120, *366* |
| Annona grandiflora / Annona palustris | *Asimina obovata* (Willd.) Nash | big flower pawpaw | 95, 97, 101, *366* |
| Annona lanceolata | *Asimina angustifolia* Raf. | slimleaf pawpaw | 64 |
| Annona muricata | *Annona muricata* L. | soursop | 120 |
| Annona triloba | *Asimina triloba* (L.) Dunal | common pawpaw | 110, 111, 231, 315, *356* |
| Anonymos azaleoides / Azalea-like shrub | *Rhododendron pilosum* (Michx.) Craven | minniebush | 256, 306 |
| Anonymos carpinoides / Carpinus / Carpinus . . . / Carpinus ostrya | *Carpinus caroliniana* Walter or *Ostrya virginiana* (Mill.) K. Koch | American hornbeam / American hop-hornbeam | 175, 176, 196, 275, 285, 294 |
| Anonymos ligustroides / new tree with opposite leaves | *Forestiera acuminata* (Michx.) Poir. or *Forestiera ligustrina* (Michx.) Poir. | swamp-privet / southern-privet | 109, 284, 285, 295 |
| Aquatic cypress / Cupressus thyoides | *Chamaecyparis thyoides* (L.) BSP. | Atlantic white cedar | 32, 33, 36, 261 |
| Aquatic plant, 3 stamens | *Elodea canadensis* Michx. | common waterweed | 176 |
| Aquilegia canadensis / Aquilegia? | *Aquilegia canadensis* L. | Canada columbine / eastern columbine | 66, 176 |
| Aralia new sp. / Aralia new, hispid | *Aralia hispida* Vent. | bristly sarsaparilla | 177, 178, 181 |
| Aralia nudicaulis | *Aralia nudicaulis* L. | wild sarsaparilla | 177 |
| Aralia racemosa / Zanthoxylum, monoecious | *Aralia racemosa* L. | spikenard | 131, 175 |

| Michaux's names for a particular species | Modern binomial (scientific name) | Common name(s) | Page(s) in this volume |
|---|---|---|---|
| Aralia spinosa | *Aralia spinosa* L. | devil's-walking-stick / Hercules'-club / prickly-ash | 101, *366* |
| Arbutus acadiensis?, leaves entire / Arbutus acadiensis / Arbutus uva ursi / Arbutus, leaves with wooly membranaceous margins | *Arctostaphylos uva-ursi* (L.) Spreng. | bearberry / kinnikinick | 175, 178, 199 |
| Arbutus andrachne | *Arbutus andrachne* L. | Greek strawberry tree | 38 |
| Arbutus, loved by bears / Vaccinium (or Arbutus) of the bears | *Gaylussachia ursina* (M. A. Curtis) Torr. & A. Gray ex A. Gray | bear huckleberry | 73, 74 |
| Arbutus unedo | *Arbutus unedo* L. | strawberry tree | 38 |
| Arctium lappa | *Arctium lappa* L. | great burdock | 130 |
| Arethusa bulbosa | *Arethusa bulbosa* L. | dragon's-mouth | 62 |
| Arethusa divaricata | *Cleistesiopsis divaricata* Pansarin & F. Barros | large spreading pogonia | 62 |
| Arethusa ophioglossoides | *Pogonia ophioglossoides* (L.) Ker Gawl. | rose pogonia | 62 |
| Aristolochia macrophilla / Aristolochia scandens / Aristolochia Sipho / Aristolochia sipho seu [or] macrophylla / Aristolochia sipho-tom | *Isotrema macrophyllum* (Lam.) C. F. Reed | pipevine / Dutchman's-pipe | 73, 145, 228, 229, 275, *359* |
| Artemisia crithmoides | *Artemisia campestris* L. subsp. *caudata* (Michx.) H. M. Hall & Clem. | caudate wormwood | 184, *369* |
| Arum triphyllum | *Arisaema triphyllum* (L.) Schott | jack-in-the-pulpit | 228 |
| Arum, with spotted stem | *Peltandra sagittifolia* (Michx.) Morong | spoonflower | 63 |
| Arundo arenaria | *Ammophila breviligulata* Fernald | American beach-grass | 184 |
| Arundo donax | *Arundo donax* L. | giant reed | 38 |
| Arundo, glumes 2-flowered | *Deschampsia cespitosa* (L.) P. Beauv. | tufted hairgrass | 184, 190, *371* |
| Arundo, glumes with one floret | *Calamagrostis canadensis* (Michx.) P. Beauv. | bluejoint | 186, *369* |
| Asarum canadense | *Asarum canadense* L. | wild ginger | 176, 228, 270 |
| Asclepias / Asclepias frutescens? | *Asclepias* sp. | milkweed | 61, 62, 100, 146 |

| Michaux's names for a particular species | Modern binomial (scientific name) | Common name(s) | Page(s) in this volume |
|---|---|---|---|
| Asclepias linifolia | *Asclepias verticillata* L. | whorled milkweed | 146 |
| Asclepias purpurea | *Asclepias purpurascens* L. | purple milkweed | 146 |
| Asclepias rubra | *Asclepias rubra* L. | purple savanna milk-weed / red milkweed | 146 |
| Asphodelle | *Aspodelus* sp. | aspodel | 152 |
| Asplenium rhizophyllum | *Asplenium rhizophyllum* L. | walking fern | 235 |
| Aster grandiflorus | *Symphyotrichum novae-angliae* (L.) G. L. Nesom | New England aster | 315 |
| Astragalus | *Astragalus* sp. or *Astragalus villosis* Michx. or *Astragalus michauxii* (Kuntze) F. J. Herm. | milkvetch / bearded milkvetch / sand-hills milkvetch | 62 |
| Astragalus / Astragalus new sp. | *Astragalus canadensis* L. | Canada milkvetch | 138, 146 |
| Atraphaxia | *Polygonella polygama* (Vent.) Engelm. & A. Gray | common October-flower | 250 |
| Avena paniculata | *Schizachne purpurascens* (Torr.) Swallen | false medic / purple oatgrass | 189 |
| Azalea / Azalea viscosa | *Rhododendron viscosum* (L.) Torr. | swamp azalea / swamp honeysuckle | 199 |
| Azalea alba / Azalea, found along rivers with white flowers / Azalea rosea | *Rhododendron arborescens* (Pursh) Torr. | sweet azalea | 257, 272 |
| Azalea coccinea lutea / Azalea flava / Azalea fulva / Azalea lutea / Azalea, leaves with glandular tip / Azalea, yellow flowers, style very long | *Rhododendron calendulaceum* (Michx.) Torr. | flame azalea | 138, 148, 256, 257, 272, 306, *362* |
| Azalea glauca | *Rhododendron canadense* (L.) Torr. | rhodora | 177 |
| Azalea nudiflora | *Rhododendron periclymenoides* (Michx.) Shinners | pinxterflower / pinx-terbloom azalea | 61, 272, *362* |
| Azalea octandra [PA] | *Rhododendron* sp. | azalea | 227 |
| Azalea, scarlet flower / Azalea coccinea / Azalea, found along rivers with brilliant red flowers / Azalea with the color of fire | *Rhododendron flammeum* (Michx.) Sarg. | Oconee azalea | 51, 61, 272, *362* |
| Baccharis | *Baccharis halimifolia* L. | groundsel tree | 106 |
| Balsamina | *Impatiens balsamina* L. | garden balsam | 150 |

| Michaux's names for a particular species | Modern binomial (scientific name) | Common name(s) | Page(s) in this volume |
|---|---|---|---|
| Bartsia pallida | *Castilleja septentrionalis* Lindl. | Labrador Indian-paintbrush | 190 |
| Befaria | *Bejaria racemosa* Vent. | tarflower | 163, *366* |
| Betula . . . | *Betula michauxii* Spach | Newfoundland dwarf birch | 187 |
| Betula alba | *Betula papyrifera* Marshall or *Betula minor* (Tuck.) Fernald | paper birch / canoe birch / dwarf white birch | 190, 192, 193 |
| Betula alba seu papyrifera / Betula papyrifera | *Betula papyrifera* Marshall | paper birch / canoe birch | 175, 176, 178, 186 |
| Betula australis / Betula australis or spuria / Betula spuria | *Betula nigra* L. | river birch | 176, 285, 295, *356* |
| Betula nigra* [used in error] | *Betula lenta* L. | black birch / sweet birch | 113, 193 |
| Betula papyrifera* [used in error] | *Betula nigra* L. | river birch | 61, 62 |
| Betula pumila | *Betula glandulosa* Michx. or *Betula pumila* L. | resin birch / bog birch | 184, 190, 191, 193 |
| Bignonia crucigera | *Bignonia capreolata* L. | cross-vine | 108, 248, 249, 250, 285, *356* |
| Bignonia pentaphylla | *Tabebuia bahamensis* (Northr.) Britton | Bahamian trumpet tree | 120 |
| Bignonia radicans | *Campsis radicans* (L.) Seem. ex Bureau | trumpet-creeper | 228, 229, 249 |
| Bignonia sempervirens, small climbing shrub | *Gelsemium sempervirens* (L.) St.-Hil. | Carolina jessamine | 62, 89, 108, 148, 248, 249, 250, 285, *367* |
| Bixa / Bixa orellana | *Bixa orellana* L. | lipstick tree | 149, 151 |
| Black oaks | *Quercus velutina* Lam. or *Quercus marilandica* Münchh. | black oak / blackjack oak | 68, 70, 137, 247, 276 |
| Bonducs / Guilanda? / Guilandina dioica | *Gymnocladus dioicus* (L.) K. Koch | Kentucky coffee-tree | 232, 235, 279, 283, 284, 286, 314 |
| Bryonia, plant monoecious calyx 5-parted, corolla 5-parted, male flowers spicate, axillary, female flowers each axillary, fruits with harmless spines | *Echinocystis lobata* (Michx.) Torr. & A. Gray | wild-cucumber | 230 |
| Bursera gummifera | *Bursera simaruba* (L.) Sarg. | gumbo limbo | 120, *367* |
| Buxus balearica | *Buxus balearica* Lam. | balearic boxwood | 38 |
| Cacalia | *Prenanthes* sp. | rattlesnake-root | 228 |

| Michaux's names for a particular species | Modern binomial (scientific name) | Common name(s) | Page(s) in this volume |
|---|---|---|---|
| Cacalia atriplicifolia | *Arnoglossum atriplicifolium* (L.) H. Rob. | pale Indian-plantain | 78 |
| Cacalia hastata | *Prenanthes trifoliolata* (Cass.) Fernald | gall-of-the-earth | 193 |
| Cacalia incana | *Prenanthes racemosa* Michx. | rattlesnake root | 193, *360* |
| Calceolaria? | *Calceolaria multiflora* Cav. | slipper flower | 120 |
| Calla palustris | *Calla palustris* L. | wild calla / water-arum | 177, 179 |
| Callitriche americana | *Callitriche heterophylla* Pursh var. *heterophylla* | twoheaded water-starwort | 308 |
| Calycanthus | *Calycanthus floridus* L. | sweet-shrub | 65, 66, 109, 110, 111, 126, 148, *357* |
| Camelia japonica | *Camellia japonica* L. | camellia | 9, 38 |
| Campanula, many flowered, not campanulate / Campanula . . . | *Campanula americana* L. | tall bellflower | 314 |
| Campanula . . . | *Campanula rotundifolia* L. | bluebell / harebell | 175 |
| Cane / Reeds or canes up to 2–3 inches diameter, 25–30 feet high | *Arundinaria tecta* (Walter) Muhl. or *Arundinaria gigantea* (Walter) Muhl. | switch cane / small cane / giant cane / river cane | 101, 303 |
| Carduus virginicus | *Cirsium virginianum* (L.) Michx. | Virginia thistle | 253, 254 |
| Carex | *Carex* sp. | sedge | 97, 175, 176, 193 |
| Carica papaya | *Carica papaya* L. | papaya | 91 |
| Carolina myrica | *Morella carolinensis* (Mill.) Small or *Morella cerifera* (L.) Small or *Morella pumila* (Michx.) Small | evergreen bayberry / common wax-myrtle / southern bayberry / dwarf bayberry / dwarf wax-myrtle | 249 |
| Carpinus . . . / Carpinus, in fruit / Carpinus ostrya / Carpinus ostrya americana / Horn-beam | *Ostrya virginiana* (Mill.) K. Koch | American hop-hornbeam | 111, 175, 176, 196, 275, 294 |
| Cashew nut | *Anacardium occidentale* L. | cashew | 151 |
| Cassia chamaecrista | *Chamaecrista fasciculata* (Michx.) Greene | common partridge-pea | 78, *357* |
| Cassia marylandica | *Senna marilandica* (L.) Link | Maryland wild senna | 228 |
| Cassia nictitans | *Chamaecrista nictitans* (L.) Moench | sensitive-plant | 78 |

| Michaux's names for a particular species | Modern binomial (scientific name) | Common name(s) | Page(s) in this volume |
|---|---|---|---|
| Cassine | *Ilex montana* Torr. & A. Gray ex A. Gray | mountain holly | 256 |
| Catalpa / Bignonia catalpa | *Catalpa speciosa* (Warder) Warder ex Engelm. | northern catalpa | 249, 285, 288 |
| Catesbaea spinosa | *Catesbaea spinosa* L. | prickly apple / Spanish guava | 120 |
| Ceanothus / Ceanothus americanus | *Ceanothus americanus* L. | New Jersey tea | 67, 175, 197, 198 |
| Ceanothus floridanus / Ceanothus new sp. / Unknown shrub | *Ceanothus microphyllus* Michx. | redroot / little-leaf buckbush | 109, 154 |
| Celastrus / Celastrus scandens / Celastrus ... | *Celastrus scandens* L. | American bittersweet | 172, 175, 196 |
| Celtis occidentalis | *Celtis occidentalis* L. | northern hackberry | 228, 270, 286, 288 |
| Cephalanthus / Cephalanthus occidentalis | *Cephalanthus occidentalis* L. | buttonbush | 129, 175, 196, 229 |
| Cerasus | *Prunus* sp. | cherry | 176 |
| Cerasus azarero | *Prunus lusitanica* L. | Portuguese cherry-laurel | 38 |
| Cerasus corymbosus, petioles glandular / Cerasus corymbosus | *Prunus pensylvanica* L. f. | fire cherry / pin cherry | 186, 193 |
| Cerasus, fruit black, petioles non-glandular / Cerise de Sable | *Prunus pumila* L. var. *depressa* Pursh | prostrate dwarf-cherry / northern sand cherry | 186 |
| Cerasus racemosa, petioles glandular / Padus virginiana / Padus virginiana minor | *Prunus virginiana* L. | choke cherry | 176, 186 |
| Cercis canadensis | *Cercis canadensis* L. | eastern redbud | 228, 229, 274, 288 |
| Chamaerops | *Sabal palmetto* (Walter) Lodd. ex J. A. Schult. & J. H. Schult. | cabbage palmetto | 85, 93, 99 |
| Chamaerops acaulis / Chamaerops of Carolina | *Sabal minor* (Jacq.) Pers. | dwarf palmetto | 63, 250 |
| Chamaerops monosperma, leaves acute, dentate, rootstock prostrate / Chamaerops repens / Chamaerops, saw toothed / Palm, Chamaerops recurvata caule | *Serenoa repens* (W. Bartram) Small | saw palmetto | 63, 89, 92, 95, *363* |
| Chelone glabra / Chelone glabra, white flowers | *Chelone glabra* L. | white turtlehead | 64, 176, 183 |
| Chelone hirsuta | *Penstemon hirsutus* (L.) Willd. | northeastern beard-tongue / hairy beardtongue | 176 |

| Michaux's names for a particular species | Modern binomial (scientific name) | Common name(s) | Page(s) in this volume |
|---|---|---|---|
| Chene blanc / Quercus alba / Quercus pinnatifida / White oak | *Quercus alba* L. | white oak | 68, 135, 239, 274, 277, 279, 299 |
| Chestnut oak / Quercus prinus / Quercus prinus saxosa / Quercus prinus-glauca | *Quercus michauxii* Nutt. or *Quercus montana* Willd. | swamp chestnut oak / rock chestnut oak | 199, 256, 274, 275, 277, 279, 283, 285, 294 |
| Chicasaw plum / Cherokee plum | *Prunus angustifolia* Marshall | chickasaw plum / sand plum | 43, 158, 275 |
| Chinquapin / Fagus chinquapin / Fagus pumila / Fagus pumila (Chinquapin) | *Castanea pumila* (L.) Mill. | common chinquapin | 104, 105, 133, 144, 238, 239 |
| Chionanthus | *Chionanthus virginicus* L. | fringe-tree / old man's beard | 62, 105, 133, 147, *357* |
| Chrysocoma new sp. | *Eupatorium capillifolium* (Lam.) Small | dog-fennel | 120 |
| Chrysophyllon glabrum | *Chrysophyllum argenteum* Jacq. | bastard redwood / smooth star apple | 149 |
| Cinna arundinacea | *Cinna arundinacea* L. | sweet woodreed / common woodreed | 314 |
| Circea canadensis | *Circaea canadensis* (L.) Hill subsp. *canadensis* | Canada enchanter's-nightshade | 175, 195, 229 |
| Cistus canadensis | *Crocanthemum canadense* (L.) Britton | Canada frostweed | 176 |
| Claytonia virginica | *Claytonia virginica* L. | spring-beauty | 270, 304, 305 |
| Clematis erecta | *Clematis ochroleuca* Aiton | curlyheads | 137, 146 |
| Clematis semineparvo | *Clematis* sp. | undetermined clematis | 146 |
| Clethra | *Clethra alnifolia* L. | coastal sweet-pepperbush / coastal white-alder | 108 |
| Clethra buxifolia? / Ledum? buxifolium / Ledum buxifolium | *Kalmia buxifolia* (P. J. Bergius) Gift, Kron & Stevens | sand-myrtle | 255, 257, 259 |
| Clethra new sp., Clethra montana | *Clethra acuminata* Michx. | mountain sweet pepperbush | 73, 256, *357* |
| Climbing fern | *Lygodium palmatum* (Bernh.) Sw. | American climbing fern | 237, 303 |
| Clinopodium incanum | *Pycnanthemum incanum* (L.) Michx. | mountain-mint | 228 |
| Clinopodium vulgare / Ziziphora new sp. | *Clinopodium vulgare* L. | wild basil | 228 |
| Coccoloba | *Coccoloba* sp. | undetermined | 120 |
| Coffea | *Psychotria nervosa* Sw. | wild coffee | 99 |

| Michaux's names for a particular species | Modern binomial (scientific name) | Common name(s) | Page(s) in this volume |
|---|---|---|---|
| Coix? | *Coix lacryma-jobi* L. | Job's-tears | 152 |
| Collinsonia canadensis | *Collinsonia canadensis* L. | northern horse-balm / richweed | 175 |
| Collinsonia tuberosa | *Collinsonia tuberosa* Michx. | stoneroot | 260 |
| Comarum / Comarum palustre | *Comarum palustre* L. | purple marshlocks | 177, 184, *370* |
| Convallaria?, berries blue [Quebec] | *Clintonia borealis* (Aiton) Raf. | bluebead-lily | 186 |
| Convallaria, 2-leaved | *Maianthemum canadense* Desf. | false lily-of-the-valley / Canada mayflower | 175, 178, 186, 256, 271 |
| Convallaria, 3-leaved / Convallaria stellata | *Maianthemum stellatum* (L.) Link | starry Solomon's-plume / starry false lily-of-the-valley | 178, 186, 193 |
| Convallaria bifolia / Convallaria multiflora | *Polygonatum biflorum* (Walter) Elliott | Solomon's-seal | 255, 271 |
| Convallaria majalis / Convallaria majalis, leaves entire, glabrous, flowers on one side of raceme, and blue berries / Convallaria majalis?, yellow berries | *Convallaria majalis* L. subsp. *majuscula* (Greene) Gandhi, Reveal & Zarucchi | American lily-of-the-valley | 255, 271, *356* |
| Convallaria polygonatum maximum | *Polygonatum biflorum* (Walter) Elliott var. *commutatum* (J. A. & J. H. Schultes) Morong | large Solomon's-seal | 175 |
| Convallaria racemosa | *Maianthemum racemosum* (L.) Link | false Solomon's-seal / eastern Solomon's plume | 186, 255, 271 |
| Convallaria trifolia | *Maianthemum trifolium* (L.) Sloboda | three-leaf Solomon's-seal / three-leaf false lily-of-the-valley | 178, 186, 193 |
| Convallaria umbellata, blue berries / Convallaria umbellata, leaves with entire margins and [illegible], softly hairy, flowers in umbels, blue berries / Convallaria? umbellata | *Clintonia umbellulata* (Michx.) Morong | speckled wood-lily | 255, 256, 271 |
| Convolvulus | *Convolvulus* sp. | field-bindweed | 150 |
| Convolvulus dissectus | *Merremia dissecta* (Jacq.) Hallier f. | noyau vine | 97 |
| Coreopsis verticillata, leaves ovate | *Coreopsis verticillata* L. | threadleaf coreopsis | 253 |
| Cork-tree / Quercus suber | *Quercus suber* L. | cork oak | 38, 163 |
| Cornus | *Cornus* sp. | dogwood | 175 |

*Table of Described Plants (continued)*

| Michaux's names for a particular species | Modern binomial (scientific name) | Common name(s) | Page(s) in this volume |
|---|---|---|---|
| Cornus (Red osier of the Canadians) / Cornus canad. (Cornus Osier rouge) / Cornus, stolonib. red stems / red Osier | *Cornus stolonifera* Michx. | red osier dogwood | 190, 193, *363* |
| Cornus alternifolia | *Cornus alternifolia* L. f. | alternate-leaf dogwood | 70, 111, 113, 175 |
| Cornus, branches punctate | *Cornus racemosa* Lam. | northern swamp dog-wood / gray dogwood | 174 |
| Cornus canadensis / Cornus herbacea | *Cornus canadensis* L. | bunchberry / dwarf dogwood | 175, 177, 186, 193, *357* |
| Cornus florida | *Cornus florida* L. | flowering dogwood | 199, 229, 274 |
| Cornus, branches pitted | *Cornus rugosa* Lam. | roundleaf dogwood | 175 |
| Cornus, leaves like laurel willow | *Nectandra coriacea* (Sw.) Grisb. | lancewood | 120 |
| Corylus americana | *Corylus americana* Walter | American hazelnut | 305, 306 |
| Corylus cornuta | *Corylus cornuta* Marshall var. *cornuta* | beaked hazelnut | 306 |
| Crataegus . . . / Crataegus coccinea / Crataegus lutea | *Crataegus* sp. | hawthorn | 136, 150, 176, 196 |
| Crinum americanum | *Crinum americanum* L. | string-lily | 99 |
| Crinum rubrum | *Crinum* sp. | swamp-lily | 79 |
| Crotalaria alba? | *Baptisia alba* (L.) Vent. | thick-pod white wild indigo | 228 |
| Crotalaria / Crotalaria laburnifolia | *Crotalaria* sp. or *Crotalaria laburnifolia* L. | rattlebox | 149, 151 |
| Croton benzoin | *Styrax benzoin* Dryand. | Benjamin tree | 38 |
| Croton cascarilla / Croton cascarilla / Ricinoides?, leaves of Eleagnus | *Croton cascarilla* (L.) W. Wright | eleutheria-bark / cascarilla-bark | 120, 123, 153 |
| Croton, dioecious shrub / Ricino? leaves like rosemary / Eleuthera bark | *Croton eleuteria* (L.) W. Wright | eleutheria-bark / cascarilla-bark | 153, 154 |
| Cucurbita okeechobeensis | *Cucurbita okeechobeensis* (Small) L. H. Bailey | Okeechobee gourd, wild squash | 101 |
| Cunila / Cunila pulegioides, flowers 4-parted | *Cunila origanoides* L. Britton or *Hedeoma pulegioides* (L.) Pers. | stone-mint / American pennyroyal | 228, 248 |
| Cuphea viscosa / Cuphea viscososa | *Cuphea viscosissima* Jacq. | blue waxweed | 235, 315 |
| Cupressus disticha / Cypress | *Taxodium ascendens* Brong. or *Taxodium distichum* (L.) L. C. Rich. | pond-cypress / bald-cypress | 61, 75, 89, 95, 136, 147, 261 |

| Michaux's names for a particular species | Modern binomial (scientific name) | Common name(s) | Page(s) in this volume |
|---|---|---|---|
| Cupressus fastigata / Pyramidical cypress | *Cupressus sempervirens* L. | Italian cypress | 38, 44, 154 |
| Cynoglossum | *Cynoglossum virginianum* L. | wild comfrey | 175, 179 |
| Cynoglossum officinalis | *Cynoglossum officinale* L. | garden comfrey / hound's-tongue | 175 |
| Cynoglossum, 3 species | *Cynoglossum* sp. | hound's-tongue | 229 |
| Cypripedium | *Cypripedium* sp. | lady's-slipper | 176 |
| Cypripedium calceolaria, red flower | *Cypripedium acaule* Aiton | pink lady's-slipper / moccasin-flower | 195 |
| Cypripedium calceolaria, two species / Cypripedium calceolaria, yellow flower / Cypripedium luteum | *Cypripedium parviflorum* Salisb. or *Cypripedium parviflorum* Salisb. var. *pubescens* (Willd.) Knight | large yellow lady's-slipper / small yellow lady's-slipper | 66, 256, 272 |
| Cyrilla / Cyrilla racemiflora | *Cyrilla racemiflora* L. | ti-ti | 79, 248, 249 |
| Cytire | *Cytisus scoparus* (L.) Link | scotch broom | 44 |
| Delphinium | *Delphinium exaltatum* Aiton | tall larkspur | 254 |
| Dianthera . . . / Dianthera americana / Dianthera new sp. / Justicia? | *Justicia americana* (L.) Vahl. | American water-willow | 177, 229, 231, 252 |
| Dianthus, in bloom | *Dianthus armeria* L. | deptford pink | 308 |
| Diervilla / Lonicera diervilla | *Diervilla lonicera* Mill. | northern bush-honeysuckle | 175, 186, 193, 256, 372 |
| Dionaea muscipula | *Dionaea muscipula* J. Ellis | Venus flytrap | 134 |
| Dioscorea, fruit inferior | *Dioscorea villosa* L. | common wild yam | 229 |
| Diospyros / Diospyros virginiana / Persimmons | *Diospyros virginiana* L. | American persimmon | 68, 105, 129, 247, 274, 285 |
| Dirca palustris | *Dirca palustris* L. | leatherwood / leatherbark | 61, 68, 175, 235 |
| Disette | *Beta vulgaris* L. | beet | 97 |
| Dodecatheon meadia | *Primula meadia* (L.) A. R. Mast & Reveal | eastern shooting star | 270 |
| Draba bursa-p. | *Draba arabisans* Michx. | rock draba | 176 |
| Dracocephalum virginianum | *Physostegia virginiana* (L.) Benth. | obedient-plant | 228, 229, 314 |

| Michaux's names for a particular species | Modern binomial (scientific name) | Common name(s) | Page(s) in this volume |
|---|---|---|---|
| Drosera | *Drosera anglica* Huds. or *Drosera intermedia* Hayne | English sundew / spoonleaf sundew | 179 |
| Drosera rotundifolia | *Drosera rotundifolia* L. | roundleaf sundew | 182 |
| Droseras | *Drosera* sp. | sundews | 30 |
| Echium vulgare | *Echium vulgare* L. | viper's-bugloss | 130 |
| Elm used for making wheel hubs [IL] | *Ulmus thomasii* Sarg. | cork elm / rock elm | 288 |
| Empetrum nigrum | *Empetrum nigrum* L. | black crowberry | 179, 181 |
| Epigea / Epigea procumbens / Epigea repens | *Epigaea repens* L. | trailing arbutus | 70, 109, 114, 116, 138, 175, 179, 186, 189, 190, 238, 257, 307 |
| Epilobium, linear leaves | *Epilobium* sp. | willow-herb | 189 |
| Epilobium, petals split in two | *Epilobium ciliatum* Raf. subsp. *glandulosum* (Lehm.) Hoch & P. H. Raven | fringed willow-herb | 186 |
| Epilobium, stamens bent | *Chamerion platyphyllum* (Daniels) Löve & Löve | fireweed / great willow-herb | 186 |
| Equisetum | *Equisetum hyemale* L. or *Equisetum* sp. | tall scouring-rush / horsetail | 176, 292 |
| Erigeron canadense | *Conyza canadensis* (L.) Cronquist var. *canadensis* | common horseweed | 229 |
| Eriocaulon . . . | *Eriocaulon aquaticum* (Hill) Druce | seven-angled pipewort | 184, *371* |
| Eriophorum | *Eriophorum virginicum* L. | tawny cottongrass | 253 |
| Eryngium, lanceolate leaves | *Eryngium aquaticum* L. | marsh eryngo | 253 |
| Erythrina | *Erythrina herbacea* L. | coral bean | 98, 147 |
| Erythronium dens-leonis | *Erythronium umbilicatum* Parks & Hardin | dimpled trout lily | 270 |
| Erythronium, spotted leaves | *Erythronium americanum* Ker Gawl. | trout-lily / dog-tooth violet | 305 |
| Evonimus, capsules glabrous / Evonimus latifolius?, flowers 4-merous, capsules glabrous, leaves petiolate / Evonimus, branches almost round, capsules smooth / Evonimus latifolius | *Euonymus atropurpureus* Jacq. | American wahoo / burning bush | 230, 235, 274, 315 |

| Michaux's names for a particular species | Modern binomial (scientific name) | Common name(s) | Page(s) in this volume |
|---|---|---|---|
| Evonimus / Evonimus americanus, flowers 5-merous, capsules warty, leaves subsessile / Evonmus, branches 4-angled, capsules with spines / Evonimus americanus | *Euonymus americanus* L. | hearts-a-bustin'- (with-love) / strawberry-bush | 148, 235, 274, 315, *358* |
| Eupatorium / Eupatorium perfoliatum | *Eupatorium* sp. or *Eupatorium perfoliatum* L. | throughwort / boneset | 176, 228 |
| Eupatorium aromaticum | *Ageratina aromatica* (L.) Spach | small-leaved white snakeroot | 229, 232 |
| Eupatorium celestinum | *Conoclinium coelestinum* (L.) DC. | mistflower | 229 |
| Eupatorium maculatum | *Eutrochium maculatum* (L.) E. E. Lamont | spotted joe-pye-weed | 229 |
| Eupatorium odoratum | *Ageratina altissima* (L.) King & H. Rob. | common white snakeroot | 229 |
| Euphrasia odontites | *Euphrasia nemorosa* (Pers.) Wallr. | common eyebright | 178 |
| Fagara | *Zanthoxylum piperitum* DC. | Sichuan pepper | 150 |
| Fagara / strawberries | *Fragaria virginiana* Mill. | wild strawberry / Virginia strawberry | 175, 176 |
| Fagara, roots of as remedy | *Zanthoxylum fagara* (L.) Sarg. | lime prickly-ash | 288 |
| Fagus americana / Fagus sylvatica / Fagus sylvatica americana | *Fagus grandifolia* Ehrh. | American beech | 175, 176, 196, 229, 238, 274, 277, 279, 294, 299 |
| Fagus castanea / Fagus castanea americana | *Castanea dentata* (Marshall) Borkh. | American chestnut | 199, 227, 238, 256, 274, 277 |
| Festuca | *Chasmanthium sessiliflorum* (Poir.) Yates | longleaf spikegrass | 172 |
| Fig tree | *Ficus aurea* Nutt. | strangler fig | 92 |
| Fothergilla / Fothergilla gardeni | *Fothergilla gardenii* L. | coastal witch-alder | 103, 104, 105, 145, 147, *372* |
| Franklinia / Gordonia or Franklinia | *Franklinia alatamaha* W. Bartram ex Marshall | Franklinia, Franklin tree | 41, 51, *358*, 428n34 |
| Frasera foetida | *Frasera caroliniensis* Walter | American columbo | 308 |
| Fraxinus | *Fraxinus* sp. | ash | 176, 183, 274, 285, 295, 299 |
| Fraxinus (quadrangularis) / Fraxinus, leaflets serrate, branches 4-angled / Fraxinus ramis quadrangularis | *Fraxinus quadrangulata* Michx. | blue ash | 232, 235, 314 |
| Fraxinus, etc. etc. / Fraxinus, leaflets tomentose, serrate | *Fraxinus pensylvanica* Marshall | green ash | 186 |

| Michaux's names for a particular species | Modern binomial (scientific name) | Common name(s) | Page(s) in this volume |
|---|---|---|---|
| Fraxinus, leaflets subentire | *Fraxinus americana* L. | white ash | 235 |
| Fraxinus palustris | *Fraxinus caroliniana* Mill. | Carolina ash / pop ash | 95, 106 |
| Fumaria sempervirens | *Capnoides sempervirens* (L.) Borkh. | rock harlequin | 176 |
| Fumaria vesicaria / Fumaria vesicaria scandens | *Adlumia fungosa* (Aiton) Greene ex BSP. | Allegheny-vine | 174 |
| Galium / Galium album | *Galium* sp. | bedstraw | 175, 179 |
| Gardenia (native to Bahamas) / Gardenia new sp. | *Casasia clusiifolia* (Jacq.) Urb. | seven-year apple | 120, *367* |
| Gaultheria procumbens / Gaultheria / Shrub, new, with toothed leaves / Shrub with saw-toothed leaves | *Gaultheria procumbens* L. | wintergreen / teaberry | 36, 138, 175, 176, 193, 238, 257 |
| Gentiana? / Gentiana pneumonanthe | *Gentiana linearis* Froel. | narrowleaf gentian | 184, 186, 190 |
| Gentiana pumila | *Gentiana pumila* Jacq. | (European alpine species) | 146 |
| Geranium [IL] | *Geranium maculatum* L. | wild geranium | 286 |
| Geranium [NY] | *Geranium maculatum* L. or *Geranium robertianum* L. | wild geranium / herb-Robert | 174 |
| Geranium [VT] | *Geranium bicknellii* Britton | northern cranebill | 176 |
| Gerardia, stems usually unbranched, leaves sessile, opposite, ovate, flowers axillary, purplish | *Agalinis auriculata* (Michx.) S. F. Blake | earleaf false foxglove | 280 |
| Geum . . . | *Geum virginianum* L. | cream avens | 176 |
| Ginkgo | *Ginkgo biloba* L. | ginkgo / maidenhair tree | 9, 38, 39 |
| Ginseng / Panax, 5 leaflets / Panax quinquefolium | *Panax quinquefolius* L. | ginseng / American ginseng | 70, 128, 174, 176, 215, 228, 229, 230, 314 |
| Glaux? | *Lysimachia maritima* (L.) Galasso, Banfi, & Soldano | sea-milkwort | 179 |
| Gleditsia / Gleditsia triacanthos / honey locust | *Gleditsia triacanthos* L. | honey locust | 59, 61, 62, 70, 78, 232, 235, 279, 283, 285, 286 |
| Gleditsia monosperma / Gleditsia aquatica / Gleditsia, oval one-seeded closed capsule | *Gleditsia aquatica* Marshall | water locust | 64, 79, 80, 100, 105, 145, 147, 148 |
| Gnaphalium dioicum | *Antennaria plantaginifolia* (L.) Richardson | plantain pussytoes / woman's tobacco | 176, 270, 301 |

| Michaux's names for a particular species | Modern binomial (scientific name) | Common name(s) | Page(s) in this volume |
|---|---|---|---|
| Gordonia / Gordonia lasianthus | *Gordonia lasianthus* (L.) J. Ellis. | loblolly bay | 61, 89, 93, 145, 156, 159, 250 |
| Gramen monoecious, bulbous / buffalo grass [KY] | *Bouteloua curtipendula* (Michx.) Torr. var. *curtipendula* | side-oats grama | 314 |
| Grapefruit | *Citrus x paradisi* Macfad. | grapefruit | 153 |
| Grenadier | *Punica granatum* L. | pomegranate | 150 |
| Guilandina bonduccella | *Caesalpinia bonduc* (L.) Roxb. | gray nicker | 94 |
| Guilandina moringha | *Moringa oleifera* Lam. | horseradish tree / drumstick tree | 149 |
| Guinea grass / Panicum altissimum | *Panicum maximum* Jacq. | guinea grass | 143, 164 |
| Halesia | *Halesia* sp. | silverbell | 105, 109, 147, 307 |
| Halesia diptera | *Halesia diptera* J. Ellis | two-wing silverbell | 62, 64, *368* |
| Halesia tetraptera | *Halesia tetraptera* J. Ellis | common silverbell | 111, 256 |
| Halesia tetraptera, with small flowers | *Halesia carolina* L. | Carolina silverbell / little silverbell | 89 |
| Hamamelis . . . / Hamamelis virginiana | *Hamamelis virginiana* L. | northern witch-hazel | 175, 199, 229, 256 |
| Hedisarum | *Desmodium canadense* (L.) DC. | Canadian tick-trefoil / showy tick-trefoil | 176 |
| Hedisarum | *Hedysarum* sp. | sweetvetch / sainfoin | 151 |
| Hedisarum gyrans | *Codariocalyx motorius* (Houtt.) Ohashi | telegraph-plant | 151 |
| Helianthus atrorubens | *Helianthus atrorubens* L. | Appalachian sunflower | 257 |
| Heliotropium | *Heliotropium indicum* L. | turnsole | 253 |
| Heliotropium peruvianum | *Heliotropium arborescens* L. | grey leaf heliotrope | 38 |
| Hemerocallis fulva | *Hemerocallis fulva* (L.) L. | orange day-lily | 62 |
| Heuchera . . . / Heuchera americana | *Heuchera americana* L. | American alumroot | 235, 270 |
| Hibiscus | *Hibiscus* sp. | rose-mallow | 150, 154 |
| Hibiscus sinensis | *Hibiscus mutabilis* L. | cotton rose / confederate rose | 152 |

*Table of Described Plants (continued)*

| Michaux's names for a particular species | Modern binomial (scientific name) | Common name(s) | Page(s) in this volume |
|---|---|---|---|
| Hibiscus, with hastate leaves, outer calyx subulate and laciniate, flowers pale pink | *Hibiscus laevis* All. | smooth rose-mallow | 277 |
| Hieracium paludosum? | *Hieracium kalmii* L. or *Heracium scabrum* Michx. | Kalm's hawkweed or rough hawkweed | 190, 193 |
| Holcus? | *Phalaris arundinacea* L. | reed canary-grass | 279 |
| Hopea / Hopea tinctoria | *Symplocos tinctoria* (L.) L' Hér. | sweetleaf / horse sugar | 134, 148, 248, 249 |
| Hordeum murinum | *Hordeum jubatum* L. subsp. *jubatum* | foxtail barley / mouse barley | 179 |
| Houstonia purpurea | *Houstonia longifolia* Gaertn. | longleaf bluet | 175 |
| Hydrangea (glauca) | *Hydrangea radiata* Walter | snowy hydrangea | 111 |
| Hydrangea arborescens / Hydrangea frutescens | *Hydrangea arborescens* L. | smooth hydrangea | 159, 227, 235 |
| Hydrastis | *Hydrastis canadensis* L. | golden-seal | 131 |
| Hydrocotyle reniformis | *Centella asiatica* (L. f.) Fernald | centella / coinleaf | 154 |
| Hydrophyllum canadense | *Hydrophyllum canadense* L. | mapleleaf waterleaf / Canada waterleaf | 229 |
| Hydrophyllum virginicum | *Hydrophyllum virginianum* L. | eastern waterleaf | 173 |
| Hypericum [VT] | *Hypericum canadense* L. | Canada St. John's-wort | 176 |
| Hypericum kalmianum* / Hypericum kalmianum* grandiflorum [used in error] | *Hypericum* sp. | St. John's-wort | 238, 275 |
| Hypericum new sp. | *Hypericum ascyron* L. | great St. John's-wort | 177 |
| Hypericum perforatum | *Hypericum perforatum* L. | European St. John's-wort | 130 |
| Hypericum new sp. of subshrub, calcareous and dry sites | *Hypericum spherocarpum* Michx. | roundseed St. John's-wort | 314 |
| Hypoxis erecta | *Hypoxis hirsuta* (L.) Coville | common stargrass | 175 |
| Hyppophae / Hyppophae canadensis | *Shepherdia canadensis* (L.) Nutt. | russet buffalo-berry / rabbit-berry | 175, 176, 179, 184, 198, *370* |
| Ilex | *Ilex opaca* Aiton | American holly | 137, 147 |
| Ilex aestivalis | *Ilex decidua* Walter | possum-haw | 248 |
| Ilex angustifolia / Ilex cassine [NC] | *Ilex cassine* L. | dahoon / cassena | 99, 101 |

| Michaux's names for a particular species | Modern binomial (scientific name) | Common name(s) | Page(s) in this volume |
|---|---|---|---|
| Ilex angustifolia / Ilex, with narrow and small leaves [SC] | *Ilex myrtifolia* Walter | myrtle holly | 116n, 135, 148, 250 |
| Illecebrum divaricatum, plant annual, erect | *Paronychia canadensis* (L.) Wood | Canada whitlow-wort | 315 |
| Illicium / Illicium, flowers yellow / Illicium new sp. | *Illicium parviflorum* Michx. ex Vent. | yellow anise-tree / swamp star-anise | 99, 101, 105, 106, 135, 150 |
| Illicium floridanum | *Illicium floridanum* J. Ellis | Florida star-anise | 105, 139, *368* |
| Impatiens . . . | *Impatiens capensis* Meerb. | jewelweed / spotted touch-me-not | 176 |
| Indigofera? | *Indigofera* sp. | indigo | 150 |
| Ipomoea | *Ipomoea alba* L. | moonflower | 101 |
| Iresine celesioides | *Iresine rhizomatosa* Standl. | bloodleaf | 231, 235, 315 |
| Iris coerulea | *Iris versicolor* L. | northern blue flag | 175 |
| Jatropha pinnatifolia | *Jatropha multifida* L. | coralbush / physic nut | 149 |
| Juglans hiccory / Juglans alba / Hickories, hard nuts, pungent nut, American nut, bitternut, tree with round nut | *Carya* sp. incl. *Carya aquatica* (Michx. f.) Elliott and *Carya cordiformis* (Wangenh.) K. Koch and *Carya ovata* (Mill.) K. Koch | hickory / water hickory / bitternut hickory / common shagbark hickory | 89, 196, 229, 274, 279, 283, 285 |
| Juglans nigra | *Juglans nigra* L. | black walnut | 81, 227, 274, 275, 279 |
| Juglans oblonga | *Juglans cinerea* L. | butternut | 138, 176, 179, 229, 274 |
| Juglans pecan / Pecan trees | *Carya illinoinensis* (Wangenh.) K. Koch | pecan | 279, 285, 293 |
| Juglans pignut / pignut hickory | *Carya glabra* (Mill.) Sweet | pignut hickory | 274, 285 |
| Juniper (repens) / Juniperus communis / Juniperus europea? | *Juniperus communis* L. var. *depressa* Pursh | common juniper / ground juniper | 173, 174, 179, 181, 184, 191, 198, *371* |
| Juniperus / Juniperus (cedar) / Juniperus virginiana | *Juniperus virginiana* L. | eastern red cedar | 79, 105, 176, 199, 239, 275 |
| Juniperus of the Bahamas | *Juniperus barbadensis* L. | Barbados cedar / pencil cedar | 123, 153, 154 |
| Juniperus sabina? | *Juniperus horizontalis* Moench | creeping juniper | 181, 193 |
| Kalmia repens / Kalmia hirsuta | *Kalmia hirsuta* Walter | hairy wicky | 63, 163, *368* |

| Michaux's names for a particular species | Modern binomial (scientific name) | Common name(s) | Page(s) in this volume |
|---|---|---|---|
| Kalmia angustifolia | *Kalmia angustifolia* L. | northern sheepkill / sheep laurel | 36, 134, 177, 179, 182, 187, 189, 190, 191, 193, *371* |
| Kalmia glauca | *Kalmia polifolia* Wangenh. | bog laurel / swamp laurel | 148, 177, 178, 186, 187, 188, 189, 190, 191, 193, 196 |
| Kalmia latifolia | *Kalmia latifolia* L. | mountain laurel | 61, 70, 72, 109, 110, 111, 139, 199, 227, 238, 256, *359* |
| Kalmia, new | *Kalmia cuneata* Michx. | white wicky | 253, 270, 308, *359* |
| Labrador tea / Ledum palustre | *Rhododendron groenlandicum* (Oeder) Kron & Judd | Labrador tea, bog Labrador tea | 177, 178, 179, 182, 185, 186, 187, 188, 189, 191 |
| Lamium | *Lamium* sp. | dead-nettle | 195 |
| Lapathum occidentale | *Cliftonia monophylla* (Lam.) Britton ex Sarg. | black ti-ti / buckwheat-tree | 65, 109, *367* |
| Lappa . . . | *Arctium minus* (Hill) Bernh. | common burdock | 130, 195, 196 |
| Larix / Pinus, needles clustered / Pinus larix | *Larix laricina* (Du Roi) K. Koch | eastern tamarack / eastern larch | 178, 179, 185, 186, 190, 191, 192, 193 |
| Larix orientalis | *Larix kaempferi* (Lamb) Carrière | Japanese larch | 38 |
| Laurus aestivalis | *Litsea aestivalis* (L.) Fernald | pondspice | 78, 83, 248 |
| Laurus borbonia | *Persea borbonia* (L.) Spreng. | red bay | 83, 89, 99, 156, 249 |
| Laurus indica | *Persea indica* (L.) Spreng. | viñatigo (Spanish) | 83, 120, 156 |
| Laurus persea | *Persea americana* Mill. | American avocado | 120 |
| Laurus sassafras / Sassafras | *Sassafras albidum* (Nutt.) Nees | sassafras | 27, 199, 229, 299 |
| Lawsonia inermis | *Lawsonia inermis* L. | henna | 36, 38, 151 |
| Lebanon cedar | *Cedrus libani* A. Rich. | cedar of Lebanon | 37 |
| Leontia thalictroides / Leontice thalictroides | *Caulophyllum thalictroides* (L.) Michx. | common blue cohosh | 146, 175, 228, 315 |
| Lepidium | *Lepidium* sp. | pepperweed | 176 |
| Liana [woody vine] Rajanioides / Rajania . . . dioecia 8-dria | *Brunnichia ovata* (Walter) Shinners | American buckwheat-vine | 284, 285 |

| Michaux's names for a particular species | Modern binomial (scientific name) | Common name(s) | Page(s) in this volume |
|---|---|---|---|
| Ligustrum monosper-mum / Olea americana | *Osmanthus ameri-canus* (L.) Benth. & Hook. ex A. Gray | wild olive / devilwood | 61n, 78, 80, 99, 101, 258 |
| Lilium canadense | *Lilium canadense* L. | Canada lily | 175 |
| Lilium philadelphicum | *Lilium philadelphicum* L. | wood lily | 175 |
| Limodorum | *Calopogon tuberosus* (L.) BSP. var. *tuberosus* | grass pink | 62 |
| Linnaea borealis | *Linnaea borealis* L. subsp. *americana* (Forbes) Hul-tén ex R. T. Clausen | American twinflower | 175, 176, 179, 186, 190, 193 |
| Liquidambar / Liquidambar styraciflua | *Liquidambar styraciflua* L. | sweet gum | 27, 30, 82, 136, 148, 199, 274, 275, 283, 285, 286, 294 |
| Liquidambar asplenifolium / Liquidambar peregrinum | *Comptonia peregrina* (L.) J. M. Coult. | sweet-fern | 27, 199, 261 |
| Liriodendron / Lirioden-dron tulipifera / tulip tree | *Liriodendron tulipifera* L. | tulip-tree / yel-low poplar | 27, 79, 199, 277, 278, 288, 294, 299, 300 |
| Lithospermum . . . | *Lithospermum officinale* L. | common gromwell | 196 |
| Lobelia cardinalis | *Lobelia cardinalis* L. | cardinal flower | 228 |
| Lobelia inflata | *Lobelia inflata* L. | Indian-tobacco | 228 |
| Lobelia simplex | *Lobelia dortmanna* L. | water lobelia | 184 |
| Lobelia siphilitica | *Lobelia siphilitica* L. | great blue lobelia | 176, 228, 315 |
| Lonicera (Chamaeceras) / Lonicera . . . | *Lonicera canadensis* W. Bartram ex Marshall | American fly-honeysuckle | 175 |
| Lonicera glauca / Lonic-era, glaucous, climbing | *Lonicera dioica* L. | limber honeysuckle | 175, *370* |
| Lonicera . . . | *Lonicera tatarica* L. | Tartarian honeysuckle | 175 |
| Lonicera . . . / Lonicera (Chamaeceras) / Lonicera camaecerasus, leaves tomentose | *Lonicera canadensis* W. Bartram ex Mar-shall or *Lonicera villosa* (Michx.) Schult. | American fly-honeysuckle / moun-tain fly-honeysuckle | 175, 186, *371* |
| Lucerne | *Medicago sativa* L. | alfalfa | 171 |
| Ludwigia | *Ludwigia* sp. | seedbox | 83 |

*Table of Described Plants (continued)*

| Michaux's names for a particular species | Modern binomial (scientific name) | Common name(s) | Page(s) in this volume |
|---|---|---|---|
| Lupinus hirsutus / Lupinus perennis [NC and SC] | *Lupinus perennis* L. subsp. *gracilis* (Nutt.) Dunn. | southern sundial lupine | 61, 250, 270, 308, 309, *368* |
| Lupinus hirsutus / Lupinus perennis [NY] | *Lupinus perennis* L. subsp. *perennis* | northern sundial lupine | 174 |
| Lupinus pilosus | *Lupinus diffusus* Nutt. or *Lupinus villosus* Willd. | blue sandhill lupine / sky blue lupine / pink sandhill lupine | 61, 250, 254, 308, 309 |
| Lupinus pilosus, blue flowers | *Lupinus diffusus* Nutt. | blue sandhill lupine / sky blue lupine | 89, *368* |
| Lycopodium | *Lycopodium* or *Lycopodiella* or *Huperzia* or *Dendrolycopodium* | clubmoss | 177, 178, 186 |
| Lycopodium inundatum | *Lycopodiella inundata* (L.) Holub | northern bog clubmoss / inundata clubmoss | 187 |
| Lysimachia, 4-leaved? | *Lysimachia quadrifolia* L. or *Lysimachia quadriflora* Sims | whorled loosestrife / smooth loosestrife | 175 |
| Lythrum lineare | *Lythrum lineare* L. | wand loosestrife | 154 |
| Magnolia (hastata) / Magnolia auriculata / Magnolia cordata, discovered a few years earlier / Magnolia cordata or auriculata | *Magnolia fraseri* Walter | Fraser magnolia / mountain magnolia | 73, 112, 113, 114, 126, 138, 145, 159, 256, 271, 305 |
| Magnolia acuminata / Magnolia acuminata, flowers glaucous | *Magnolia acuminata* (L.) L. | cucumber-tree | 72, 74, 75, 105, 110, 115, 138, 145, 227, 238, 256, 274, 305 |
| Magnolia acuminata, yellow flowers | *Magnolia acuminata* (L.) L. var. *subcordata* (Spach) Dandy | yellow cucumber-tree | 260 |
| Magnolia cordata, leaves glaucous / Magnolia, tomentose-glaucous, very long cordate leaves / Magnolia, with purple at base of the petals / Magnolia, with very large leaves auriculate, oblong, glaucous / Magnolia, leaves very long and cordate | *Magnolia macrophylla* Michx. | bigleaf magnolia | 126, 137, 254, 271, 275, 307, *356* |
| Magnolia glauca [NY to DE] | *Magnolia virginiana* L. | northern sweet bay | 32, 33, 36, 82, 133, *360* |
| Magnolia glauca [SC to FL] | *Magnolia virginiana* L. var. *australis* Sarg. | southern sweet bay | 93, 95, 99, 105, 147, 156 |
| Magnolia grandiflora | *Magnolia grandiflora* L. | southern magnolia | 31, 51, 61, 63, 78, 79, 89, 95, 99, 101, 147, 250, 347 |
| Magnolia tripetala | *Magnolia tripetala* (L.) L. | umbrella magnolia / umbrella-tree | 62, 115, 147, 162, 273, 275, *360* |

| Michaux's names for a particular species | Modern binomial (scientific name) | Common name(s) | Page(s) in this volume |
|---|---|---|---|
| Magnolia, leaves with 2-inch long petioles, cordate at the base, flowers with pleasant odor [GA] | *Magnolia pyramidata* W. Bartram | pyramid magnolia | 62 |
| Magnolias | *Magnolia* sp. | magnolia | 147, 318 |
| Malacea dolichos? | *Pachyrhizus tuberosus* (Lam.) Spreng. | yam bean | 152 |
| Malanga | *Xanthosoma* sp. | malanga | 152 |
| Male véronique / Veronica officinalis | *Veronica officinalis* L. | common speedwell | 97, 130, 228 |
| Mangifera, 1 seed | *Mangifera indica* L. | mango | 149 |
| Mangrove, fruits like those of Catesby's fig tree | *Laguncularia racemosa* (L.) C. F. Gaertn. | white mangrove | 94, *368* |
| Mangrove / Rhizophora mangle | *Rhizophora mangle* L. | red mangrove | 88 |
| Mangroves | *Rhizophora* sp. | mangroves | 88, 91, 93 |
| Martynia from Vera Cruz | *Martynia annua* L. | uña de gato (Spanish) / cat's claw | 151 |
| Medeola virginica | *Medeola virginiana* L. | Indian cucumber-root | 175, 195 |
| Menispermum / Menispermum canadense | *Menispermum canadense* L. | moonseed | 138, 176, 228, 229, 253 |
| Mentha, stamens longer than corolla | *Mentha arvensis* L. | field mint | 184 |
| Menyanthes? | *Menyanthes trifoliata* L. | bogbean | 151 |
| Mesembrianthemum edule | *Carpobrotus edulis* (L.) N.E. Br. | Hottentot fig | 151 |
| Mespilius | *Amelanchier* sp. or *Amelanchier obovalis* (Michx.) Ashe or *Amelanchier canadensis* (L.) Medik. | Serviceberry / coastal plain serviceberry / eastern serviceberry | 59 |
| Mespilus canadensis arborea [Quebec] | *Amelanchier hartramiana* (Trausch.) M. Roemer | Bartram's shadbush / oblong-fruited serviceberry | 192 |
| Mespilus canadensis arborea [VT] / Mespilus of the mountains [SC] | *Amelanchier arborea* (Michx. f.) Fernald | downy serviceberry / common serviceberry | 114, 176 |
| Mespilus?, very large treelet, red fruits | *Crataegus aestivalis* (Walter) Torr. & A. Gray | mayhaw / eastern mayhaw | 64 |
| Mimosa | *Mimosa* sp. | undetermined | 149, 150, 151, 152 |

| Michaux's names for a particular species | Modern binomial (scientific name) | Common name(s) | Page(s) in this volume |
|---|---|---|---|
| Mimosa . . . | *Mimosa microphylla* Dryander | eastern sensitive-briar | 134, *360* |
| Mimosa el abrisin | *Albizia julibrissin* Durazzini | mimosa | 9, 38, 39 |
| Mimosa erecta-herbacea / Mimosa mississipi | *Desmanthus illinoensis* (Michx.) MacMill. ex B. L. Rob. & Fernald | prairie mimosa / bundleflower / Illinois bundleflower | 275, 315 |
| Mimulus ringens | *Mimulus ringens* L. | Allegheny monkey-flower | 229 |
| Mirabilis clandestina or umbellata or parviflora | *Mirabilis nyctaginea* (Michx.) MacMill. | wild four-o'clock | 275 |
| Mitchella repens | *Mitchella repens* L. | partridge-berry | 299 |
| Mitella aphylla | *Mitella nuda* L. | naked miterwort | 183, 195 |
| Mitella diphylla | *Mitella diphylla* L. | two-leaved miterwort | 235 |
| Mollugo verticillata | *Mollugo verticillata* L. | carpetweed | 235 |
| Monarda coccinea / Monarda didyma | *Monarda didyma* L. | bee-balm / oswego tea | 228, 256, *361* |
| Monarda fistulosa | *Monarda fistulosa* L. | bergamot | 256 |
| Mussanda | *Pinckneya bracteata* (W. Bartram) Raf. | fever-tree | 164, *352, 372* |
| Myrica cerifera | *Morella cerifera* (L.) Small | common wax-myrtle / southern bayberry | 89, 249 |
| Myrica gale | *Myrica gale* L. | sweet gale | 176, 184, 186, 190, 192, 193 |
| Narthecium | *Narthecium* sp. | asphodel | 186 |
| Narthecium calyculatum | *Tofieldia pusilla* (Michx.) Pers. | scotch false asphodel | 178, *370* |
| Narthecium calyculatum / Narthecium, calyx with equal parts | *Triantha glutinosa* (Michx.) Baker | northern bog asphodel / sticky bog asphodel | 190, *370* |
| Nepeta altissima | *Agastache nepetoides* (L.) Kuntze | yellow giant-hyssop | 315 |
| Nymphaea / Red Nymphaea | *Nuphar* sp. or *Nymphaea* sp. | waterlily / pondlily | 152 |
| Nymphaea hastata | *Nuphar sagittifolia* (Walter) Pursh | narrowleaf pondlily | 261 |
| Nymphaea lutea / Nymphea lutea major | *Nuphar variegata* Durand in G. W. Clinton | variegated yellow pondlily | 183, 184 |

| Michaux's names for a particular species | Modern binomial (scientific name) | Common name(s) | Page(s) in this volume |
|---|---|---|---|
| Nymphaea lutea minor / Nymphaea lutea, leaves and flowers small | *Nuphar microphylla* (Pers.) Fernald | yellow pondlily | 183, 184 |
| Nymphaea nelumbo | *Nelumbo nucifera* Gaertn. | sacred lotus | 79 |
| Nyssa | *Nyssa aquatica* L. or *Nyssa sylvatica* Marshall | tupelo / water tupelo / black gum | 51 |
| Nyssa aquatica / Nyssa dentata / Nyssa, with large fruits / Nyssa, leaves acute, dentate | *Nyssa aquatica* L. | water tupelo | 59, 60, 61, 63, 75, 108, 136, 147, 249, 274 |
| Nyssa montana | *Nyssa sylvatica* Marshall | black gum / tupelo | 114, 199, 256, 283, 288, 294, 299 |
| Nyssa ogechee | *Nyssa ogeche* W. Bartram ex Marshall | Ogeechee lime / Ogeechee tupelo | 63, 162 |
| Oenothera | *Oenothera perennis* L. | little sundrops | 186 |
| Olea odorata | *Olea europaea* L. | olive | 38 |
| Orange myrthe | *Pimenta dioica* (L.) Merr. | allspice | 152 |
| Orchid whose two oval, entire, parallel veined leaves persist all winter | *Aplectrum hyemale* (Muhl. ex Willd.) Torr. | Adam-and-Eve | 301 |
| Oriza aristrala | *Leersia oryzoides* (L.) Sw. | rice cutgrass | 149 |
| Orobanche virginica / Orobanche virginiana | *Epifagus virginiana* (L.) W. P. C. Barton | beechdrops | 176, 294 |
| Osmunda / Osmunda cinnamomea / Osmunda filiculifolia | *Osmundastrum cinnamomeum* (L.) C. Presl | cinnamon fern | 63, 176, 193 |
| Osmunda regalis | *Osmunda regalis* L. | royal fern | 190, 193 |
| Osmunda / Osmunda filiculifolia | *Osmunda claytoniana* L. | interrupted fern | 176, 193 |
| Oxalis | *Oxalis* sp. | wood-sorrel | 176 |
| Oxalis new sp. | *Oxalis montana* Raf. | American wood-sorrel / white wood-sorrel | 177 |
| Oxalis stricta | *Oxalis stricta* L. | common yellow wood-sorrel | 228 |
| Padus sempervirens, Carolina laurel | *Prunus caroliniana* (Mill.) Aiton | Carolina laurel cherry | 109 |
| Paliurus | *Paliurus spina-christi* Mill. | Christ thorn / Jerusalem tree | 37, 43 |
| Panax trifoliatum | *Panax trifolius* L. | dwarf ginseng | 314 |

| Michaux's names for a particular species | Modern binomial (scientific name) | Common name(s) | Page(s) in this volume |
|---|---|---|---|
| Pancratium [Florida] | *Hymenocallis latifolia* (Mill.) M. Roem | mangrove spiderlily | 63 |
| Parietaria . . . | *Parietaria pensylvanica* Muhl. ex Willd. | Pennsylvania pellitory | 235 |
| Passiflora | *Passiflora* sp. | passionflower | 149 |
| Passiflora cuprea | *Passiflora cupraea* L. | passionflower | 120 |
| Pavia alba / Pavia spicata | *Aesculus parviflora* Walter | bottlebrush buckeye | 109, 159, *355* |
| Pavia lutea | *Aesculus flava* Sol. | yellow buckeye | 114, 129, 230, 231 |
| Peaches | *Prunus* sp. | peach | 70 |
| Pedicularis | *Pedicularis* sp. | lousewort / wood betony | 176 |
| Pedicularis canadensis | *Pedicularis canadensis* L. | eastern lousewort / wood-betony | 176 |
| Pentapetes, seeds | *Pentapetes phoenicia* L. | scarlet pentapetes | 164 |
| Penthorum sedoides | *Penthorum sedoides* L. | ditch-stonecrop | 229 |
| Phaseolus, from India | *Phaseolus* sp. | bean | 151 |
| Phaseolus or Dolichos | *Phaseolus polystachios* (L.) BSP. | wild bean / thicket bean | 94 |
| Phaseolus, hairy | *Phaseolus* sp. | bean | 152 |
| Philadelphus inodorus | *Philadelphus inodorus* L. | Appalachian mock-orange | 159, 275, 294 |
| Philirea latifolio | *Phillyrea latifolia* L. | green olive tree | 44 |
| Phlox / Phlox, with white flowers | *Phlox nivalis* Lodd. ex Sweet | pineland phlox | 253, 270, 308 |
| Phlox lanceolata, in flower | *Phlox carolina* L. or *Phlox glaberrima* L. | Carolina phlox or smooth phlox | 270 |
| Phlox, with pink flowers | *Phlox pilosa* L. | downy phlox | 270 |
| Phryma leptostachia | *Phryma leptostachya* L. | American lopseed | 228, 314 |
| Pine with two needles [FL] | *Pinus clausa* (Chapm. ex Engelm.) Vasey ex Sarg. | sand pine | 95 |
| Pine with two needles [SC] | *Pinus glabra* Walter | spruce pine | 59 |

| Michaux's names for a particular species | Modern binomial (scientific name) | Common name(s) | Page(s) in this volume |
|---|---|---|---|
| Pine, 2 needles shorter and slightly twisted, scales sharp, rough, recurved / Pinus, 2 needles, cones with rigid and prickly scales / Pinus, 2 needles | *Pinus pungens* Lamb. | table mountain pine | 238, 239, 247 |
| Pine, two needles with scales that do not recurve after the seeds have fallen out, but are only spread apart and concave, leaves long and straight, a large tree [VA] | *Pinus* sp. | pine | 247 |
| Pine, with 3 needles / Pine 3 needles, rough scales, rather long leaves / Pinus taeda [VA to FL] | *Pinus taeda* L. | loblolly pine | 89, 248, 249 |
| Pinguicula alpina? / Pinguicula | *Pinguicula vulgaris* L. | common butterwort | 190, 193 |
| Pinus | *Pinus* sp. | pine | 106, 247 |
| Pinus, 2 needles / Pinus, with 2 or 3 needles / Pinus, 2 and 3 needles inter-mingled on the same branch [PA to SC] | *Pinus echinata* Mill. | shortleaf pine | 68, 227, 248, 249 |
| Pinus abies / Pinus abies alba | *Picea glauca* (Moench) Voss | white spruce | 175, 178, 186 |
| Pinus abies balsamifera | *Abies fraseri* (Pursh) Poir. | Fraser fir | 257, 258 |
| Pinus palustris / Pinus, ex-tremely long needles | *Pinus palustris* Mill. | longleaf pine | 61, 106, 148, 248, 249 |
| Pinus strobus / Pinus, nee-dles 5 / Weymouth pine | *Pinus strobus* L. | eastern white pine | 31, 36, 73, 112, 113, 175, 179, 186, 193, 194, 199, 227, 238, 239, 256, 257, 306 |
| Pinus, 2 needles / Pinus, needles paired / Pinus, twin needles, cone ovate, smooth [VT and Quebec] | *Pinus resinosa* Aiton | red pine | 175, 176, 186, 198 |
| Pinus, 2 needles / Pinus, 2 nee-dles, smallish oblong cone, scales acute, recurved [PA to KY] | *Pinus virginiana* Mill. | Virginia pine / scrub pine | 227, 238, 299 |
| Pinus, 3 needles / Pinus, needles in 3 (pitch pine) [VT to VA] | *Pinus rigida* Mill. | pitch pine | 176, 199, 227, 238, 247 |
| Pinus, needles paired / Pinus, needles in 2s / Pinus, needles in 2s, short [Quebec] | *Pinus banksiana* Lamb. | jack pine | 179, 186, 188, 189, 191, 192, 193 |
| Piper? | *Piper* sp. | pepper | 152 |
| Pisonia* / Pisonia baccifera* / Piso-nia inermis* [used in error] | *Sageretia minutiflora* (Michx.) C. Mohr | small-flowered buckthorn | 162, 163, 250 |
| Pistatia nut | *Pistacia vera* L. | pistachio nut | 43 |
| Plant, 12 feet in height, annual | *Amaranthus australis* (A. Gray) J. D. Sauer | southern water-hemp | 91 |

| Michaux's names for a particular species | Modern binomial (scientific name) | Common name(s) | Page(s) in this volume |
|---|---|---|---|
| Plant, annual, erect opposite four-angled branches, leaves ovate, 3-nerved, subsessile | *Trichostema dichotomum* L. | common blue curls | 260 |
| Plant, bulbous, flower with usually 2 spathes / Pancratium mexicanum | *Hymenocallis crassifolia* Herb. | spring-run spiderlily | 59, 63 |
| Plant with solitary fruit on a stalk / Podophyllum diphyllum / Jeffersonia | *Jeffersonia diphylla* (L.) Pers. | twinleaf | 127, 128, 271, 274, 304 |
| Plantago maritima | *Plantago maritima* L. var. *juncoides* (Lam.) A. Gray | seaside plantain / goose tongue | 178 |
| Platanus occidentalis / Platanus | *Platanus occidentalis* L. | sycamore | 61, 179, 197, 229, 231, 274, 275, 279, 283, 285, 286, 295 |
| Plum trees, common | *Prunus americana* Marshall | wild plum | 157 |
| Plum, wild | *Prunus* sp. | plum | 70 |
| Poa, glumes with 4 florets | *Poa palustris* L. | fowl bluegrass / fowl meadow-grass | 186 |
| Podophyllum peltatum / Podophyllum umbellatum | *Podophyllum peltatum* L. | may-apple | 176, 229, 271, 274, 314 |
| Podophyllum, flowers . . . , blue berries | *Diphylleia cymosa* Michx. | umbrella-leaf | 256 |
| Polyg[onum?] necess. | *Polygonum* sp. | knotweed | 270 |
| Polygala | *Polygala* sp. | milkwort | 154 |
| Polygala rosea | *Polygala incarnata* L. | pink milkwort / procession flower | 62 |
| Polygala senega | *Polygala senega* L. | seneca snakeroot | 176 |
| Polygala viridescens | *Polygala sanguinea* L. | blood milkwort / field milkwort | 176 |
| Polygonum . . . / Polygonum amphybium | *Persicaria amphibia* (L.) S. F. Gray | water smartweed | 229 |
| Polygonum aviculare / Polygonum aviculare, stamens 5, styles 3 | *Polygonum aviculare* L. | knotweed | 195, 229, 280 |
| Polygonum aviculare, larger, stamens 5, styles 3 | *Polygonum ramosissimum* Michx. | bushy knotweed | 280 |
| Polygonum hydropiper | *Persicaria hydropiper* (L.) Opiz | common smartweed / marshpepper smartweed | 229 |
| Polygonum scandens | *Fallopia scandens* (L.) Holub | common climbing buckwheat | 229 |
| Polymnia canadensis / Polymnia canadensis, ray florets very small . . . | *Polymnia canadensis* L. | white-flowered leafcup | 231 |

*Table of Described Plants (continued)*

| Michaux's names for a particular species | Modern binomial (scientific name) | Common name(s) | Page(s) in this volume |
|---|---|---|---|
| Polymnia uvedalia, ray florets large | *Smallanthus uvedalia* (L.) Mack. ex Small | bearsfoot / leafcup | 315 |
| Polypodium scolopendroides | *Blechnum serrulatum* Rich. | swamp fern | 93 |
| Populus / Populus (fastigatus?) / Populus caroliniana | *Populus deltoides* W. Bartram ex Marshall | eastern cottonwood | 196, 285 |
| Populus balsamifera | *Populus balsamifera* L. | balsam poplar | 176, 179, 185, 193 |
| Populus caroliniana / Populus heterophylla | *Populus heterophylla* L. | swamp cottonwood | 64, 105 |
| Populus fastigiata | *Populus nigra* L. | Lombardy poplar | 38 |
| Potamogeton | *Potamogeton* sp. | pondweed | 183 |
| Potamogeton, etc. | *Stuckenia pectinata* (L.) Börner | sago-pondweed | 179 |
| Potentilla | *Potentilla* sp. | cinquefoil | 176 |
| Potentilla | *Potentilla anserina* L. | silverweed | 176 |
| Potentilla fruticosa | *Dasiphora fruticosa* (L.) Rydb. | shrubby-cinquefoil | 188, 192, 193 |
| Potentilla nivea | *Potentilla norvegica* L. | rough cinquefoil / strawberry-weed | 179, 184, 186, *370* |
| Potentilla tridentata | *Sibbaldiopsis tridentata* (Aiton) Rydb. | mountain-cinquefoil | 194, 257 |
| Pride of China | *Melia azedarach* L. | chinaberry / pride of India | 43 |
| Prinos glaber | *Ilex glabra* (L.) A. Gray | little gallberry / inkberry | 36, 67 |
| Prinos verticillatus | *Ilex verticillata* (L.) A. Gray | winterberry | 36, 175, 184, 196 |
| Prinos new sp. | *Ilex* sp. | holly | 131, 147 |
| Protea argentea | *Leucadendron argenteum* (L.) R. Br. | silver tree | 152 |
| Protea conocarpus | *Leucospermum conocarpodendron* (L.) H. Buek | grey tree pincushion | 152 |
| Psydium | *Psidium guajava* L. | guava | 120 |
| Pteris / Pteris aquilina | *Pteridium aquilinum* (L.) Kuhn subsp. *latiusculum* (Desv.) Hultén | eastern bracken | 186, 190, 193 |
| Pteris lineata | *Vittaria lineata* (L.) J. Smith | shoestring fern | 93 |

| Michaux's names for a particular species | Modern binomial (scientific name) | Common name(s) | Page(s) in this volume |
|---|---|---|---|
| Pteris, new sp. | *Pellaea atropurpurea* (L.) Link | purple cliff brake | 235 |
| Pulmonaria maritima | *Mertensia maritima* (L.) S. F. Gray | oysterleaf / seaside-bluebell | 179 |
| Pyrola | *Pyrola* sp. | shinleaf / wintergreen | 175 |
| Pyrola umbellata? | *Chimaphila umbellata* (L.) W. P. C. Barton var. *cisatlantica* Blake | prince's pine | 175 |
| Pyrola uniflora | *Moneses uniflora* (L.) A. Gray | single delight / one-flowered wintergreen | 179 |
| Quasia amara | *Quassia amara* L. | bitterwood / amargo | 149 |
| Quercus . . . Chêne chataignier / Quercus castaneaefolio | *Quercus muehlenbergii* Engelm. | chinquapin oak | 199, 227 |
| Quercus alba, known as mountain white oak | *Quercus stellata* Wangenh. | post oak | 301 |
| Quercus cerroides / Quercus cerroides, known by the French as curly oak, by the Americans as overcup white oak / Quercus cerroides–curly oak–overcup white oak / Quercus, leaves lyrate, underneath tomentose, cups large with laciniate margins, generally enclosing an acorn, overcup white oak / Quercus, with large acorns, fruit enclosed, called overcup white oak | *Quercus macrocarpa* Michx. | bur oak / mossy-cup oak | 196, 275, 277, *278, 279*, 283, 299, *350* |
| Quercus integrifolia / Quercus latifolia / Shingle oak / Ram's oak | *Quercus imbricaria* Michx. | shingle oak | 280, 283 |
| Quercus nigra | *Quercus marilandica* Münchh. | blackjack oak | 135, 301, 429n53 |
| Quercus palustris, deltoid leaves | *Quercus nigra* L. | water oak | 136, 249 |
| Quercus phellos | *Quercus phellos* L. | willow oak | 79, 89, 275 |
| Quercus phellos latifolia / Quercus salicifolia | *Quercus laurifolia* Michx. | laurel oak | 136 |
| Quercus praemorsa | *Quercus lyrata* Walter | overcup oak | 274, 275 |
| Quercus prinus humilis | *Quercus prinoides* Willd. | dwarf chinquapin oak / dwarf chestnut oak | 274 |
| Quercus rubra | *Quercus rubra* L. | northern red oak | 136, 239, 274, 275, 277, 279, 283, 285 |
| Quercus rubra, much branched | *Quercus pagoda* Raf. | cherrybark oak | 283, 284 |

| Michaux's names for a particular species | Modern binomial (scientific name) | Common name(s) | Page(s) in this volume |
|---|---|---|---|
| Quercus tomentosa | *Quercus bicolor* Willd. | swamp white oak | 274, 275 |
| Quercus glauca | *Quercus* sp. | oak | 114, 138 |
| Ranunculus . . . / Ranunculus reptans leaves narrow, stems creeping | *Ranunculus flammula* var. *reptans* (L.) E. Mey. | creeping spearwort | 183, 196 |
| Red elm / Ulmus viscoa / Ulmus viscosa fungosa / Ulmus viscosa, twin golden flowers with 4–5 or 6 stamens and purple stigmas / Ulmus . . . | *Ulmus rubra* Muhl. | slippery elm | 175, 274, 275, 304, 305, 307 |
| Red oak with long petioles, acorns short, sessile and thick | *Quercus falcata* Michx. | southern red oak | 137, 253, *351* |
| Rhamnus | *Rhamnus* sp. | buckthorn | 23, 150, 234 |
| Rhamnus (dioicus) | *Rhamnus alnifolia* L'Hér. | alder-leaf buckthorn | 175 |
| Rhexia | *Rhexia* sp. | meadow-beauty | 83, 150, 234 |
| Rhexia | *Rhexia virginica* L. | Virginia meadow-beauty / handsome Harry | 253 |
| Rhexia, large Carolina | *Rhexia alifanus* Walter | smooth meadow-beauty | 249 |
| Rhinanthus crista-galli | *Rhinanthus groenlandicus* Chabert | arctic rattlebox | 193 |
| Rhinanthus, plant similar to, 4 feet high | *Aureolaria* sp. | oak-leach / false-foxglove | 315 |
| Rhododendron / Rhododendron maximum | *Rhododendron maximum* L. | great laurel, rosebay rhododendron | 32, 33, 36, 71, 72, 112, 113, 139, 164, 227, 238, 256, 305 |
| Rhododendron minus / Rhododendron new sp. | *Rhododendron minus* Michx. | gorge rhododendron / punctatum | 109, 110, 111, 112, 114, 148, 159, 164, 272, 306, *372* |
| Rhus | *Rhus* sp. | sumac | 175 |
| Rhus . . . / Rhus pumilum | *Rhus michauxii* Sarg. | Michaux's sumac / dwarf sumac | 254, 308 |
| Rhus glabra / Rhus glabrum | *Rhus glabra* L. | smooth sumac | 175, 228, 254 |
| Rhus radicans / Rhus toxicodendron | *Toxicodendron radicans* (L.) Kuntze var. *radicans* | eastern poison ivy | 228 |
| Rhus toxicodendron | *Toxicodendron pubescens* Mill. | poison oak | 228, 315 |
| Rhus typhina | *Rhus typhina* L. | staghorn sumac | 228 |

*Table of Described Plants (continued)*

| Michaux's names for a particular species | Modern binomial (scientific name) | Common name(s) | Page(s) in this volume |
|---|---|---|---|
| Rhus vernix | *Toxicodendron vernix* (L.) Kuntze | poison sumac | 228 |
| Rhus, with winged leaves, dioecious / Rhus copallinum / Rhus, winged between the leaflets | *Rhus copallinum* L. | winged sumac | 228, 254, 277 |
| Ribes / Ribes (miquelon) | *Ribes* sp. | currant / gooseberry | 175, 190, 193 |
| Ribes cynosbati | *Ribes cynosbati* L. | prickly gooseberry / dogberry | 175, 186 |
| Ricinus | *Ricinus communis* L. | castor-bean | 152 |
| Rivina humilis | *Rivina humilis* L. | rouge-plant | 100 |
| Robinia | *Robinia hispida* L. | bristly locust | 31, 65, *355* |
| Robinia chamlaga | *Caragana sinica* (Buc'holz) Render | Chinese pea shrub | 38 |
| Robinia pseudoacacia | *Robinia pseudoacacia* L. | black locust | 229, 235, 256, 283 |
| Robinia viscosa | *Robinia viscosa* Vent. | clammy locust | 256 |
| Rosa . . . | *Rosa* sp. | rose | 176 |
| Round grains, malvaceous sp., Curcuma | *Curcuma longa* L. | turmeric | 152 |
| Rubus arcticus | *Rubus arcticus* L. subsp. *acaulis* (Michx.) Focke | stemless arctic bramble / dwarf raspberry | 186, 192, 193, *358* |
| Rubus arcticus | *Rubus pubescens* Raf. | dwarf blackberry / dwarf red raspberry | 176, 229 |
| Rubus canadensis | *Rubus canadensis* L. | smooth blackberry | 176 |
| Rubus hispidus | *Rubus hispidus* L. | swamp dewberry / bristly dewberry | 176 |
| Rubus occidentalis [Quebec] | *Rubus idaeus* L. var. *strigosus* (Michx.) Maxim. | red raspberry | 186, 192, 193 |
| Rubus occidentalis [VT] | *Rubus occidentalis* L. | black raspberry | 176, 229 |
| Rubus odoratus | *Rubus odoratus* L. | flowering raspberry | 176, 193, 229, 256 |
| Rudbeckia | *Rudbeckia* sp. | black-eyed susan / yellow coneflower | 257 |
| Rushes | *Juncus roemerianus* Scheele | blackneedle rush | 60 |
| Sagittaria sagittifolia | *Sagittaria latifolia* Willd. | broadleaf arrowhead | 176 |

| Michaux's names for a particular species | Modern binomial (scientific name) | Common name(s) | Page(s) in this volume |
|---|---|---|---|
| Salicornia . . . | *Salicornia* sp. | glasswort | 196 |
| Salix | *Salix exigua* Nutt. var. *sericans* (Nees) G. L. Neesom | sandbar willow | 295 |
| Salix capreoides / Salix, stipules leaf-like | *Salix eriocephala* Michx. | heart-leaved willow | 306 |
| Salix pumila | *Salix humilis* Marshall | prairie willow / upland willow | 301 |
| Salix sericea | *Salix pellita* (Andersson) Bebb | satiny willow | 186 |
| Salix, stipules leaf-like | *Salix cordata* Michx. | sand dune willow | 186 |
| Salsola . . . | *Salsola kali* L. | northern saltwort | 196 |
| Sambucus | *Sambucus* sp. | elderberry | 175 |
| Sambucus canadensis, black fruit | *Sambucus canadensis* L. | common elderberry | 229 |
| Sambucus, red fruit, tomentose leaves | *Sambucus racemosa* L. var. *pubens* (Michx.) Koehne | red elderberry | 178, 229, *360* |
| Sanguinaria / Sanguinaria canadensis | *Sanguinaria canadensis* L. | bloodroot | 176, 229, 274, 304, 305, *362* |
| Sanicula | *Sanicula* sp. | snakeroot | 175 |
| Sanicula marylandica / Sanicula marylandica, called root of Becquel | *Sanicula marilandica* L. | Maryland sanicle / Becquel root | 228, 280 |
| Sapindus / Sapindus saponaria | *Sapindus saponaria* L. | soapberry | 97, 98, 99 |
| Sapodilla from Saint-Domingue / Sapodilla from Curacao | *Manilkara zapota* (L.) Royen | sapodilla | 149 |
| Sarothra gentianoides | *Hypericum gentianoides* (L.) BSP. | pineweed / orange-grass | 247 |
| Sarracenia lutea | *Sarracenia flava* L. | yellow pitcher-plant / trumpets | 59, *362* |
| Sarracenia new sp. | *Sarracenia psittacina* Michx. | parrot pitcherplant | 162, *369* |
| Sarracenia purpurea | *Sarracenia purpurea* L. | northern purple pitcherplant | 35, 36, 178, 189, 190, 191, 193 |
| Sarracenia tubifolia | *Sarracenia minor* Walter | hooded pitcherplant | 75 |
| Saururus cernuus | *Saururus cernuus* L. | lizard's-tail | 164 |
| Saxifraga nivalis* | *Micranthes virginiensis* (Michx.) Small | early saxifrage | 173 |

| Michaux's names for a particular species | Modern binomial (scientific name) | Common name(s) | Page(s) in this volume |
|---|---|---|---|
| Saxifraga pennsylvanica | *Micranthes pennsylvanica* (L.) Haw. | swamp saxifrage | 172 |
| Saxifrage | *Tiarella cordifolia* L. | foamflower | 304 |
| Scheuchzera . . . | *Scheuchzeria palustris* L. | pod-grass / rannoch-rush | 177 |
| Scirpus | *Scirpus* sp. | bullrush | 97 |
| Scrophularia | *Scrophularia* sp. | figwort | 176 |
| Scrophularia marylandica / Scrophularia, new sp., 3 leaves, opposite, ternate, pinnatifid | *Scrophularia marilandica* L. | eastern figwort | 229, 315 |
| Sedum, with lower leaves dentate and upper leaves entire | *Hylotelephium telephioides* (Michx.) H. Ohba | Allegheny live-for-ever | 256 |
| Serratula?, new genus, phyllaries imbricate, outer ones largest, plant showy | *Liatris aspera* Michx. | rough blazing-star | 315 |
| Serratula [Bahamas] | *Liatris garberi* A. Gray | Garber's gayfeather | 154 |
| Serratula praealta | *Vernonia noveboracensis* (L.) Michx. | New York ironweed | 232 |
| Serratula, 2 unknown species | *Liatris spheroidea* Michx. and *Liatris squarrosa* (L.) Michx. | spherical blazing-star / scaly blazing-star | 235 |
| Serratula, plumose pappus | *Liatris squarrosa* (L.) Michx. | scaly blazing-star | 315, *359* |
| Shrub, with the aspect of an Erica | *Ceratiola ericoides* Michx. | Florida rosemary / sandhill rosemary | 109 |
| Shrub, fruit in the shape of a pear | *Pyrularia pubera* Michx. | buffalo-nut | 72 |
| Shrub, leguminous with ternate leaves | *Dalea carnea* (Michx.) Poir. | pink-tassels / whitetassels | 89 |
| Shrub, resembling Calycanthus, with red root | *Nestronia umbellula* Raf. | nestronia | 126, 254, 308, *361* |
| Shrubby tree with blackish wood / Laurus benzoin | *Lindera benzoin* (L.) Blume | spicebush | 27, 78, 199, 294, 299, 301 |
| Sicyos angulata | *Sicyos angulatus* L. | bur-cucumber | 228 |
| Sida | *Sida acuta* Burm. f. | wireweed | 253 |
| Sideroxilon tomax / Sideroxylon | *Sideroxylon tenax* L. | tough bumelia / tough buckthorn | 61, 78, 79, 147, 159, *361* |
| Sigesbeckia . . . | *Verbesina occidentalis* (L.) Walter | southern crownbeard / yellow crownbeard | 228 |

*Table of Described Plants (continued)*

| Michaux's names for a particular species | Modern binomial (scientific name) | Common name(s) | Page(s) in this volume |
|---|---|---|---|
| Silene . . . flowers pinkish / Silene . . . | *Silene caroliniana* Walter | catchfly | 308 |
| Silene virginica | *Silene virginica* L. | fire-pink | 62, 270, *363* |
| Silphium connatum / Silphium trifoliatum | *Silphium integrifolium* Michx. | prairie rosinweed | 279 |
| Silphium laciniatum | *Silphium laciniatum* L. | compass-plant | 279 |
| Silphium perfoliatum | *Silphium perfoliatum* L. | common cup-plant | 279 |
| Silphium pinnatifidum | *Silphium pinnatifidum* Elliott | tansy rosinweed / cutleaf prairie-dock | 279 |
| Silphium, leaves very large | *Silphium terebinthinaceum* Jacq. | prairie-dock | 279, *363* |
| Sisyrinchium bermudiana | *Sisyrinchium montanum* Greene | strict blue-eyed grass | 179 |
| Smilax | *Smilax* sp. | greenbrier | 72 |
| Smilax herbacea . . . / Smilax herbacea erecta | *Smilax herbacea* L. | common carrionflower | 176, 270 |
| Smilax laurifolia | *Smilax laurifolia* L. | bamboo-vine / laurel greenbrier | 248, 253 |
| Smilax pseudochina | *Smilax pseudochina* L. | coastal carrionflower | 285 |
| Smilax sarsaparilla | *Smilax glauca* Walter | cat greenbrier / wild sarsaparilla | 288 |
| Smilax with red berries | *Smilax walteri* Pursh. | coral greenbrier | 248 |
| Smilax, spiny and deciduous [IL] | *Smilax hispida* Raf. | bristly greenbrier | 288 |
| Sophora occidentalis | *Sophora tomentosa* L. | yellow necklacepod | 93 |
| Sophora sinica | *Sophora japonica* L. | Japanese pagoda tree | 38 |
| Sophora, with yellow flowers | *Baptisia* sp. | undetermined | 253 |
| Sophora, flowers blue / Sophora, leaflets in 3s stipulate broadly lanceolate, flowers blue standard petal shorter than corolla / Sophora, perennial, blue, cool sites, submerged during winter | *Baptisia australis* (L.) R. Br. | blue wild indigo | 229, 314 |
| Sophora, known as yellow lupine | *Baptisia cinerea* (Raf.) Fernald & B. G. Schub. | Carolina wild indigo | 249 |
| Sophora, new | *Cladrastis kentukea* (Dum. Cours.) Rudd | yellow-wood | 292, 303, 304, *353, 364* |

| Michaux's names for a particular species | Modern binomial (scientific name) | Common name(s) | Page(s) in this volume |
|---|---|---|---|
| Sorbus aucuparia | *Sorbus americana* Marshall or *Sorbus aucuparia* L. | mountain-ash / European mountain-ash | 38, 178, 190, 256, 257, *365* |
| Sorbus aucuparia americana | *Sorbus decora* (Sarg.) C. K. Schneid. | showy mountain-ash / northern mountain-ash | 192, 193 |
| Spanish chustnuts [Chestnuts] | *Castanea sativa* Mill. | Spanish chestnut | 43 |
| Sparganium / Sparganium erectum / Sparganium natans | *Sparganium* sp. | bur-reed | 176, 179, 184 |
| Sphagnum / Sphagnum palustre | *Sphagnum* sp. | sphagnum | 30, 35, 36, 182, 189, 190, 191, 253 |
| Spigelia | *Samolus ebracteatus* Kunth or *Samolus parviflorus* Raf. | limewater brookweed / water-pimpernel | 93 |
| Spigelia marylandica | *Spigelia marilandica* (L.) L. | Indian pink / pinkroot | 108, 145, *363* |
| Spiraea (paniculata) trifoliata | *Gillenia trifoliata* (L.) Moench | mountain Indian-physic / bowman's-root | 256, 286, *359* |
| Spiraea / Spiraea tomentosa | *Spiraea tomentosa* L. | steeplebush | 196, 259, 307 |
| Spiraea new sp. / Spirea, pink flowers in panicles | *Filipendula rubra* (Hill) B. L. Rob. | queen-of-the-prairie | 129, 147 |
| Spiraea opulifolia | *Physocarpus opulifolius* (L.) Maxim. var. *opulifolius* | eastern ninebark | 196 |
| Spiraea salicifolia | *Spiraea latifolia* (Aiton) Borkh. | broadleaf meadowsweet | 186 |
| Spiraea, dioecious / Spiraea | *Aruncus dioicus* (Walter) Fernald | goat's-beard | 122 |
| Staphylea trifoliata | *Staphylea trifolia* L. | bladdernut | 175, 228 |
| Stewartia / Stewartia malacodendron | *Stewartia malacodendron* L. | silky camellia | 55, 80, 104, 105, 147, 249 |
| Stewartia pentagyna / Stewartia? | *Stewartia ovata* (Cav.) Weath. | mountain camellia | 254, 307 |
| Stillingia | *Stillingia aquatica* Chapm. | water toothleaf / corkwood | 147 |
| Stillingia / Stillingia herbacea / Stillingia silvatica / Stillingia sylvatica | *Stillingia sylvatica* Garden ex L. | queen's-delight | 61, 96, 147, 250, 312 |
| Styrax | *Styrax* sp. | snowbell | 55, 67, 104, 105, 129, 150 |
| Styrax angustifolia | *Styrax americanus* Lam. | American snowbell | 59, 145 |
| Styrax latifolia | *Styrax grandifolius* Aiton | bigleaf snowbell | 59, 145 |

*Table of Described Plants (continued)*

| Michaux's names for a particular species | Modern binomial (scientific name) | Common name(s) | Page(s) in this volume |
|---|---|---|---|
| Styrax tomentosum | *Styrax americanus* var. *pulverulentus* (Michx.) Perkins ex Rehder | downy American snowbell | 312 |
| Supple jack / Ziziphus scandens | *Berchemia scandens* (Hill) K. Koch | supplejack / American rattan | 79, 89 |
| Swertia corniculata | *Halenia deflexa* (Sm.) Griseb. | American spurred gentian | 182 |
| Tagetoides | *Dyssodia papposa* (Vent.) Hitchc. | dogweed / fetid marigold | 292 |
| Talipot | *Corypha umbraculifera* L. | talipot palm | 151 |
| Tamarindus indica | *Tamarindus indica* L. | tamarind | 120 |
| Taxus / Taxus monoica / Taxus monoicus | *Taxus canadensis* Marshall | Canada yew | 175, 176, 186 |
| Teucrium canadense | *Teucrium canadense* L. | Canada germander | 228, *365* |
| Thalictrum dioicum [VT] | *Thalictrum dioicum* L. | early meadowrue | 176 |
| Thalictrum dioicum / Thalictrum purpurascens [Quebec] | *Thalictrum pubescens* Pursh | common tall meadowrue | 176 |
| Thea viridis | *Camellia sinensis* (L.) Kuntze | tea plant | 9, 38 |
| Thesium umbellatum | *Comandra umbellata* (L.) Nutt. | bastard-toadflax | 175 |
| Thuya / Thuya occidentalis | *Thuja occidentalis* L. | American arborvitae | 129, 175, 176, 178, 193, 199, 238, 239 |
| Thuya orientalis | *Platycladus orientalis* (L.) Franco | oriental arborvitae | 38 |
| Thymbra new sp. | *Thymbra* sp. | mint | 154 |
| Tilia / Tilia americana | *Tilia americana* L. | basswood | 66, 176, 228, 229, 279 |
| Tradescantia umbellata, flowers pink | *Cuthbertia rosea* (Vent.) Small | common roseling | 63 |
| Tragia volubilis / Tragia, monoecious plant, fruits like Euphorbias | *Tragia cordata* Michx. | heartleaf noseburn | 234, 315 |
| Trientalis | *Trientalis borealis* Raf. | northern starflower / maystar | 173, 175, 196 |
| Trifolium . . . | *Trifolium* sp. | clover | 196 |
| Trifolium lagopus | *Trifolium arvense* L. | rabbitfoot clover | 130 |
| Trifolium rubens | *Trifolium pratense* L. | red clover | 176 |

| Michaux's names for a particular species | Modern binomial (scientific name) | Common name(s) | Page(s) in this volume |
|---|---|---|---|
| Trifolium? pentandrum majus | *Orbexilum onobry- chis* (Nutt.) Rydb. | lanceleaf scurfpea | 280 |
| Trifolium? pentandrum, flowers purple | *Orbexilum peduncula- tum* (Nutt.) Rydb. | Sampson's-snakeroot | 280 |
| Triglochin / Triglochin palustre | *Triglochin maritima* L. or *Triglochin palustris* L. | marsh arrowgrass / seaside arrowgrass | 177, 184 |
| Trillium, capsule angulate vi- olet / Trillium erectum | *Trillium erectum* L. | red trillium | 175, 186, 227 |
| Trillium cernuum [GA] | *Trillium rugeli* Rendle | southern nod- ding trillium | 66 |
| Trillium cernuum [NJ, PA, Quebec] | *Trillium cernuum* L. | northern nod- ding trillium | 173, 229 |
| Trillium sessile | *Trillium maculatum* Raf. or *Trillium reliquum* Freeman | mottled trillium / relict trillium | 66 |
| Trillium sessile | *Trillium sessile* L. | sessile toadshade / sessile trillium | 274 |
| Trillium, capsule ovate, red / Tril- lium cernuum, erect with red fruits | *Trillium undulatum* Willd. | painted trillium | 186, 256, *356* |
| Triosteum [SC] | *Triosteum angustifolium* L. | lesser horse-gentian | 137 |
| Triosteum / Triosteum perfoliatum | *Triosteum perfoliatum* L. | perfoliate horse-gentian | 196, 228 |
| Triticum spelt | *Triticum spelta* L. | wheat | 171 |
| Typha / Typha altissima | *Typha angustifolia* L. or *Typha latifolia* L. | common cattail / narrowleaf cattail | 176, 179 |
| Tythymalis | *Euphorbia* sp. | spurge | 61 |
| Tythymalus, resembling Stillinga | *Ditrysinia fruticosa* (W. Bartram) Go- vaerts & Frodin | Sebastian-bush | 59 |
| Ulmoides, anonymous / Tree from the Santee River, new, elm-like leaves, rough capsule, 1 subovate seed | *Planera aquat- ica* J. F. Gmel. | planer-tree / water-elm | 270, 285 |
| Ulmus | *Ulmus* sp. | elm | 175, 196, 229, 285 |
| Ulmus americana / Ulmus, white elm / White elm | *Ulmus americana* L. | American elm | 157, 175, 186, 295 |
| Ulmus, said to be from Siberia | *Ulmus pumila* L. | dwarf elm | 38 |
| Umbellifer, little bulbous | *Erigenia bulbosa* (Michx.) Nutt. | harbinger-of-spring / pepper-and-salt | 303, 304 |
| Uniola maritima | *Uniola paniculata* L. | sea oats | 312 |

| Michaux's names for a particular species | Modern binomial (scientific name) | Common name(s) | Page(s) in this volume |
|---|---|---|---|
| Uniola montana | *Chasmanthium latifolium* (Michx.) Yates | river oats | 312 |
| Urtica | *Urtica* sp. | nettle | 176 |
| Urtica divaricata / Urtica inermis | *Laportea canadensis* (L.) Wedd. | wood-nettle | 228, 229, 315 |
| Uvularia / Uvularia perfoliata [VT] | *Uvularia* sp. or *Uvularia perfoliata* L. | bellwort / perfoliate bellwort | 175, *364* |
| Uvularia? [SC] | *Uvularia* sp. or *Uvularia floridana* Chapm. | Bellwort / florida bellwort | 270 |
| Vaccinium arboreum | *Vaccinium arboreum* Marshall | sparkleberry | 248, 250, 312 |
| Vaccinium atoca / Vaccinium atoca, 8 stamens / Vaccinium repens / Vaccinium . . . cranberry / Atoca, edible red fruit | *Vaccinium oxycoccos* L. | small cranberry | 36, 173, 179, 182, 184, 186, 189, 193 |
| Vaccinium atoca / Vaccinium coccineum / Vaccinium cranberry related to Oxicoccus | *Vaccinium macrocarpon* Aiton | cranberry / large cranberry | 30, 179, 257 |
| Vaccinium coccineum / Vaccinium, leaves with ciliate margins, reticulate on the surface, axillary peduncles with single flower, corolla revolute, 4-parted, stamens 8, fruit inferior, a pear-shaped berry, red, 4 locules | *Vaccinium erythrocarpum* Michx. | mountain cranberry | 256 |
| Vaccinium corymbosum / Vaccinium corymbosum minus / Vaccinium corymbosum, 10 stamens | *Vaccinium angustifolium* Aiton or *Vaccinium boreale* I. V. Hall & Aalders or *Vaccinium corymbosum* L. or *Vaccinium myrtilloides* Michx. | northern blueberry / northern lowbush blueberry / smooth highbush blueberry / sourtop / velvetleaf blueberry | 175, 177, 186, 187, 191, 193 |
| Vaccinium repens | *Pieris phillyreifolia* (Hook.) DC. | climbing fetterbush | 162 |
| Vaccinium resinosum | *Gaylussacia baccata* (Wangenh.) K. Koch | black huckleberry | 175, 182, 256, 312 |
| Vaccinium stamineum | *Vaccinium stamineum* L. | deerberry | 162, 175, 256, *364* |
| Vaccinium uliginosum | *Vaccinium uliginosum* L. | bog-bilberry | 188, 190 |
| Vaccinium vitis idaea / Vaccinium vitis idaea, 8 stamens / Vaccinium, leaves with glandular tip | *Vaccinium vitis-idaea* L. or *Vaccinium vitis-idaea* L. subsp. *minus* (Lodd.) Hultén | lingonberry | 181, 187, 190, 193, *369* |
| Vaccinium, with solitary fruits in the leaves' axils, fruit bluish, calyx with 5 segments / Vaccinium, dwarf with 10 stamens | *Vaccinium cespitosum* Michx. | dwarf bilberry | 189, 193 |

| Michaux's names for a particular species | Modern binomial (scientific name) | Common name(s) | Page(s) in this volume |
|---|---|---|---|
| Vaccinium, creeping, stamens 8 / Vaccinium hispidulum or Attoca, with white fruit / Vaccinium niveum / Vaccinium niveum, 8 stamens / Vaccinium, white fruit / Vaccinium, creeping, with white berry | *Gaultheria hispidula* (L.) Muhl. ex Bigelow | creeping snowberry | 173, 179, 186, 190, 193 |
| Vaccinium, evergreen leaves, prostrate stems, black fruits / Vaccinium repens, black fruit, etc. / Vaccinium sempervirens | *Vaccinium crassifolium* Andrews | creeping blueberry | 249, 250 |
| Valeriana | *Valerianella radiata* (L.) Dufr. | corn salad | 270 |
| Veratrum / Veratrum viride album | *Veratrum viride* Aiton | white-hellebore / green hellebore | 176, 255, 256 |
| Veratrum luteum | *Chamaelirium luteum* (L.) A. Gray | devil's bit / fairy wand | 256 |
| Verbascum album | *Verbascum lychnitis* L. | white mullein | 130 |
| Verbascum blattaria | *Verbascum blattaria* L. | moth mullein | 174 |
| Verbascum nigrum | *Verbascum nigrum* L. | black mullein | 130 |
| Verbascum thapsus | *Verbascum thapsus* L. | wooly mullein | 175 |
| Verbena aublatia | *Glandularia canadensis* (L.) Nutt | rose vervain / rose verbena | 60, 309 |
| Verbena carolinana | *Verbena carnea* Medik. | Carolina-vervain | 61 |
| Verbena utricifolia, stem erect, panicles spreading, bracts shorter than the flowers, flowers white | *Verbena urticifolia* L. | white vervain | 279 |
| Verbena, creeping stems, leaves pinnatifid, very long bracts | *Verbena bracteata* Lag. & Rodr. | prostrate vervain / big-bracted vervain | 279 |
| Verbena, stem erect, panicle with clustered branches, bracts and calyx pilose, flowers purplish blue | *Verbena hastata* L. | common vervain / blue vervain | 279 |
| Verbena, stems erect, panicles erect, leaves ovate, tomentose, twice-serrate | *Verbena stricta* Vent. | hoary vervain | 279 |
| Veronica . . . | *Veronica* sp. | speedwell | 228 |
| Veronica virginica | *Veronicastrum virginicum* (L.) Farw. | Culver's-root | 286, 288 |
| Viburnum | *Viburnum lantanoides* Michx. | hobblebush | 175 |
| Viburnum | *Viburnum* sp. | viburnum | 126, 138, 146, 162 |

| Michaux's names for a particular species | Modern binomial (scientific name) | Common name(s) | Page(s) in this volume |
|---|---|---|---|
| Viburnum / Viburnum, trilobed leaves | *Viburnum acerifolium* L. | maple-leaf viburnum | 137, 175 |
| Viburnum cassinoides | *Viburnum obovatum* Walter | small-leaf viburnum | 89, 164 |
| Viburnum dentatum | *Viburnum dentatum* L. | arrow-wood | 105 |
| Viburnum nudum | *Viburnum cassinoides* L. | northern wild raisin / withe-rod | 175, 186 |
| Viburnum opulus, petiole glandular | *Viburnum edule* (Michx.) Raf. | squashberry | 175, 186 |
| Viburnum opulus, petiole glandular | *Viburnum opulus* L. var. *americanum* Aiton | highbush-cranberry / cranberry-tree | 175 |
| Viburnum tinus | *Viburnum tinus* L. | laurustinus | 164 |
| Viburnum, leaves dentate, fruit compressed, new species | *Viburnum molle* Michx. | soft arrow-wood / Kentucky viburnum | 315 |
| Vicia [SC] | *Vicia caroliniana* Walter or *Vicia sativa* L. | Carolina vetch or common vetch | 270 |
| Vinca lutea | *Pentalinon luteum* (L.) B. F. Hansen & Wunderlin | Catesby's vine | 120 |
| Vinca lutea? / Apocynum cannabinum | *Trachelospermum difforme* (Walter) A. Gray | climbing dogbane | 62n, 64 |
| Viola [VT] | *Viola canadensis* L. | Canadian white violet / tall white violet | 176 |
| Viola, yellow, scapose, leaves hastate | *Viola hastata* Michx. | halberd-leaf violet / spearleaf violet | 73, 306, 307, *364* |
| Violet, reniform, dentate leaves, hairy petiole, flowers yellow | *Viola rotundifolia* Michx. | round-leafed yellow violet / early yellow violet | 306 |
| Viscum parasite | *Phoradendron serotinum* (Raf.) M. C. Johnst. | American mistletoe | 294 |
| Vitis | *Vitis* sp. | grape | 175 |
| Vitis rubra or monosperma / Vitis, crenate acuminate glabrous leaves, stems prostrate / Vitis repens / Vitis riparia | *Vitis riparia* Michx. | riverbank grape | 277, 278, 284, 285 |
| Vitis vulpina / Vitis, reticulate glabrous leaves, small berries, winter grape | *Vitis vulpina* L. | frost grape | 229 |
| Vitis, glabrous leaves, berries very large, muscadine | *Vitis rotundifolia* Michx. | muscadine / scuppernong | 277 |
| Vitis, lower surface of leaves tomentose, large berries, fox grape | *Vitis labrusca* L. | fox grape | 256, 277 |
| Vitis, tomentose leaves, smaller berries, fruiting around the 10th of September, called summer grape | *Vitis aestivalis* Michx. | summer grape | 277 |

*Table of Described Plants (continued)*

| Michaux's names for a particular species | Modern binomial (scientific name) | Common name(s) | Page(s) in this volume |
|---|---|---|---|
| Willows | *Salix* sp. | willow | 295 |
| Winterania / Winterania canella | *Canella winterana* (L.) Gaertn. | pepper cinnamon / wild cinnamon | 120, 149, *367* |
| Xanthoxillum / Zanthoxylum clava herculis | *Zanthoxylum clava-herculis* L. | Hercules'-club / toothache tree | 105, 288 |
| Yucca aloifolia | *Yucca aloifolia* L. | Spanish dagger | 148 |
| Yucca filamentosa | *Yucca filamentosa* L. | curlyleaf yucca | 148 |
| Yucca gloriosa | *Yucca gloriosa* L. | mound-lily yucca / Spanish bayonet | 148 |
| Zamia / Zamia pumila | *Zamia floridana* (L.) var. *umbrosa* (Small) D. B. Ward | coontie / Florida arrowroot | 88, 89 |
| Zanthoriza or Marboisia | *Xanthorhiza simplicissima* Marshall | yellowroot | 66, 70, 74, 111 |
| Zanthoxilon | *Zanthoxylum* sp. | prickly-ash / toothache tree | 147 |
| Zizania / Zizania palustris | *Zizania aquatica* L. | southern wild-rice | 63, 64, 104, 146 |

# Table of Described Animals

The following table lists the animals André Michaux described in his North American journals and letters. It is organized according to Michaux's names for the species, and the modern common name is provided for each species.

| Michaux's names for a particular species | Modern binomial (scientific name) | Common name(s) | Page(s) in this volume |
|---|---|---|---|
| Alligator / caiman | *Alligator mississipiensis* | Alligator | 66, 68, 95, 100 |
| American buffalo / buffalo | *Bison bison* | Bison, buffalo | 51, 232, 276, 283, 294 |
| American fisher-martin | *Ceryle alcyon* | Belted kingfisher | 286 |
| American pelican | *Pelecanus erythorhynchos* | American white pelican | 286 |
| Aquatic birds including ducks and wild geese | Undetermined | Undetermined | 93 |
| Banded snake (yellow-black-red) | *Lampropeltis triangulum elapsoides* | Scarlet kingsnake | 61 |
| Bear / ours | *Ursus americanus* | American black bear | 73–74, 114, 192–93, 201, 276, 280, 284–85, 294 |
| Beaver | *Castor canadensis* | Beaver | 187–88, 192, 201, 276, 294 |
| Bird, affinity to magpie | *Perisoreus canadensis* | Gray jay | 192 |
| Bird, body ash-colored | *Bombycilla cedorum* | Cedar waxwing | 247 |
| Bird sent to Bosc | *Aphelocoma coerulescens* | Florida scrub jay | 445n49 |
| Black snake that kills rattlesnakes | *Drymarchion cooperi* | Eastern indigo snake | 61 |
| Black squirrel | *Sciurus niger* | Fox squirrel | 68 |
| Bluebird | *Sialia sialis* | Eastern bluebird | 247 |
| Blue jay | *Cyanocitta cristata* | Blue jay | 294 |
| Canada goose / outarde | *Branta canadensis* | Canada goose | 286 |
| Carcajou | *Gulo gulo* | Wolverine | 179, 192 |
| Cardinal / Carolina cardinal | *Cardinalis cardinalis* | Northern cardinal | 247, 276, 294 |
| Chaffinch | *Carduelis tristis* | American goldfinch | 68 |
| Coachwhip | *Masticophis flagellum* | Eastern coachwhip | 69 |

*Table of Described Animals (continued)*

| Michaux's names for a particular species | Modern binomial (scientific name) | Common name(s) | Page(s) in this volume |
|---|---|---|---|
| Crossbills | *Loxia leucoptera, Loxia curvirostra* | White-winged and red crossbills | 242 |
| Crow | *Corvus brachyrhynchos* | Crow | 294 |
| Deer / dwarf deer | *Odocoileus virginianus* | Whitetail deer | 50, 72, 80, 112, 118, 144, 237, 276, 280, 294 |
| Elk | *Cervus canadensis* | Elk, Illinois deer | 276, 283 |
| Goose, with white head | *Anser albifrons* | Greater white-fronted goose | 287 |
| Gray squirrel | *Sciurus carolinensis* | Gray squirrel | 238, 276, 294 |
| Green parrot with yellow head | *Conuropsis carolinensis* | Carolina parakeet | 294 |
| Grouse | *Bonasa umbellus* | Ruffed grouse | 276 |
| Kingbird/ Lanius tyrannus | *Tyrannus tyrannus* | Eastern kingbird | 276 |
| Linx | *Lynx canadensis* | Canadian lynx | 192 |
| Loutre / otter | *Lutra canadensis* | River otter | 191–92, 294 |
| Marmotte | *Marmota monax* | Woodchuck | 192 |
| Martes | *Martes americana* | American marten | 192 |
| Moccasin | *Agkistrodon piscivorus* | Eastern cottonmouth | 61 |
| Moor-hen | *Fulica americana* | American coot | 286 |
| Moquer | *Mimus polyglottos* | Northern mockingbird | 247 |
| Muscicapa / flycatchers | *Empidonax* sp. | Flycatcher | 276 |
| Muskrat | *Ondatra zibethica* | Muskrat | 276, 294 |
| Opossum | *Didelphis virginiana* | Opossum | 287, 294 |
| Owl, large species | *Bubo virginianus, Tyto alba,* or *Strix varia* | Great horned owl, barn owl, or barred owl | 69, 294 |
| Parus / mésanges | *Poecile carolinensis, P. atricapilla* | Carolina chickadee, black-capped chickadee | 247 |
| Parus americanus | *Baeolophus bicolor* | Tufted titmouse | 247 |

| Michaux's names for a particular species | Modern binomial (scientific name) | Common name(s) | Page(s) in this volume |
|---|---|---|---|
| Perdix / Tetrao lagopus | *Lagopus lagopus* | Willow ptarmigan | 193 |
| Picus | Undetermined | Woodpecker | 276 |
| Pigeon, passenger | *Ectopistes migratorius* | Passenger pigeon | 214 |
| Pigeon / turtledove | *Zenaida macroura* | Mourning dove | 214 |
| Porpoise, white | *Delphinapterus leucas* | Beluga whale | 179 |
| Raccoon | *Procyon lotor* | Raccoon | 287, 294 |
| Rattlesnake | *Crotalus adamanteus* | Eastern diamond-back rattlesnake | 61 |
| Raven / corbeau | *Corvus corax* | Raven | 194, 237 |
| Reindeer / renne | *Rangifer tarandus* | Caribou | 190, 192 |
| Renard | *Vulpes fulva* | Red fox | 192 |
| Ring-necked goose | *Branta bernicula* | Brant | 190 |
| Robin | *Turdus migratorius* | American robin | 276 |
| Salamander / mountain alligator | *Cryptobranchus alleganiensis* | Eastern hellbender | 494n99 |
| Sparrow | *Spizella arborea* | American tree sparrow | 247 |
| Striped squirrel, Sciurus striatus | *Tamias striatus* | Eastern chipmunk | 276, 505n80 |
| Summer ducks | *Axi sponsa* | Wood duck | 50, 55–56, 78–80, 201, 287, 330, 332–33 |
| Turkey | *Meleagris gallopavo* | Wild turkey | 50, 56, 69, 72, 75n, 114–15, 161, 230, 276, 286, 293 |
| Turtle with claws | Undetermined | Undetermined | 213 |
| Wolf | *Canis lupus* | Gray wolf | 276 |
| Woodpecker | Undetermined | Woodpecker | 68, 276 |
| Woodpecker, red throat | *Melanerpes erythrocephalus* | Red-headed woodpecker | 294 |
| Woodpecker, small, black body white spots | *Picoides pubescens* or *P. villosus* | Downy or hairy woodpecker | 192 |

# Notes

## Introduction

1. Charles S. Sargent, director of the Arnold Arboretum of Harvard University, recognized the importance of Michaux's work and was instrumental in publishing a transcription of Michaux's diary.

2. Lafayette to George Washington, September 3, 1785, in Gottschalk 1976, 187. The original letter is in the archives of Lafayette College, Easton, PA. Lafayette to James McHenry, September 3, 1785, in *Autograph Letters of Washington, J. Adams, Hamilton and Lafayette presented to the Southern Relief Association, Baltimore, April 2, 1866* (Baltimore, MD: S. S. Mills, 1866), Manuscripts Department, MSS5:6 W4418.1, Virginia Historical Society, Richmond. McHenry, a signer of the US Constitution, was secretary of war from 1796 to 1800. His namesake Fort McHenry is where a large American flag flying during the War of 1812 inspired the United States national anthem, "The Star-Spangled Banner."

3. Régis Pluchet (2004) provides historian Maroteaux's estimate of around 150 hectares (ca. 370 acres) for the Michaux farm. See also Maroteaux 2000. The 1805 English translation by Charles König and John Sims of Deleuze's 1804 "Notice historique sur André Michaux" reported the farm's size as "500 acres" (Deleuze 1805). König and Sims, however, did not take into consideration that the area stated by Deleuze was 500 *arpents*. Because an *arpent* was only 0.85 of an acre, the farm's area would have been more accurately translated as 425 acres.

4. Pluchet 2004; Pluchet, unpublished data on Michaux family genealogy; Joslin 2004.

5. Lamaute 1981, 12, 217; Michaux's first biographer, Deleuze, reported Michaux's date of birth as March 7, 1746, and this earlier date has been widely repeated.

6. Pluchet, unpublished data on Michaux family genealogy; Joslin 2004.

7. Joseph Philippe François Deleuze (1753–1835) was assistant naturalist and librarian at the Muséum national d'histoire naturelle, Paris.

8. Deleuze 1805; Deleuze (1805) 2011.

9. Deleuze 1805, 322.

10. Physician Louis-Guillaume Lemonnier first studied physics but then turned to medicine. Upon meeting Claude Richard (1704–84), he became interested in exotic plants. He soon met Louis XV and received training from Bernard de Jussieu. He encouraged the king to develop a garden in Versailles at the Petit Trianon with de Jussieu in charge. Lemonnier's chief interest was the introduction and naturalization of useful plants from abroad. With the

support of Louis XV and later Louis XVI, Lemonnier was instrumental in sending to various parts of the world at least a dozen trained individuals to collect and transport useful plants from one colony to another and to France. He established some of these plants in his garden in Montreuil near Versailles. His house still stands today although his garden is much reduced. Without Lemonnier's help, Michaux would have remained a farmer (Robbins 1958; Pluchet 2004, 229; Deleuze 1805, 323–24).

11. André Thouin was the oldest son of the head gardener at the Jardin du Roi. Upon his father's death, Buffon appointed him, then seventeen, to take his place. He became an authority on naturalization of foreign plants, grafting, and horticulture. He trained Michaux as well as many other plant explorers. He was a good friend of Thomas Jefferson, whom he met in 1785, when the latter was ambassador to France. They corresponded, and Thouin sent seeds to Monticello for many years. In 1793, upon the reorganization of the Jardin du Roi, Thouin became first professor of horticulture at the Jardin des Plantes. Oscar Leclerc, Thouin's nephew, published his uncle's *Cours de culture et naturalization des vegetaux* (1827) after his death (Letouzey 1989).

12. Antoine-Laurent de Jussieu was born to a family of eminent pharmacists and botanists in Lyon. He finished his medical studies in Paris and worked alongside his uncle. He is remembered for publishing the first natural classification of plants, *Genera Plantarum* (1789), where he classified plants into one hundred orders (equivalent to families today) based on a large number of hierarchical morphological characters. Most of his families are recognized today. This book was an elaboration of the work of his uncle, who had arranged plants at the Petit Trianon in Versailles according to this natural system. Antoine-Laurent participated in the reorganization of the Jardin du Roi into the Jardin des Plantes and Muséum national d'histoire naturelle in 1793, during the revolution, and was appointed its director in 1798 (Stevens 1994).

13. That Michaux had even earlier contact with England is indicated by a 1778 record that Dr. John Fothergill (1712–80), noted patron of John and William Bartram, had sent Michaux seeds of twenty species from India. *Journal des envois de plantes, arbres et graines faits au jardin du Roy par ses correspondents, avec ceux faits par le jardin dudit jardin du Roy aux correspondents d'icelui*, 1772–an V (1797), 117 fols. (transcribed by James E. McClellan III), MS 1327, Archives bibliothèque centrale du Muséum national d'histoire naturelle, Paris (MNHN).

14. André Michaux (AM) to Sir Joseph Banks, FRS [Fellow of the Royal Society], March 15, 1784, Add. MS. 8096.137, Banks Letters, British Library (BM), London; Deleuze 1805, 324–25.

15. R. Williams 2001, 57–70; Deleuze 1805, 325–26.

16. Deleuze 1805, 326.

17. Michaux 1781.

18. AM to Count d'Angiviller, April 15, 1791. Documents on his botanizing in the United States, 1785–1807. Manuscripts Department, American Philosophical Society Library (APS), MSS film 330: three microfilm reels, originals in the Bibliothèque nationale de France (BNF) and Archives nationales (AN), Paris.

19. Pluchet 2014; Al-Zein and Musselman 2004, 201–2.

20. Pluchet 2010.

21. Reveal 2004, 30–31.

22. AM to François André Michaux, March 21, 1784. [André Michaux], Letters and Papers, 1783–1801, Manuscripts Department, APS.

23. AM to Louis-Guillaume Lemonnier, May 23, 1785, and June 2, 1785 (transcribed by Pluchet), Pap. Lemonnier, Archives MNHN; Chastellux 1963, 611.

24. Deleuze 1805, 330; Reveal 2004, 32.

25. Charles Claude Flahaut de la Billarderie, Count d'Angiviller had a brilliant military career under Louis XV. In 1774, when Louis XVI took the throne, he was named director of buildings, arts, gardens, academies, and royal manufactures. He was energetic and enterprising in promoting architecture and the acquisition of art from abroad. The count personally acquired a large mineral collection that he donated in 1780 to the Jardin du Roi. During the revolution, he was accused of embezzlement of public funds, and all his possessions were confiscated. He fled first to Russia and subsequently to Germany. As Michaux's immediate supervisor, in 1785–1791, he was sent at least forty-five letters by the botanist, reporting on the latter's activities in the New World. The last letter that we have from this correspondence is dated Charleston, April 15, 1791; the count had already left France (De Sacy 1953).

26. Count d'Angiviller's instructions for André Michaux and response, August 18, 1785 (APS MSS film 330). AM to Monsieur, September 8, 1785, and AM to Count d'Angiviller, September 2, 1785 (APS MSS film 330).

27. Otto was a brilliant diplomat who was to serve in Austria, England, and Germany, as well as in the United States. Of German origin, Otto studied foreign languages and law at the University of Strasbourg. He came to the United States in 1779 as secretary to the French minister and served as chargé d'affaires from 1785 to 1788. Otto counseled Michaux well and also helped by intervening with Governor Livingston and the New Jersey legislature so that a foreigner could purchase land in that state. Otto was exceptionally well connected; he soon married the governor's niece Eliza Livingston, and after her early death he married the daughter of Crèvecoeur. In 1792, the Girondins named him head of the division of external affairs (secretary of foreign affairs), but when the Jacobins came to power in May 1793 and the Reign of Terror ensued, Otto was imprisoned and nearly met the guillotine (Chinard 1943; Allen and Asselineau 1987, 159).

28. Louis-Guillaume Otto to Count d'Angiviller, October 27, 1785; and November 25, 1785 (APS MSS film 330).

29. AM to Count d'Angiviller, December 1785 (APS MSS film 330); AM to Louis-Guillaume Lemonnier, December 9, 1785 (reprinted in Duprat 1957, 241–43).

30. Robbins and Howson 1958, 354–55.

31. AM to Count d'Angiviller, May 13, 1786; AM to Cuvillier (Count d'Angiviller's secretary), May 12, 1786 (APS MSS film 330).

32. AM to Cuvillier, May 12, 1786 (APS MSS film 330).

33. Robbins 2007, 45.

34. AM to André Thouin, January 19, 1786 (transcription by Régis Pluchet). Archives MNHN. Extensive excerpts from this letter are reprinted in the next chapter.

35. AM to Cuvillier, May 12, 1786 (APS MSS film 330).

36. AM to Count d'Angiviller, July 15, 1786 (APS MSS film 330); André Michaux's expense report to Count d'Angiviller, April 1, 1787 (hereafter cited as Michaux Expenses Journal), "Jun. 1786" from "Journal de Dépenses Aug. 1785 to Mar. 1787" (APS MSS film 330); Sargent 1915.

37. Fry and Jefferson (1755) 2000, 84–85. Some Virginia maps of the eighteenth century indicate modern Winchester, seat of Frederick County, as Frederick Town.

38. Michaux Expenses Journal, "Aug. 1786."

39. Berkeley and Berkeley 1982, 301–2.

40. The Woodlands became a horticultural showplace that contained the largest collection of foreign fruit and shade trees in the United States. William Hamilton was the first in America to grow the ginkgo, *Ginkgo biloba* L., and the Lombardy poplar, *Populus nigra* L. (Spongberg 1990, 60–63, 84–86). Michaux's New Jersey gardener, Saunier, is credited with spreading

the latter so widely in New York and New Jersey that many believed that he and Michaux had introduced it into the United States (Rusby 1884, 89; Robbins and Howson 1958, 366). Most of Hamilton's correspondence has been lost, but the APS has a letter to his private secretary, Lancaster, dated October 3, 1789, where he mentions Michaux's help. Hamilton was a friend of Jefferson, and the latter gave him seeds to germinate from the Lewis and Clark expedition. At different times, two noted botanists, John Lyon (ca. 1765–1814) and Frederick Pursh (1774–1820) were employed as gardeners at the Woodlands. Today a portion of the property including the house remains as part of the Woodlands Cemetery.

41. Heitzler 2005, 1:167–69; Michael J. Heitzler, pers. comm.

42. AM to Count d'Angiviller, November 12, 1786; and April 2, 1787 (APS MSS film 330).

43. AM to Count d'Angiviller, September 30, 1786; November 12, 1786; March 10, 1787; March 26, 1787; and April 2, 1787 (APS MSS film 330); Michaux Expenses Journal, "Nov. 1786–Mar. 1787."

44. AM to André Thouin, February 9, 1790. Facsimile provided by Régis Pluchet from the copy owned by a private individual in Vauréal, France. François André returned to France in February 1790 and may have personally delivered a collection of seeds to Thouin.

45. United States Bureau of the Census 1908b, 12:36; white people were counted by age and gender, but enslaved people were reported only as a total number.

46. Michael J. Heitzler, pers. comm.

47. In both February and March 1787, Michaux reported the services of three black people rented from their owners at the rate of sixteen shillings per month (Michaux Expenses Journal, "Feb. and Mar. 1787").

48. Grégoire 1810, 159.

49. André Michaux, memorandum to Edmund Charles Genet, May 21, 1793, trans. Henry Savage Jr., "Abridged Memoir Concerning My Journeys in North America" (hereafter cited as Michaux Abridged Memoir), Henry Savage Papers, folder 150, Caroliniana Library, University of South Carolina, Columbia (SCL).

50. Wadström 1798, 88.

51. AM to Count d'Angiviller, October 18, 1789 (APS MSS film 330).

52. AM to Count d'Angiviller, February 5, 1790 (APS MSS film 330).

53. Abbé Nolin to Count d'Angiviller, April 30, 1790 (APS MSS film 330).

54. AM to Count d'Angiviller, February 23, 1790; and October 4, 1790 (APS MSS film 330); André Michaux, *Journal de mon Voyage*, end of cahier 9, "Year 1790, Seeds sown in the garden in Carolina near Charleston," Manuscripts Department, APS. The seed list was not reproduced in the transcription of the journal edited by Sargent and published by the APS in 1889.

55. Michaux and Boott 1826. This paper is an English translation of a report on his North American travels that André Michaux drafted for the minister of the interior, Pierre Bénézech, in February 1797. Boott obtained the document from François André Michaux in 1819. The identity of the translator is unknown, but Boott added introductory remarks and arranged for the publication. The original in Michaux's hand is archived at the Royal Botanic Garden, Kew, UK (KG).

56. Michaux and Boott 1826, 129.

57. Agricultural Society of South Carolina records, 1785–1860, 2 vols., 2 folders, Manuscripts Department, South Carolina Historical Society, Charleston (SCHS). The members of the Agricultural Society included some of the wealthiest and most prominent planters in Charleston. Both George Washington and Thomas Jefferson were already honorary members when Michaux was named honorary member in 1792.

58. Michaux and Boott 1826, 129–30; AM to Louis Bosc, April 1, 1793, Fonds patrimonial,

dossier André Michaux, Bibliothèque municipale de Versailles (VERS); AM to François André Michaux, December 30, 1792, in *Esprit des journaux français et étrangers français et étrangers*, 22 année, vol. 5, Mai 1793 (transcriptions by Régis Pluchet).

59. *Dictionary of American Biography* (*DAB*), s.v. "Pond, Peter."

60. Robbins 2007, 85–95.

61. Michaux and Boott 1826, 130; AM to François André Michaux, December 30, 1792.

62. Sargent 1889, 75.

63. Savage and Savage 1986, 124–39.

64. Michaux and Boott 1826, 131; Michaux Abridged Memoir. This episode involving Jefferson and the APS is documented and analyzed in the digital collections of the Library of Congress and in Catanzariti 1990, 1992, and 1995. See especially the editorial note (Catanzariti 1995, 26:75–83).

65. Michaux had joined this organization, which championed the French Revolution and its ideas, prior to his first meeting with Genet (Kennedy 1976).

66. Turner 1898.

67. Williams 2004; Lowitt 1948.

68. Sargent 1889, 108–14.

69. Sargent 1889, 114–27; Deleuze 1805, 344–45.

70. Sargent 1889, 127–40; Michaux and Boott 1826, 131–32.

71. AM to André Thouin, October 14, 1796; AM to Charles Louis L'Hertier de Brutelle, November 7, 1796, trans. Henry Savage Jr., Henry Savage Papers, folder 226, SCL; Deleuze 1805, 346; Michaux and Boott 1826, 131.

72. Michaux and Boott 1826, 130; Deleuze 1805, 346.

73. Deleuze 1805, 347–48.

74. MacPhail 1981, 2–3; Reveal 2004, 38. The archives at KG have what appears to be an original draft of the manuscript of *Histoire des chênes de l'Amérique* along with some proof pages annotated by François André Michaux.

75. Hyde 2013, 93–96.

76. Reveal 2004, 40.

77. McMillan 2007.

78. James Edward Smith of London purchased Linnaeus's herbarium, library, and other natural history collections from his widow after the deaths of Linnaeus (1707–78) and their only son Carl (1741–83). For more information, see the Linnaean Society website at http://www.linnean.org.

79. Berkeley and Berkeley 1963, 173–74; Wulf 2008, 200–204.

80. Ward 2007; Simpson, Moran, and Simpson 1997, 9–10.

81. Michaux (1803) 1974, xii–xv; this reprint edited by Ewan includes the first published English translation of François André Michaux's Latin introduction to the *Flora*.

82. Soon after de Jussieu's book was published, Thouin wrote from Paris to Michaux that he was sending him a copy of it, but Michaux never mentioned whether he received it (André Thouin to AM, November 21, 1789, Archives, Huntington Library, San Marino, CA [HU]).

83. Jandin 1995, 7.

84. Michaux's reasons for leaving Baudin's expedition are more fully explored in the epilogue.

85. Deleuze 1805, 348–50; Dorr 2004.

86. Pluchet 2004, 228, 231; Deleuze 1805, 350; Dorr 2004. Michaux and the gardener Chapelier are buried in adjacent tombs overlooking Ivondro River a few miles south of Toamasina (Tamatave) on the east coast of Madagascar.

87. In addition to the complete and more widely known König and Sims translation (1805) of Deleuze's tribute to Michaux, the *City Gazette* of Charleston began a three-part publication of its own slightly abridged English translation of this document on July 20, 1804 (Dorr 2004).

88. The APS owns the often-reproduced portrait of François André Michaux by Rembrandt Peale. A less well-known portrait of François André by an unknown artist in the Hôtel de Ville, Pontoise, France, is reproduced in Chinard (1957). The portrait collection at KG includes a small painting that closely resembles the Pontoise portrait and was once owned by Francis Boott (1792–1863), François André's London friend. The Gray Herbarium, Harvard University, has photoprints from the daguerreotype taken in 1851 for Asa Gray as well as engravings made from the Rembrandt Peale portrait.

89. Jangoux 2013, 405.

90. Dorr 2004.

91. Pluchet 2014, 18.

92. Charles Kuralt, "André Michaux Talk at Grandfather Mountain, Aug. 28, 1994," Manuscript, Appalachian Collection, Appalachian State University, Boone, NC.

## Chapter One

1. APS MSS film 330.

2. Abbé Pierre-Charles Nolin (1717–96), controller general of French nurseries from 1772 until 1792, frequently corresponded with Michaux.

3. Reprinted in Duprat 1957, 241–43.

4. The small, shrubby tree with blackish wood and scarlet fruit is probably *Lindera benzoin* (L.) Blume, spicebush.

5. APS MSS film 330.

6. Michaux expenses journal, "Nov. 1785–Dec. 1785."

7. APS MSS film 330.

8. APS MSS film 330.

9. Michaux had chosen a garden site that, although not large, supported an impressive diversity of plants because of its varied topography.

10. MNHN, facsimile provided by Régis Pluchet.

11. Lorient (L'Orient) on the Brittany coast was the home port of the French East India Company, and it had a botanical garden where plants sent from India and China were sometimes planted to be acclimatized before being forwarded to the Jardin du Roi in Paris.

12. Chrétien-Guillaume de Lamoignon de Malesherbes (1721–94) was a lawyer, magistrate, government minister, and amateur botanist. Malesherbes defended Louis XVI at his trial and was subsequently executed during the Terror.

13. Jacques-Phillippe Martin Cels (1740–1806) was a gifted horticulturalist who devoted himself to cultivating the new plants that were being introduced by French explorer-botanists, including Michaux. Charles Louis l'Héritier de Brutelle (1746–1800) was magistrate and supervisor of the waters and forests of Paris and an amateur botanist. In a 1788 publication, l'Héritier described and named a new genus of Middle Eastern plants *Michauxia* in honor of Michaux. Dantic was Louis-Augustin-Guillaume Bosc (1759–1828), a botanist, entomologist, and mineralogist who had been a friend and colleague of Michaux's since their days in Paris together at the Jardin du Roi before Michaux's Persian journey. The son of Paul Bosc d'Antic, a physician to the king, Bosc suppressed the second part of his name, "d'Antic," during the revolution. He was friends with political figures who were executed during the Terror, and he left Paris for his own safety.

14. Louis Jean Marie Daubenton (1716–99) was a member of the Académie des sciences, a zoologist, and the assistant to Count de Buffon at the Jardin du Roi.

15. Michaux's inclusion here of *Gaultheria* (a woody subshrub with minutely toothed, aromatic leaves) is something of an enigma. Note that only two paragraphs later in this same letter he asks for a *Gaultheria* leaf to aid him in identification of this species. See also the discussion of Michaux's discovery of *Shortia galacifolia* on June 13, 1787, as well as his elaborate directions for locating an unknown aromatic shrub with toothed leaves on December 9 and 11, 1788. Many have believed the plant Michaux described on December 9 and 11, 1788, to be *Shortia*. Nonetheless, the accumulation of evidence since 1886, when C. S. Sargent proposed that the plant was *Shortia*, suggests that the aromatic shrub Michaux encountered in December 1788 was instead *Gaultheria procumbens*, and that he did not immediately identify the species when he first encountered it in South Carolina. Michaux made a correct identification before he shipped fifty of the plants to France in January 1789 (AM to Count d'Angiviller, from Charleston, January 26, 1789, APS MSS film 330).

16. At his own expense, l'Héritier had published the first part of his *Stirpes novae aut minus cognitae* (Plants either new or little known; Paris, 1784–85), illustrated by Pierre Joseph Redouté. With the coming of the French Revolution, he never completed his *Stirpes novae*.

17. *Atlas américain septentrional* by Georges-Louis Le Rouge, engineer and geographer to the king, Quai des Augustins, Paris, 1778.

18. *Dictionnaire des particules anglaises, précédée d'une grammaire raisonnée* by Jean-François Lefebre de Cillebrune, Paris, 1774.

19. *Eléments de la langue anglaise, ou méthode pratique pour apprendre facilement cette langue* by Louis-Pierre Siret, Paris, 1780.

20. Jose Celestino Mutis (1732–1808) collected plants in Colombia, South America, during the 1760s. Although his "Flora de Bogota" remained unfinished, the specimens that he sent to Linnaeus were described in a supplement to *Species Plantarum* in 1781.

21. Pluchet 2004, 230.

22. Note that Michaux refers to all these plants as "trees," *arbres*. We would refer to some of them as "shrubs."

23. Rusby 1884; Michaux's source for the Lombardy poplar could have been William Hamilton's estate in Philadelphia, which he first visited in June 1786 (Spongberg 1990, 60).

24. AM to "Citizen Minister," Paris, January 25, 1799, trans. Henry Savage Jr., Henry Savage Papers, folder 229, SCL.

25. For a discussion of Michaux's ornamental plant introductions into the United States, see Cothran 2004.

26. Robbins and Howson 1958, 355.

27. Michaux Expenses Journal, "Jan.–Apr. 1786."

28. APS MSS film 330.

29. Michaux Expenses Journal, "June 1786."

30. APS MSS film 330.

31. Trans. in Taylor and Norman 2002, 187.

32. Details about Michaux's visits to Baltimore are lacking. He mentions only a sojourn in the city and botanizing in Maryland during his return trip; however, another of the letters he carried was an introduction from the Marquis de Lafayette to James McHenry (1753–1816), who had served as a Continental army officer on Lafayette's staff in 1780–81. Now a prominent and politically active Baltimore citizen, McHenry would be a representative from Maryland at the Constitutional Convention in 1787 and later serve as secretary of war. Fort McHenry in Baltimore is named for him (*DAB*, s.v. "McHenry, James").

33. Reprinted in Sargent 1915.

34. Armand Louis Gontant, Duke de Lauzun (1747–93), a colorful and dashing leader of volunteer cavalry, served in the French army in America from 1780 to 1783 and was well known to Washington.

35. Gottschalk 1976, 197.

36. Sargent 1915.

37. APS MSS film 330.

38. APS MSS film 330.

39. Michaux Expenses Journal, "June 1786–Sept. 1786."

40. APS MSS film 330.

41. APS MSS film 330.

42. Archives, HU. Michaux's letter is undated but would have been sent after August 19. "Young Archibal" is likely Archibald MacMaster, an experienced Scottish gardener who arrived in France ten years earlier under the sponsorship of the landscape architect Thomas Blaikie (R. Pluchet, pers. comm.).

43. APS MSS film 330.

44. Charleston was then the capital of South Carolina, and Gen. William Moultrie (1730–1805) was serving the first of his two terms as governor.

45. Jean Baptiste Petry was the consul in charge of the French consulate in Charleston.

46. John Jay (1745–1829), later the first chief justice of the United States, was secretary of foreign affairs under the Articles of Confederation; his French counterpart was Charles Gravier, Count de Vergennes (1719–87).

47. The southern garden that André Michaux established was located ten miles north of the city of Charleston in Goose Creek Parish on the neck of land between the Ashley and Cooper Rivers, where the main road from the interior entered the city (Coker 1911). The garden became known locally as the "French Garden." In 2008 a South Carolina State Historical Marker was erected at 2330 West Aviation Avenue in the city of North Charleston adjacent to the Charleston Air Force Base to mark the site. An older granite marker erected by the Garden Club of Charleston in 1954 was relocated in 2017 to a highly visible place outside the main terminal where Michaux and his era in Charleston are depicted in a magnificient new mural. Heitzler (2005) provides detailed information on the Goose Creek area's history and residents.

48. APS MSS film 330.

49. Twenty sous equals one livre. Michaux suggests here that he could have his boxes constructed on the plantation for approximately a quarter of the retail cost.

50. Some early botanists believed that the signature plant discovery of John and William Bartram, *Franklinia alatamaha* W. Bartram ex Marshall, Franklin tree or Franklinia, was a *Gordonia*. Fry (2000) recounts the *Franklinia* story in fascinating detail.

51. APS MSS film 330.

52. Michaux Expenses Journal, "Dec. 1786."

53. APS MSS film 330.

54. Michaux Expenses Journal, "Jan. 1786–Mar. 1786."

55. AM to Count d'Angiviller, from Charleston, February 2, 1787, and March 10, 1787 (APS MSS film 330).

56. APS MSS film 330.

57. AM to Count d'Angiviller, April 2, April 8, and April 15, 1787; AM to Cuvillier, April 2, April 8, and April 15, 1787 (APS MSS film 330).

58. APS MSS film 330.

## Chapter Two

1. Michaux's travel routes in coastal South Carolina and Georgia may be examined on period maps, including William Faden's *Map of South Carolina and a Part of Georgia*, London, 1780, which is available digitally through the Library of Congress at www.loc.gov/item/74692514. It is cited throughout this work as the "1780 Faden map." Even more detailed maps of the entire state of South Carolina may be found in *Mills' Atlas of South Carolina: An Atlas of the Districts of South Carolina in 1825*, compiled by Robert Mills ([1825] 1979) and cited as "1825 *Mills' Atlas*."

2. Michaux began this journey south toward Georgia from the plantation he had acquired ten miles outside the city of Charleston. He first crossed the Ashley River at the Ashley Ferry, which was later known as Bee's Ferry and which is the site of a modern railroad bridge.

3. Throughout his early travels, Michaux only made notes in his journal about the relatively few plants that interested him the most. He had already mentioned finding *Styrax*, a flowering understory tree, in his letter to Count d'Angiviller of February 2, 1787 (APS MSS film 330), and a plant in the genus *Sarracenia* was included in one of his first shipments to France. His *S. lutea* here is *S. flava* L., yellow pitcher plant. His two *Styrax* species are *S. americanus* Lam., American snowbell, and *S. grandifolius* Aiton, bigleaf snowbell, while the *Nyssa aquatica* L., water tupelo, was a species he encountered often and readily recognized (APS MSS film 330).

4. Michaux's route from Ashley Ferry to Parker's Ferry on the Edisto River may be examined on period maps, including the 1780 Faden map and the 1825 *Mills' Atlas*. He first followed the route of modern Bee's Ferry Road north of Charleston until it joined the track of modern US Highway 17, leading southward through modern Colleton and Beaufort Counties to Jasper County, where Michaux left the route of US 17, following what are now state and county roads to reach the Savannah River.

5. Our plant identifications are often documented with reference to the Michaux herbarium on microfilm published by the Inter Documentation Co. (IDC), followed by number references to specific fiche and location on the fiche. Michaux's observation here of *Mespilus* could include species from the modern genera *Amelanchier*, serviceberry; *Crataegus*, hawthorn; or *Aronia*, chokeberry. Possible species based on Michaux's *Flora*, his herbarium, and the present-day distribution are *Amelanchier obovalis* (Michx.) Ashe, coastal plain serviceberry (IDC 65-21); and *A. canadensis* (L.) Medik., eastern serviceberry.

6. The plant resembling *Tythymalus* was *Ditrysinia fruticosa* (W. Bartram) Govaerts & Frodin, Sebastian bush; the pine with two needles was *Pinus glabra* Walter, spruce pine.

7. Ashepoo Ferry was located where the US Highway 17 bridge now crosses the Ashepoo River in Colleton County.

8. The Combahee Ferry crossing site, located on US Highway 17 adjacent to the modern Harriet Tubman Bridge, was an important crossing of the Combahee River. At some times in the eighteenth century, this crossing had a bridge, and at other times a ferry. Traveling this same route a year earlier, Italian naturalist Luigi Castiglioni (1757–1832) reported that the causeway approaching the Combahee was nearly a mile long and in poor condition, and that the water in the river was almost level with the land, subjecting the causeway to frequent flooding (Castiglioni 1983, 122).

9. Michaux refers here to John Days or, more likely, John Deas. His estate, indicating the ownership of 120 enslaved persons, is listed in the 1790 US Census of South Carolina with the spelling *Days*; however, this is probably the census taker's spelling error. A name pronounced "Days" would typically be spelled *Deas* in this region of South Carolina, where Deas was an old and prominent family name.

10. This description fits *Hymenocallis crassifolia* Herb., spring-run spiderlily.

11. The route and distance traveled on April 23 brought Michaux across three tidal rivers: the Pocotaglio, Tullifinny, and Coosawhatchie. Extensive tidal marshes of *Spartina* sp., cordgrass, are found in the area today. The town of Coosawhatchie lies on the south side of the Coosawhatchie River. It appears that Michaux used the ellipsis to abbreviate this difficult Native American name in his journal.

12. *Glandularia canadensis* (L.) Nutt., rose vervain or rose verbena.

13. The distances given for the journeys of April 24 and April 25 suggest that Michaux's camping place was near Coosawhatchie. He did not follow the most direct route from Coosawhatchie to Savannah through the town of Purrysburg, South Carolina. Michaux instead continued along a more northerly course (perhaps the same one taken in April 1775 by William Bartram) toward Two Sisters Ferry, more than fifteen miles upstream on the Savannah River from Purrysburg. This northern route from Coosawhatchie to the Savannah River crosses a large sand ridge known today as the Tillman Sand Ridge. It is shown on the Beaufort District Map of the 1825 *Mills' Atlas* and is likely the stretch of sandy and hilly soil that Michaux describes on April 25.

14. Two Sisters Ferry on the Savannah River, two miles east of present-day Clyo, Georgia, was a well-traveled crossing with a ferry house where William Bartram had dined in 1775. Settled by French Huguenots in the 1730s, Purrysburg (later abandoned) was upstream from Abercorn, located fifteen miles upstream from the original settlement of Savannah on the Georgia side of the river. The Abercorn settlement was short lived; the ten families who settled there in 1733 were gone by 1737, but the Abercorn boat landing continued in use (Jones [1878] 1997, 137–40). The statement that the ferry house was "situated equidistant between Purrysburg and Abercorn" is not correct for the house at Two Sisters Ferry, which was located upstream of and not between these settlements.

15. Michaux's *Lupinus pilosus* is *Lupinus villosus* Willd., pink sandhill lupine (IDC 85-20). In this instance, Michaux's specimen included his species name, location, and date; many specimens lack some or all of this information.

16. The two species here would likely be *Asimina triloba* (L.) Dunal, common pawpaw; and *A. parviflora* (Michx.) Dunal, small-flowered pawpaw.

17. *Betula papyrifera* is not found at this latitude. The species observed here is instead *B. nigra* L., river birch. Michaux realized his error in 1792 when he observed *B. papyrifera* in Vermont.

18. Local residents' observations of the death struggle of two reptiles no doubt produced the imaginative interpretation reported here by the botanist. The black snake mentioned as a predator of the rattlesnake best fits the eastern indigo snake. Michaux's observation is from the historic range of this large reptile, but it is now federally listed as threatened and found only in a much smaller range. It is a large, heavy-bodied snake, black or blue-black on all sides, equipped with powerful jaws (Gibbons and Dorcas 2005, 129–32). Michaux had a healthy fear of snakes and killed any snake he found.

19. Michaux observed *Dirca palustris* L., leatherbark, several times. His specimen (IDC 53-8) mentions Lake Champlain, Kentucky, Pennsylvania, Georgia, and the Carolinas.

20. A specimen in the Michaux herbarium labeled *Azalea calendulacea* β *flammea* collected near Two Sisters Ferry is likely the *Azalea coccinea* mentioned here. It is now recognized as *Rhododendron flammeum* (Michx.) Sarg., Oconee azalea.

21. *Azalea nudiflora* was a deciduous azalea that Michaux would sometimes compare to the new azalea species he encountered; it is recognized today as *Rhododendron periclymenoides* (Michx.) Shinners, pinxterbloom azalea. Linnaeus had first named the plant *Azalea lutea*, then later changed it to *A. nudiflora*, perhaps because the shrubs are bare when the flowers

appear. It had been introduced into England as early as 1734 and was among the first American azaleas to be grown in Europe (Galle 1987, 73).

22. Michaux's description fits the rare *Magnolia pyramidata* of William Bartram, pyramid magnolia.

23. In Michaux's era, *Andromeda arborea* was the correct name for the tree we recognize today as *Oxydendrum arboreum* (L.) DC., sourwood. He often recorded observing this species in his travels and measured the circumference of several large specimens.

24. The first settlement named Ebenezer, founded by Lutheran Saltzburgers escaping religious persecution in Europe, was found to be so inaccessible that in 1736 the town moved to its present location on the Savannah River. The "Ebenezer" mentioned here is probably the abandoned town because period maps show that the road from Two Sisters passed through it. New Ebenezer, as the later settlement nearby was known, had brick buildings, including a church that is still used on special occasions today (Jones [1878] 1997, 11–44).

25. *Bignonia sempervirens* is *Gelsemium sempervirens* (L.) St.-Hil., Carolina Jessamine. The other small climbing shrub having much in common with it (woody vine with opposite lanceolate leaves, yellow flowers) is *Trachelospermum difforme* (Walter) A. Gray, climbing dogbane. Observing this plant again on May 12, Michaux guessed that it might be a *Vinca* "lutea" (with yellow flowers).

26. Michaux's herbarium has two species of *Astragalus* from this region, and he could be referring to either *A. villosus* Michx., bearded milkvetch, southern milkvetch (IDC 88-14), or *A. michauxii* (Kuntze) F. J. Herm., (formerly *A. glaber* Michx.), sandhills milkvetch, Michaux's milkvetch (IDC 88-12).

27. These three terrestrial orchids are all in different genera today. *Arethusa divaricata* is *Cleistesiopsis divaricata* Pansarin & F. Barros, large spreading pogonia. *A. ophioglossoides* is *Pogonia ophioglossoides* (L.) Ker Gawl., rose pogonia. The *Limodorum* is *Calopogon tuberosus* (L.) BSP., grass pink.

28. On April 28 or April 29, Michaux, following the road from Ebenezer to Savannah, crossed a portion of Morton Hall Plantation. Walker (1996) reports that *Rosa laevigata* Michx., Cherokee rose, was first cultivated in Georgia at Morton Hall just prior to the revolution, when this plantation was owned by Nathaniel Hall. A merchant in Savannah, Hall had connections with a number of ship captains. Walker proposes that Hall obtained the seeds of the Cherokee rose from them and grew it on his plantation, where Michaux found and collected the plant during his first journey into Georgia.

29. The palm that is "different from the *Chamaerops* of Carolina" (*Sabal minor* [Jacq.] Pers., dwarf palmetto) was probably *Serenoa repens* (W. Bartram) Small, saw palmetto.

30. Savannah was laid out by Georgia's founder, James Oglethorpe, following plans drawn in England with rectilinear streets crossing at right angles. In this symmetrical layout, numerous squares were reserved for public parks, and subsequent generations followed Oglethorpe's plans. Savannah was the capital of Georgia throughout the colonial era.

31. *Tradescantia umbellata* is in the *Flora* (1:193) as *T. rosea* (IDC 44-15, "from Carolina to Florida") and recognized today as *Cuthbertia rosea* (Vent.) Small, common roseling.

32. Period maps, including the 1780 Faden map, indicate the Ogeechee Ferry on the main road south from Savannah. This road was a coastal route that led to Darien with roads branching off east to Sunbury and west to Fort Barrington.

33. Sunbury was an important colonial-era seaport located south of Savannah where the Medway River empties into St. Catherines Sound. After the Revolutionary War, Sunbury declined in importance. The discussion about finding a way to St. Augustine confirms that Michaux's intention was to continue south into Spanish Florida on this journey. No doubt hostilities with the Creek Indians influenced the decision to turn north.

34. Michaux does not mention in his journal that the extended foray to the banks of the Altamaha made by his son, Fraser, and a servant was prompted by the search for *Franklinia alatamaha*, a rare tree discovered there in 1765 by John and William Bartram. Michaux had seen and admired this tree in Bartram's Garden the year before. Now, while Michaux was traversing the southern wilds, Abbé Nolin in Paris wrote and specifically asked him to find it (Abbé Nolin to AM on June 7, 1787, AN O/1/2112): "Try to obtain l'Altamaha to which the name *Franklinia* has been applied. The seeds that I bought several times have always been sent before they are mature and did not grow. I have the drawing that was sent from Philadelphia and gives one the desire to have the tree."

The year following Michaux's journey, a letter from the Charleston gardener and nurseryman John Watson to Humphry Marshall in Philadelphia, dated April 8, 1788 (now at Clements Library, University of Michigan, Ann Arbor [CLL]), explains: "there is not a plant of the *Franklinia* to be found[;] the French King's Botanist have been several times looking for it where it is said to grow, also a Gentelman from London but found non[e.]"

*Franklinia* is listed in Michaux's *Flora* (2:42) under the binomial *Gordonia pubescens* Lam., and there are three specimens of it in his herbarium (IDC 83-1, 83-2, 83-3), the last one labeled "Franklenia," but none of his specimens has a locality listed.

35. This seems to indicate that he was attempting to use the system of de Jussieu, who used the term *orders* for what we consider families today.

36. Michaux's party retraced its route into Savannah and then turned northwest once again, traveling on the road called the Augusta Road or River Road. A Georgia historical highway marker near modern Rincon in Effingham County reminds travelers of the historic taverns on the old road to Augusta without specifically mentioning Fifteen Mile House. The roads near the Savannah River that Michaux followed are shown prominently on period maps, including the 1780 Faden map.

37. "Capt. Prevott" and "the captain, Maj. Revots" mentioned on May 17 are the same prominent local citizen Abraham Ravot (d. 1795), a captain of the Effingham County, Georgia, militia (McCall [1941] 1968–69, 3:145). Ravot was a son of Gabriel Frances Ravot, one of the initial Huguenot settlers of New Bordeaux, South Carolina, in 1764, and Louise Catherine Malette, whose family initially lived in the French settlement of Purrysburg (Hirsch [1928] 1999, 84).

38. The mention of Beaver Dam Creek indicates that Michaux turned left, away from the Savannah River, at the fork of this road just south of Black Creek that are shown on the 1780 Faden map and other period maps. The left fork was the Quaker Road and led to the community of Wrightsboro, approximately thirty-five miles west of Augusta. Beaver Dam Creek is a tributary of the much larger and longer Brier Creek. The confluence of these creeks near the point where Brier Creek empties into the Savannah River is another feature shown on period maps.

39. We see here that sixteen-year-old François André Michaux has already developed a discerning eye for plants on this, his first extensive trip into the field with his father. Rehder (1923) compares *Lapathum occidentale*, now *Cliftonia monophylla*, to Tournefort's *L. orientale*.

40. At this time the Creek Indians occupied the lands west of Georgia's Oconee River, about two-thirds of the territory of the modern state of Georgia.

41. McMillan (2007) pointed out that no *Trillium* would be in bloom near Augusta in late May, but the mottling of the leaves of either *T. maculatum* Raf., spotted trillium or *T. reliquum* Freeman, relict trillium would draw Michaux's attention.

42. Alligators may readily be observed today in the Phinzy Swamp Natural Area just south of Augusta.

43. This stream could be either Rocky Creek or, perhaps, Butler's Creek.

44. Virginia native Col. Lee Roy Hammond (1726–90) moved to the area in 1765 where he soon became a leading citizen. Hammond resided across the Savannah River in South Carolina, and he represented his district in the South Carolina General Assembly several times. He saw active service as an officer in the Revolutionary War, sometimes campaigning with Gen. Andrew Pickens. His plantations, New Richmond and Snow Hill (the latter being where he died), are shown on the Edgefield District map of the 1825 *Mills' Atlas*. Michaux's hostess was Hammond's wife, Mary Ann Tyler, also a native Virginian. Michaux did obtain a letter of introduction to Gen. Pickens from Col. Hammond, so apparently the botanist made another, later visit that is not mentioned in the journal.

45. Ninety-Six, a small courthouse town in modern Greenwood County, South Carolina, had been the scene of several Revolutionary War battles. The numerical name, said to be in use by 1730, perhaps came from the mileage to the Cherokee town of Keowee.

46. *Pavia (spicata)* is *Aesculus parviflora* Walter, bottlebrush buckeye; Michaux's erstwhile traveling companion John Fraser also visited this disjunct population on the bluffs of the Savannah River opposite Augusta (Wyatt 1985) where he collected the type specimen for the species (Rembert 1984).

47. Battles fought there during the Revolutionary War almost destroyed Augusta; nonetheless the rebuilt town was the capital city of Georgia from 1785 to 1795. Despite Michaux's difficulty purchasing supplies there, Augusta was an important trading center located at the fall line of the Savannah River, where large boats carried the region's products downriver to Savannah and returned with European goods.

48. These observations are as interesting for what they reveal about Michaux, as for the criticism of the residents of Augusta who, as noted earlier, were in the process of rebuilding after the destruction of the Revolutionary War. We are reminded that Michaux is a sober, hardworking man who does not like to pass his time in what he sees as unproductive leisure activity.

49. Michaux refers to botanist and nurseryman John Fraser (1750–1811) as both a Scotsman and as an Englishman; he was the former. Fraser left Charleston with the Michauxs and traveled with them to Georgia. When the party's horses were stolen, Fraser managed to recover his animals. The two botanists then separated, although each traveled to the mountains. Michaux's low opinion of the Scotsman was somewhat unearned. Fraser returned from this journey to assist South Carolina planter and botanist Thomas Walter, who was producing a regional flora that he then carried to London and arranged for the publication of in 1788 as *Flora Caroliniana*. This book included a number of new plants that Fraser collected on his journey with Michaux and that would otherwise have been Michaux's own discoveries. The most spectacular new plant in the book was William Bartram's *Magnolia auriculata*, his new mountain magnolia, which was instead named *M. fraseri*, the name accepted today.

50. Backcountry Georgia, with its fertile lands, was especially attractive to settlers from other states whose farmlands were worn out. Coulter (1965) describes the Virginians who settled this area of Georgia at the end of the Revolutionary War. Many in this elite group became successful planters, and the transplanted Virginians provided many future political leaders.

51. The location of Scott's Ferry on the Savannah is shown on the Edgefield District map of the 1825 *Mills' Atlas*. Today, US Highway 221 crosses the Savannah River from Clark's Hill, South Carolina, into Georgia at this location, where the Savannah has been dammed to form J. Strom Thurmond Lake.

52. He is perhaps speaking about an American goldfinch here.

53. In this locale the pines with two needles were probably *Pinus echinata* Mill., shortleaf pine, and the black oaks *Quercus marilandica* Münchh., blackjack oak.

54. The torrent is probably Little River, which Michaux had to cross in order to travel through the New Bordeaux community of French settlers described next.

55. Founded in 1764, New Bordeaux was a settlement of French Huguenots located on Long Cane Creek three and a half miles upstream from its confluence with the Savannah River. In a page of notes not reproduced in Sargent's 1889 transcription of Michaux's journal, the botanist mentions three people in New Bordeaux: M. de la How, doct.; M. Langavin; and M. Guillaumeau, propr. Dr. John de la Howe (ca. 1710–97) was a physician who came to South Carolina in 1764 and settled in New Bordeaux in the 1780s. His legacy was a school that continues in operation and bears his name today. M. Langavin is likely Pierre Engevine (1727–1805), a Charleston merchant who married Jeanne Boutiton Gibert, the widow of New Bordeaux founder Jean Louis Gibert (1722–73). After Gibert's death in Charleston, Engevine moved with his stepchildren to her New Bordeaux property, which later became known as Badwell. The last name mentioned was André Guillebeau (1739–1814), who married Mary Jane Rocquemore and served in the Revolutionary War. The site of his home and family cemetery is marked with a South Carolina Historical Marker. His sturdy home (listed on the National Register of Historic Places) has been relocated to nearby Hickory Knob State Park and is available for rental (Andrews 1998, 154, 156).

56. This was the Lower Long Cane (or Hopewell) Presbyterian Church (founded 1760 and closed 1950), located twelve miles southwest of modern Abbeville, South Carolina. Other observers have remarked on the large number of worshippers in this congregation. As the area became more densely populated, other Presbyterian churches were formed nearby from this first early congregation. These settlers were descendants of Scottish Presbyterian dissenters against the established Church of England who had lived in Northern Ireland for a generation or two and had then moved to America for better opportunities.

57. Squire "Coohm" was Patrick Calhoun (1737–96). A South Carolina Historical Marker outside Abbeville in Abbeville County marks his family graveyard and homesite (Andrews 1998, 1). Calhoun was the leading man and spokesman of the South Carolina backcountry districts referred to as the "Upcountry" (the area along the coast near Charleston was the "Lowcountry"). In 1756, Calhoun led the first group of Scots settlers into this region west of the Saluda River. They found the *Arundinaria gigantea* (Walter) Muhl., giant cane or river cane, growing along the creeks and rivulets to be from five to thirty feet high (an indicator of fertile soil) and gave the area the name "Long Canes." They were briefly driven out of the area by the Cherokee in 1760, but they soon returned and assisted in the establishment of two other groups nearby, the French at New Bordeaux and the Germans on Hard Labor Creek. Calhoun became the first representative from the Upcountry to the provincial legislature and served in the South Carolina state legislature during and after the revolution. He was the father of US senator and champion of the Old South John C. Calhoun (1782–1850).

58. Gen. Andrew Pickens (1739–1817) married Rebecca Calhoun, a niece of Patrick Calhoun, in 1765 and became one of the leading men in the South Carolina Upcountry. An Elder of the Presbyterian Church and widely respected as an honorable, upright man, Pickens served with distinction in both the French and Indian and Revolutionary Wars. After the revolution he was a military commander and negotiator with the native tribes, and he served in the US House of Representatives. Both his son Andrew Pickens Jr. (1779–1838) and his grandson Francis Wilkinson Pickens (1805–69) became governors of South Carolina. It was natural that Michaux would visit him, not only because he was prominent in the area, but also because Pickens knew the Cherokee well and the botanist planned to enter their territory. The Pickens plantation located at the site of modern Abbeville is the home that Michaux visited, however, during this period Pickens also built a new home on the Seneca River that

he called "Hopewell," which is extant and now owned by Clemson University (Andrew 2017; Waring 1962).

59. Capt. Middle, also referred to by Michaux in his footnote to his June 22 entry as Capt. Vedle, was Revolutionary War veteran Capt. Moses Liddell (d. 1802), who married Elizabeth Johnson in 1784 and is listed in the 1790 US Census of South Carolina. The Abbeville District map of the 1825 *Mills' Atlas* indicates "Liddle's" seven miles northwest of Pickens. The precise location of Liddell's 1775 land grant on Park (Parks) Creek is mapped at the website *Mapping Colonial Abbeville*, http://www.bfthompson.com/abbeville_colonial_plats/AbbePlats_home. htm. (J. Blythe, pers. comm.; Moss 1983, 568).

60. Only one Thomas Lee is listed in the 1790 US Census of South Carolina as living in this area. His household included one woman and six enslaved persons. Rocky River drains into the Savannah; the area where Michaux crossed is probably inundated by Lake Secession today.

61. The minerals mentioned here are common components of pegmatite and indicate that Michaux's knowledge of the natural world extended beyond his knowledge of plants. He recognized and remarked on these minerals again when traveling in the Piedmont of North Carolina near Morganton on November 18, 1789.

62. Deep Creek is downstream from the confluence of Twenty Six Mile Creek and Twenty Three Mile Creek. In discussing William Bartram's journey through this area, Sanders (2002) points out that both Bartram and Michaux crossed Deep Creek at Porter Shoals, which is Porter Marina on Lake Hartwell today. Both botanists then followed the high ground from Seneca to Fort Rutledge, remaining all the while on the east side of the Keowee (now the Seneca River submerged beneath Lake Hartwell). They passed through Pendleton, which was being settled about this time.

63. This creek would be Eighteen Mile Creek.

64. Col. Andrew Williamson's South Carolina militia constructed Fort Rutledge in 1776 during the campaign against the Cherokee. The fort has disappeared, but there is a large stone marker on the site. The Cherokee town of Seneca, or Esseneca, is spelled several ways in the record, but "Seneca," which Michaux uses, is the accepted spelling today. The two Cherokee villages, usually called Little Esseneca and Esseneca, lay on either side of the river of the same name. The sites are largely inundated beneath the waters of Lake Hartwell at Clemson University today. William Bartram also visited these villages, which he called "Sinica."

65. Lewis (Louis) D. Martin is listed in the 1790 US Census of South Carolina as a resident of this area. He was a deputy surveyor of South Carolina involved in laying off and mapping unclaimed lands being purchased. Near his own land holdings on the West Fork of Cane Creek, a tributary of the Keowee, Martin mapped for Michaux three tracts totaling 2,888 acres (Savage and Savage 1986, 73).

66. The Tugaloo River is the name still in use today for the west fork of the Savannah. Zahner and Jones (1983) point out that the east fork, the Keowee River of Michaux's journal, includes three rivers as one proceeds upstream today. Michaux's Keowee River is named the Seneca River at its junction with the Tugaloo; it is called the Keowee upstream from Twelve Mile Creek at Clemson to the junction with the Toxaway in the Jocassee Valley, where it is called the Whitewater River. Today, a series of dams has altered this landscape. The entire bed of the Seneca River is completely submerged in the Lake Hartwell impoundment. The Keowee River bed is beneath Lake Keowee, and the lower Whitewater River bed and its junction with the Toxaway River are beneath Lake Jocassee.

67. Michaux crossed to the west side of the Keowee River and began his journey upstream along this river on established paths. The Mouzon map of 1775 indicates a path running north and south on the west side of the Keowee River, which on this map is called by a Native

American name, Isundigaw River. The river that Michaux crossed on slippery rocks was the Little River, a western tributary of the Keowee now largely inundated beneath Lake Keowee. See this map in Henry Mouzon and William Patterson Cumming, *An Accurate Map of North and South Carolina with Their Indian Frontiers. North Carolina in Maps, Plate VIII* (Raleigh: North Carolina Department of Archives and History, 1966), or digitally through the American Memory Project of the Library of Congress, memory.loc.gov. For modern features, including the topography of the area now underwater, see *Lake Hartwell Recreation and Fishing Guide, Enlarged Version* (Marietta, GA: Atlantic Mapping, 2005).

68. June 8 is meant; on May 8 Michaux was in Sunbury recuperating from the insect bite.

69. Zahner and Jones (1983) report that Michaux continued to travel on poor trails up the west bank of the Keowee River past the abandoned Cherokee towns of Keowee, Kulsage (or Sugar Town), and Toxaway to camp overnight on June 12 at a point several miles downstream of the present Jocassee Dam.

70. This is the type locality for this shrub, *Pyrularia pubera* Michx., buffalo-nut (IDC 123-10, 123-11), discovered here (Uttal 1984, 62).

71. Although Michaux does not mention it in his journal, research in the modern era has shown that he first encountered and collected the plant we know as *Shortia galacifolia* Torr. & A. Gray, Oconee bell, on June 13, 1787, here in the Jocassee Valley, not in December 1788 as has often been reported. The interesting and puzzling story concerning this plant, which stymied some of the best nineteenth-century botanists, can be found in Davies (1956); Zahner and Jones (1983); and Williams, Norman, and Aymonin (2004). Sanders (2007) reprints all three studies and also includes a summary by Williams, "The 'Lost Shortia,' a Botanical Mystery."

72. The "pine or sapinette" (fir tree) is Michaux's initial observation of *Tsuga canadensis* (L.) Carrière, eastern hemlock.

73. Zahner and Jones (1983; see also Zahner 1994) carefully checked Michaux's distances on the ground and freshly interpreted his route. They concluded that on June 13, Michaux, his young interpreter, and his two Cherokee guides continued upstream through the Jocassee Valley (now inundated beneath Lake Jocassee), then climbed along the path of the Whitewater River to reach its headwaters and a more level camping site west of Chimney Top Mountain in Cashiers Valley, where they spent the evening of June 14. The place where the Keowee River begins is surrounded by present Chimney Top, Terrapin, and Sassafras Mountains just south of present Cashiers, North Carolina (Zahner and Jones 1983, 171; US Department of the Interior, Geological Survey [USGS], "Cashiers Quadrangle, N.C.–S.C.–GA," and "Highlands Quadrangle, N. C.–GA," 7.5-minute series topographic [maps] 1:24,000 [Reston, VA: USGS, 1946]).

74. Zahner and Jones (1983) interpret Michaux's reference to the Tugaloo River at this point in his narrative as the headwaters area of both the Chattooga and Tugaloo Rivers.

75. Both *Clethra acuminata* Michx., mountain sweet pepperbush (IDC 59-16, 59-18), and the violet, *Viola hastata* Michx., halberd-leaf violet (IDC 106-15), were discovered here and are mentioned in Michaux's letter to Count d'Angiviller of August 2, 1787 (APS MSS film 330).

76. As interpreted by Zahner and Jones (1983), the next day, June 15, 1787, Michaux's party left Cashiers Valley through Wildcat Gap in the Cowee Mountains, crossed the Highlands Plateau, continued on the Blue Ridge Divide through present-day Scaly, North Carolina, and then moved southwest through Webster Gap, Georgia, north of Rabun Bald Mountain, to the Little Tennessee River at present-day Dillard, Georgia. In the vicinity of Dillard and Rabun Gap, they picked up the main Cherokee Trail. After two more days of travel, they returned to their starting point, the Cherokee village of Seneca (not the modern town of Seneca, but rather located nearer present-day Clemson, South Carolina). Their route through Georgia led

along Warwoman Creek and crossed the Chattooga River into South Carolina at Earl's Ford (Seaborn 1976, 24–35).

77. McMillan (2007) reinterprets this shrub, the favorite of bears, as *Gaylussacia ursina* (M. A. Curtis) Torr. & A. Gray ex A. Gray, bear huckleberry, which is abundant in the lower elevations, while *Vaccinium erythrocarpum* Michx., bearberry, is restricted to the mountaintops in this area.

78. The azalea with yellow flowers is *Rhododendron calendulaceum* (Michx.) Torr., flame azalea.

79. This was the Cherokee village of Tomassee (Tamassee) Town near modern Tamassee in Oconee County, South Carolina.

80. There are two Cane Creeks in this region; Michaux refers to the Cane Creek that flows east through modern Walhalla in Oconee County.

81. Hard Labor Creek in the Londonborough Township, which was the site of a 1765 settlement of German Lutherans. A South Carolina Historical Marker for this settlement, near Bradley in Greenwood County, helps us establish Michaux's route of travel (Andrews 1998, 121).

82. Michaux returned to Charleston, not by way of Augusta as he had come, but on more direct roads through the interior of South Carolina. Turkey Creek in modern Edgefield County is a waypoint given in the 1784 legislation for a new road from Charleston to the Long Canes section that ran south of the South Fork Edisto River. The older route had passed to the north through Orangeburg and Ninety-Six. Michaux's route may be followed using the Abbeville, Barnwell, Colleton, Charleston, Edgefield, and Ninety-Six District maps of the 1825 *Mills' Atlas* (McCord 1841, 9:282).

83. Michaux notes the transition from the Piedmont into the Sandhills region, probably in eastern Edgefield County.

84. The mention of the Edisto River and the plans to return during the following winter confirm that Michaux traveled the new route defined in the 1784 legislation, which he would also follow the next winter. This route approached the south fork of the Edisto, traversing modern Bamberg, Barnwell, and Colleton Counties; Michaux crossed the Edisto at Givhan's Ferry and returned to the plantation near Charleston via the Dorchester Road.

85. Pierre-François Barbé de Marbois Jr. (b. 1754), vice-consul in Philadelphia.

86. Michaux referred to the New Jersey garden by several different names.

87. Louis-Guillaume Otto remained chargé d'affaires in New York, while Antoine-René-Charles-Mathurin de La Forest was vice-consul general.

88. Roland had submitted reports to the French navy about naval timber in American forests (Furstenburg 2014, 464).

89. An indication that Michaux did not initially plan to relocate his base to Charleston permanently is that he had left his herbarium and some of his important books behind.

90. Michaux's letter to Count d'Angiviller of August 2, 1787 (APS MSS film 330), provides a brief summary of his trip to the southern mountains in the most positive language and calls his supervisor's attention to the quantity of a "very pretty shrub named *Fothergilla*," which he is sending as live plants along with a box of seeds.

91. Michaux's letter to Count d'Angiviller of March 10, 1787 (APS MSS film 330), describes a very large shipment of twenty-one cases of plant material on the ship *Adriana*, Capt. Clark in command. Unfortunately the *Adriana* ran aground in Charleston Harbor; the shipment had to be off-loaded, and the plants suffered damage. The ship arrived in Bordeaux, and the plants suffered further damage when transported overland to Rambouillet. Abbé Nolin, controller general of French nurseries who reported to the count, was critical of Michaux's

work on this occasion and perhaps others. Michaux vented his frustrations to his friend and colleague André Thouin in a letter of November 6, 1787, that is excerpted later in this chapter.

92. In addition to the discussion of funds for his operations in America, Michaux points out in his letter to Count d'Angiviller of August 29, 1787 (APS MSS film 330), that there are no regular ships sailing between Charleston and New York except sloops, which have wet decks and therefore cannot be used to send ducks or other animals. He also again requests that a French ship be sent to Charleston to transport his collections to France.

93. Michaux's letter to Count d'Angiviller of September 25, 1787 (APS MSS film 330), announces a shipment of two boxes of seeds and a cage with eight summer ducks. The letter says that Mr. Paul Nerac or Nariax is the *négociant à Bourdeaux* 'the merchant involved in this transaction.'

94. All the species mentioned occur in the vicinity of Charleston, so these are short, local trips.

95. The Réaumur thermometer readings from October 30 through November 5 listed below were omitted from the 1889 Sargent transcription:

Oct 30, morning, 4 degrees [41°F, 5°C]; noon, 14 degrees [64°F, 10°C]
Oct 31, morning, 6 degrees above zero [46°F, 8°C]; noon, 17 degrees [70°F, 21°C]
Nov 1, morning, ice, 2 degrees above zero [37°F, 3°C]; noon, 9 degrees [52°F, 11°C]
Nov 2, morning, 1 degree [34°F, 1°C]; noon, 6 degrees [46°F, 8°C]
Nov 3, morning, 0 degrees [32°F, 0°C]; noon, 16 degrees [68°F, 20°C]; evening, 5 degrees [43°F, 6°C]
Nov 4, morning, 1 degree above zero [34°F, 1°C]; noon, 16 degrees [68°F, 20°C]; evening, 4 degrees [41°F, 5°C]
Nov 5, morning, 2 degrees above zero [37°F, 3°C]; noon, 17 degrees [70°F, 21°C]; evening, 7 degrees [48°F, 9°C]
Nov 6, morning, 5 degrees [43°F, 6°C]; noon, 14 degrees [64°F, 18°C]

96. Strawberry Ferry was on the Cooper River downstream from Moncks Corner in Berkeley County. Michaux's biographers Henry Savage Jr. and Elizabeth J. Savage (1986) erred in reporting that *Nymphaea nelumbo* collected here was *Nelumbo lutea* Willd., American lotus. It was in fact the Asiatic species now recognized as *Nelumbo nucifera* Gaertn., sacred lotus. This exotic species had already been introduced into South Carolina by way of the West Indies and was known locally as the Jamaica lily (Drayton 1943, 59). Supplejack is *Berchemia scandens* (Hill) K. Koch.

97. John Watson (d. 1789) was an English gardener and nurseryman who moved to Charleston in 1755. For a time Watson worked for Henry Laurens, a wealthy merchant friend of both John and William Bartram (Sarudy 1992, 6–10).

98. The history of *Crinum* as a garden plant is a long one. This lily was among the first flowers to be deliberately crossed by early breeders. A list of hybrids containing nearly thirty varieties was compiled in 1837 (Ogden 2007, 192).

99. Because the French packet boats only sailed between French ports and New York, any shipment of plant material from France to Michaux in Charleston had to be transferred to another ship at New York. It is easy to speculate, therefore, that this shipment originated in France and included some of the plants that Michaux had requested in his letter of January 19, 1786 (excerpted in chapter 1).

100. Moncks Corner was a small community about a day's journey from Goose Creek at a road junction north of Charleston near Biggin Bridge over the Cooper River.

101. European deer are larger than the eastern North American species, and Michaux repeatedly refers to the American whitetail deer as a dwarf deer.

102. AM to Count d'Angiviller, from Charleston, December 12, 1787 (APS MSS film 330), describes this shipment.

103. AM to Count d'Angiviller, from Charleston, December 27, 1787 (APS MSS film 330), describes this shipment. The captain refused to take the deer; his ship was a brig with the decks exposed to the weather. M. Limozin, from Le Havre (then Havre-de-Grâce), is mentioned in other documents. He is the merchant who received Michaux's shipments when they arrived in Le Havre and forwarded them to their destinations in France.

104. Translated in and reprinted from Taylor and Norman (2002, 199–203). The original is now in the archives of the Huntington Library (HU), HM 71889.

105. Michaux's observation was likely correct. None of these three species remains in the genus *Laurus* today.

106. Michaux chafed under Count d'Angiviller's directive to send almost everything he shipped to the nursery at Rambouillet as he had friends and colleagues whom he wished to supply with seeds from America.

## Chapter Three

Note: This chapter is based to a large extent on Walter Kingsley Taylor and Eliane M. Norman's 2002 volume *André Michaux in Florida: An Eighteenth Century Botanical Journey*, pp. 64–120.

1. Michaux sent a letter to Governor Zéspedes requesting permission to make explorations in Florida (AM to Governor and Captain General, St. Augustine, March 8, 1788, Library of Congress, Washington, DC [LC], East Florida Papers 51).

Dr. André Michaux, subject of His Most Catholic Majesty and professor of the botanical sciences, respectfully informs you of his desire to conduct botanical observations in this province which undoubtedly will lead to the advancement of botanical science.

I respectfully request that Your Excellency grant me the authority to travel in the province accompanied by two assistants and a negro [word unclear] who will substantiate my observations. Your kindness in granting me this favor will eternally be remembered. I am hopeful of making a discovery that will be worthy of bearing your name as a symbol of my eternal respect for you.

2. St. Augustine was then the capital of Spanish East Florida.

3. It is more likely that Americans cut down the palmettos during the Revolutionary War. Sullivan's Island was the site of the palmetto log fort now known as Fort Moultrie. The patriot militia commanded by Col. (later Gen.) William Moultrie (1730–1805) manning this fort decisively repulsed the British navy's first attack on Charleston, June 28, 1776.

4. Packet boats sailed from New York to France on a regular schedule and often carried shipments from Michaux and Saunier.

5. Michaux omitted from his journal and letters any descriptions and impressions of St. Augustine and the immediate area surrounding the town.

6. The third *Andromeda* specimen is *Lyonia ferruginea* (Walter) Nutt., crookedwood, staggerbush (IDC 57-11, "No 3. *Andromeda* leaves revolute, rusty beneath; St. Augustine"), *Flora* 1:252.

7. The most likely reading source for Michaux's description of Florida was the book *A Concise Natural History of East and West Florida* by Bernard Romans ([1775] 1999). Later in his journal, the botanist mentioned the dwelling of Capt. Roger, called the Indian River the Aïsa Hatcha, and described present-day Mosquito Lagoon and Indian River as "Arms of the Sea." Romans used the same descriptors for these sites. The map that Michaux used is not known, but it could be one from Romans's book or one of William Gerard De Brahm's maps.

8. Deleuze (1805) described Michaux's canoe as about twenty-two feet long, three feet wide, two and one-half feet deep, made from a hollowed-out bald cypress trunk equipped with a sail. Two persons could not sit side by side. These dugout canoes were commonly used for riverine transport throughout the Southeast, including by the Bartrams during their 1765–66 explorations on the St. Johns River.

9. Michaux's host on St. Anastasia Island (also called Fish's Island) was Jesse Fish, one of the most colorful and oldest inhabitants of the St. Augustine area. Fish's home on the island, called "El Verge" (the Garden), was known for its elaborate gardens with olive, date, lemon, and sweet orange trees (Forbes [1821] 1964; Mowat [1943] 1964). The citrus groves on Anastasia Island were world famous (Siebert 1929). François André Michaux, fifteen years later, reminisced about Fish's beautiful plantation, which he had seen in 1788, and about the quality of the oranges that he remembered to be large, sweet, and thin-skinned (F.-A. Michaux [1805] 1904, 305).

10. *Rhizophora mangle* L., red mangrove, is no longer widespread on Anastasia Island because of past freezes. It was thought to have been extirpated, but it has been recently rediscovered at Fort Matanzas on the southern tip of the island (Zomlefer, Judd, and Giannasi 2006). While not in the *Flora*, a Michaux specimen of this species from Florida is in the general collection of the Paris herbarium. The other plant mentioned is *Zamia floridana* var. *umbrosa* (Small) D. B. Ward., Florida arrowroot or coontie (IDC 125-2; *Flora* 2:242). It is recognized by some botanists as *Z. pumila* L., and it still grows on the island.

11. Michaux's North West River is today's Pellicer Creek, which borders St. Johns and Flagler Counties.

12. Michaux used the name *Juglans hiccory* (a name of no botanical standing) for different hickory trees in different parts of America. Here in Florida, he was probably referring to *Carya aquatica* (Michx. f.) Elliott, water hickory.

13. The leguminous shrub with leaves in threes, no. 17, is *Dalea carnea* (Michx.) Poir., pink-tassels (IDC 84-19). Sargent (1889) was incorrect in suggesting that it was instead *Erythrina herbacea* L. The unknown shrub, no. 18, is *Ceanothus microphyllus* Michx., redroot or littleleaf buckbush (IDC 37-5, "March 16, arid areas near the Northwest River"; *Flora* 1:154). It is the type for the species (Uttal 1984, 25), and it occurs today at Faver-Dykes State Park located on Pellicer Creek.

14. *Andromeda formosissima* is *Agarista populifolia* (Lam.) Judd, Agarista, pipe-plant (IDC 57-10; *Flora* 1:253).

15. Michaux probably followed Kings Road south to the Tomoka River, which he called Tomoka Creek. The plantation that Michaux mentioned was known as Rosetta and had formerly belonged to Florida Lt. Governor John Moultrie (1729–98), Loyalist brother of Gen. William Moultrie of South Carolina. Savage and Savage (1986, 377) erroneously referred to this plantation as John Moultrie's elaborate and more famous Bella Vista, located four miles south of St. Augustine near Moultrie Creek of today.

16. The abandoned habitation on the left bank of the lagoon was undoubtedly Mount Oswald. This plantation was owned by Richard Oswald, a wealthy Scottish merchant, West Indian planter, and British envoy at the Paris negotiations in 1782. The identity of "Orange Island" four miles from the abandoned plantation of James Penman has not been determined.

17. *Carica papaya* L., papaya, is native to tropical America and was introduced early in Florida where it has escaped from cultivation. William Bartram located the plant in 1774 growing on the banks of the St. Johns River south of Mud Lake.

18. Michaux's descriptions of Dr. Andrew Turnbull (1718–92) and his ill-fated New Smyrna colony are similar to Romans's ([1775] 1999) account, another indication that Michaux had read Romans's book.

19. A white frost in late March is unusual in coastal Florida.

20. The plantation in ruins where the Michaux party stopped was Mount Plenty, which formerly belonged to Capt. Robert Bisset, an active planter in East Florida during the period of British occupation; Bisset moved his operations to the West Indies when Florida reverted to Spain. The swamp was located west of Mount Plenty.

21. The specimen labeled *Pancratium* (IDC 43-20) has a single flower and appears to be *Hymenocallis latifolia* (Mill.) M. Roem., mangrove spiderlily, commonly found today in the swales and mangrove swamps of Florida's east coast. *Amaranthus australis* (A. Gray) J. D. Sauer, southern water-hemp, is common today in wet areas of Florida's east coast, where individual specimens grow to twelve feet tall or more. *A. australis* is likely the species labeled *Acnida rusocarpa* on plate 50 of the *Flora*.

22. Mount Tucker, known today as Turtle Mound and located about eight miles south of New Smyrna Beach, Volusia County, is a prominent Indian shell mound on the Florida east coast and featured on several early maps, including William Bartram's map in *Travels of William Bartram* (1958). Many tropical plants reached their northernmost distributions at or near Turtle Mound (Norman 1976), but in subsequent years many of these plants have vanished because of severe freezes (Norman and Hawley 1995).

23. Michaux, like Romans ([1775] 1999), remarked that Roger's sugarcane plantation was the southernmost plantation that the English had established in Florida, but this was incorrect because British East Florida absentee owner Peter Elliot owned the southernmost plantation. A Scot, John Ross, was one of Elliot's managers, and probably Romans, and then Michaux, confused Ross with Roger.

24. The swampy area where Michaux crossed was undoubtedly the southern extension of the Great Swamp at the head of the Indian River and located on the Bisset Plantation. Romans also mentioned both Aïsa Hatcha and Rio d'Ais for present-day Indian River. Camp for March 28 was probably along Mosquito Lagoon. In 1837, Dr. Jacob Rhett Motte (1953, 160) described the southern part of Mosquito Lagoon as "an immense extent of flat land, overflowed by the sea, giving it the appearance of a long lake, the water nowhere exceeding six feet in depth."

25. Michaux calls present-day Mosquito Lagoon a "canal." The narrowest place between the Indian River and Mosquito Lagoon is the area known as the Crossing Place or Haulover and is shown on early maps of this area, such as William Gerard De Brahm's 1769 map. It was a place where Indians dragged or "rolled" their canoes over felled trees across the narrow strip of land between Mosquito Lagoon and the Indian River. The Haulover Canal, a man-made canal, connects these two bodies of water today. Hordes of saltwater mosquitoes are still a loathsome nuisance, and one can appreciate the desire of the explorers to leave Mosquito Lagoon. Many bodies of water along the east Florida coast today have thick stands of *Juncus roemerianus* Scheele, black needle rush.

26. *Ficus aurea* Nutt., strangler fig (IDC 106-10), is in Michaux's herbarium. The new *Sophora* was likely a plant in the modern genus *Baptisia* and probably not the *S. tomentosa* L., yellow necklacepod, that Michaux found in flower on April 2. Sargent's (1889) suggestion that the two "*Sophoras*" were the same species is likely incorrect.

27. A one-mile passage of dragging or rolling the canoe through scrub and saw palmetto is about the correct distance across the land following the old Haulover Canal from Mosquito Lagoon to the east bank of the Indian River. Writing in his journal in 1837, Motte (1953) described the narrow strip of land between Mosquito Lagoon and Indian River as "one unbroken expanse of scrub-saw-palmetto from three to four feet high" and reported that the Haulover consisted of a crude Indian path connecting Mosquito Lagoon with the Indian River.

28. The central part of the growing point of cabbage palms is still eaten today. All the ferns that Michaux listed in the journal are still common along the east Florida coast: *Vittaria lineata* (L.) J. Smith, shoestring fern (IDC 129-10; *Flora* 2:261); *Blechnum serrulatum* Rich., swamp fern or marsh fern (*Flora* 2:264), which grows on the stems and branches of cabbage palms; and *Acrostichum danaeifolium* Langsd. & Fisch., giant leather fern (IDC 130-18). *A. danaeifolium* was not distinguished from *A. aureum* L. until after Michaux's era (Morton 1967).

29. The explorers crossed the Indian River to the west bank. The blooming *Sophora* that Michaux found here, the *S. tomentosa* L., yellow necklacepod, is still common along Florida's east coast, but it is not found in either Michaux's herbarium or his *Flora*.

No species of *Spigelia* occurs today in the area where Michaux was botanizing. The plant that Michaux identified as *Spigelia* was probably *Samolus ebracteatus* Kunth, limewater brookweed (IDC 27-6). In the Michaux herbarium, it is mounted on the same sheet as *S. parviflorus* Raf., water-pimpernel. Both species of *Samolus* occur in the area today.

30. The new species of *Annona* that Michaux describes with very large white flowers was probably *Asimina obovata* (Willd.) Nash, big flower pawpaw (IDC 73-22), and the other *Annona* mentioned was likely *Asimina pygmaea* (W. Bartram) Dunal, dwarf pawpaw (*Flora* 1:330).

31. Although Michaux did not encounter any Native Americans during the entire trip, he decided not to venture to the Cape of Florida as originally planned because of possible conflicts.

32. It is difficult to determine exactly the route that Michaux traveled in the Merritt Island–Cape Canaveral area. In a letter written and dated April 24, 1788, from St. Augustine, Michaux told Count d'Angiviller that he went a little past latitude 28°15'. Judging from the mileage that Michaux gave in his journal along with the latitude mentioned, the explorers probably went just past the present-day town of Bonaventure, Brevard County, Florida. This is probably farther south than William Bartram traveled along the east Florida coast in winter 1766.

33. The return trip was made at a quicker pace; most areas were revisited, and the written accounts in the *Journal* are briefer. The island that Michaux visited on April 6 may be today's Pardon Island, the largest island in Mosquito Lagoon in the vicinity of Haulover Canal. *Guilandina bonduccella*, now *Caesalpinia bonduc* (L.) Roxb., gray nicker, was listed in a shipment of seeds from Florida sent to Count d'Angiviller. The mangrove is *Laguncularia racemosa* (L.) C. F. Gaertn., white mangrove. The legume with large fruits is probably *Phaseolus polystachios* (L.) BSP., wild bean, thicket bean (IDC 86-19, "undetermined *Phaseolus* or *Dolichos*"; *Flora* 2:60).

34. Pelican Island is likely the unnamed island where they made camp.

35. *Annona palustris* with large, white flowers was probably *Asimina obovata* (Willd.) Nash, big flower pawpaw. The pond apple (*Annona glabra* L.) that Michaux thought might be a variety of *Asimina triloba* could be *Asimina parviflora*; it is very similar to *A. triloba*. Today, the pond apple occurs considerably farther south of the Tomoka River. The only *Annona glabra* L. in the Michaux herbarium (IDC 73-18) was collected during his trip to the Bahamas. The species is not in the *Flora*. The *Pinus* with two leaves is surely *P. clausa* (Chapm. ex Engelm.) Vasey ex. Sarg., sand pine. Three species of ashes are in the *Flora* and IDC, but none has the locality as Florida. The most likely ash on the Tomoka River would be *Fraxinus caroliniana* Mill., Carolina ash or pop ash.

36. Here and in other places Michaux was distraught with the annual burning practice of the Indians and others to stimulate new growth of grasses. We know today that periodic controlled burning is a natural process and healthy for fire-based communities.

37. *Stillingia sylvatica* Garden ex L., queen's-delight (IDC 119-15; *Flora* 2:213), is fairly common in the area.

38. John Leslie was one of the owners of Panton, Leslie and Company, which operated upper and lower stores on the St. Johns River and traded merchandise with the Indians, but he was not an Indian agent.

39. The *disette* was one of the varieties of *Beta vulgaris*, beet; and *véronique* was probably *Veronica officinalis* L., common speedwell, a medicinal plant used widely at that time for digestive and respiratory ailments.

Although Michaux did not mention so in his journal, he also planned to send to St. Augustine a male specimen of *Phoenix dactylifera* L., date palm, so that the lone female palm there could bear fruit (Deleuze 1805, 336). Michaux had become well acquainted with the reproduction of this plant during his travels in the Middle East (Pluchet 2014, 108).

40. The notebooks that contained the listing of 105 species found by Michaux in Florida and mentioned in the journal have not been found.

41. The same party of five left St. Augustine for the St. Johns River.

42. Job Wiggins, a friend of William Bartram, lived on the east bank of the St. Johns River near present-day East Palatka, in Putnam County. Wiggins accompanied Bartram on the trip to the Alachua Savanna, today's Paynes Prairie, where Bartram served as a delegate to the Indians at Cuscowilla, Florida (Bartram 1958).

43. The store was Spalding's Lower Store, located on the west side of the St. Johns at present-day Stokes Landing, southwest of Palatka, Putnam County, and it was a frequent stopping place used by William Bartram. The party probably camped near today's Welaka Spring.

44. The Indian camp was probably Mount Royal, near Fruitland in Marion County, a sand mound named by John Bartram when he and William made their trip up the St. Johns River in 1765.

45. The island was Drayton Island and Michaux's "alligator point" is today's Lake George Point.

46. *Illicium* refers to *I. parviflorum* Michx. ex Vent., yellow anise-tree, swamp star-anise (IDC 73-3 "Illicium, yellow flowers, Florida"; *Flora* 1:326). The type specimen was collected here (Uttal 1984, 40).

47. The explorers entered Lake George and proceeded to the west bank. At the outfall of today's Salt Spring Run, Michaux observed the famous crystal fountain of William Bartram, who wrote that the fountain exceeded two to three feet in height. This is the site where Bartram discovered the yellow anise in 1766 (Taylor and Norman 2002, 115).

48. The "southern point of the bay" is Lisk Point. The "hill of oranges" where they camped was a shell mound; the fairly wide river is known today as Silver Glen Spring.

49. Although Michaux gave their navigation to be eight miles, the exact place of their camp is uncertain. A good candidate is Zinder Point, a shelly bluff that was covered with orange trees and located near the junction of Lake George and the St. Johns River.

50. The place frequented by Indians was probably Spalding's Upper Store. Bluffton or Orange Bluff was most likely the site of the large "orange orchard" that Michaux encountered. Bluffton was a large shell mound covering about seventeen acres. The hill where they camped was probably Bartram's Mound, also called Little Orange Mound or Idlewild Dock. The shell mound, built on the edge of a swamp, was located about two miles from Bluffton on the west side of the St. Johns River.

51. Michaux did not distinguish between alligator and caiman. His description of the alligator is similar to that in William Bartram's *Travels of William Bartram* (1958). Both Bartram and Michaux encountered large numbers of alligators in the Lake Dexter area of the St. Johns. Bartram called the place "Battle Lagoon."

52. The source of the river that came out of the ground is Blue Spring, located on the east

side of the St. Johns River in modern Orange City, Volusia County. The remote habitation where the party ate lunch was the ruins of the plantation house that belonged to Lord Beresford, located about four miles from Blue Spring. *Cucurbita okeechobeensis* (Small) L. H. Bailey, wild squash, Okeechobee gourd, is rare in Florida, but it still can be found in this area today. William Bartram had already seen this species near the same site (Bartram 1958).

53. Based on a straight-line distance of Michaux's 11 miles from Blue Spring, the explorers could have reached High Banks, Volusia County, before returning. Had he continued about five additional miles, he would have entered Lake Monroe.

54. Camp was made the next day at the Indian Mound at Silver Glen (the "hill of orange trees").

55. The Michaux party returned to Salt Springs Run. The cane is *Arundinaria gigantea* (Walter) Muhl., giant cane or river cane. The *Annona* mentioned here is unlikely to be *Asimina obovata*, which would not be found with illiciums, but it might be *A. parviflora*.

56. The party returned to the residence of Job Wiggins and picked up their horses for the return trip.

57. Michaux's host for this meal, Capt. Carlos Howard, was Governor Zéspedes's main assistant.

58. Relatively little is known of Michaux's religious beliefs (Brasher 2004), but in Charleston in 1788, he could not attend Mass because no Catholic congregation was established in Charleston until 1791 (Woods 2011). Here in Spanish Catholic Florida, the Feast of Corpus Christi was observed, and Michaux participated in the solemn procession of this religious festival.

59. That Michaux respected Governor Zéspedes and appreciated his service is indicated in his published *Flora* (2:70), where the genus *Lespedeza* was named in the governor's honor. It has been suggested that the spelling discrepancy between *Zespedeza* and *Lespedeza* was either an error made by the printer of the *Flora* or a spelling alteration on the part of Michaux (Hochreutiner 1934; Ricker 1934). Regardless of the discrepancy, it is clear from Michaux's letter to the governor that he had planned, before the Florida exploration even began, to name a new plant in his honor.

60. Michaux's detailed account of his Florida observations that he gave to the Spanish government has not been found.

61. Twenty Mile Post (or Twenty Mile House) was located twenty miles north of St. Augustine where travelers changed horses and rested for the night. The Post of St. Vincent, a military outpost, was forty miles from St. Augustine and twelve miles upstream on the St. Johns River, where it occupied high ground and guarded the approach from the sea. It was known as San Vicente Ferrer by the Spanish and St. Johns Bluff by the British.

62. Michaux sailed this small boat from San Vincente Ferrer through the channels and sounds between the barrier islands and the mainland along the coast to Savannah.

63. Michaux camped opposite the bar of the Nassau River, approximately ten miles south of the Georgia border.

64. Camden County is Georgia's southernmost coastal county. People of European descent had settled only along the coast and along the Savannah River. The much larger interior section of present-day Georgia remained lands of the Creek Indians (Coulter 1960).

65. Georgians were at war with the Creeks after the Continental Congress repudiated treaties between Georgia and the Creek Nation. The portion of Georgia north from the St. Marys to the Altamaha River where Michaux was traveling was disputed territory. The war between Georgia and the Creeks was incessant, low-intensity, raiding warfare with results such as Michaux described.

66. Both the Great Satilla and Little Satilla Rivers empty into St. Andrews Sound, which lies between Cumberland Island to the south and Jekyll Island to the north.

67. St. Simons Sound lies between Jekyll and St. Simons Islands. The Turtle River (whose name changes to the Brunswick River at the Port of Brunswick today) empties into St. Simons Sound.

68. Frederica was located on the western side of St. Simons Island, where Georgia founder James Oglethorpe (1696–1785) established a fortified town to control the inner passage between Spanish Florida and the Georgia settlements. Oglethorpe defeated a Spanish attack on Frederica in 1742, and thereafter Georgia was secure for the English colonists.

69. James Spaulding (1734–94) was a prominent planter and merchant of St. Simons Island who operated trading stores in Florida and had assisted William Bartram during his southern travels in the 1770s. Spalding's wife, Margery McIntosh (1754–1818), was a niece of Gen. Lachlan McIntosh (1727–1806) a commander of Georgia Continentals who was at Valley Forge with Washington (Heitman [1914] 1982; Coulter 1960).

70. Michaux probably refers here to Altamaha Sound or Duboy Sound, which lies between Sapelo and St. Catherines Islands.

71. *Fothergilla gardenii* L., a coastal plain species of pocosins and savannas, is known from several counties near Charleston. In his August 2, 1787, letter to the count, Michaux singled out *Fothergilla* as a small tree that produced very desirable flowers, and he included seeds and plants of the species in several shipments to France.

72. Leyritz, a well-educated officer returning to France, was prepared to advance funds to Michaux and carried the botanist's letters to Abbé Nolin and Count d'Angiviller.

73. Because relatively few ships sailed directly from Charleston to French ports, Michaux was often obliged to ship cargo to New York, where it would then be transferred to the French packet boats that sailed from New York to French ports on a regular schedule during this period.

74. Christopher Williman (d. 1813) owned White House plantation on the Ashley River, and his large landholding (2,153 acres at his death) adjoined Michaux's Goose Creek plantation (Smith 1919, 7–8; Joyce 1988, 10; 2009, 4).

75. The source of the box of trees is not known; we speculate that it may have come from the Bartrams of Philadelphia.

76. Dorchester, a community on the Ashley River, was less than a day's journey from the plantation.

77. Michaux's "Alaterne from Carolina" was probably the Carolina buckthorn, now recognized as *Frangula caroliniana* (Walter) A. Gray, a species lacking both scientific description and name until the 1788 publication of Thomas Walter's *Flora Caroliniana*. It likely reminded him of *Rhamnus alaternus* L., Italian buckthorn, native to the Mediterranean.

78. Michaux was probably sending Capt. Marshall a special container for transporting plants.

79. One of the problems in the new United States was the absence of a paper currency accepted throughout the country. In this instance, Michaux needed money for a trip to Georgia because South Carolina paper money was not accepted outside South Carolina.

## Chapter Four

1. Michaux traveled through St. George's Dorchester Parish, now Dorchester County, and through the town of Old Dorchester (today's Old Dorchester State Park), then crossed the upper Ashley River and continued to Givhan's Ferry (formerly Wort's Ferry) on the Edisto River.

2. Michaux began this journey along the route that he had followed when returning to Charleston from his trip to the Georgia and South Carolina mountains in 1787. Appropriate roads showing this route are indicated in the 1825 *Mills' Atlas*; once across the Ashley River, the route generally followed the track of modern SC Highway 61 to its intersection with US

Highway 78 in Bamberg County, and then it continued northwest. South Carolina legislation enacted in 1778 and revised in 1784 and 1788 established this route as a public road to the mountain district of South Carolina and named Henry Peoples as one of the road commissioners (McCord 1841, 9:282–84, 9:315). He, or another of his family, is likely the Peoples indicated here. French-born John d'Antignac (1748–1827), a Revolutionary War officer and member of the South Carolina legislature, owned land on the Edisto River; the ferry was likely on his property (Culler 1995, 274). The surnames Stanley and Bruton are found more than once in the 1790 US Census of this district, while only one Chester, David Chester (located in what is now upper Barnwell County), is listed, and he is known to have hosted other travelers (Culler 1995, 186).

3. This innkeeper was Walter Robinson (or Robison), who kept a tavern in old Winton County, all of whose territory is now part of other counties (Holcomb 1978, 140). White Pond is in modern Aiken County and remains a major road junction today. There US Highway 78 leads northwest into Aiken and road connections with US Highway 25 to Edgefield, on the route to the old Long Canes settlement (modern Abbeville County), while a connector road leads to US Highway 278, the direct route to Augusta.

4. "Col. Stallion" was Col. James Stallings, who was appointed colonel of the Richmond County, Georgia, militia on February 10, 1785. He served as a captain in the Legionary Corps of the Georgia Continental Line during the Revolutionary War (Knight [1920] 1967, 15, 403). Stallings was active in local affairs, and his thousand-acre landholding was located north of the town of Augusta on the modern boundary between Richmond and Columbia Counties near the Savannah River.

5. This *Rhododendron* is the new species, *R. minus* Michx., gorge rhododendron or punctatum, that Michaux mentions collecting in this area near Augusta on November 19. No other evergreen *Rhododendron* species is found in this locality today.

6. The unknown tree with opposite leaves is probably *Forestiera ligustrina* (Michx.) Poir. ex Lam., southern privet. It occurs with *Aesculus parviflora* Walter, bottlebrush buckeye (Michaux's *Pavia spicata*), on the Savannah River bluffs in North Augusta, Aiken County, South Carolina (T. Jones, pers. comm.). Although the common names are similar, this uncommon native shrub should not be confused with the invasive Asian shrubs in the genus *Ligustrum* also called privet that have escaped from horticultural plantings.

7. The shrub resembling an *Erica* is *Ceratiola ericoides* Michx., Florida rosemary.

8. Lambert's, whose location is indicated on the 1780 Faden map and other period maps, was a landmark tavern along the Quaker Road near Odom's Ferry over Brier Creek, approximately six miles south of the new town of Waynesboro in old Burke County, Georgia. Italian naturalist Luigi Castiglioni also stayed at Lambert's in 1785 (Castiglioni 1983, 140), and George Washington dined there during his 1791 southern tour (Jackson and Twohig 1979, 6:140–43).

9. Bel Taverne and Mr. Bel refer to the location where Michaux found lodging. Families with the surname Bell were living in old Burke County at this time, and some were prominent in local affairs.

10. Michaux shipped the plants that he had collected in the vicinity of Augusta by boat downriver to Savannah, where they could be sent to Charleston by sea (note his journal entry for December 31, 1788, later in this chapter).

11. Bedford was a small community in Richmond County northwest of Augusta. It appears on the 1822 Young and Delleker Georgia map, which can be accessed online at http://dlg.galileo .usg.edu/hmap/id:hmap1822y6.

12. Col. John Graves (1748–1824), born in Culpepper County, Virginia, served in Washington's army and then with Gen. Nathaniel Greene in the campaign in the South. After the

revolution, he was a lieutenant colonel in the Georgia militia (McCall [1941] 1968–69, 3:76).

13. Wilkes County is one of the original eight counties of the state of Georgia. When established in 1777, it was the northwest frontier of the state. Wilkes County was nicknamed the "Hornet's Nest" because of the area's successful resistance to British domination. During the war, the state government met at different sites in Wilkes. The town of Washington, named for George Washington, was chartered in 1780 during the dark days of the war.

14. The French physician was Dr. Antoine Poullaine (d. 1794), sometimes spelled Anthony Pullian, who lived and practiced medicine in the town of Washington. He is said to have served with Lafayette during the revolution, and in roughly 1787, he married Sara Garland Wingfield in Hanover County, Virginia. The couple then moved to Wilkes County, as did Sara's father and brother. Dr. Thomas Noel Poullaine (1792–1889) was the child of Antoine and Sara Wingfield Poullaine (McCall [1941] 1968–69, 3:141; Davidson 1932, 1:273–74).

15. French-born Daniel Terondet (or Terrandett) and his wife Nancy lived in the town of Washington. A large landowner in the region, Terondet died in 1794 or 1795 with a will revealing that many people owed him money (Davidson 1932, 1:255; Hudson 1988, 91).

16. "Col. Stablerfield" was Virginia native Peter Stubblefield (1751/52–94), the youngest of the four sons of the elder George Stubblefield (d. 1751); all four sons served as officers in the Revolutionary War. Stubblefield and his wife Peggy Apperson moved to Wilkes County, Georgia, in 1785 (Stewart 1926, 55–56). He became commissary general for the Second Regiment of the Wilkes County militia in 1788 ("Appointment of Field Officers of the Militia," *Georgia State Gazette or Independent Register* [Augusta, GA], November 10, 1788). The Stubblefields were among the wealthiest citizens of the town of Washington, and they also owned hundreds of acres in both Wilkes and Franklin Counties. Peter Stubblefield made his will in neighboring Columbia County in 1794 (Hudson 1988, 91).

17. Daniel Gaines (1745–1803) moved from Virginia to Wilkes County after the Revolutionary War that included his service under Lafayette at Yorktown. Col. Gaines married twice to Virginians, first Mary Hudson and then Mary Gilbert. A lover of the classics, he named three of his children after famous ancient Greeks (McCall [1941] 1968–69, 3:71). In two 1784 transactions, Gaines purchased one thousand acres located on the south side of the Broad River adjacent to the mouth of Chickasaw Creek (Davidson 1932, 119).

18. With this crossing of the Broad River, Michaux entered modern Elbert County. He would continue his journey north into modern Hart County, before crossing the Tugaloo River and entering South Carolina.

19. This is likely a reference to James Tate, a prominent local citizen of old Wilkes County who owned more than twenty-five hundred acres of land in Wilkes and Franklin Counties. He served as a justice of the peace in 1786.

20. Lt. Col. John Cunningham (d. 1829) performed most of his wartime service as an officer with the South Carolina militia. His widow, Ann Davis Cunningham, received a Revolutionary War widow's pension from 1838 until her death in 1849 (McCall [1941] 1968–69, 3:56).

21. The lovely lady was probably the wife of James Freeman, who served in Capt. Bailey's district of the Wilkes County militia. The Freeman plantation was in today's Hart County and is now submerged beneath the waters of Lake Hartwell (Hudson 1988, 11, 117).

22. Seaborn (1976, 65) reports that Larkin Cleveland lived on the Georgia side of the Tugaloo River, while Rev. John Cleveland (ca. 1740–ca. 1825), the first Baptist minister in the area, lived on the South Carolina side. The Georgia–South Carolina state boundary was in dispute when the Clevelands moved into the area. After the 1787 Beaufort Convention between the two states located the state line at the Tugaloo River, the Clevelands lived in different states.

23. In this locale, *Abies* spruce is probably *Tsuga canadensis* (L.) Carrière, eastern hemlock.

24. Michaux's expanded description of this *Carpinus* as *Carpinus fructo lupulo* when he shipped the plants to France in January 1789 confirms the identification as *Ostrya virginiana* (Mill.) K. Koch, American hop-hornbeam.

25. Most of the Cherokee people had been driven from their villages in South Carolina before Michaux's visits. He found several abandoned villages on his travels through the area in 1787. Michaux met and dined with the headman of one of the small Cherokee bands remaining in the area.

26. Michaux was introduced to rockahominy, Indian corn parched without burning and ground to a powder, a staple provision of wilderness travel in early America. The moisture having been removed, rockahominy was lighter, easier to carry than corn kernels, and not readily spoiled by moisture.

27. The two species of *Rhododendron* found in this locality are *R. maximum* L., great laurel, and *R. minus* Michx., gorge rhododendron, punctatum.

28. William Bartram (1739–1823) was the first botanist to visit these mountains, and he found many new plants, including this new mountain magnolia that he called *Magnolia auriculata*. Michaux met Bartram in Philadelphia in 1786 and learned of this new species, but apparently Michaux could not remember the name that Bartram had given the species when he himself encountered it a year later. During this observation of the new species, Michaux thus mentioned *M. cordata* as well as *M. auriculata* as a possible name for the mountain magnolia of Bartram. *M. auriculata* was Bartram's only name for the tree. Michaux's memory lapse has been a source of confusion since that time because, unfortunately, in his *Flora*, *Magnolia cordata* is the name assigned to an entirely different species of magnolia. Further complicating matters, Bartram's delay in publishing the name *M. auriculata* enabled botanist Thomas Walter (ca. 1740–88) to publish the name *M. fraseri* Walter for the same magnolia in honor of his partner, John Fraser (1750–1811). Walter's name is the accepted binomial today.

29. C. S. Sargent (1889, 46) wrote that this plant with toothed leaves was *Shortia galacifolia* Torr. & A. Gray, Oconee bell, but additional information about the plant in Michaux's entry for December 11, and other new information, indicates that Sargent's identification was incorrect; see also note 35.

30. The Keowee River is formed by the junction of the Toxaway and Whitewater Rivers. This confluence is submerged beneath the waters of Lake Jocassee today.

31. Lower Whitewater Falls is the last of a series of cascades where the sixty-foot-wide Whitewater River descends seven hundred feet in half a mile. Michaux continued up the mountain past Upper Whitewater Falls to the headwaters of this river near modern Cashiers, North Carolina.

32. This *Magnolia cordata* is *M. fraseri* Walter; see note 28.

33. Michaux could observe the six-thousand-foot Balsam Mountains in North Carolina from this region.

34. Here André Michaux first noted that the plant species he found in the southern mountains were often the same as those he found farther north. His son later expanded on this observation to suggest that the southern mountains were higher than the mountains in the northeastern states, an insight confirmed when the elevations of the southern mountains were accurately measured several decades later. As Michaux made fresh explorations into new areas, especially Canada, he continued to improve his understanding of plant geography, and his *Flora* incorporated much of this knowledge on plant distribution. It was the first book on American plants to include distributional information.

35. These are the most elaborate directions for finding a plant given anywhere in Michaux's journal. In the 1880s, C. S. Sargent reported that these directions were Michaux's instructions

for locating the plant we know as *Shortia galacifolia* Torr. & A. Gray, Oconee bell. With the new information that has come to light since the 1950s, including the discovery of a previously unknown Michaux-collected specimen of *S. galacifolia* in Paris in 2004 (Williams, Norman, and Aymonin 2004), it is clear that Sargent was mistaken. The plant with good-tasting, aromatic leaves that Michaux described on December 11, 1788, was, as Asa Gray realized at the time, *Gaultheria procumbens* L., wintergreen or teaberry (Sargent 1886, 473). See chapter 2, note 71, which relates to Michaux's entry for June 13, 1787, the probable collection date for his *Shortia* specimens.

36. Apparently Michaux was able to ship part of his collection of plants to Charleston on this wagon.

37. Michaux's letters to Count d'Angiviller detailing the contents of his shipments of January 5 and January 26, 1789, show that he included a total of ninety-six plants of *Rhododendron minus* Michx., gorge rhododendron or punctatum, and an unknown quantity of seeds of that species in these two shipments. At this time he refers to the species as *Rhododendron new species*.

38. "*Mespilus* of the mountains" is *Amelanchier arborea* (Michx. f.) Fernald, downy serviceberry.

39. Michaux retraced the route that he had followed when returning to Charleston in 1787; he did not return through Georgia. Rocky River is a South Carolina tributary of the Savannah River.

40. This is Michaux's third visit with Gen. Andrew Pickens, one of the leading men of the district, who at this time lived at the present site of Abbeville, South Carolina (see also note 58 in chapter 2, for Michaux's June 5, 1787, journal entry).

41. Possibly the John Randall (b. 1744) originally from Virginia who, while a resident of the Edgefield District of South Carolina in 1778, enlisted for service under Col. Lee Roy Hammond and subsequently served under several other commanders. Randall was at different times a private, a lieutenant, and a captain (Moss 1983, 800).

42. On his outbound journey, Michaux had visited the Robertson Tavern at White Pond on November 8 and Chester house on November 7.

43. Michaux also stayed with Peoples the night of November 6, 1788.

44. Michaux here refers to the house at his garden near Charleston.

45. Michaux brought the last items that he had collected in the mountains with him, but plants collected earlier in the journey were shipped to Charleston separately. While some arrived on Dec. 31, 1788, another shipment did not arrive until Jan. 24, 1789, and there were possibly other shipments that he did not note in his journal.

46. With a one-page letter to Count d'Angiviller dated December 29, 1788, Michaux reported that he had returned from the mountains of western Carolina with a grand collection of new trees.

47. Antoine-René-Charles-Malthurin de La Forest, a French consular official; and Pierre-Paul Saunier, the gardener in charge of Michaux's New Jersey garden; both previously identified.

48. This shipment of ten cases to Le Havre is described in a letter to Count d'Angiviller dated January 5, 1789.

49. Dantic, usually known as Bosc, was an old friend and colleague of Michaux's (see chapter 1, note 13). Claude-Jean Rigoley, Baron d'Ogny (1725–93), was the royal postmaster general of France, and Bosc was secretary of the postal service. This shipment of birds is likely to have included the Florida scrub-jay, *Aphelocoma coerulescens*, of which Bosc became the authority.

50. This refers to amateur botanist Charles Louis l'Héritier de Brutelle (see chapter 1, note 13).

51. The plants included in this shipment of four cases are named in the list attached to Michaux's letter to Count d'Angiviller dated January 26, 1789. The third and fourth cases contained a variety of woody plants collected on Michaux's recent journey to the mountains. Michaux's list includes fifty plants that he identified as *Gualtheria? procumbens* (*Gaultheria procumbens* L., wintergreen, teaberry). The misspelling and the question mark indicate that Michaux did not immediately identify the species and remained somewhat uncertain whether or not his identification was correct.

52. Michaux was obtaining personal funds from his brother-in-law in France, Jean Charles Desaint.

53. The *Alnus* is *A. serrulata* (Aiton) Willd., tag alder. Michaux examined this species again on January 30, 1791.

54. New Providence is the island in the Bahamas where the city of Nassau is located.

55. John Murray, fourth earl of Dunmore (1732–1809) was governor of the Bahamas from 1787 to 1796, but he is better known in American history as the last royal governor of Virginia.

56. In this locale, *Vinca lutea* would refer to *Pentalinon luteum* (L.) B. F. Hansen & Wunderlin, Catesby's vine.

57. The *Cornus* is *Nectandra coriacea* (Sw.) Grisb., lancewood.

58. The new species of *Gardenia* is *Casasia clusiifolia* (Jacq.) Urb., seven-year apple.

59. This new species was likely *Chrysocoma capillacea*, today known as *Eupatorium capillifolium* (Lam.) Small ex Porter & Britton, common dog-fennel (M. Vincent, pers. comm.).

60. Michaux listed seeds of five species of *Passiflora* from the Bahamas, including this one, in the plant list attached to his letter to Count d'Angiviller dated April 28, 1789.

61. No species in the genus *Anacardium*, cashew, is known to occur in the native flora of the Bahamas.

62. This is probably Francis Robinet or Robinett, a cooper of French heritage with an interest in natural history, who was a Charleston friend of Michaux's. He is mentioned in a January 15, 1793, letter from Michaux to Consul Mangourit as one of several individuals in Charleston who might have received and held mail for Michaux while the botanist was away traveling in Canada.

63. Lucayas was sometimes used as a name for the entire island group that we call the Bahamas today, but it was also used to identify the Abaco Islands within the Bahamas group. Michaux visited other islands within the Bahamas in addition to New Providence.

64. This is probably the island that we know as Grand Bahama.

65. The dioecious *Spiraea* was *Aruncus dioicus* (Walter) Fernald, goat's-beard.

66. APS MSS film 330.

67. Frenchman Charles Plumier (1646–1704), Dutchman by birth and Austrian by choice Nicholas Joseph Jacquin (1727–1817), and Englishman Sir Hans Sloane (1660–1753) were pioneers in the study of the botany of the West Indies. Michaux indicated here that he was familiar with their works.

68. BM Add MS 8097.334.

69. Michaux thought that the juniper was *Juniperus virginiana* L., eastern red cedar, but the Bahamian plant was the closely related *J. barbadensis* L., Barbados cedar, pencil cedar.

70. *Cascarilla* is the diminutive of *cascara*, the Spanish name for the rind or bark of trees, and Eleuthera was the island in the Bahamas where it was chiefly gathered. Considerable confusion existed among seventeenth- and eighteenth-century botanical and medical writers concerning cascarilla or Eleuthera-bark and the plant yielding it, which is now recognized as *Croton eluteria* (L.) W. Wright.

## Chapter Five

1. This is the first of Michaux's five visits to Camden, in north-central South Carolina; it is the oldest town in the interior of the state. Camden had been the scene of bitter fighting during the American Revolution, but it was recovering from the ill effects of the war when Michaux visited (Edgar 1998, 163, 232–44; Buchanan 1997, 80, 157–72, 397).

2. Mecklenburg County and its seat of Charlotte, North Carolina, were named in honor of George III's Queen Charlotte of Mecklenburg-Strelitz, a patron of botany (Dwelle 1968, 1–3). Her generous support for the inexperienced botanist Thomas Young Jr. led indirectly to John Bartram becoming George III's botanist in North America and being given the resources to explore Florida (Berkeley and Berkeley 1982, 215–17). Notwithstanding the names honoring royalty, Mecklenburg County had been a hotbed of patriot resistance during the American Revolution.

3. Michaux's routes of travel throughout North Carolina can be followed on a contemporaneous state map that shows many of the roads he followed and landmarks he mentioned: Jonathan Price and John Strother, *This First Actual Survey of the State of North Carolina* (Philadelphia: 1808), available online at http://lib.unc.edu/dc/ncmaps (hereafter referred to as the 1808 Price and Strother map).

4. The new species that Michaux and his son observed here near the modern town of Stanley in Gaston County was *Magnolia macrophylla* Michx., bigleaf magnolia. The "*M. cordata* discovered a few years earlier" was Michaux's faulty memory of the name that William Bartram had applied to the tree that we now recognize as *Magnolia fraseri* Walter, Fraser magnolia or mountain magnolia. Earlier researchers who misinterpreted the identity of the tree encountered here either overlooked, or failed to understand, the geography in François André Michaux's account of the discovery of *Magnolia macrophylla* in his *North American Sylva* (1817–19, 2:25–26) and who did not comprehend that the binomial *Magnolia cordata* referred to different trees when used in Michaux's journal and his *Flora* (Williams 1999).

5. Although it is seldom found today in any of the counties where Michaux encountered it, the unknown shrub, which resembled *Calycanthus floridus* L., sweet-shrub, was *Nestronia umbellula* Raf., nestronia, a low, colonial shrub with opposite leaves.

6. Burke Courthouse is modern Morganton, North Carolina. In Michaux's time, Burke County included far more territory than it does today, and many of the highest mountains in the Appalachians were found in Burke (Corbitt 1950, 42–48; Phifer [1977] 1982, 3–10).

7. This is the first of five visits with Col. Waightstill Avery (1741–1821) that Michaux recorded in his journal. Avery's home, Swan Ponds, was located on a tract of land along the Catawba River a few miles west of Morganton (Furr 2007). A Connecticut native and Princeton graduate, he was one of the firebrand Mecklenburg County leaders who early advocated total separation from Britain. During the war Avery became the state's first attorney general, served in the militia, and later moved to Burke County, where he practiced law and was elected to the North Carolina legislature. Avery was in a unique position to assist the botanist with his work because he possessed first-hand knowledge of the people, geography, and routes of travel through the mountains. He routinely traveled across the mountains to the courts in what is now east Tennessee. It is likely that Avery was personally acquainted with everyone whom Michaux met in this thinly settled region. North Carolina's Avery County is named in his honor (Avery 1979; Deyton 1947, 432, 435; Newsome 1934, 309–10).

8. All of Michaux's *Magnolia cordata* observations in this locale west of Lincolnton refer to the mountain magnolia of William Bartram, recognized today as *M. fraseri* Walter.

9. Col. William Wofford (ca. 1728–1823) served in the South Carolina militia and was a

founding partner in Wofford's Iron Works in Spartanburg County, South Carolina. The iron-works was the scene of fighting during the Revolutionary War and was destroyed in 1780, about the time that Col. Wofford moved to a farm in Turkey Cove on the upper Catawba in what is now McDowell County, North Carolina, along the main route over the mountains into Tennessee. On September 29, 1780, one wing of the army of frontier militiamen, known to history as the "Overmountain Men," camped there on the way to the Battle of King's Mountain. Col. Wofford later moved to Habersham County, Georgia. Wofford College in Spartanburg is named for his nephew Rev. Benjamin Wofford (Wait, Wofford, and Floyd 1928, 42–46; Draper [1881] 1996, 182–83; Alderman 1986, 86).

10. Turkey Cove is located at the foot of the Blue Ridge Mountains, north of modern Marion, North Carolina, and is indicated on the US Geological Survey's modern topographic maps. The 1808 Price and Strother map gives the name of the creek flowing into the North Fork of the Catawba River as Turkey Creek, but it is known as Armstrong Creek today. (USGS, "Little Switzerland Quadrangle, N.C.," 7.5 minute series topographic [map] 1:24,000 [Reston, VA: USGS, 1960, 1979]).

11. James Ainsworth was among the earliest white settlers of this area. His large household of two adult men, four male children under age sixteen, three women, and eight enslaved persons is the only Ainsworth household listed in the North Carolina 1790 US Census for this region. The relationship between the Wofford and Ainsworth families has not been determined.

12. This journey to the Black Mountain was a true wilderness trip for Michaux. He traveled with his guides on difficult hunting trails and was apparently unable to estimate distances with any accuracy. Until the 1850s, the fifteen-mile-long Black Mountain range rising thirty miles north of Asheville, North Carolina, was referred to as a single mountain. Today, the range includes twelve named peaks that rise over 6,200 feet above sea level and 3,500 to 4,000 feet above neighboring valley floors, including the 6,684-foot Mount Mitchell, the highest peak east of the Mississippi River. The Black Mountains were not recognized as higher than neighboring Roan and Grandfather Mountains until the pioneering nineteenth-century work of Elisha Mitchell (1793–1857), who lost his life measuring the height of the mountain that bears his name today. Three forest zones are present on these mountains: an oak hickory forest at lower elevations, a northern deciduous forest from about 4,500 to 5,500 feet, and a spruce-fir forest at the highest elevations (Schwarzkopf 1985, xiii, 14–34).

13. The entire ridge complex encompassing Roan Mountain—topped by Roan High Knob (6,286 feet) and three lower-elevation peaks now named Yellow Mountain, Little Yellow Mountain, and Big Yellow Mountain—was marked on state maps as simply "Yellow Mountain" until the middle of the nineteenth century. Nonetheless, the name Roan Mountain for the highest of this complex of peaks was in use locally much earlier and was the name used by John Strother (d. 1815), who led the party surveying the North Carolina–Tennessee state line through this area in 1799 (Williams 1920). The top of the mountain is an open area where trees do not grow, but it is home to a celebrated display of *Rhododendron catawbiense* Michx., Catawba rhododendron, pink laurel. The note on the specimen of this species in Michaux's herbarium says "on the summit of the high mountains of North Carolina," and it was almost certainly collected on this journey.

14. André Michaux's route from Turkey Cove over the Roan Highlands into Tennessee between June 23 and June 28, 1789, cannot be accurately determined from the information in his journal. The loss of his original journal notebook for this period may have deprived us of useful landmarks, but Michaux was clearly accompanied by a very knowledgeable guide or guides. In his journal for March 20, 1796, Michaux later noted that Col. Waightstill Avery was with him in 1789 at Maj. Landon Carter's house (his next overnight stop, Elizabethton, Tennessee), and

it is plausible that Michaux and Avery made this journey over the mountains together (Fulcher 1998). The "hike of several miles on the highest mountains, called the Blue Ridges" suggests that Michaux was traveling along the open mountaintops that run almost due east from Roan High Bluff on the track of the modern Appalachian Trail through the Roan Highlands. These open mountaintops, well below the elevation of the tree line, but where trees do not grow, are a signature feature of this region and are called *balds*. The party's failure to find any settlers living along their route until they reached Peter Parkinson's dwelling indicates that they were not following Bright's Trace, a route over these mountains that Michaux would follow in 1795, but a different rough, steep route that required, as Michaux noted, frequent travel on foot for safety and trail-clearing with tomahawks for the men and horses to pass.

15. Revolutionary War veteran Capt. Peter Parkinson or Parkeson (ca. 1751–92) lived near modern Hampton, Tennessee, and was an ally of Col. John Tipton in the conflict to maintain the authority of North Carolina's government over the state's transmontane counties of Washington and Sullivan, where a fierce political dispute raged between 1784 and 1788. While Col. Tipton led the settlers who supported North Carolina's claims, Col. John Sevier (1745–1815) led a group of rival settlers who attempted to break away and form the new state of Franklin. In winter 1788, this dispute led to actual bloodshed when Sevier and a group of his followers besieged Tipton at his farm. Parkinson led a company of men to Tipton's relief. In both 1795 and 1796, Michaux would find hospitality with Col. Tipton. When North Carolina ratified the US Constitution in 1789, the Franklin issue waned because North Carolina ceded its western lands to the national government. The counties that had attempted to form the state of Franklin then became part of the new Territory of the United States South of the Ohio, and in 1796, part of the new state of Tennessee (Williams [1924] 1993, 337; Fischer 1996, 235; Griffey 2000, 320).

16. Maj. Landon Carter (1760–1800) was one of the most prominent men of what is now east Tennessee. He campaigned actively in the Revolutionary War with John Sevier and Francis Marion. Carter was an active proponent of the state of Franklin and other public affairs of the upper east Tennessee country. Carter County, Tennessee, is named for him, and its seat Elizabethton for his wife. Their home, on the bank of the Watauga River, built by his father John (d. 1781), is a Tennessee state historic site today (Williams [1924] 1993, 299–301; *DAB* s.v. "Carter, Landon").

17. When leaving Carter's, Michaux first crossed the Watauga River and then the South Fork of the Holston River in the vicinity of modern Kingsport before reaching the Block House, where the road that he was following intersected the road to Kentucky.

18. The Block House, at the head of Carter's Valley in Scott County, Virginia, was erected in 1777 by Col. John Anderson (d. 1817). For ten years after the Revolutionary War, the Block House was the chief point for settlers to assemble before heading west into the wilderness of Kentucky (Kincaid 1947, 159). A historical plaque and a large stone chimney are found at the site today.

19. The route through Virginia that Michaux followed was used by thousands of settlers coming from the opposite direction into the western lands that are now the states of Kentucky and Tennessee. Many of the towns and landmarks that Michaux mentions along this route are given with the distances between them in Filson (1784, 113–14). The roads that Michaux followed and most of the towns that he traveled through in Virginia are shown on the period map by James Madison, *A Map of Virginia Formed from Actual Surveys* (Richmond: James Madison, William Prentis, and William Davis, 1807; reprinted in Stephenson and McKee 2000), cited hereafter as the 1807 Bishop Madison Virginia map. On modern maps, Virginia's Highway 11 closely follows Michaux's route for more than three hundred miles from Bristol in the southwest corner of the state to Winchester in the northwest corner.

20. Abingdon, the seat of Washington County, Virginia, was first settled between 1765 and 1770. Black's Fort located there provided refuge during the Cherokee uprising in 1776, and the name became Abingdon in 1778. Some of the buildings that Michaux observed and may have visited in Abingdon remain standing today (Virginia Writer's Project 1940, 439–41).

21. At this site in eastern Smyth County, Michaux collected live specimens of this plant, known today as *Jeffersonia diphylla* (L.) Pers., twinleaf (based on *Podophyllum diphyllum* L.). He gave plants to the Bartrams, who grew it in their Philadelphia garden, where Benjamin Smith Barton (1766–1815) observed the plant and named it in honor of Thomas Jefferson (Barton 1793; Ewan and Ewan 2007, 276).

22. Filson (1784) lists Stone Mill, on the outskirts of modern Wytheville, as a waypoint on the Wilderness Road, 11 miles southwest of Fort Chiswell and 93 miles northeast of the Block House.

23. Kincaid (1947, 64) reports that Fort Chiswell was constructed by Col. William Byrd III during the French and Indian War and named for his friend Col. John Chiswell. The Montgomery County seat is now Christiansburg, but the first courthouse for the larger, original Montgomery County was constructed at Fort Chiswell in 1778. In 1790, Wythe County was formed from the southern part of Montgomery that included Fort Chiswell, and Wytheville (then called Evansham) became the seat of Wythe County (Kegley 1989, 40–41).

24. Filson (1784) reports the distance from Fort Chiswell to the New River as twenty-eight miles.

25. Michaux's mentions of lodgings and mileage appear to best fit crossing the New River at Ingles Ferry near modern Radford, Virginia. This ferry, first licensed to John Ingles in 1762, was the oldest and most important ferry on the New River. The story of John Ingles's wife, Mary Draper Ingles (1732–1815), who was captured by the Shawnee during the French and Indian War and taken to Ohio but escaped and rejoined her husband in Virginia, is one of the epics of the frontier (Kincaid 1947, 54–65).

26. Many of the rivers in the southern Appalachians, including the New River, eventually drain into the Gulf of Mexico. Soon after crossing the New River, Michaux reached the watershed of the Roanoke River, a stream that drains into the Atlantic Ocean instead. Continuing along the road north of Ingles Ferry, shown on the 1807 Bishop Madison Virginia map, Michaux passed through the site of modern Christiansburg, over the mountain ridge identified on that map as the Allegheny Mountains, and then followed a road along the South Fork of the Roanoke River before passing through modern Salem and continuing north along the road to the James River.

27. Natural Bridge is a ninety-foot bridge of stone spanning a 215-foot gorge cut by Cedar Creek, a small tributary of the James River.

28. Lexington, established in 1777 and located near the south end of the Shenandoah Valley, is now the seat of Rockbridge County.

29. Staunton, one of the first two towns established in the Shenandoah Valley, was the seat of Augusta County. This bustling trading and service town was located at the crossroads of the north–south valley road and an east–west road from the Tidewater area through Charlottesville that continued into the mountains (Hendricks 2006, 88–99).

30. The *Spiraea* was likely *Filipendula rubra* (Hill) B. L. Rob., queen-of-the-prairie.

31. The North River rises on the eastern slopes of the Shenandoah Mountain and flows across the upper Shenandoah Valley in Augusta and Rockingham Counties, where it joins the South Fork of the Shenandoah River. The nearest modern city to the river crossing on Michaux's route is Harrisonburg in Rockingham County, then popularly known as Rocktown.

32. New Market is located at the foot of the western slope of Massanutten Mountain,

the ridge dividing the North and South Forks of the Shenandoah River. The town formed around an inn and store established at this crossroads in 1761 by the young John Sevier before he moved west to Tennessee and became a transformative leader (Gilmore [1887] 1997, 11–14; Fischer and Kelly 2000, 146–47).

33. Filson (1784) correctly indicated that Strasburg, his Stover's-town, was only eighteen miles south of Winchester, not forty miles.

34. This is Michaux's second visit to Winchester; he referred to the town as Frederick Town in a letter to the Count d'Angiviller dated July 15, 1786. Hendricks (2006, 88–95) noted that Winchester, the seat of Frederick County, was both the first incorporated town west of the Blue Ridge, and, at the end of the colonial era, the largest, with a population of about fifteen hundred people.

35. Charles Town later became the seat of Jefferson County and is now in West Virginia. The town was laid out in 1786 by Charles Washington (1738–99), the youngest brother of George Washington.

36. Here in the Virginia backcountry far from any seaport, Michaux documented several weedy plants of European origin no doubt accidentally introduced by settlers.

37. Other travelers of the period, including Italian naturalist Luigi Castiglioni, who visited in 1785, also commented on the attractiveness of Frederick, Maryland, a settlement of industrious German immigrants (Castiglioni 1983, 205–6).

38. Michaux was mistaken in believing that the boundary between Maryland and Pennsylvania was the Susquehanna River and that York was in Maryland. York, laid out in 1741, is the seat of York County and the oldest town in Pennsylvania west of the Susquehanna River. The long boundary dispute between Maryland and Pennsylvania led to the famous survey of Mason and Dixon in 1763–68. The Continental Congress met in York in 1777 and 1778; while there, it ratified the Articles of Confederation and the treaties with France (Beyer 2000, 385, 391).

39. The Philadelphia Turnpike connected Lancaster, a center of German settlement in Pennsylvania, to Philadelphia. Lancaster was the capital of Pennsylvania from 1799 to 1812. The ferry across the Susquehanna River between Lancaster and York was Wright's Ferry, established about 1726 (Beyer 2000, 210, 212, 223).

40. Michaux's high opinion of Philadelphia is found in his letter to Cuvillier (Count d'Angiviller's secretary) dated July 15, 1786, excerpted in chapter 1. Philadelphia would become the national capital of the newly formed United States for ten years beginning in 1790.

41. This refers to Pierre-François Barbé de Marbois Jr., vice-consul for Pennsylvania and Delaware since 1785 (Rice 1936–37, 370).

42. The monoecious *Zanthoxylum* is *Aralia racemosa* L., spikenard.

43. As he had no doubt done in spring 1786, Michaux rode the stagecoach from Philadelphia to New York. One of Michaux's countrymen left a detailed account of his own 1788 journey on this route, where stagecoaches traveled every day of the week. The ninety-six-mile trip could be made in either one or one and a half days depending on the line chosen. The horse-drawn stages were open four-wheeled wagons with curtains of leather and wool cloth to protect travelers from the elements. Stages were equipped with four benches seating three each, while heavy luggage was carried at the rear (Brissot de Warville 1964, 152–59).

44. Hector St. John de Crèvecour (1731–1813) was consul for New York, New Jersey, and Connecticut from 1783. When he returned to France on furlough, he requested that Antoine-René-Charles Mathurin de La Forest (1756–1846), then consul in Charleston, move to New York to perform his consular duties. Louis-Guillaume Otto served as chargé d'affaires from 1785 through 1788 (Allen and Asselineau 1987; Nasatir and Monell 1967; Robbins and Howson 1958, 353).

45. On William Hamilton (1745–1813), see introduction, note 40.

46. Wilmington, Delaware, is located at the confluence of the Christina and Delaware Rivers. It is named for Spencer Compton, earl of Wilmington, a friend of proprietor Thomas Penn.

47. The road connecting Philadelphia and Baltimore passed through the towns of Wilmington and Christiana Bridge in Delaware before entering Maryland, where it crossed the Elk River (Carey 1796, 11).

48. The rapidly growing city of Baltimore was the commercial capital of Maryland, but Annapolis was the political capital.

49. Charles-Francois-Adrien d'Annemours (1743–1808) had been Lafayette's officer and had served as consul for Maryland since 1778 (Rice 1936–37, 369–70).

50. Working as a young surveyor's assistant in 1749, George Washington helped lay out the streets of Alexandria, Virginia, located just below the fall line of the Potomac River. Four decades later, Alexandria had become a major seaport of the upper Chesapeake region.

51. Three years earlier, on June 19, 1786, Michaux had paid a visit to Washington at Mount Vernon and delivered a gift of special seeds and plants (Sargent 1915).

52. Colchester, a small tobacco port, was founded in 1753 at the ferry crossing the Occoquan River on the main post road between Boston and Charleston, at the junction with Ox Road leading west to the Blue Ridge and Winchester (Arnold 2007, 15).

53. Dumfries, chartered in 1749, was a substantial town of the period. Located where Quantico Creek empties into the Potomac River, it was an important port for the shipment of tobacco in the mid-eighteenth century, and the tonnage shipped through Dumfries rivaled that of Alexandria, Baltimore, and New York (Arnold 2007, 26).

54. Fredericksburg, located at the head of navigation of the Rappahannock River, where a public ferry had been in operation as early as 1722, was a major town and the prosperous urban center of a fertile agricultural region where agricultural products were delivered for export (Virginia Writers Project 1940, 216–17; Arnold 2007, 28).

55. Located at the head of navigation on the James River, Richmond became the capital of Virginia in 1779, and it soon grew to become the largest city in the most populous state in the United States. Dramatic events of the revolutionary period took place in Richmond, including Patrick Henry's famous "give me liberty, or give me death" speech. High on a hill overlooking the James River, the imposing Virginia State Capitol designed by Thomas Jefferson, although not completed until 1798, was already being used by the legislature when Michaux visited the city (Virginia Writers Project 1940, 284–90).

56. Petersburg, a trading center from the earliest days of settlement, was on the site of Fort Henry at the falls of the Appomattox River. The Virginia colonial legislature created the town of Petersburg in 1748 by amalgamating four adjacent communities under one name (Virginia Writers Project 1940, 273–74).

57. *Hopea* is *Symplocos tinctoria* (L.) L'Hér., sweetleaf, horse sugar.

58. Modern Emporia is composed of two old towns on opposite sides of the Meherrin River: Belfield and Hicksford (or Hicksville). The latter grew up at the site of a ford named for Capt. Robert Hix (Hicks), the commander of nearby Fort Christanna in 1717, and an associate of Virginia's lieutenant governor Alexander Spotswood (Brown 1968, 31–38, 81–86).

59. Halifax on the Roanoke River, the seat of Halifax County, was one of the principal towns of North Carolina, and was often the meeting place of the legislature during the Revolutionary War.

60. Enfield was the seat of Edgecombe County from 1745 until the county was divided into the separate counties of Halifax and Edgecombe in 1759. The old courthouse building in

Enfield remained standing as a landmark for many years thereafter (Sharpe 1954–65, 3:1311).

61. Dorchester Bridge over Swift Creek in Edgecombe County, Duncan Lemon's Ferry across the Tar River in Nash County, and Peacock's Bridge over Contentnea Creek in Wilson County are all shown on the 1808 Price and Strother map. These landmarks indicate that Michaux's route turned slightly west through the vicinity of the modern cities of Rocky Mount and Wilson as he continued traveling south.

62. Whitfield's Ferry over the Neuse River in southeast Wayne County near the boundary with Lenoir County is indicated on the 1808 Price and Strother map.

63. North Carolina historical highway markers near modern Wallace in Duplin County commemorate the 1756 organization of Rockfish Presbyterian Church and a 1781 battle fought between the British army and the North Carolina militia where the road from Wilmington crossed Rockfish Creek, a tributary of the North East Cape Fear River (Hill 2001, 76–77). Two miles downstream from the confluence of the creek and the river is the former site of the river town of South Washington, then in northern New Hanover, now Pender County. South Washington relocated in the nineteenth century to be on the new railroad line rather than to remain on the North East Cape Fear River and also changed its name to Hiawatha, later shortened to Watha (Powell and Hill 2010, 496). Both South Washington and Rockfish Creek are shown on the 1808 Price and Strother map. Savage and Savage (1986, 104) were mistaken when they wrote that Michaux traveled east and passed through Washington, a town in Beaufort County, to reach Wilmington.

64. *Dionaea muscipula* J. Ellis, Venus flytrap, a unique carnivorous plant native only to this area of North Carolina and adjacent South Carolina. The generic name refers to Dione, mother of Venus, the love goddess, because of the leaf blade's supposed resemblance to female genitalia; the specific epithet means "mousetrap." John Bartram referred to the plant as "tipitiwitchet," a euphemism for the vulva that was apparently in use at the time, and his efforts to send the plant to Peter Collinson in England are described in Berkeley and Berkeley (1982).

65. Wilmington, the seat of New Hanover County, is located thirty miles upstream from the Atlantic Ocean at the confluence of the Northeast Cape Fear and Cape Fear Rivers. An old settlement that had become the major seaport between the Chesapeake and Charleston, Wilmington had a resident French consul, G. J. A. Ducher (Rice 1936–37, 370).

66. Old maps indicate that the community of Town Creek was on the creek of the same name in Brunswick County, on the west side of the Cape Fear River, downstream from Wilmington on the east side. Michaux crossed both the main channel of the Cape Fear River that flowed past the Wilmington waterfront as well as the channel between the Brunswick County mainland and Eagle Island, which is marked on modern maps as the Brunswick River. Eagle Island is shown as two distinct islands with a channel in between on some maps, but this additional channel may have been Michaux's third river. See the following, contemporaneous North Carolina maps: William Barker, Jonathan Price, and John Strother, *A Map of Cape Fear River and Its Vicinity from the Frying Pan Shoals to Wilmington* (Philadelphia: William Barker, 1807); and John Lodge and John Bew, *Cape Fear River with the Counties Adjacent, and the Towns of Brunswick and Wilmington . . .* (London: J. Bew, 1781). These maps can be viewed online at http://lib.unc.edu/dc/ncmaps. The 1808 Price and Strother map also shows the road that Michaux is likely to have taken toward his next landmark, Lockwood Folly. This road is directed south toward the town of Brunswick well inland from the river, then it continues its inland path, curving to follow the contour of the coastline below Brunswick.

67. Lockwood Folly is the name given to an inlet and a tidal river in Brunswick County. The unusual name is derived from a resident of Barbados named Lockwood, whose early attempt at settlement was destroyed by the Native Americans after he ignored warnings and

foolishly built on their land (Sprunt [1914] 2005, 36–37). The Little River is a tidal river beginning at Little River Inlet just across the state line in South Carolina. See the Horry District map in the 1825 *Mills' Atlas*.

68. Michaux's mileage and descriptions for September 12 and 13 indicate that he traversed the area now known as the Grand Strand of South Carolina, including Myrtle Beach, where he would have walked and ridden directly on the sandy beach at low tide for several miles. He subsequently first reached not the Santee but the broad lower Waccamaw River ("Wackama" on the 1780 Faden map) south of Georgetown where it empties into Winyah Bay, an arm of the sea. In his 1794 journey through this same area, Michaux described not only the ferry crossing of the Waccamaw but also the Santee, and he remarked on how difficult and dangerous these crossings were.

69. Although Michaux does not mention landmarks along his route from the Santee River to the Cooper River, his note that it was only five miles from the Cooper to his plantation allows us to determine that he followed the more inland route to Clement's Ferry, which crossed the Cooper upstream closer to his plantation than the Hobcaw Ferry directly opposite the city of Charleston.

70. Michaux's early biographer J. P. F. Deleuze describes the successful plantings of *Illicium parviflorum* Michx. ex Vent., yellow anise-tree, in Charleston a year earlier.

71. Even in these restrained notes, we see Michaux's concern for his son's welfare. François André would soon leave for France where he would begin his medical education.

72. Moncks Corner, a backcountry commercial center that grew from one store in 1753 to five stores and four well-kept taverns less than twenty years later, owed its growth to its location at the fork of the main road leading north from Charleston (Edgar 1998, 163).

73. Only one Jackson, Thomas, is listed as a head of household in this district, St. John's Berkeley, according to the US Census of 1790 for South Carolina.

74. Early botanists left a thicket of nomenclatural confusion as they struggled to describe and classify the oaks. Michaux used Mark Catesby's names, *Quercus nigra* and *Q. aquatica*, for two familiar oaks in this list. Catesby's *Q. nigra* is now recognized as *Q. marilandica* Münchh., blackjack oak. One of Michaux's correspondents in France, Lamarck, published the name that Michaux used here for the water oak, *Q. nigra aquatica*, but we recognize this species today as *Q. nigra* L., the binomial established by Linnaeus (Michaux 1801, species nos. 11 and 12; Weakley 2011, 516).

75. *Quercus salicifolia* is not a recognized binomial but rather appears to be Michaux's descriptive phrase meaning oak with willow-like leaves. While the most likely candidate is *Q. phellos* L., willow oak, Michaux had often recognized this species earlier. Perhaps *Q. laurifolia* Michx., laurel oak, another oak with willow-like leaves native to this area, was meant on this occasion.

76. Eutaw Springs, now beneath the waters of Lake Marion, a hydroelectric lake, was a limestone springs near Nelson's (or Neilson's) Ferry on the Santee River. It was also the site of the last major battle between Gen. Nathaniel Greene's American army and the British army in September 1781.

77. The High Hills of Santee are the highly eroded remnants of an ancient seashore north of the Santee River and east of the Wateree River in western Sumter County. The hills are twenty-five miles long and five miles wide at their widest point. Their greatest elevation rise is three hundred feet above the Wateree. By the Revolutionary War era, the High Hills of Santee were relatively populous and had gained a reputation as a healthy place to spend the summer (Edgar 2006, 443).

78. Dr. Isaac Alexander (ca. 1750–1812), a leading citizen and elder of the Presbyterian

Church, was twice elected to serve in the state legislature as well as to serve as intendant (mayor) of Camden. From 1801 to 1804, he was one of the initial thirteen trustees of the University of South Carolina. When Edmund Charles Genet, minister of France, visited Camden on April 26, 1793, Dr. Alexander was one of the dignitaries who welcomed him to the celebration held in his honor and who proposed a toast (Kirkland and Kennedy 1905, 317–19, 342–43; Edgar, Watson, and Bailey 1981, 32–33).

79. Capt. William Nettles (1742–1833), who moved from Virginia to the Camden district before the revolution, served several tours in the militia from 1776 until the end of the war, ultimately rising in rank to captain. Nettles was a stalwart patriot who continued the fight despite being involved in several disastrous engagements, and he went on to enjoy a long life as an honored citizen of the region. He lived on Sanders Creek, six miles north of Camden (Moss 1983, 724).

80. A local landmark, Hanging Rock was the site of Revolutionary War fighting. Bear Creek rises southeast of modern Lancaster, seat of Lancaster County, and feeds the town's reservoir today.

81. The fork in the road just north of Hanging Rock Creek is shown on the Mouzon map of 1775, and this section of South Carolina is also shown on the 1808 Price and Strother map of North Carolina. Michaux indicates that he took the right fork, which is the center road on the Price and Strother map. This route across Lancaster County, following modern county roads, leads to the modern town of Waxhaw in Union County, North Carolina. Robert Bartley, listed in the South Carolina 1790 US Census in the Camden District of Lancaster County, was born in 1762 in Ireland and arrived in South Carolina at the age of seven. He volunteered for military duty in September 1780, serving under Gen. Thomas Sumter and several other officers. He later moved to Kentucky, to Tennessee, and eventually to Alabama (Moss 1983, 51).

82. The next two creeks that Michaux mentions crossing are Camp Creek at six miles from the Johnson house, and Cane Creek at ten miles. See the Lancaster District map in the 1825 *Mills' Atlas*, as well as USGS "Lancaster: SC–NC," [map] 1:100,000 metric topographic (Reston, VA: USGS, 1986).

83. This is likely John Crye (or Cry), who with his wife, Catherine, lived on a five-hundred-acre farm in what is now Union but was then Mecklenburg County, North Carolina. He is mentioned in county court records and left a will that identifies his surviving children and their spouses (Ferguson 1993, 21; 1995, 7, 158, 161, 194, 196, 203, 228).

84. Tuckasege Ford (Tuckesege) on the Catawba River at modern Mount Holly in Gaston County is now submerged beneath the waters of Lake Wylie, a hydroelectric lake. Michaux also mentioned crossing this ford traveling west in 1794 and traveling east 1796; it was on the main route from Charlotte in Mecklenburg County to Lincolnton in Lincoln County. The ford and road network between Charlotte and Lincolnton are indicated on the 1808 Price and Strother map. Modern spellings of the name vary slightly from county to county (Tuckasegee, Tuckaseege, or Tuckaseegee).

85. Peter Smith (1761–1838) was a settler of German descent who lived on Mauney Creek in Gaston County, near the modern town of Stanley. Eight years after Michaux's visit, Smith acquired property from his neighbor Benjamin Smith (no known relation) with a population of *Magnolia macrophylla* Michx., bigleaf magnolia. Four times between 1803 and 1809, Peter Smith hosted botanist John Lyon (ca. 1765–1814), who came from Philadelphia and collected large quantities of the young magnolia trees on Smith's property. Lyon collected so many young trees at Smith's (3,600 in 1809) that it is probable that he had an arrangement with Smith to grow the plants from seed (Ewan and Ewan 1963, 48; Williams 1999, 4–9).

86. This is a fuller description of *Magnolia macrophylla* Michx., first mentioned in this area

on June 10, 1789 (see note 4 in this chapter). Michaux's biographers Henry Savage and Elizabeth Savage (1986, 106, 379) erred when they concluded that this long Latin description referred to *M. virginiana*, L., southern sweet bay.

87. Lincolnton, North Carolina, incorporated in 1785, is the seat of Lincoln County. The first log courthouse was replaced in 1788 by a plank building painted bright red.

88. George Heinrich Weidner (1717–92) married in Pennsylvania in 1749 and obtained a thousand-acre grant of land in North Carolina the next year. He was one of the earliest settlers of modern Catawba County, and his home stood on the Henry's Fork (named for him) of the South Fork of the Catawba River. During the French and Indian War, Weidner narrowly escaped a Cherokee war party, but he returned to reclaim his land after hostilities ended. When he fled the Cherokee, he took refuge with a Robinson family living in what was then Anson County, North Carolina. When they grew to adulthood, one of the sons in the Robinson family, Jesse, married Weidner's daughter Mollianna, and Weidner gave this son-in-law his homeplace on Henry's Fork. In later journeys, Michaux refers to this same stop as "Robertsons, formerly Watner's" (Preslar 1954, 414–21; McAllister and Sullivan 1988, 1–7, 16–17; Freeze 1995, 22–30).

89. North of Turkey Cove, the North Fork of the Catawba River descends gently through a narrow valley called the North Cove. The western slope of Linville Mountain rises more than two thousand feet above North Cove, and the Blue Ridge rises even higher above Turkey Cove (USGS, "Linville Falls Quadrangle" 15 minute series topographic [map], 1:62,500 [Washington, DC: USGS, 1956]).

90. From Turkey Cove, Michaux would first climb the Blue Ridge then travel to the west and southwest in order to reach the Black Mountain.

91. There are numerous cascades but no major waterfalls on the South Toe River where it flows through the Black Mountains (Hairr 2007, 95).

92. This note appears at the top of the last page of Michaux's fifth notebook. The handwriting, although André Michaux's, is slightly different from the next, final entry in this notebook, dated December 9, 1789. This note was not written in 1789 but in 1791 during André Michaux's visit to Camden County, Georgia, and seems to have been inadvertently placed in this notebook. Michaux's other notes from his 1791 visit to Georgia are found in the sixth notebook. The people named lived not in North Carolina but in Georgia. The key figure is Mr. Seagrove. The Seagrove brothers, James and Robert (d. 1795), were prominent local citizens of Camden County, Georgia. James Seagrove (ca. 1750–1812) was especially prosperous and active in local and state affairs, a founder of the town of St. Marys, and later agent for the Creek Indians. In Camden County, Seagrove was closely associated with the Stafford brothers, Robert and Thomas (d. 1800). Moreover, Seagrove was an unusual name in the United States in 1790 and is not found in the North Carolina or any other extant reports of the 1790 US Census; however, the 1790 US Census of Georgia is lost (Christian 1990, 20–26; Vocelle 1914, 43–46).

93. Although Michaux provides no details of his return journey, it is almost certain that, encumbered with his plant collections, he returned to Charleston along the same route through Lincolnton, Charlotte, and Camden that he had taken on the two earlier journeys to these mountains in June and November.

94. HU, HM 71892, transcription by Régis Pluchet.

95. Count d'Angiviller opposed the revolution and thus had many enemies. He was accused in 1790 of grossly inflating his expenses, and the next year the National Assembly ordered the seizure of all his property. He left for Germany and then Russia, and he finally returned to Germany, where he died in a monastery in 1809.

96. M. Mauduit was a physician and specialist in ornithology with an interest in naturalization of species.

# Chapter Six

1. APS MSS film 330.

2. Thomas Walter (ca. 1740–89) and his partner, John Fraser, believed that they had discovered an important new cereal crop, a grass they called *Agrostis cornucopiae*, which is now recognized as *Agrostis perennans* (Walter) Tuck., bentgrass. After his return to Britain, Fraser attempted without success to interest English agriculturalists in the grass. He then traveled to France, where he again failed to sell the plant, but Count d'Angiviller was curious enough to ask Michaux to obtain this grass for him. Fraser's promotional pamphlet on the plant also includes his sketchy report of the 1787 journey with Michaux and his explorations after they separated (Fraser 1789; Rembert 1980, 19).

3. The other grass mentioned here, *Panicum maximum* Jacq. (also *Megathyrsus maximus* [Jacq.] B. K. Simon & S. W. L. Jacobs, guinea grass), is a forage plant introduced earlier to America that is native to Africa.

4. APS MSS film 330.

5. APS MSS film 330.

6. Archives MNHN, transcription by Régis Pluchet.

7. Banks's letter from London dated April 6, 1790, is in the Banksian Correspondence, KG (Kew: B.C. 2. 7).

8. BM Add. Ms 8098.72–73.

9. Banks sent his letter to François André Michaux in France, who then forwarded it to his father in Charleston.

10. René Godard was chancellor of the French consulate in Charleston when Jean Baptiste Petry was consul.

11. This is a news report of the unsuccessful American military campaign led by Gen. Josiah Harmar (1753–1813) in October 1790 against Indians in present-day Ohio and Indiana. Michaux was no doubt interested in news from this region because he hoped to explore it.

12. Archives MNHN, transcription by Régis Pluchet.

13. Maj. Ephraim Mitchell (d. 1792) was a Continental Army veteran and member of the Society of the Cincinnati. He resided outside Charleston at a plantation in the parish of St. John's Berkeley, the parish immediately north of St. James Goose Creek, where Michaux lived. Mitchell was surveyor general of the state of South Carolina (Moss 1983, 687; Webber 1920).

14. Murray's Ferry on the Santee River and a network of roads on both sides of the river are shown on early maps. The lower section along the river, called the French Santee, was settled by industrious French Huguenots and enjoyed a period of great prosperity that had come to an end before Michaux visited (Savage 1956, 105–6).

15. Michaux previously mentioned the Strawberry Ferry on November 4–5, 1787. The ferry crossed the western branch of the Cooper River several miles downstream from Moncks Corner and provided more direct access from the Goose Creek plantations north of Charleston to Lenud's Ferry on the Santee River.

16. Correspondence with France had become increasingly difficult. Michaux's complaints about poor mail service can be found in his letters to Count d'Angiviller dated February 5 and October 4, 1790. In France, François André's attempts to get mail delivered to his father in America are reported in a letter dated April 7, 1791. Count d'Angiviller was fully aware of the problems and also commented on the difficulties with mail in his letter to Michaux of April 9, 1791. The problem was widespread, and others also reported mishandling and long delays with the mail sent on the packet boats between France and New York (Allen and Asselineau 1987, 151, 155).

17. At this time the responsibility for construction and maintenance of the public roads lay

with the citizens who lived along the route. Road overseers, who were appointed public officials, directed the work in their local areas.

18. This is André Michaux's last report of personal contact with John Fraser.

19. *Andromeda* was a large genus established by Linnaeus, but it has since been divided into several genera, including *Agarista*, *Chamaedaphne*, *Eubotrys*, *Lyonia*, *Oxydendrum*, and *Zenobia*. *Andromeda polifolia* L., bog rosemary, the sole remaining member of the genus, is a plant of colder climates. Michaux must have obtained it from New Jersey or Pennsylvania.

20. Early in his training Michaux had been tutored in the natural system of plant classification being developed by Bernard de Jussieu and carried forward by his nephew Antoine-Laurent de Jussieu. Here Michaux appears to be employing something of the organization found at the Trianon Garden, Versailles, where he studied botany with Bernard de Jussieu.

21. Further evidence of Michaux's extensive correspondence, now largely lost. John Bartram Jr., his brother William Bartram, and William Hamilton were men whom Michaux had last visited during his August 1789 sojourn in Philadelphia.

22. Autograph Coll., vol. 4, p. 17, Gray Herbarium, Harvard University, Cambridge, MA.

23. University-educated physician Moses Bartram Jr. (1768–91) was the gifted and precocious grandson of John Bartram. Brilliant by the standards of any age, he unfortunately died from a fever soon after opening a medical practice near Charleston (Baird 2003).

24. APS MSS film 330.

25. No copy of this letter has been located, but much of the content was likely repeated in d'Angiviller's letter to Michaux dated April 9, 1791, which survives in the French archives (AN 0 2113a-61bis, transcription by Régis Pluchet). It deals with many of the same subjects: difficulties with the mail service, disposition of the Charleston establishment, the question of whether Michaux wished to remain in America, and Saunier's fitness to be in charge of the New Jersey garden.

26. Christopher Williman owned the plantation adjacent to the garden outside Charleston where Michaux lived. See also chapter 3, note 74.

27. For a period map available digitally that indicates many of the place-names used in Michaux's narrative of his exploration of Camden County, Georgia, see: Henry Schenck Tanner, *Georgia and Alabama* (Philadelphia: H. S. Tanner, 1823), available at http://georgiainfo.galileo. usg.edu/histcountymaps/camden1823map.htm.

28. The initial Georgia settlements at the mouth of the St. Marys River were located on Cumberland Island, but a need for a town on the mainland led to the founding of St. Marys in 1787. James Findley, the surveyor of Camden County, laid out the town on the north bank of the St. Marys River (Knight 1914, 2:614–15).

29. Brothers Robert and Thomas Stafford (d. 1800) were among the earliest settlers of St. Marys, having come to Cumberland Island from Florida in the 1780s before the founding of the town of St. Marys. Thomas was the more prominent and was elected delegate to the 1789 state convention that met in Augusta to consider ratification of the US Constitution. His son Robert (1790–1872) became a prominent planter (Bullard 1995, 8–20).

30. The name James Moore appears in the early Camden County land grant books and on the reconstructed 1800 US Census of Camden County (Christian 1995).

31. A John Crawford and four men with the surname Wright were heads of households living in Camden County at this time (Christian 1995).

32. The journey from Capt. Stafford's that Michaux describes between April 24 and May 2, 1791, is a loop. First, the botanist traveled north by water through the Cumberland Sound and River as well as the St. Andrew Sound to enter the Satilla River. He continued west, traveling upstream on the Satilla River and visiting settlers, until, probably reaching the old

settlement of Burnt Fort, he left the Satilla River to travel south overland to Coleraine on the St. Marys River. A road between Burnt Fort and Coleraine appears on somewhat later maps but had likely existed for many years. Once on the St. Marys River, Michaux traveled east downstream for two more days before reaching the town of St. Marys and then Cumberland Island on May 2. An important Indian trading post was located at Coleraine, and in 1796 a peace treaty between the Creek Indians and the United States was signed there (Vocelle 1914, 44–46).

33. *Sarracenia psittacina* Michx., parrot pitcher plant (IDC 70-16; *Flora* 1:311). The type specimen was likely collected here (Uttal 1984, 39).

34. *Pisonia* is a tropical genus with two species known from southern Florida today, but it is not found this far north nor is it mentioned in Michaux's *Flora*. Instead, Michaux had found the rare coastal shrub *Sageretia minutiflora* (Michx.) C. Mohr, small-flowered buckthorn, which is in his *Flora* (1:154) as *Rhamnus minutiflorus* (S. Hill, pers. comm.)

35. A reminder to himself that Michaux inserted out of its geographic context. Although the Middletons had many plantations, the reference to three miles from Dorchester (South Carolina) indicates that this note refers to the family seat on the Ashley River, now Middleton Place National Landmark (Doyle, Sullivan, and Todd, 2008). The word *cork-tree* is written in English in the manuscript. It is likely that the Middletons, agriculturalists and gardeners of a high order, had a specimen of *Quercus suber* L., cork oak, acquired through commerce. Although native to southern Europe and North Africa, *Quercus suber* could succeed in this climate. An interesting coincidence, perhaps, is that this species appears on Michaux's 1786 list of the species he wished to "multiply in America."

36. Michaux named this new plant, commonly known as the fever-tree, *Pinckneya pubens* in his *Flora* (1:103). The scientific name recognized today, *Pinckneya bracteata* (W. Bartram) Raf., combines Michaux's generic name, *Pinckneya*, honoring Charles Cotesworth Pinckney (1748–1825), while reverting to the specific epithet, *bracteata*, coined by William Bartram, who observed this small tree earlier near Fort Barrington, Georgia. The St. Marys River locale given in Michaux's *Flora* suggests that he collected the type specimen (IDC 1:105) about April 30, 1791.

37. The names Belin and Bleym (or Belvin) refer to the same person. Peter Belin (1725–ca. 1795), a Huguenot rice planter from the Santee region of South Carolina, lived on Cumberland Island from 1788 to around 1794. Belin invented machinery used on rice plantations and was related to the Lynch family whose estate, jointly with Gen. Nathanael Greene's, owned Cumberland Island (Bullard 1995, 11–14).

38. Francis P. Fatio was a Swiss who came to Florida in 1771 during the British occupation and remained when the Spanish returned. He was a wealthy landholder in Spanish Florida, having a ten-thousand-acre plantation named New Switzerland on the eastern bank of the St. Johns River (Taylor and Norman 2002, 62).

39. The Crooked River is a short tidal river flowing into the Cumberland Sound north of the St. Marys River.

40. Michaux was interested in the pharmaceutical properties of plants, and similar notes are often found in his journals.

41. *Viburnum tinus* is an evergreen *Viburnum* native to the Mediterranean region of Europe. Michaux had developed a unique approach to grafting very early in his career (Deleuze 1805, 325). The Michaux graft is illustrated on plate 51C of Thouin and Leclerc (1827).

42. Cow-ford on the St. Johns River is now Jacksonville, Florida (Sanders 2002, 101). These notes about Cow-ford and the Altamaha are most likely for a journey that Michaux contemplated making at a later time.

43. Michaux and Boott 1826, 129–30; translation is Boott's.

44. Archives MNHN, transcription by Régis Pluchet.

45. Pierre Marie Auguste Broussonet (1761–1807), ichthyologist and botanist friend of Michaux, was a deputy of the legislative assembly during the Girondin period and later became director of the Montpellier Botanic Garden, France's oldest botanic garden.

## Chapter Seven

1. The contents of this page are not included in the 1889 transcription of Michaux's journals.

2. *Capsicum annum* L., pepper, continues to have many uses as a medicinal plant. For example, capsaicin is the active ingredient in several topical arthritis medications.

3. Called *catsup* by the English, this Chinese product was made of the brine from pickled walnuts and mushrooms. The much sweeter tomato catsup known today as *ketchup* was a nineteenth-century American invention.

4. Michaux refers to the arrangement with his friend John J. Himely concerning the Charleston garden that is described more fully in the introduction and the previous chapter.

5. As noted earlier, Michaux had difficulty sending and receiving mail while he was in Charleston, but on this occasion he was able both to send mail to France and to receive mail from France. Arnaud de La Porte (1737–92) was intendant of the civil list, managing and dispensing the king's private funds to individuals. He was the official who had directed Michaux to sell the Charleston garden and in normal times would have arranged for Michaux to receive the funds promised to him.

6. French consul general Antoine-René Charles Mathurin, Count de La Forest, was stationed in Philadelphia (then the capital of the United States). The reference to de La Forest in this entry suggests that it was written after Michaux arrived in Philadelphia. The information is restated on April 26.

7. Michaux was no longer able to obtain money from the drafts provided by the government and now had to borrow money using his property in France as collateral.

8. Jean Baptiste Petry continued to serve as consul at Charleston at this time. He was replaced in September 1792 by Michel-Ange-Bernard Mangourit (1752–1829), who was more supportive and sympathetic to Michaux (Alderson 2008, 48)

9. James and Shoemaker was a mercantile business at 25 North Water Street in Philadelphia.

10. Jean Baptiste Ternant (1751–1833) was appointed French minister to the United States in August 1791 and served until Genet succeeded him in May 1793. Ternant had served in the American Revolution and was recommended by Lafayette.

11. William Gerard De Brahm (1717–99) was a military engineer, geographer, climatologist, and naturalist, as well as the author of navigation tables and pilot guides. De Brahm drew detailed maps of South Carolina, Georgia, and the Atlantic coast of Florida (*DAB*, s.v. "DeBrahm, William Gerard"). William Bartram worked for him as a draftsman in 1766, after being shipwrecked as he attempted to leave Florida (Sanders 2002, 99). De Brahm's deputy surveyor for a short period, Bernard Romans (ca. 1720–84), wrote *A Concise Natural History of East and West Florida* ([1775] 1999), a book that Michaux appears to have studied prior to his own 1788 travels in Florida.

12. Ralph Izard (1741–1804) of Charleston, South Carolina, was one of the wealthiest men in the country, with large land holdings devoted to the cultivation of rice and indigo. Izard was educated in England and married Alice De Lancey, whose family was prominent in colonial politics in New York. In 1792, Izard was living in Philadelphia and serving as US senator from South Carolina. He was a signer of Thomas Jefferson's "Exploration of the West"

document, pledging twenty dollars toward Michaux's proposed exploration of the trans-Mississippi West. This important document will be examined later (*DAB*, s.v. "Izard, Ralph").

13. Educated in Scotland and England, Philadelphia physician Benjamin Rush (1745–1813) was probably the best-known American physician of his day. He was a member of the Continental Congress and a signer of the Declaration of Independence (*DAB*, s.v. "Rush, Benjamin").

14. Thomas Pinckney (1750–1828) was the brother of Charles Cotesworth Pinckney. After soldiering in the American Revolution, he served two one-year terms as governor of South Carolina (1787–89) and then was appointed minister to Great Britain in 1792. As special envoy to Spain in 1795, he successfully negotiated the Treaty of San Lorenzo ("Pinckney's Treaty"), which granted Americans free navigation on the Mississippi and access to the port of New Orleans (fulfilling the wishes of the Kentuckians who supported Genet). Pinckney was in Philadelphia in spring 1792, but he was preparing to leave to take up his post in Great Britain (*DAB*, s.v. "Pinckney, Thomas").

15. Rush and Corner 1948, 220. May 1, 1792, was a Tuesday; Rush mistakenly labels it as a Thursday.

16. Dr. Alexander Baron (1745–1819) was a prominent Charleston physician interested in natural history who had arrived from Scotland in 1769 (Raven 2002).

17. *Triticum spelta* is an ancient species of cultivated wheat that grows even in poor soils. Lucerne is another name for alfalfa, *Medicago sativa* L.

18. Sir John Chardin (1643–1713) authored *The Travels of Sir John Chardin into Persia and the East-Indies; to which is added, the coronation of this present king of Persia, Solyman the Third* (London, 1686), which was also published in French as *Des voyages du Chevalier Chardin en Perse et à d'autres lieux de l'Orient.*

19. This entry refers to William Hamilton (1745–1813); see introduction, note 40.

20. Physician and naturalist Benjamin Smith Barton (1766–1815) was at the center of the Philadelphia scientific establishment and a nephew of astronomer David Rittenhouse, president of the APS when Michaux's exploration of the West was being planned. At Jefferson's request, Barton visited Michaux and negotiated with him the terms of the agreement to explore the West that Barton described in his letter to Jefferson of January 3, 1793, reprinted in the next chapter (Ewan and Ewan 2007, 498–99).

21. Botanist Ambroise Marie François Joseph Palisot de Beauvois (1752–1820) was a French nobleman who worked in Africa and the Caribbean, but who lost most of his collections. He came to America as a refugee from the revolt on Saint-Domingue (Haiti) and settled in Philadelphia, where he assisted Charles W. Peale at his museum from 1793 to 1798 with the preparation of a museum catalog that was never completed. At the direction of the consul general, he visited Michaux's Charleston garden in 1796 and submitted a rambling but critical report about Michaux's operations that totally ignored the successful garden in New Jersey. This malicious attempt to undermine Michaux and deprive him of funding was supported, at least temporarily, by the nobleman's powerful friends in the botany community in France and caused Michaux difficulty after his return to Paris. More details are given in chapter 14.

22. Stagecoaches ran each day between Philadelphia and New York City at low fixed prices. The most often described route ran through Trenton and Princeton, but period maps also indicate a route through Bordentown and Cranbury to Amboy.

23. The *Festuca* (IDC 17-20) is now recognized by some botanists as *Chasmanthium sessiliflorum* (Poir.) Yates, and by others as *Chasmanthium laxum* (L.) Yates var. *sessiliflorum* (Poir.) Wipff & S. D. Jones, longleaf spikegrass. This herbarium sheet was annotated by both Asa Gray and A. S. Hitchcock.

24. The North American range of *Micranthes nivalis* (L.) Small, alpine saxifrage, encompasses only the northern part of Quebec. This entry probably refers to *M. virginiensis* (Michx.) Small, early saxifrage (*Flora* 1:269).

25. *Vaccinium hispidulum* is *Gaultheria hispidula* (L.) Muhl. ex Bigelow, creeping snowberry (illustrated in the *Flora*, plate 23).

26. Peter Pond (1740–1807) was a soldier and explorer who entered the fur trade after service in the French and Indian War. He was one of the first traders to open the Athabasca region in what is today the extreme north region of the provinces of Saskatchewan and Alberta, where a lake is now named for him. Pond was one of the original shareholders of the North West Company, commercial rival to the Hudson Bay Company (*DAB*, s.v. "Pond, Peter").

27. It is likely that Michaux made this round-trip to New Haven and Milford, Connecticut, by stagecoach.

28. The *Cornus* with branches punctate is probably *C. racemosa* Lam., northern swamp dogwood, gray dogwood (IDC 24-2, 24-8), although similar attributes are also found in *C. amomum* Mill., silky dogwood, and *C. rugosa* Lam., roundleaf dogwood, which grow in this area.

29. The two Geraniums are likely *G. maculatum* L., wild geranium (IDC 82-7), *Flora* 2:38, and *G. robertianum* L., herb robert, (IDC 82-6 "mountains north of the Hudson River, 15 mi from New York"), although not in the *Flora*.

30. West Point, where the Hudson is narrow and makes two almost right-angle turns, was a key fortification during the Revolutionary War. To prevent British warships from sailing up the Hudson River, a great chain protected by artillery batteries was stretched across the river.

31. Poughkeepsie, a thriving river port at the time of Michaux's visit, is approximately midway between New York City and Albany.

32. Kingston, an old Dutch settlement on the west bank of the Hudson where Rondout Creek enters the river, was first named Esopus.

33. Albany, settled by the Dutch in 1615, was a large inland town (1790 population 3,498), strategically located on the west bank of the Hudson near the junction of the water routes to the Great Lakes and Montreal. It became the state capital of New York in 1797 (Faragher 1990, 11).

34. Lansingburgh is a river town a few miles north of Albany on the east bank of the Hudson just north of the confluence of the Mohawk and Hudson Rivers.

35. Mount Rafinesque (elevation 1,020 feet), also called Bald Mountain, is about four miles east of Lansingburgh and can be seen for some miles. It is named for botanist C. S. Rafinesque (1783–1840), who visited the area in 1833. While there, he gave popular lectures at nearby institutions such as Rensselaer School and the Troy Lyceum, and he met his friend, the well-known botanist and geologist Amos Eaton (Boewe 2011).

36. *Fumaria* (*scandens*) *vesicaria* (IDC 85-2) is *Adlumia fungosa* (Aiton) Greene ex BSP., Allegheny-vine.

37. Saratoga, between Lake Champlain and Albany, was on the British army's Revolutionary War invasion route from Canada, and it was the location of a 1778 American victory that influenced France to come to the aid of the rebellious Americans. Saratoga National Historical Park occupies the site today.

38. Whitehall, New York, located at the southern terminus of Lake Champlain was then sometimes called Skeensboro, the town's name prior to the Revolutionary War. Michaux more fully describes the route between Saratoga and Whitehall on his return journey from Canada, mentioning stops at Fort Ann and Fort Edward, French and Indian War fortifications on this route.

39. *Linnaea borealis* L., twinflower, is in the herbarium (IDC 22-5, IDC 22-6) from this locality.

40. The only *Arbutus* in the herbarium, *A. uva ursi* (IDC 55-21), is recognized as *Arctostaphylos uva-ursi* (L.) Spreng., bearberry. It is noted again on June 25–27.

41. Burlington is the largest city in Vermont and home to the University of Vermont, chartered in 1791. Mount Mansfield, the highest mountain in Vermont, rises twenty miles east of Burlington to a 4,393-foot summit, which is above the treeline.

42. Mention of Cumberland Head, a prominent peninsula just north of modern Plattsburgh, New York, indicates that Michaux crossed the lake to the western shore.

43. In 1797, Michaux described this 1792 journey from New York to Quebec as one continual plant collection (Michaux and Boott 1826, 130). The following list of plants (which he put in paragraph style to save paper) forms the most extensive plant list that he made anywhere in his journal and covers five and one-half pages written in a shorthand unique to Michaux. Many of the plants that he encountered had been described and named earlier by European botanists. Michaux usually, but not always, recognized these species and provided the binomials for them. His lists combined binomials that were correct at the time, correct genera without specific epithets, and brief descriptions of plants that he did not recognize. Sometimes he mistakenly thought that he recognized a European species, and he provided the correct binomial for it, whereas the American species is different enough to be recognized as such. Also, a few of his two-word descriptions look like binomials, but they are his use of a genus name with an added descriptive word (or phrase) that is not the established specific epithet of the species being observed.

44. *Pinus abies canadensis* is *Tsuga canadensis* (L.) Carrière, eastern hemlock. In this locality a pine with needles in fascicles of twos is probably *Pinus resinosa* Aiton, red pine. *P. strobus* L. is the eastern white pine. A pine with needles inserted on all sides is probably *Picea*, spruce.

45. Although *Betula nigra* L. is the accepted modern binomial for the river birch, this plant is not found in the wild in the region of Lake Champlain (Gilman 2010). Instead, Michaux is using the binomial *B. nigra* here to indicate the black birch, the common name of *B. lenta* L. Michaux also used this binomial in the southern mountains on December 11, 1788, for a birch that was certainly *B. lenta* L., black birch, sweet birch.

46. Michaux's white elm is *Ulmus americana* L., American elm. Red elm is *U. rubra* Muhl., slippery elm (IDC 39-14).

47. A Michaux identification of *Carpinus* can mean either *C. caroliniana* Walter, American hornbeam, or *Ostrya virginiana* (Mill.) K. Koch, American hop-hornbeam; both were included in Linnaeus's genus *Carpinus*.

48. *Lonicera diervilla* is now *Diervilla lonicera* Mill., northern bush-honeysuckle. While Michaux used the genus *Lonicera* in his journal, the genus given on herbarium specimens and in his *Flora* is *Xylosteon*. One of the plants identified only by genus here is most likely *Lonicera canadensis* W. Bartram ex Marshall, American fly-honeysuckle, or possibly *L. villosa* (Michx.) Schult., mountain fly-honeysuckle (IDC 26-15). The other *Lonicera* mentioned only by genus is probably the exotic invasive species *L. tartarica* L., Tartarian honeysuckle, which he collected at Lake Champlain (IDC 26-13), while *L. glauca* is probably *L. dioica* L., limber honeysuckle.

49. Michaux provided a specific epithet for only one of the four *Viburnum* species that he noted here in his journal, *V. nudum* L. This binomial is now recognized only for the southern wild raisin, while a plant here in Vermont would be the northern wild raisin, *V. cassinoides* L., although some authorities refer to it as *V. nudum* var. *cassinoides* (L.) Torr. & A. Gray (Gilman 2010). Michaux's herbarium enables us to propose the identities of the other three: *V. lantanoides* Michx., hobblebush (IDC 40-12); *V. acerifolium* L., maple-leaf viburnum (IDC 41-9, 41-10); and *V. opulus* L., cranberry-tree (IDC 41-11). These specimens have notes indicating that they were found at Lake Champlain.

50. Both *Actaea pachypoda* Elliott, white baneberry, doll's eyes, and *A. rubra* (Aiton) Willd., red baneberry, occur at Lake Champlain (and in Quebec). The Michaux herbarium includes specimens of each labeled, respectively, *Actaea spicata alba* (IDC 70-1) and *Actaea spicata rubra* (IDC 70-2).

51. *Vaccinium stamineum* L., deerberry, is a plant that Michaux had seen in South Carolina and thus would be expected to recognize, but it is not known to occur in northern Vermont near Lake Champlain today, only in Bennington County in the southwestern corner of the state (Gilman 2015). At this latitude, it is likely that Michaux's *V. corymbosum* is *Vaccinium corymbosum* L., smooth highbush blueberry, and his *V. resinosum* is *Gaylussacia baccata* (Wangenh.) K. Koch, black huckleberry (Gilman 2010).

52. Michaux's herbarium provides a possible identity for this grass, specimen (IDC 12-1), labeled by Michaux as "*Panicum muricatum* Lake Champlain." It is likely *Echinochloa muricata* (P. Beauv.) Fernald, barnyard grass.

53. None of Michaux's herbarium specimens of *Houstonia* include New York, Lake Champlain, or Canadian locality information. Michaux's *H. purpurea* (IDC 21-17), lacking any locality data, has been annotated as *H. longifolia* Gaertn., longleaf bluet, which is a known part of the local flora (Gilman 2010).

54. *Galium album* Mill., white bedstraw, is not native to North America. It is likely that Michaux refers here to a *Galium* with white flowers. His herbarium contains two *Galium* specimens with white flowers and northern ranges: *G. asprellum* Michx., rough bedstraw (IDC 20-7), and *G. tinctorium* (L.) Scop., stiff marsh bedstraw (IDC 20-8).

55. Several specimens of *Cornus* in Michaux's herbarium include location information for Lake Champlain. Three species that we can be certain he observed there are *C. alternifolia* L. f., alternate-leaf dogwood (IDC-24-9); *C. canadensis* L., bunchberry, dwarf dogwood (IDC 23-6, "Lake Champlain, Hudson Bay"); and *C. rugosa* Lam., roundleaf dogwood (IDC 23-7).

56. *Fragaria virginiana* Mill., wild strawberry, Virginia strawberry (IDC 68-10), is the only *Fragaria* species in Michaux's herbarium.

57. *Cynoglossum virginianum* L., wild comfrey, is a native plant found in the Michaux herbarium (IDC 31-6) and alternately known to some botanists as *Andersonglossum virginianum* (L.) J. I. Cohen. *Cynoglossum officianale* L., garden comfrey, hound's tongue, is native to Eurasia. *Symphytum officianle* L., common comfrey, is naturalized from Europe.

58. This four-leaved *Lysimachia* is *L. quadrifolia* L., whorled loosestrife.

59. This refers to *Campanula rotundifolia* L., bluebell, harebell (IDC 26-20).

60. Michaux's list on August 22, 1792 (in chapter 8), adds that the leaves of *Lonicera* (Chamaceras) were tomentose (hairy), suggesting that the plant was either *L. villosa* (Michx.) Schult., mountain fly-honeysuckle, or *L. oblongifolia* (Goldie) Hook., swamp honeysuckle.

61. Probably one of the two unidentified *Asclepias* was *A. syriaca* L., common milkweed (IDC 28-15).

62. One of the two undetermined *Uvularia*, bellwort, varieties observed here was *U. sessifolia* L., wild-oats (IDC 45-19).

63. The limits of the genus *Convallaria* in the eighteenth century were much broader than today, and Michaux included in it species found today in genera such as *Polygonatum*, *Maianthemum*, and *Clintonia*. *Convallaria* is now limited to one species with three subspecies (Gandhi, Reveal, and Zarucchi 2012). *C. polygonatum maximum* is possibly *Polygonatum biflorum* var. *commutatum* (J. A. & J. H. Schultes) Morong, large Solomon's-seal. *C.* "bifolia" (with two leaves) is *Maianthemum canadense* Desf., Canada mayflower (IDC 46-20).

64. *Andromeda paniculata* in this region is *Lyonia ligustrina* (L.) DC. var. *ligustrina*, northern maleberry.

65. *Padus virginiana* is *Prunus virginiana* L., choke cherry. The undetermined plants in Michaux's genus *Cerasus* are cherries.

66. *Mespilus canadensis arborea* is a tree-like *Amelanchier*, probably *A. arborea* (Michx. f.) Fernald var. *arborea*, downy serviceberry, common serviceberry. *M. canadense frutescens* is shrubby and possibly *A. canadensis* (L.) Medik., eastern serviceberry.

67. These names remain valid in the genus *Rubus*, however, *R. arcticus* L. is not found in Vermont; this is possibly a reference to the closely related *R. pubescens* Raf., dwarf black-berry, dwarf red raspberry (Gilman 2010). Michaux's "*Rubus saxitalis americana*, Canada, end of Lake Champlain" (IDC 67-14) is *R. saxatilis* var. *canadensis* Michx., in the *Flora* (1:298), and now recognized as a synonym for *R. pubescens* Raf.

68. This entry probably refers to *Potentilla anserina* L., silverweed (IDC 68-12).

69. Michaux's herbarium specimen (IDC 68-6) locates *Geum virginianum* L., cream avens, at Lake Champlain, although it is not part of the local flora today (Gilman 2010).

70. *Thalictrum purpurascens* (IDC 72-11), now recognized as *T. pubescens* Pursh, common tall meadowrue, and *T. dioicum* L., early meadowrue, are part of the local flora (Gilman 2010).

71. Here, Michaux refers to *Chelone glabra* L., white turtlehead, while *Chelone hirsuta* (IDC 78-21) was later determined by F. W. Pennell to be *Penstemon hirsutus* (L.) Willd., northeastern beardtongue, hairy beardtongue.

72. The undetermined *Scrophularia* could be either *S. lanceolata* Pursh, eastern figwort, or *S. marylandica* L., eastern figwort, both of which occur in the area (Gilman 2010).

73. *Draba bursa-p* is *Draba arabisans* Michx., rock draba (IDC 80-15, "collected on rocks at Lake Champlain"). This specimen is the type collection of this species (Uttal 1984, 43).

74. This plant was likely *Hedysarum canadense* (IDC 90-18), today known as *Desmodium canadense* (L.) DC., Canadian tick-trefoil.

75. By *Trifolium rubens* Michaux probably means simply "red clover" here. If so, a likely species is *Trifolium pratense* L., red clover. A variety of this European plant was widely culti-vated and had freely naturalized.

76. *Hypericum canadense* L., Canada St. John's-wort (IDC 92-1).

77. *Gnaphalium dioicum* (IDC 101-2) is *Antennaria plantaginifolia* (L.) Richardson, plantain pussytoes, woman's tobacco.

78. *Viola* probably refers to *Viola canadensis* L., Canadian white violet, tall white violet (IDC 107-1, "Lake Champlain and mountains of Carolinas").

79. In this locality the pine with needles in twos is probably *Pinus resinosa* Aiton, red pine. The pine with needles in threes is probably *P. rigida* Mill., pitch pine. A pine with needles in fives is *P. strobus* L., eastern white pine. A pine with the needle apex shallowly notched is probably *Abies balsamea* (L.) Mill., balsam fir. A pine with needles denticulate (having minute teeth) is probably *Picea mariana* (Mill.) BSP., black spruce, while a pine with needles in fas-cicles is likely *Larix laricina* (Du Roi) K. Koch, eastern tamarack, eastern larch. A pine with needles inserted on all sides is probably a *Picea*, spruce.

80. *Populus balsamifera* L., balsam poplar (IDC 125-11).

81. Point au Fer is a peninsula extending from the western shore of Lake Champlain in New York, approximately four miles south of the Canadian border.

82. Leaving Lake Champlain, Michaux crossed into Canadian territory entering the Richelieu River, the lake's outlet stream, which empties into the St. Lawrence River down-stream from Montreal at the town of Sorel.

83. This was the most direct route to Montreal. La Prairie lies across the St. Lawrence on the opposite bank from Montreal.

84. Mount Royal, from which the name of the city of Montreal comes, is today a park and a cemetery within the city limits.

85. The Twenty-Sixth Regiment of Foot (Cameronians) saw service in Canada and in the northern colonies during the American Revolution. It redeployed to Canada in 1787, where various units of the regiment manned several posts in the Montreal District, protecting the Canadian frontier from Indian uprisings (or US incursions). Capt. Hugh Scott joined the regiment in 1779 as an ensign and was promoted to captain in 1786 (Gareth Cheeseman, assistant museums officer, South Lanarkshire Museums, Hamilton, UK, to Lt. Col. George H. Sutcliffe, US Army, ret., March 23, 2006; hereafter cited as 2006 Cheeseman to Sutcliffe letter).

86. The family *Typhaceae* in eastern North America has only two genera, *Sparganium* and *Typha*.

87. The "No. 2 genus" is *Elodea canadensis* Michx., common waterweed (IDC 5-17, "environs of Montreal") It is the type specimen for this species (Uttal 1984, 7).

88. The only *Alisma* for Canada in the herbarium is *A. triviale* Pursh, northern waterplantain (IDC 48-21, "lakes & rivers Quebec").

89. Wealthy fur trader and entrepreneur Joseph Frobisher (1740–1810) was a member of the North West Company and associate of Peter Pond. Frobisher was a leading citizen of Montreal; he occupied a place of honor in the Beaver Club of Montreal, a select group made up of the wealthiest fur traders in Canada who were renowned for their hospitality (*Dictionary of Canadian Biography* [*DCB*], s.v. "Frobisher, Joseph").

90. Alexander Henry (1739–1824), a fur trader, merchant, militia officer, and author, was an associate of Pond and the Frobishers; he was also a member of the Beaver Club. After his last visit to England in 1781, Henry sent Sir Joseph Banks a detailed plan for an expedition to find an overland route to the Pacific. Henry's memoir, *Travels and Adventures in Canada and the Indian Territories, between the years 1760 and 1776* (New York: 1809), is considered an adventure classic (*DCB*, s.v. "Henry, Alexander").

91. This area, now part of Montreal, is called, tongue-in-cheek, "Lachine" because a series of rapids here prevented explorer Jacques Cartier's further passage upstream in 1535, as he sought the fabled northwest passage to China.

92. The *Dianthera* is recognized today as *Justicia americana* (L.) Vahl., American water-willow (IDC 2-7, "Ohio, upper St. Lawrence"; *Flora* 1:7). The new *Hypericum* is *H. ascyron* (L.) subsp. *pyramidatum* (Aiton) N. Robson., great St. John's-wort (IDC 91-18 "Montréal"; in *Flora* 2:82 as *H. macrocarpon*).

93. Maj. Patrick Murray served with the Sixtieth Royal American Regiment of the British army. The regiment campaigned in the American South during the Revolutionary War, and two of its four battalions were stationed in Canada in 1792 (Col. I. H. McCausland, ret., Regimental Museum Archives, the Royal Green Jackets, Winchester, Hants, UK, to Lt. Col. G. H. Sutcliffe, US Army, ret., June 1, 2006).

94. Sorel is located where the Richelieu River enters the south bank of the St. Lawrence.

95. Lac St. Pierre is the name applied to the wide section of the St. Lawrence extending approximately from Sorel to Trois-Rivières.

96. In his Charleston garden in 1791, Michaux planted seeds from this species, recognized today as *Chamaedaphne calyculata* (L.) Moench, leatherleaf, cassandra, suggesting that he may have collected it earlier when visiting one of the disjunct populations in the Carolina coastal plain.

97. Trois-Rivières, founded as a fortified trading post in 1634, lies on the north bank of the St. Lawrence, like Batiscan, the community mentioned next. It is where the St. Lawrence

begins to be affected by the Atlantic tides. After Sorel there are no further mentions of points on the south bank.

98. *Ledum palustre* L. (now recognized as *Rhododendron tomentosum* Harmja) only grows far north of Lake Mistassini in the northernmost parts of Quebec, which were not visited by Michaux. All of Michaux's references to *Ledum palustre* are *Rhododendron groenlandicum* (Oeder) Kron & Judd, Labrador tea.

99. *Comarum* refers to *C. palustre* L., purple marshlocks (IDC 68-11, "Hudson Bay, Quebec").

100. *Triglochin* refers to either *T. maritima* L., seaside arrowgrass, or *T. palustris* L., marsh arrowgrass, the herbarium specimen of *Triglochin* (IDC 47-14, "Lake Mistassini, St. Laurence") includes both species on the same sheet.

101. *Scheuchzeria* refers to *S. palustris* L., pod-grass (IDC 47-16, "Batiscan, Canada; New England, Connecticut").

102. The St. Lawrence River narrows, and navigation becomes difficult for ocean-going vessels, at the city of Quebec, the site of an ancient Native American settlement first permanently settled by the French under Champlain in 1608. The city was the administrative center of Lower Canada when Michaux visited (Morison 1972, 102–8).

103. Maj. Gen. Sir Alured Clarke (1745–1832) served as commander of British forces in North America, and as civil governor of the new province of Lower Canada from 1791 to 1793 (*DCB*, s.v. "Clarke, Sir Alured").

104. Michaux's "*Oxalis acetosella* Québec" (IDC 82-9) is recognized as a synonym for *O. montana* Raf., American wood-sorrel. The same herbarium sheet also has a specimen of *O. corniculata* L., creeping lady's-sorrel.

105. Archibald Charles Dodd (ca. 1740/45–1831) was a lawyer and judge on Cape Breton Island and later judge of the Supreme Court (*DCB*, s.v. "Dodd, Archibald Charles").

106. *Aconitum uncinatum* L. is the southern blue monkshood or wild monkshood, whose present-day distribution does not reach north to Quebec. Tisavyanne is a common name in Canada for *Coptis trifolia* (L.) Salisb., goldthread or goldenroot, a plant found in Quebec. It is an Algonquin word meaning "skin dye." Native Americans used the yellow rhizomes (underground stems) of these plants to make a dye that they applied to prepared animal skins. A decoction was also used for mouth and eye sores and other maladies (Gerard 1885).

107. Printer Samuel Neilson (1771–93) then published the weekly newspaper *Quebec Gazette / La Gazette de Québec*. In September 1792, he began publication of the first illustrated periodical in Quebec, the sixty-four-page bilingual *Quebec Magazine / Le Magazin de Québec*. Unfortunately Neilson died of tuberculosis in January 1793, and his magazine ceased publication in May 1794 (*DCB*, s.v. "Neilson, Samuel").

108. *Convallaria stellata* is *Maianthemum stellatum* (L.) Link, starry Solomon's-plume, starry false lily-of-the-valley (IDC 46-15, "*Convallaria stellata* ? Rivers N of Quebec, Lake Champlain"). *Convallaria* with three leaves is *Maianthemum trifolium* (L.) Sloboda, three-leaf Solomon's-seal, while *Convallaria* with two leaves is *Maianthemum canadense* Desf., Canada mayflower.

109. At this latitude, this *Sorbu*s is likely *S. americana*. There is no indication that Michaux separated *S. americana* Marshall and *S. decora* (Sarg.) C. K. Schneid., North American species of mountain-ash, from the European species *S. aucuparia* L. (IDC 65-12, "*Sorbus aucuparia*, Roan Mt., Black Mt., Grandfather Mt., Quebec, Lake Mistassini").

110. *Narthecium calyculatum* is *Triantha glutinosa* (Michx.) Baker, northern bog asphodel (IDC 47:18; in *Flora* 1: 210 as *Narthecium glutinosum*).

111. *Euphrasia odonitites* from the journal is in the herbarium (IDC 77-20, "*E. officinalis*,

River St. Laurence toward Quebec") and was determined to be *E. nemorosa* (Pers.) Wallr., common eyebright.

112. In addition to *Actaea rubra* (Aiton) Willd., red baneberry, and *A. pachypoda* Elliott, white baneberry, *A. rubra* forma *neglecta* (Gillman) B. L. Rob., a white-fruited form of the red baneberry, is also found in the area.

113. John Mervin Nooth (1737–1828) was a scientist and inventor as well as the most respected physician in Canada. Nooth corresponded with Sir Joseph Banks and sent him samples of wild rice, *Zizania aquatica* (Rousseau 1931). He served with the British army in New York during the American Revolution and was sent to Quebec in 1788 as superintendent general of hospitals, where he remained until 1799 (*DCB*, s.v. "Nooth, John Mervin").

114. This large waterfall, known in English as Montmorency Falls and fed by the Montmorency River, is located in a park within the modern city of Quebec.

115. *Picea glauca* (Moench) Voss, white spruce, is the spruce commonly found at Montmorency Falls today.

116. Brunet (1864) says the right bank was wooded. The River St. Charles is now within the modern city of Québec; it empties into the St. Lawrence just north of the much smaller old city of Québec that Michaux visited.

117. L'Ancienne-Lorette was northwest of the old walled city of Quebec; it is part of the larger city of Quebec today. Linnaeus's disciple Peter Kalm also botanized at Lorette during his 1749 visit to Quebec.

118. Cape Tourmente and Cape Brulé are on the north bank of the St. Lawrence, downstream from the city of Quebec.

119. Baie-Saint-Paul is on the north bank of the St. Lawrence opposite Isle-aux-Coudres, which Cartier named in recognition of the many hazelnut trees (*coudriers*) found on the island.

120. The white porpoise may have been a beluga whale, *Delphinapterus leucas*; the species still occurs in the St. Lawrence.

121. *Potentilla nivea* L. has a generally more northern range, but another three-leafed cinquefoil grows in this area, *P. norvegica* L., rough cinquefoil, strawberry-weed. Michaux's "*Potentilla hirsuta*, Tadoussac, Malbaye" (IDC 68-13) is a good specimen of *P. norvegica* L.

122. This variety of *Vaccinium* is likely either *V. oxycoccos* L., small cranberry, or *V. macrocarpon* Aiton, cranberry, large cranberry; both grow in this area.

123. Michaux's "*Drosera anglica*, Batiscan, Canada" (IDC 43-14) is either *D. anglica* Huds., English sundew, or *D. intermedia* Hayne, spoonleaf sundew.

124. Both *Typha angustifolia* L., narrowleaf cattail, and *T. latifolia* L., common cattail, broadleaf cattail, occur here.

125. *Sparganium erectum* L., bur-reed, is found from western Europe to Siberia but not in Quebec.

126. Michaux's *Flora* (1:101–2) includes six species of *Potamogeton*, pondweed. The St. Lawrence River is given as habitat for only one, *P. marinum* L., which matches Michaux's specimen (IDC 26-1, "*Potamogeton marinus*, Québec"). This species name is now accepted as a synonym of *P. pectinatus* L. (Haynes 1986, 570–71), and the species is recognized today as *Stuckenia pectinata* (L.) Börner, sago pondweed.

127. La Malbaie is a picturesque spot on the north bank of the St. Lawrence. Champlain visited in 1608 and found the tidal anchorage deceptive and thus gave it the name Malbaie (literally "evil bay") when he found that the receding tide had grounded his ships.

128. *Pulmonaria maritima* (IDC 31-14, "*Pulmonaria parviflora*, Malbaie") is recognized today as *Mertensia maritima* (L.) S. F. Gray, oysterleaf.

129. *Glaux maritima* L. is the only *Glaux* in the *Flora* (1:114) and specimen in the herbarium

(IDC 28-4, 28-5, 28-6). Today, the genus *Glaux* is subsumed into the genus *Lysimachia*, and this plant is recognized as *L. maritima* (L.) Galasso, Banfi & Soldano, sea-milkwort.

130. The only *Sisyrinchium* species to be found at this latitude is *S. montanum* Greene, strict blue-eyed grass.

131. Michaux's "*Pyrola uniflora* Canada, Saguenay River" (IDC 56-9) is *Moneses uniflora* (L.) A. Gray, single delight, one-flowered wintergreen.

132. This species is *Empetrum nigrum* L., black crowberry (IDC 127-15, "borders of the sea in Canada").

133. The Saguenay River is over eight hundred feet deep with steep rocky shores rising more than fifteen hundred feet from the water. Subject to dramatic tides, it is a true North American fjord and is now home to the Saguenay–St. Lawrence Marine Park.

## Chapter Eight

1. André Michaux's account of his voyage into Canada and his description of the vegetation that he found were extensively studied by Brunet (1864). Cayouette (2014) and Asselin, Cayouette, and Mathieu (2017) observed that when Michaux arrived in 1792, few plant collections had been made from the part of Canada north of the St. Lawrence. Most plant identifications for this study of Michaux's Quebec journey have been determined by Cayouette.

2. *DCB*, s.v. "Albanel, Charles."

3. Tadoussac is located on the north shore of the Saguenay where it enters the St. Lawrence. Explorer Jacques Cartier visited Tadoussac in 1535, and a trading post was first established there in 1600. The village became the center of the fur trade in the region. Today's chief enterprise is tourism, especially observing marine mammals.

4. Michaux's three Native American guides also took their dog with them on the trip (Michaux and Boott 1826, 130).

5. Chicoutimi was the end of deep-water navigation.

6. Many times throughout this journey to Lake Mistassini, Michaux notes *Ledum palustre* and *Andromeda calyculata*. At this latitude the former is now *Rhododendron groenlandicum* (Oeder) Kron & Judd, Labrador tea, while the latter is now *Chamaedaphne calyculata* (L.) Moench, leatherleaf, cassandra.

7. *Swertia corniculata* (IDC 25-3) is recognized today as *Halenia deflexa* (Sm.) Griseb., American spurred gentian.

8. There is longstanding discord among plant taxonomists concerning the species within the genus *Nuphar* (Padgett, Les, and Crow 1999). Michaux's "*Nymphea lutea* River Chicoutimi" (IDC 70-18) is recognized as *Nuphar variegata* Durand in G. W. Clinton, variegated yellow pondlily, and his "*Nymphea lutea minor* on river Chicoutimi, Lake Champlain, and Canada" (IDC 70-21) is recognized now as *Nuphar microphylla* (Pers.) Fernald, yellow pondlily.

9. The only *Abies* found in this area is *Abies balsamea* (L.) Mill., balsam fir.

10. *Alisma subulata* is *Sagittaria graminea* Michx., grassy arrowhead. The specimen gathered on this trip (IDC 113-5) is the type specimen (Uttal 1984, 59).

11. Michaux's "*Polygonum amphibia* Lac Saint-Jean" (IDC 54-11) is recognized today as *Persicaria amphibia* (L.) S. F. Gray, water smartweed.

12. One herbarium sheet (IDC 17-11, "on the bank of the St. Lawrence River, Lac Saint-Jean, Riv. Saguenay") contains specimens of two species, but only the *Ammophilia breviligulata* Fernald, American beach-grass, occurs in Lac Saint-Jean.

13. François Panet, helped by his brother Thomas, managed the trading post at Pointe-Bleue until 1802, when Thomas lost his life in the lake, and his brother lost his sanity as a result of the accident. Earlier researchers have been divided as to whether the post mentioned by

Michaux was Pointe-Bleue. Rousseau (1948) wrote that this trading post was not Pointe-Bleue but rather one situated at the mouth of the Metabechouan River, because he believed that Michaux overestimated distances in this portion of the journey by one-fourth to one-third. Brunet (1864), however, called the trading post Pointe-Bleue.

14. This species was also observed on September 5 and collected near Lake Mistassini. Michaux's "*Aira ambigua* Lake Mistassini" (IDC 15-2) is recognized today as *Deschampia cespitosa* (L.) P. Beauv., tufted hairgrass.

15. The species that Michaux calls *Pinus abies* is either *Picea glauca* (Moench) Voss, white spruce, or *Picea mariana* (Mill.) BSP., black spruce. *Picea rubens* Sarg., red spruce, is very rare in this area.

16. This waterfall and rapids are known as the Larges Rapides and locally as Kissegau (Rousseau 1948).

17. *Vaccinium corymbosum* L., smooth highbush blueberry, is found only in southern Quebec. This and other mentions of a *Vaccinium corymbosum* on this part of Michaux's journey would refer to either either *V. angustifolium* Aiton, northern lowbush blueberry, or *V. myrtilloides* Michx., sourtop, velvetleaf blueberry.

18. *Cerasus racemosa* may be *Prunus virginiana* L., choke cherry; it occurs in the Lac Saint-Jean area.

19. Michaux's "*Cerasus corymbosa* Quebec, Lake Mistassini" (IDC 64-4) is *Prunus pensylvanica* L. f., fire cherry, pin cherry.

20. The type specimen of *Cornus stolonifera* Michx., red osier dogwood, is IDC 23-17 (Uttal 1984, 18).

21. *Cornus rugosa* Lam., roundleaf dogwood, is the most likely of the *Cornus* species found in this area with pitted branches.

22. *Convallaria* with blue berries is *Clintonia borealis* (Aiton) Raf., bluebead-lily (IDC 46-17).

23. The three-leaved *Convallaria* is either *Maianthemum stellatum* (L.) Link, starry Solomon's-plume, or *M. racemosum* (L.) Link., eastern Solomon's plume, false Solomon's-seal. The two-leaved *Convallaria* is *M. canadense* Desf., false lily-of-the-valley, Canada mayflower. The third *Convallaria* could be either *M. racemosum* (L.) Link, eastern Solomon's-plume, false Solomon's-seal, or *M. stellatum* (L.) Link, starry Solomon's-plume, starry false lily-of-the-valley.

24. Following Linnaeus ([1753] 1957–59), Michaux usually lumped fir (*Abies*), larch (*Larix*), spruce (*Picea*), and hemlock (*Tsuga*), as well as pines, into the genus *Pinus*. Some of his descriptive names look like binomials but are not recognized as such. Michaux's *Pinus larix* is probably *Larix laricina* (Du Roi) K. Koch, eastern tamarack, eastern larch; *Pinus balsamea* is *Abies balsamea* (L.) Mill., balsam fir; *Pinus abies alba* is *Picea glauca* (Moench) Voss, white spruce; *Pinus abies nigra* is *Picea mariana* (Mill.) BSP., black spruce; while *Pinus strobus* is always *P. strobus* L., eastern white pine. *Pinus* twin leaves, cone ovate, smooth, is *P. resinosa* Aiton, red pine; and *Pinus* needles in twos, short, is probably *P. banksiana* Lamb., jack pine.

25. This item is perhaps *Salix pellita* (Andersson) Bebb, satiny willow (it is not *Salix sericea* Marshall, silky willow).

26. This *Salix* with leaf-like stipules is likely either *S. cordata* Michx., sand dune willow, or *S. eriocephala* Michx., Missouri willow, heart-leaved willow.

27. Michaux's "*Arundo canadensis* Hudson to Canada" (IDC 17-10) is the type specimen of *Calamagrostis canadensis* (Michx.) P. Beauv., bluejoint (Uttal 1984, 15).

28. This undetermined species is not *Ribes cynosbati* L., prickly gooseberry; it does not occur in the Lac Saint-Jean area.

29. *Spiraea salicifolia* L. does not occur in Quebec. This plant is probably *S. latifolia* (Aiton) Borkh, broadleaf meadowsweet.

30. Michaux's "*Oenothera pusilla* Lake Mistassini" (IDC 50-1) is recognized today as *O. perennis* L., little sundrops.

31. *Thalictrum dioicum* L., early meadowrue, does not occur here. This plant is probably *T. pubescens* Pursh, common tall meadowrue.

32. This *Actaea* was likely either *A. pachypoda* Elliott, white baneberry, doll's-eyes, or *A. rubra* (Aiton) Willd. forma *neglecta* (Gillman) B. L. Rob., white baneberry.

33. Brunet (1864) estimated the height of Monte-à-Peine to be only eight hundred to nine hundred feet, but he agreed that the ascent was difficult and dangerous.

34. Because of the range and the bog habitat, the very small birch that Michaux noted here and collected as *Betula nana* (IDC 111-17) was probably the species now known as *B. michauxii* Spach, Newfoundland dwarf birch (Fernald 1950).

35. According to Rousseau (1948), the portage from Monte-à-Peine leads toward Lac des Cygnes (Swan Lake), and the little river seems to be the discharge from this lake, which empties into the Mistassini River at about latitude 49°54′.

36. *Picea rubens* Sarg., red spruce, does not occur here; if this is a spruce, it is *P. mariana* (Mill.) BSP., black spruce.

37. Michaux's *Avena striata* (IDC 17-7; *Flora* 1:73) later became *Melica striata* (Michx.) Hitchc. and is recognized today as *Schizachne purpurascens* (Torr.) Swallen, false medic, purple oatgrass.

38. *Narthecium calyculatum* is probably either *Tofieldia pusilla* (Michx) Pers., scotch false asphodel, or *Triantha glutinosa* (Michx.) Baker, northern bog asphodel, sticky bog asphodel. Uttal (1984, 30) identifies IDC 47-18 as the type collection of *Narthecium glutinosum* Michx. Later placed in the genus *Tofieldia*, it is recognized today as *Triantha glutinosa*.

39. There are a number of potential species here, among the most likely are "*Epilobium oliganthum* environs of Mistassini" (IDC 49-17), recognized as *E. palustre* L., marsh willow-herb, and "*E. angustifolium* Lake Mistassini" (IDC 49-16), recognized as *Chamerion platyphyllum* (Daniels) Löve & Löve, great willow-herb, fireweed.

40. At almost one hundred miles long and twelve miles wide, Lake Mistassini is the largest natural lake in Quebec; however, it is irregularly shaped with a series of rocky island ridges virtually dividing it into two lakes. Dutilly and Lepage (1945) paddled from Hudson Bay to Lake Mistassini to establish the botanist's correct route and refute suggestions that Michaux came to Mistassini by way of Lake Albanel. They concluded that Michaux reached the southern extremity of the lake (arriving by way of a river that runs north) and followed the southwest coast of the lake as far as the Rupert River.

41. Michaux's "*Gentiana pneumonanthe* Canada, Hudson Bay" (IDC 40-4) is *G. linearis* Froel., narrowleaf gentian.

42. Rousseau (1948) concluded that the Goelands River must be the western arm of the Rupert River, called *Natastan chipi* by the Indians.

43. This ring-necked goose was a brant, *Branta bernicla*.

44. While *Betula pumila* is probably *B. pumila* L., bog birch or dwarf birch, Michaux also described and collected the type specimen of *B. glandulosa* Michx., resin birch (IDC 111-19; *Flora* 2:180), in the vicinity of Lake Mistassini, where it is a dominant shrub, and it might be the species meant here instead (Uttal 1984, 58).

45. Michaux consistently used the name of the European species *Sorbus aucuparia* L. for each *Sorbus* that he encountered. The most likely *Sorbus* at this latitude is *S. decora* (Sarg.) C. K. Schneid., northern mountain-ash, showy mountain-ash.

46. The Michaux herbarium has four specimens of *Ribes* with either Mistassini or the probable outlet river for the lake mentioned on their labels. Two specimens (IDC 27-7 and IDC 27-10) are *R. triste* Pallas, swamp red currant, wild red currant; the other two are *R.*

*lacustre* (Pers.) Poir., bristly black currant, spiny swamp currant (IDC 27-8), and *R. glandulo-sum* Graver, skunk currant (IDC 27-11).

47. *Pinguicula alpina* L. is a Scandinavian species; only *P. vulgaris* L., common butterwort, is found at Lake Mistassini today.

48. *Hieracium paludosum* L., marsh hawksbeard, now recognized as *Crepis paludosa* (L.) Moench, is a European species not in the North American flora. This note could possibly refer to either of two *Hieracium* species that Michaux collected at Lake Mistassini: *H. kalmii* L., Canadian hawkweed (IDC 93-8), or *H. scabrum* Michx., rough hawkweed (IDC 93-9), and if the latter, would be its type location (Uttal 1984, 48).

49. Rousseau (1948) offered the opinion that both the Atchouke and Goelands Rivers are part of the Rupert River (longitude 74°20' west, latitude 51°12' north).

50. Michaux and his party ceased traveling north and began their return journey. They had reached latitude 51° north and were perhaps 150–175 miles east of James Bay, the southern arm of Hudson Bay.

51. The muskrat is recognized as *Ondatra zibethicus* today.

52. *Rubus arcticus* L. subsp. *acaulis* (Michx.) Focke, stemless arctic bramble, dwarf raspberry. This is the type location for this species (Uttal 1984, 38).

53. The small woodpecker was most likely a downy woodpecker, *Picoides pubescens*, or a hairy woodpecker, *Picoides villosus*; both are common species there.

54. This bird is the gray jay, *Perisoreus canadensis*.

55. The *Vaccinium* dwarf with ten stamens may be *V. cespitosum* Michx., dwarf bilberry (IDC 52-14); this is the type specimen for this species (Uttal 1984, 33).

56. *Euphrasia odontites* is probably "*Euphrasia officinalis* Mistassini River, St. Laurence around Quebec" (IDC 77-20), recognized today as *E. nemorosa* (Pers.) Wallr., common eyebright.

57. The native *Rhinanthus* growing there is *R. groenlandicus* Chabert, arctic rattlebox.

58. *Cacalia hastata* is probably *Prenanthes trifoliolata* (Cass.) Fernald, gall-of-the-earth; *Cacalia incana* is probably *Prenanthes racemosa* Michx., rattlesnake root (IDC 92-11). Uttal (1984, 47) notes that this is the type collection for this species.

59. *Convallaria* is probably *Maianthemum trifolium* (L.) Sloboda, three-leaf Solomon's-seal, three-leaf false lily-of-the-valley (IDC 46-21 *C. trifolium* "Lake Mistassini").

60. The identity of these birds is subject to speculation. Michaux was near the southern edge of the present-day winter range of the willow ptarmigan, *Lagopus lagopus*, whose behavior most closely matches his description. This species exhibits clustering behavior in winter and eats a diet of tree buds, seeds, and berries.

61. Rupert River is also referred to as Goelands River in Michaux's narrative.

62. The Chicoutimi River is not an outlet for Lac Saint-Jean.

63. Contemporary maps show the Ashuapmushuan River emptying into Lac Saint-Jean a short distance south of the Mistassini River's own discharge into the lake.

64. Although Michaux does not mention it, his route to Lake Kenogami from Lac Saint-Jean would have begun on the Belle Rivière, which had been his route in the opposite direction on August 15–16.

65. By *Lappa*, Michaux likely meant *Arcticum lappa* L., now a synonym for *A. minus* (Hill) Bernh., common burdock.

66. Michaux's overestimation of the distance here perhaps reflects the difficulty of this part of his journey, where several portages were necessary.

67. A number of *Ranunculus* species occur here. Michaux's "*Ranunculus reptans* Hudson, St. Lawrence, Lake Mistassini" (IDC 72-3), now recognized as *R. flammula* L. var. *reptans* (L.) E. Mey., creeping spearwort, is perhaps the most likely candidate.

68. *Trifolium repens* L., white clover (IDC 86-6), is a possibility, but it is not the only one in this area, where there are a number of introduced *Trifolium* species.

69. Pointe aux Trembles is now Neuville, a small community on the north bank of the St. Lawrence located between Quebec and Donnacona.

70. The Jacques-Cartier River enters the north bank of the St. Lawrence at Donnacona about twenty miles upstream from the city of Quebec, while the Sainte Anne River enters on the north bank at Sainte-Anne-de-la-Pérade, which is just downstream from Batiscan, also on the north bank of the St Lawrence.

71. Michaux's observations precede the division of Linnaeus's genus *Juglans* into *Juglans* and *Carya* by C. S. Rafinesque in the nineteenth century. Michaux uses *Juglans hiccory*, a name that has no legitimacy as a botanical name, to refer to at least three different species now included in the genus *Carya*. In this locality, Michaux's *Juglans hiccory* is possibly *Carya cordiformis* (Wangenh.) K. Koch, bitternut hickory.

72. *Quercus alba* L., white oak, does not grow here, but *Q. macrocarpa* Michx., mossycup oak, bur oak, also in the white oak group, does.

73. Berthierville is on the north bank of the St. Lawrence approximately opposite Sorel on the south bank.

74. Repentigny, in the parish of L'Assomption, is only a few miles northeast of Montreal.

75. John Dease (ca. 1744–1801), who was educated in both Ireland and France and trained as a physician, had served as the Indian agent at Mackinac Island, now a part of the state of Michigan. Dease was then living in Montreal (*DCB*, s.v. "Dease, John").

76. Probably George Selby (1760–1835), an eminent Montreal physician and surgeon who counted among his intimate friends such prominent men as Joseph Frobisher, another acquaintance of Michaux's in Montreal (*DCB*, s.v. "Selby, George").

77. Probably Hubert-Joseph Lacroix (1743–1821), a merchant, militia officer, politician, and justice of the peace in the Montreal District after 1791. Lacroix's father had been a surgeon and botanist prior to becoming a merchant (*DCB*, s.v. "Lacroix, Hubert-Joseph").

78. Andrew Gordon (brevet), colonel of the Twenty-Sixth Regiment of Foot (Cameronians) in 1790, was promoted to lieutenant general in 1801 (2006 Cheeseman to Sutcliffe letter).

79. Capt. Erskine Hope of the Twenty-Sixth Regiment of Foot rose to lieutenant colonel in 1804 (2006 Cheeseman to Sutcliffe letter).

80. Shelburne, Vermont, is located on a protected bay of Lake Champlain immediately south of Burlington.

81. This enormous rock on a peninsula on the New York shore just south of Essex is an impressive natural landmark. It was often recognized as a dividing line between warring factions, and the 1713 Treaty of Utrecht recognized it as the boundary between New France and New England.

82. Basin Harbor is on the Vermont shore just north of today's Button Bay State Park.

83. Crown Point commands a very narrow point on Lake Champlain about nine miles north of Fort Ticonderoga. It is the site where Samuel de Champlain fought the Iroquois Indians in 1609 and began the enmity between the French and the nations of the Iroquois Confederacy. The French constructed a substantial stone fort there in 1731, and it remained in their hands until the French and Indian War. In 1749, Pehr (Peter) Kalm, the botanist sent by Linnaeus to North America, also traveled this route to Canada, and he described the fort as having walls of thick black limestone and mounting forty cannon (Robbins 2007, 91).

84. Fort Ann on Wood Creek south of Whitehall and Fort Edward farther south on the upper Hudson, both French and Indian War–era fortifications, enable us to verify Michaux's route south from Lake Champlain toward Saratoga (Kammen 1975, 316, 375).

85. Capt. Hezekiah Baldwin Sr. (1732–1803), whose wife was Abigail Peet, served in the Revolutionary War and is listed in the 1790 US Census of New York (Daughters of the American Revolution 2003, 1:120).

86. *Fagus castanea americana* is *Castanea dentata* (Marshall) Borkh., American chestnut.

87. The Saratoga battlefields are on the western side of the Hudson, while the community of Easton is on the eastern side.

88. Tarrytown was the site of an important Revolutionary War event. In 1780, local militiamen captured British officer Maj. John André and found the plans to the fortifications at West Point hidden in his boot. Their action exposed Benedict Arnold's treason and foiled his plot to deliver West Point into British hands.

89. Rush and Corner 1948, 303.

90. Transcription provided by Régis Pluchet, from *L'Esprit des journaux français et étrangers* 5 (May 1793).

## Chapter Nine

1. Kennedy 1976.

2. It is likely that Michaux had learned of the APS's interest in exploration west of the Mississippi River during his visit to Philadelphia in April and May 1792 prior to his journey to Canada. Several of the people whom he mentioned visiting in Philadelphia at that time were members of the APS, and the topic was then under discussion (Catanzariti 1992, 25:76–77). Michaux was not the only person who aspired to lead such a voyage. Jefferson reported that Meriwether Lewis, then eighteen years old, "warmly solicited me to obtain for him the execution of that object" (Bergh 1907, 17:145).

3. Catanzariti 1990, 24:687–88.

4. Historians believe that the guide to Kaskaskia mentioned here is Jean Baptiste Ducoigne, a Kaskaskia chief who arrived in Philadelphia in December 1792 with a party of other western Native Americans for a peace conference with President Washington (Catanzariti 1992, 25:18).

5. This is in reference to vaccination against smallpox. Jefferson understood the importance of this procedure and had himself, his family, and those he enslaved vaccinated (Leavell 1977).

6. Kaskaskia, an old French settlement on the Mississippi, later became the first capital of Illinois. In the nineteenth century the mighty river changed course and gradually destroyed the town.

7. Catanzariti 1992, 25:17–18.

8. Catanzariti, 71–72.

9. Catanzariti, 77–78.

10. APS Archives.

11. Catanzariti 1992, 25:82–83.

12. Bibliothèque municipale de Versailles, Versailles (VERS), 2US2, transcription by Régis Pluchet.

13. Prominent planter Peter Smith (1754–1821) was elected treasurer of the Agricultural Society of Charleston in 1785. His fellow planter Arnoldus Vanderhorst (1748–1815) was governor of South Carolina from 1794 to 1796 and owned Kiawah Island, where his home has been preserved. Charles Cotesworth Pinckney (1746–1824) was a hero of the Revolutionary War for his conduct while a prisoner of the British. Gen. Pinckney's extensive public service included being one of the South Carolina representatives at the US Constitutional Convention in 1787, serving an appointment as US minister to France in 1796, and twice being the Federalist Party candidate for US president.

14. Watchmaker J. J. Himely was the close friend of Michaux's who purchased the

Charleston garden at auction in 1792 and who arranged for the botanist to retain possession of the property. Merchant P. G. Chion had offices at 39 Bay Street in Charleston. Michaux wrote to Chion from Holland on November 30, 1796, after he was shipwrecked on the Dutch coast. Francis Robinett was a cooper on Cochran's Wharf in Charleston; he may have returned to Europe after March 1793 for reasons of health. Michaux wrote to him from the Bahamas on March 9, 1789. May was possibly either the same John May described in the deed record as a "merchant lately from Charleston," at whose home Michaux stayed on April 23, 1795, when traveling through Lancaster County, South Carolina, or a relative of this John May's. Mme Bars has not been identified.

15. "Homassell" likely refers to the mercantile firm of Mazurie and Homassell at 28 South Front Street in Philadelphia. Michaux mentioned Homassell again in another letter to Mangourit, dated March 5, 1793.

16. Archives des Affaires etrangères (AE), consular and commercial correspondence, Charleston, vol. 2, p. 8.

17. Abbé Friard was a model of simplicity, the patron saint of farmers and winemakers.

18. Alderson 2008, 48.

19. Stadelman (1971) describes many of the problems encountered when attempting to preserve early natural history specimens, and Michaux's have unfortunately not been preserved. Correspondence with several curators at the MNHN, Paris, indicates that the Muséum apparently has only a single animal skin collected by Michaux (the type collection of the fox squirrel subspecies *Sciurus niger rufiventer*, ZM NO-2000-598) and none of his insect collections. Inquiries regarding Michaux's bird specimens to the following US institutions: Museum of Comparative Zoology, Harvard University, Cambridge; Peabody Museum, Yale University, New Haven (which has some holdings from the old Peale Museum); and the Academy of Natural Sciences, Philadelphia; as well as the Museum of Zoology, University of Uppsala, Sweden, also proved negative.

20. Gustaf Adam von Nolken (1733–1813) was the Swedish ambassador in London from 1764 through 1793. Pehr von Afzelius, professor of medicine at Uppsala in Sweden, had visited Paris in 1785, where he met André Michaux. In this letter Michaux recommends that Afzelius correspond with him using the naturalist Rev. Nicholas Collin, minister of the Swedish Church in Philadelphia, as an intermediary. Pehr Afzelius was the brother of Adam Afzelius (1750–1835), a Swedish botanist working (with the support of Sir Joseph Banks) on the West African coast in the British settlement of Freetown, Sierra Leone.

21. Archives Uppsala University Library, Uppsala (UP), Thunberg archives, signature G 300s.

22. Uppsala is at latitude 60° north.

23. François André Michaux (FAM) to Auguste Broussonet, June 12, 1796, transcription by Régis Pluchet, courtesy of William Reese Company, New Haven, CT.

24. Archives municipales de Reims, coll. P. Tarbé, carton n°XX, n°102, transcription by Régis Pluchet.

25. VERS, Bosc file, transcription by Régis Pluchet.

26. Louis XVI was executed on January 21, 1793. France had declared war on Austria and Prussia on April 20, 1792, and then defeated a combined army of invading Austrians, Prussians, and French émigrés at Valmy on September 20, 1792. Michaux must be referring to the more recent news of the declarations of war against Great Britain and Spain, declarations that soon embroiled France in war with most of the countries of Europe.

27. Digital collection, Library of Congress (LC); see also editorial notes and commentary, Catanzariti 1992, 25:75–81.

28. Savage and Savage (1986, 131).

29. Michaux's attitude toward keeping a part of his collections for his own use is clearly stated in his letter to André Thouin of August 10, 1787 (HU 71888).

> Seeing the dangers that I lived through to gather them [plants and seeds from the Middle East], I would reproach myself if I had not kept a set for myself, thus I don't think anyone can object. I will always conduct myself according to these principles. I will divide my collections or discoveries, and what I reserve for myself will be for me.

30. Jefferson wrote to James Madison, Philadelphia, April 28, 1793: "We expect Mr. Genest [*sic*] here in a few days" (Catanzariti 1992, 25:619).

31. SCL, Henry Savage Papers, folder 224, translated by Henry Savage Jr. and Elizabeth Savage.

32. AM to Genet, May 21, 1793, "Abridged Memoir Concerning My Journeys in North America," SCL, Henry Savage Papers, folder 150, translated by Henry Savage Jr. from the original (LC).

33. In his notebook, Michaux's journal entry for May 29 immediately follows his entry for April 24. His entries for April 30 and May 10 are marked by Michaux as having been inserted after he had written the entry for June 9, and the entry dated May 10 clarifies that the shipment that he sent to Bosc on May 10 did not actually leave until a month later. Michaux's next entry, a note about the arrival of Minister Genet, while undated, is followed by entries about submitting memos to Genet on May 18 and 22.

34. Madame Desaint is Michaux's sister Marie-Victoire, who married the printer Jean Charles Desaint in Paris in 1779.

35. Lipscomb and Bergh, 1903–4, 18:144–45.

36. Williams 2004, 98–106.

37. Ewan and Ewan 2007, 499.

38. Williams, 2004, 101–6.

## Chapter Ten

1. Gen. George Rogers Clark (1752–1818) enjoyed great military prestige, especially in the country west of the Appalachians, for his successful Revolutionary War campaigns leading a small command of Virginia militia against the British and their Native American allies north of the Ohio River. He remained a leader in the struggle against the tribes who continued to menace Kentucky settlers immediately after the war. Clark was a charismatic fighting general, a master of the bold offensive stroke. His military success enabled a major redrawing of western boundaries in the 1783 treaty ending the war (James 1928, 417–27; Turner 1898; *DAB*, s.v. "Clark, George Rogers").

2. American scholars have published numerous studies of the Genet Affair; see especially Turner (1898) and Ammon (1973). Michaux's role in these events is most fully described in Lowitt (1948) and Williams (2004).

3. Catanzariti 1995, 26:394–95.

4. Williams 2004, 101–2.

5. AN, F10/392, dossier 1, #118. Certificate from Genet recognizing Michaux as botanist and political agent, July 13, 1793:

> In the name of the French people, I, Edmond Charles Genet, minister plenipotentiary of the French Republic to the United States and adjunct colonel general of its armies, certify that as a result of deliberation of the Executive Council on January 1, 1793, year 2

of the French Republic: Citizen André Michaux, political agent and botanist, has been granted an annual stipend of 2,400 pounds as well as 3,000 pounds for his travel expenses, and we authorize him in our official capacity and not otherwise to avail himself of the French legation in America so that he will be able to obtain from the legation 500 pounds per quarter, and for his expenses 750 pounds for which he will have to deliver a receipt. Written in Philadelphia, July 13, 1793, second year of the French Republic.

[signed] Minister plenipotentiary of the French Republic. Genet

6. Ten days prior to his own departure, Michaux paid Samuel Hodgdon $25.13 to ship 560 pounds of baggage from Philadelphia to Pittsburgh (Roy Rosenzweig Center for History and New Media, George Mason University, Fairfax, VA, papers of the War Department: 1784–1800, accessed through http://wardepartmentpapers.org/ [RRCHMN]).

7. Hugh Henry Brackenridge (1748–1816) was one of the leading men of western Pennsylvania. No doubt the reason that Michaux carried a letter to Brackenridge was that he was known as an outspoken supporter of the French Revolution. Brackenridge's actions during the Whiskey Rebellion the following year drew suspicion, but he was cleared of the charges of treason made against him, and his political career prospered. He ultimately became chief justice of the Pennsylvania Supreme Court as well as a successful writer of fiction about the American frontier (*DAB*, s.v. "Brackenridge, Hugh Henry").

8. Capt. Gilbert Imlay (1793, 141–47), a Revolutionary War veteran who visited Kentucky about this time, observed in his book *Description of the Western Territory of North America* that the way from Pittsburgh to Kentucky by river was so tedious and dangerous that those who did not carry much baggage found the way through the great wilderness via the Wilderness Road and Cumberland Gap preferable. Thomas Speed (1886, 11–12) remarked in *Wilderness Road* that "Even as late as 1793, when Imlay wrote, there was no such convenience as a regular business of carrying passengers and their luggage down the Ohio, but at Pittsburg or Old Fort a flat-boat or passenger boat might be obtained, according to the good luck of the traveler." In any event, the newly established regular mail service following the route down the Shenandoah Valley of Virginia, through the Cumberland Gap, and over the Wilderness Road, made the journey to Kentucky faster than the river route taken by Michaux, as evidenced by the letter that Gen. Logan received from Senator Brown before Michaux arrived. The route was a factor in Michaux's slow journey to Kentucky.

9. In 1793, Maj. (not "Maj. Sgt.") John Stagg Jr. (1758–1803) was the chief clerk in the War Department and as such acted as Secretary of War Henry Knox's adjutant, maintaining correspondence with the officers in the field. In this capacity he wrote to the army officers Maj. Isaac Craig and Capt. John Pratt to inform them that the French botanist would be traveling west (RRCHNM). Maj. Craig (ca. 1742–1826) was deputy quartermaster for the troops garrisoned in Pittsburgh. He superintended construction of fortifications and arranged for supplies and transport for the army as it campaigned against the hostile Native Americans in the old northwest. Maj. Craig commanded Fort Pitt in 1781 and cooperated with George Rogers Clark in an abortive campaign against Detroit (Craig 1854). Capt. Pratt (1753–1824) of Middleton, Connecticut, commanded a detachment of one hundred army recruits on the march from Trenton to Pittsburgh in June 1793 (RRCHNM).

10. Isaac Shelby (1750–1826) was an outstanding frontier political and military leader. He was governor of Kentucky during two especially difficult periods in the state's history, from 1792 to 1796 and again from 1812 to 1816. Shelby first gained distinction in Lord Dunmore's 1774 campaign against the Shawnee. During the Revolutionary War, Shelby was especially active leading frontier militia units in the Carolinas during the fierce conflict in 1780–81, and

he was one of the commanders at the pivotal battle of King's Mountain in 1780. After the war he was both a civil and military leader during Kentucky's drive to statehood. His wife, Susannah Hart, was a daughter of Capt. Nathaniel Hart, a member of the Transylvania Company, the founders of Kentucky. The episode with Genet was only one of many delicate and difficult matters that Shelby dealt with successfully during his first term as governor of the new state (Henderson 1920; Cooke 1963; *DAB*, s.v. "Shelby, Isaac").

11. Alexander Orr (1761–1835) became a member of the Virginia legislature in 1790 while Kentucky was a part of Virginia. When Kentucky achieved statehood in 1792, Orr became one of the new state's two US representatives, serving three terms (Biographical Directory of the US Congress, 1774–2005 [BDUSC]).

12. Lawyer James Brown (1766–1835), younger brother of US senator John Brown (1757–1837), became Kentucky secretary of state in Governor Isaac Shelby's administration. He later moved to Louisiana, where he attained great wealth, and he would eventually represent that state in the US Senate. He served as US minister to France from 1823 to 1829 and befriended Lafayette (Sprague 1973; *DAB*, s.v. "Brown, James").

13. Benjamin Logan (1743–1802) served in the French and Indian War and in Lord Dunmore's 1774 campaign against the Shawnee. In 1775 he moved to Kentucky, where he was in the forefront of the struggle against the hostile tribes. Logan was the ranking militia officer, a courageous, trusted, and influential leader throughout the Revolutionary War and thereafter. As governor, Isaac Shelby appointed him brigadier general of the Kentucky militia (Talbert 1962, 270–77; *DAB*, s.v. "Logan, Benjamin").

14. Revolutionary War veteran James Speed (1739–1811) moved his family to Danville, Kentucky, after the war. He became a member of the influential Political Club of Danville during the years when Kentucky was approaching statehood, when free navigation of the Mississippi River was a major issue. Speed's son-in-law was the physician Dr. Adam Rankin of Danville, another recipient of one of Michaux's letters of introduction (Speed 1894, 66–68).

15. Gen. James Wilkinson (1757–1825) served in the Revolutionary War and settled near Louisville in 1784. He took an active part in the movement toward Kentucky's statehood and engaged in a variety of business ventures that left him deeply in debt. It was only later revealed that, beginning in 1787, Wilkinson had intrigued with and spied for the Spanish governors of Louisiana. In the meantime, he received a commission as a lieutenant colonel in the US Army and moved up the ranks. He advanced to command the army after the death of Gen. Anthony Wayne in 1796, all the while secretly remaining on the payroll of Spain. His ultimate downfall came not from discovery of his treason but from his military failure during the War of 1812 (Linklater 2009; *DAB*, s.v. "Wilkinson, James").

16. Barthélemi Tardiveau (d. 1801) and his brother Pierre (d. 1835) were well-connected French merchant-entrepreneurs living in the western country. Barthélemi was involved in several enterprises in the Ohio Valley and was established at Kaskaskia on the Mississippi. In 1790 he was appointed judge of probate as well as lieutenant colonel of militia in St. Clair County in the Illinois settlements. Pierre had been a member of the Political Club of Danville for a time, and he later actively assisted Michaux and Clark (Rice 1938, 40; Speed 1894, 93).

17. Michaux's route across the state of Pennsylvania followed well-traveled roads and can readily be followed on the map of Pennsylvania published in *Carey's General Atlas, Improved and Enlarged* (Philadelphia, PA: M. Carey, 1814). His party averaged a respectable twenty-eight miles per day on this initial leg of the journey.

18. Humeau and LeBlanc were the two French artillerymen who accompanied Michaux to Kentucky. Michaux identified them and requested that their salaries be paid in his letter to Clark from Philadelphia dated December 27, 1793. It is likely that Genet provided Michaux

with this escort because, in addition to secret messages, the botanist was carrying about $750 in cash to Clark. This was a substantial sum for the time, but it was not nearly enough to fund Clark's expedition.

19. Lancaster, first laid out as a town in 1730, was on the principal route leading from Philadelphia to the western settlements and was the capital of Pennsylvania from 1799 to 1812. Despite three journeys through Lancaster, there is no indication in Michaux's journal of either awareness of, or visits with, Dr. Henry Ernest Muhlenberg (1753–1815), the botanist who lived there, although Michaux's son did visit Muhlenberg in 1802. Trained in Germany, Muhlenberg was pastor of the Holy Trinity Church in Lancaster and the first president of Franklin College (later Franklin and Marshall University). The grass genus *Muhlenbergia* was named in his honor (*DAB*, s.v. "Muhlenburg, Gotthilf Henry Ernest").

20. Carlisle and Shippensburg were important frontier communities on the route between Philadelphia and Pittsburgh. German traveler John David Schöpf (1752–1800) made favorable comments on the appearance of Carlisle (Schöpf [1911] 1968, 1:214).

21. This was the small community of Upper Strasburg in Franklin County, approximately eleven miles west of Shippensburg, not the Strasburg in Lancaster County. The main road toward Pittsburgh no longer follows a route through Upper Strasburg (Wallace 1998, 142–43).

22. *Pinus* with two leaves is either *P. echinata* Mill., shortleaf pine, or *P. virginiana* Mill., Virginia pine. *Pinus* with three leaves is *P. rigida* Mill., pitch pine. *P. strobus* L. is the eastern white pine.

23. The Juniata, a tributary of the Susquehanna, flows approximately southwest to northeast in these mountains. Passing this way at the same season of the year in 1802, François André Michaux noted that the river was easily forded when the water was low (F. A. Michaux [1805] 1904, 143).

24. Bedford was the site of an early trading post and of Fort Bedford (erected 1758), the base for expeditions against Fort Duquesne (Pittsburgh) in the French and Indian War, which occasioned road building between the two points.

25. Michaux followed the route of modern US Highway 30, from Bedford through Greensburg into Pittsburgh (Swetnam and Smith 1976).

26. Scholars have commented on Michaux's extensive botanizing on his trip to Kentucky, but the period of botanical study began after he delivered Genet's message to Hugh Brackenridge and continued while he was delayed waiting for a boat to take him down the river to Kentucky. Michaux could not abide wasting his time, and throughout this journey he engaged in botanical research while waiting for transportation.

27. *Eupatorium* has been split into several genera. While *Eupatorium perfoliatum* L., boneset, is unchanged, *Eupatorium maculatum* is now *Eutrochium maculatum* (L.) E. E. Lamont, spotted joe-pye weed; *Eupatorium odoratum* is probably *Ageratina altissima* (L.) King & H. Rob., common white snakeroot (IDC 96-2); and *Eupatorium celestinum* is *Conoclinium coelestinum* (L.) DC., mistflower.

28. Col. Francis Vigo (1747–1836) was a successful fur trader in Saint Louis with commercial interests in Kaskaskia and Vincennes. Vigo's assistance to George Rogers Clark during his campaign to secure the old northwest was pivotal to Clark's success and only the first of many important services that Vigo performed for his adopted country. A statue of Vigo sits on the bank of the Wabash River in Vincennes, Indiana; and Vigo County, Indiana, is named in his honor (Roselli 1933; Thompson 1917).

29. Pittsburgh merchant Pierre Audrain (1725–1820) was part of the small network of traders serving the western region and a close business associate of Pierre Charles Dehault Delassus de Luzières (1738–1806) in New Bourbon, Missouri, and of Barthélemi Tardiveau in

Kaskaskia, Illinois. A lifelong supporter of the Bourbon monarchy of France, who fled during the French Revolution, de Luzières lived in Pittsburgh with his family before moving into Spanish territory. There he was on excellent terms with Carondelet, the governor general of Louisiana, who, although a Spanish official, was like de Luzières a native of Flanders (Ekberg 2011, 26–27).

30. Lucas de Pentareau is now known by his English name, John Baptiste Charles Lucas (1758–1842). A native of Normandy, Lucas immigrated to the United States about 1784 and settled in Pittsburgh. Michaux was correctly informed about his political leanings. He served in the Pennsylvania legislature from 1792 to 1798 and in the US Congress as a Jeffersonian Republican from 1803 to 1805. Lucas moved to Saint Louis, where he served as district judge for the northern district of Louisiana (which became the Missouri Territory in 1812) until 1820 (BDUSC).

31. After the failure of separate campaigns led by Gen. Josiah Harmar and Gen. Arthur St. Clair against the native tribes in the old northwest in 1790 and 1791, Gen. Anthony Wayne had taken command of the American army and was training and equipping his forces for what became the successful campaign of the following year.

32. The *Bryonia* is in the *Flora* (2:217) as *Sicyos lobata*, and it is recognized today as *Echinocystis lobata* (Michx.) Torr. & A. Gray, wild-cucumber. This fast-spreading vine is a Michaux discovery that is now a widespread invasive species in Europe.

33. Buffalo Creek is modern Wellsburg in Brooke County, West Virginia.

34. Marietta was the site of the first permanent settlement in Ohio and became the territorial capital when Gen. Arthur St. Clair, governor of the Northwest Territory, arrived in 1788.

35. Belleville, Wood County, West Virginia; and Belpre, downstream on the Ohio in Washington County, Ohio, remain small communities today.

36. *Polymnia canadensis* L. is the white-flowered leafcup (IDC 105-19, 105-20).

37. The first settlers of Gallipolis were a group of French people who immigrated to the United States in 1790. A massive fraud prevented these settlers from obtaining the titles to the land that they had paid for and settled. Moravian missionary Johann Heckewelder, who kept a detailed journal of his journey down the Ohio in 1792, noted that the group included skilled tradesmen and remarked that they laid out gardens in the formal European style, but that they lacked the more basic skills needed to succeed in the Ohio wilderness. Heckewelder also reported that Dr. Petit, both a physician and a judge, was the most prominent man in the settlement (Tolzmann 1988, 36–37).

38. *Iresine celosioides* is recognized today as *Iresine rhizomatosa* Standl., bloodleaf (IDC 126-2 "Ohio and Mississippi rivers").

39. Although Michaux mentions nothing about dangers at this point, Heckewelder related that the confluence with the Scioto River was the most dangerous point on the Ohio in 1792 and that about 250 settlers had been killed or captured by hostile Native Americans there in the preceding two years (Tolzmann 1988, 38).

40. The settlement was Massie's Station (later Manchester, Ohio), a fort on the north bank of the river a few miles downstream from three sandy islands, a passage used by the Native Americans to cross into Kentucky (Tolzmann 1988, 38–39).

41. Limestone, the pioneer river gateway to Kentucky, was settled in 1784 and soon changed its name to Maysville (Wells 2002, 6).

42. Michaux carried a letter for Orr, who represented Kentucky in the US Congress.

43. Once again Michaux botanized while waiting for transportation. His initial observations in Kentucky include *Guilandina dioica* (IDC 24-19), recognized as *Gymnocladus dioicus* (L.) K. Koch, Kentucky coffee-tree, and the new species that he named *Fraxinus quadrangulata*

Michx., blue ash. His herbarium specimen for the latter (IDC 128-6), perhaps collected here, is the type collection for the species (Uttal 1984, 64).

44. All species found in North America of the genus *Crepis*, hawksbeard, are introduced species. *Crepis sibirica* L. is not known from North America, and it is not clear what species Michaux observed here.

45. Henry Lee (1757–1845) was a cousin of Henry "Light Horse Harry" Lee of Virginia. He also served in the revolution, and afterward moved to Kentucky, where in 1785 he established Lee's Station in Mason County. Col. Lee became a prominent leader and landowner in the region and served in the Kentucky state legislature and other offices. He was promoted to general of the Kentucky militia in 1798 (Lee 1903).

46. Washington, named for George Washington, is the seat of Mason County and was founded in 1785 by Arthur Fox, a surveyor from Virginia, and William Wood, a Baptist minister. With a population of 462 in 1790, it was the second largest town in Kentucky after Lexington (Best 1936, 2–3: Clift 1936, 55.).

47. This is the birthplace of American vertebrate paleontology and the site where wooly mammoth and ground sloth fossils were discovered in 1739. John Filson (1784) had both marked the location now known as Big Bone Lick State Park on his 1784 map of Kentucky and provided details in his *The Discovery, Settlement and Present State of Kentucke*.

48. Lexington, the seat of Fayette County, was a bustling frontier community at the time of Michaux's visit. Needham Parry, a Quaker who visited in 1794, described the town this way: "Lexington is a fine stirring town, containing about 350 houses, throngly [densely] inhabited; and is the greatest place for dealing I ever saw" (Beckner 1948, 232–33).

49. Paris, Kentucky, originally named Hopewell, is the seat of Bourbon County, one of Kentucky's original counties at the advent of statehood in 1792.

50. James Brown (Kentucky's secretary of state) and Joseph Simpson were the two persons in Lexington named on the earlier list of letters of introduction.

51. Danville, although located in the southeast corner of the settled part of the state, was the first center of government in Kentucky and had the first post office. Danville's position at the hub of a network of roads is clearly shown on John Filson's 1784 map of Kentucky. Crab Orchard, near Danville, was the end of the dangerous portion of the Wilderness Road from southwestern Virginia. It was also the point where roads led off to the Cumberland settlements (Nashville, Tennessee) and toward the falls of the Ohio (Louisville) and the Ohio River (Lexington and Maysville). The first postal delivery was made on November 3, 1792. Michaux used the mail to communicate with Genet while he was in Kentucky (Price 1940).

52. The Barbee family was wealthy and influential in Kentucky. Lt. Col. Thomas Barbee (1752–97) was a Revolutionary War veteran, later a general. One of five brothers, he was Danville's first postmaster (Heitman [1914] 1982, 86; Price 1940; RRCHNM). His younger brother Joshua (1761–1839) was a member of the influential Political Club of Danville and a captain in a militia company organized by James Brown in 1791, and he served in the state legislature (Speed 1894, 83–84).

53. Peter Tardiveau (d. 1835) and his partner Jean A. Honoré had stores in early Danville and Lexington. Before Spain's 1784 closing of the Mississippi to commerce, the Tardiveau brothers and Honoré made trading voyages down the river to New Orleans, where they established contacts. For a time Peter was a member of the Political Club of Danville (Dupre 1941a; Burnett 1976; Rice 1938, 3, 46; Speed 1894, 93–94).

54. Senator John Brown (1757–1837), then in Philadelphia, was the older brother of Kentucky secretary of state James Brown. A lawyer, he moved to Kentucky before 1784, becoming a leader in Kentucky's drive to statehood. He represented Kentucky in the Virginia legislature

and in the Confederation Congress, where he strongly championed free navigation of the Mississippi River and Kentucky's other interests. Senator Brown provided letters of introduction for Michaux to present to both Gen. Clark and Governor Shelby (American Historical Association 1897, 982–83; Warren 1938, 1962; *DAB*, s.v. "Brown, John").

55. Michaux's letters of introduction for Shelby from both Jefferson and Senator John Brown were carefully worded to say that Michaux "had the confidence of the French Minister Genet" and that he was an able botanist (American Historical Association 1897, 982–84). If Michaux told Governor Shelby of the true nature of his visit to Kentucky, he does not reveal it in his September 13 journal entry. Moreover, Shelby, when later advised of the plot by Jefferson in his capacity as secretary of state, responded that he was not aware of a plot. When Jefferson provided Shelby with the names of the French agents in Kentucky, he did not include Michaux on his list (Henderson 1920, 453–54).

56. American Historical Association 1897, 984.

57. *Tragia cordata* Michx., heartleaf noseburn (IDC III-4), is probably the type collection of the species (Uttal 1984, 58).

58. One of the oldest cities in Kentucky, Bardstown appears on John Filson's 1784 map. William Bard planned and surveyed the town in 1780, and the first log courthouse was built in 1785 (Wells 2002, 197).

59. This route is not a detour; the Salt River lies in the direct route from Louisville to Bardstown, as shown on the Filson Map.

60. Michaux's *Serratula* specimens from Kentucky are in the genus *Liatris* in his *Flora*: *Serratula squarrosa* (IDC 93-20) is *Liatris squarrosa* (L.) Michx., scaly blazing-star, and *Serratula scariosa* β *montana* (IDC 94-12) is *L. spheroidea* Michx., spherical blazing-star.

61. USDA dendrologist Elbert Little Jr. wrote a note dated November 30, 1956, on a *Fraxinus* that he observed in the Michaux herbarium that all the *Fraxinus* specimens were either *F. americana* L., white ash, or *F. quadrangulata* Michx., blue ash.

62. James Hogan petitioned the Virginia General Assembly in 1785 to operate a ferry where Hickman's Creek entered the Kentucky River, a key point on the road between Lexington and Danville. Traveling between Danville and Lexington, Michaux crossed at this point several times. US Highway 27 crosses there today (Robertson 1914, 87).

63. Isham Prewitt Sr. (ca. 1725–1806) and four of his children moved from Virginia to the Quirk's Run section of what is now Boyd County, Kentucky, about 1785. Quirk's Run is a tributary of the Salt River just west of Danville. Prewitt's son Isham Prewitt Jr. (1753–1825) is known to have provided food and lodgings for travelers in 1795 and is likely to have been the Prewitt with whom Michaux had lodgings during his stay in the Danville area (Prewitt 1976?).

64. Michaux visited Clark in Louisville on September 17, 19, and 20, delivering Genet's messages and becoming acquainted with the general; however, Clark neither immediately accepted the commission that Genet offered nor immediately agreed to carry out his earlier proposal. Clark's acceptance came two weeks later in separate letters to Michaux and Genet dated October 3 (American Historical Association 1897, 1009–10). In his letters Clark confirmed that the plan was no longer secret and that more money would be needed, and he then reminded Genet that French warships must block any Spanish reinforcements to New Orleans. Michaux's letters to Clark dated October 7 and October 10 (American Historical Association 1897, 1010, 1012) informed the general that James Brown was not only aware of the expedition but had assisted Michaux in raising funds by speaking to Lexington merchants on his behalf. Clark's letter to Michaux dated October 15 (American Historical Association 1897, 1013–14) stated that only a few hundred dollars were needed to begin building the boats without which the expedition could not embark and also asked Michaux to request that artillery pieces be

sent to Kentucky. Clark's letter to Genet dated October 25 (American Historical Association 1897, 1016), which Michaux carried to Philadelphia and delivered in person, repeated the importance of finding more money for the expedition.

65. George Nicholas (ca. 1754–99) is remembered as the person most responsible for the Kentucky State Constitution of 1792. Nicholas served as US attorney while Kentucky was still part of Virginia and became the state attorney general after Kentucky achieved statehood. He was a prominent member of the Democratic Society of Lexington in 1793 and had long been a proponent of the free navigation of the Mississippi (Dupre 1941b).

66. Travelers heading east gathered at Crab Orchard and formed temporary groups to travel together more securely through this dangerous country, where hostile Native Americans often attacked travelers. The famous militia leader Col. William Whitley (1749–1813) lived at Crab Orchard and often led militiamen against those who attacked travelers along the road. Today Whitley's strong two-story brick house, completed in 1794, is a Kentucky state historic site, and Whitley County is named for him.

67. The station, a type of defensible residential site, was common during Kentucky's early settlement. Both Langford's (Landford's) Station, located near the headwaters of the Dicks (Dix) River in Rockbridge County, and Modrel's (Moddrel's) Station, southeast of London on Little Laurel River in Laurel County, are shown on the Elihu Barker Map of Kentucky reproduced in the 1795 *Carey's American Atlas*. Small militia units occupied these stations to offer protection to traveling parties on the road. The need for such military strongpoints continued until after Gen. Wayne defeated the hostile tribes at Fallen Timbers on August 20, 1794. In his diary, Needham Parry related that on the road between Langford's and Modrel's Stations, Native Americans attacked and put to flight a party of forty-eight men on May 15, 1794 (Beckner 1948, 240). The original log structure of Langford's Station was preserved and later moved to the grounds of the Rockbridge County Courthouse in Mount Vernon, Kentucky (Hammond 2002).

68. A companion station to Modrel's Station (commanded by Lt. Robert Modrel), Middleton Station's garrison of twenty or so militiamen was commanded by Lt. Walter Middleton and located on Turkey Creek, southeast of modern Barbourville. Middleton's Station is indicated on the 1804 map by Dupuis and Sons bound into the French (1804) edition of François André Michaux's *Travels to the West of the Alleghany Mountains* (Kincaid 1947, 179; Hammond 2002, 574).

69. This fern was *Lygodium palmatum* (Bernh.) Sw., American climbing fern.

70. Stinking Creek and Flat Lick are the first two waypoints north of the ford of the Cumberland River listed by Filson (1784). The names continue in use today.

71. Davis's Station indicates the party had reached the relative safety of Cumberland Gap. It was located at the north base of the Cumberland Mountain, near the confluence of Davis Creek and Yellow Creek (Hammond 2002, 579).

72. This description indicates that he did not follow the route of modern US Highway 58 east from Cumberland Gap but rather continued southward generally along the route of modern US Highway 25E, crossing the Powell and Clinch Rivers and climbing and descending a series of rugged ridges before descending into the valley of the Holston River to the junction of the road from Cumberland Gap and the road running between Knoxville and Abingdon, Virginia. Michaux's Holston Station is Bean's Station, founded in 1776 at this junction. The original site is now submerged beneath a hydroelectric lake (Forester 1996, 7, 100).

73. The sturdy stone home of Capt. Thomas Amis (1744–97) near Rogersville (then Hawkins Courthouse), Tennessee, was a regular stop for travelers on the road after he completed it in 1784. Amis served in the North Carolina Continental Line and later as a commissary officer

before moving to North Carolina's western lands, which later became Tennessee. The house was built to withstand attack and is still owned and occupied by his descendants, who operate a restaurant and rent a cabin there today (Price 1996, 56–59; J. Jacobs, pers. comm.).

74. Michaux spent the night in the vicinity of modern Kingsport, Tennessee, then known as the Boat Yard because it was at the head of navigation on the Holston River (Spoden 1991, 156).

75. Anderson's Block House near the North Fork of the Holston River in Scott County, Virginia (Kincaid 1947). From this point Michaux traveled a route very similar to the one he had followed to Philadelphia in 1789.

76. Seven Mile Ford, on the river now named the Middle Fork of the Holston River, was seven miles downstream from noted militia leader Col. Arthur Campbell's home, Royal Oak. The modern town of Marion, seat of Smyth County, occupies this site today (Arnold 2007, 310).

77. Wythe Courthouse is modern Wytheville, but it was called Evansham on the 1807 *Bishop Madison Map of Virginia*.

78. *Hypericum kalmiana* L., Kalm's St. John's-wort, is a northern and midwestern species not known from Virginia today.

79. Peppers Ferry was located five miles downstream on the New River from Ingles Ferry, where Michaux had crossed in 1789. The botanist not only crossed the New River on a different ferry in 1793 but also followed a different route from the New River to the James than the one he had taken in 1789. The road from Peppers Ferry to Catawba Creek shown on the 1807 *Bishop Madison Map of Virginia* passes through the future site of Blacksburg before reaching Catawba Creek, which flows into the James River about twelve miles northeast of Fincastle (Botetourt Courthouse). Michaux's note for the following day, that he traveled 30 miles to Botetourt Courthouse, indicates that he spent the night near the headwaters of Catawba Creek.

80. Botetourt Courthouse is modern Fincastle, seat of Botetourt County.

81. Michaux also visited Lexington on July 8–9, 1789. The northern fork of the James River is now named the Maury River after hydrographer, meteorologist, and Virginia Military Institute professor Matthew Fontaine Maury (1806–73).

82. This is most likely young Col. James McDowell (1770–1835), who lived at his plantation Cherry Grove near modern Fairfield in Rockbridge County, Virginia. He distinguished himself in the War of 1812, while his first cousin, Kentucky physician and surgeon Ephraim McDowell (1771–1830) is remembered as the father of abdominal surgery. (Dr. McDowell later practiced with Dr. Adam Rankin, a recipient of one of the letters that Michaux carried to Kentucky.) Col. McDowell's son James Jr. (1795–1851) was governor of Virginia from 1843 to 1845 (Nicholson 1914; Brock [1888] 1973).

83. Michaux also visited Staunton on July 10, 1789.

84. Although it was sometimes called Rocktown in the eighteenth and nineteenth centuries, the town's name on the 1807 *Bishop Madison Map of Virginia* is Harrisonburg, the current name.

85. Woodstock in Shenandoah County, on the North Fork of the Shenandoah River, was a town founded before the revolution. In Woodstock in 1776, Peter Muhlenberg, local minister and brother of Pennsylvania botanist-clergyman Henry Muhlenberg (see note 19 in this chapter), recruited a Virginia regiment largely from his congregation after accepting a colonel's commission in the Continental Army (Virginia Writers Project 1940, 420–21).

86. The pine with two leaves and cones having rigid and prickly scales is *Pinus pungens* Lamb., table mountain pine.

87. Newtown, also called Stephensburg, was in Frederick County, nine miles south of Winchester. It is now a historic district on the National Register of Historic Places in Stephens City.

88. Michaux also visited Winchester in June 1786 and on July 14, 1789.

89. Michaux also visited Charles Town, now the seat of Jefferson County, West Virginia, on July 15, 1789.

90. Michaux also visited Frederick, Maryland, on July 16, 1789.

91. Michaux's Woodberry (not to be confused with the historic Baltimore suburb of Woodberry) was the community of Woodsborough, eleven miles north of Frederick, Maryland. Woodsborough was first laid out in 1786 and appears on the Frederick map of the 1865 *Martenet's Atlas of Maryland*, available online through the Maryland State Archives, www.msa.md.gov/msa/mdslavery/html/mapped_images/frmap.html. The town of Littlestown, Pennsylvania, is thirty-five miles from Frederick, Maryland, and seven miles from Hanover, Pennsylvania. Founded in 1760 and first named Petersburg, Littlestown was renamed as such in 1795.

92. Hanover was originally the site of Richard McAllister's tavern and store, a two-story log building where Benjamin Franklin was a guest in 1755 (Beyer 2000, 391).

93. Michaux continued to mistakenly believe that the boundary between Maryland and Pennsylvania was the Susquehanna River. See also chapter 5, note 38.

94. Although Michaux notes Harris's Ferry twice, both here in his narrative and in his recapitulation of distances traveled, he did not cross the Susquehanna at Harris's Ferry on this journey. It was not the most direct route, and his distances indicate that he traveled from York to Lancaster by the most direct route, crossing the Susquehanna at Wright's Ferry between the modern towns of Columbia and Wrightsville. On his journey west to Kentucky in July 1793, Michaux would indeed have crossed the Susquehanna at Harris's Ferry when he traveled between Lancaster and Carlisle on his journey west, but he made no mention of it then. Harris's Ferry, at modern Harrisburg, is forty miles from Lancaster and twenty-six miles from York. Michaux reports his distance traveled from York to Lancaster as twenty-four miles (Beyer 2000, 141, 212).

95. Thomas Jefferson was vice president of the APS, and David Rittenhouse (1732–96), an instrument maker, astronomer, and mathematician, was its president.

96. The community of Lincoln, Kentucky, is indicated in the vicinity of the confluence of the Hanging Fork and the Dix River, Lincoln County, on Alexander Anderson's "Map of the State of Kentucky with the Adjoining Territories," found in John Reid's 1796 *American Atlas*. Michaux's only mention of this community is in this recapitulation of distances from Kentucky to Philadelphia.

97. Dr. Nicolas Collin (1746–1831) was minister of the Swedish Church in Philadelphia. He had served as chair of the APS committee to raise money for Michaux's proposed western expedition. Collin, a close friend of Benjamin Franklin's, was with him in his dying hours and is included in a painting entitled *Death of Benjamin Franklin* owned by the American Swedish Historical Museum in Philadelphia, but which vanished several years ago (R. Goodman, pers. comm.). Collin collected seeds as well as insects, birds, and animal hides on behalf of Karl Thunberg of Sweden and other naturalists. He kept phenomenal records of weather, was interested in plants as materia medica and for their uses in commercial dyes, soaps, vegetable tallow, and fiber, and was curious about Native American uses of plants. He apparently was first to introduce rutabagas to this country (Johnson 1936). The Royal Swedish Academy of Sciences has five letters (April 1794–November 1796) from Michaux to Collin that deal with the shipment of seeds for Collin and Thunberg and also with birds destined for C. W. Peale's museum (see the following note).

98. Charles Willson Peale (1741–1827) was a noted painter and had a keen interest in natural history, which led to his friendship with Michaux. He founded the Philadelphia Museum,

housed at first in Independence Hall. This museum contained a diverse collection of botanical, zoological (especially birds), and archeological material. It was the first site to display mastodon bones. Michaux is listed as a contributor to the annual subscription to the Peale Museum for 1794, and Peale's son Rembrandt painted a portrait of François André Michaux now owned by the APS (*DAB*, s.v. "Peale, Charles Willson").

99. Charles François Bournonville came to the United States with Genet to serve as one of the secretaries of the French legation in Philadelphia, 1793–94.

100. Michaux combined his knowledge of agriculture with observations made during his recent journey to Kentucky to prepare a report about American grain production that he delivered to Genet on January 10, 1794. In this short document, Michaux concluded that the American wheat harvest would be poor and thus France should not rely on shipments of surplus American wheat in 1794 (AN F/17/1225, Citizen Michaux, botanist, to Citizen Genet, minister, Philadelphia, December 27, 1793).

101. Two species of crossbill are possible: *Loxia leucoptera*, white-winged crossbill, and *Loxia curvirostra*, red crossbill; however, unlike his plant collections, none of Michaux's collections of birds are known to have survived in any museum.

102. Genet made Michaux his messenger to Consul Mangourit in Charleston in early 1794. The consul's correspondence indicated that he was expecting Michaux to bring him dispatches. Mangourit was in charge of the expedition about to be undertaken against the Spanish in East Florida. The projected invasion of Florida was further advanced than the attack that George Rogers Clark in Kentucky planned against the Spanish in Louisiana (Turner 1903, 2:610–11 [Mangourit in Charleston to Genet in Philadelphia, January 31, 1794], 617–18 [Mangourit to Genet, March 16, 1794]).

103. Linnaeus outlined the classification of animals in his *Systema Naturae* as he had done for plants in his *Species Plantarum*. Barton had met with Michaux a year earlier while acting on behalf of Jefferson in discussing with Michaux the proposed explorations west of the Mississippi River (Ewan and Ewan 2007, 498–99).

104. This document refers to Michaux's commission as political and botanical agent of the French Republic, signed by Genet and dated July 13, 1793.

105. John Brown, US senator from Kentucky, met Michaux before he left to confer with Gen. Clark and provided him with a letter of introduction for Governor Shelby. Col. Alexander D. Orr was one of Kentucky's two representatives in the US Congress. The botanist had previously met Col. Orr in August 1793, when Orr had accompanied Michaux from Maysville [then Limestone] to Lexington. It is difficult to imagine how either Brown or Orr could have been unaware of the true nature of Michaux's 1793 mission.

106. During winter 1793, months before his first meeting with Genet, Michaux joined an organization of Frenchmen in Philadelphia who championed the French Revolution and its ideas (Kennedy 1976).

107. American Historical Association 1897, 1024–26.

108. This is a reference to Revolutionary War veteran Col. James Morrison (1755–1823). He moved to Pittsburgh after the war, and in 1792 he began a highly successful career as a merchant in Lexington, Kentucky, where he became an early benefactor of Lexington's pioneering Transylvania University (Holley 1823).

## Chapter Eleven

1. L'Abbé Grégoire (1810), a noted abolitionist, wrote that Michaux had purchased a young black man in Philadelphia whom he later brought back to France and still later took to the South Seas. If this is correct, it is likely that he would have purchased him at this time as this

would be his last stay in Philadelphia. More will be said about this young man, Merlot, in the epilogue.

2. While he visited Philadelphia, Michaux had boarded his horse with the Bartrams, who lived in the nearby township of Kingsessing (now a part of Philadelphia).

3. This is likely a reference to either *Poecile carolinensis*, Carolina chickadee, or *Poecile atricapilla*, black-capped chickadee. The ranges of these birds overlap in the area where Michaux was traveling.

4. The bird is a cedar waxwing, *Bombycilla cedorum*.

5. The area was rich in iron ore, and Maryland had begun to encourage an iron industry early in the eighteenth century.

6. As he had done on his most recent journey across Maryland in August 1789, Michaux traveled directly from Baltimore to Bladensburg. Today, Bladensburg is a suburb of Washington, DC, but the former is at least six decades older than Washington and was a busy port for the shipment of tobacco in the mid- to late eighteenth century.

7. Michaux also crossed the Potomac at Alexandria in June 1786 and again in August 1789. In 1789 he continued south on this same route through Dumfries and Fredericksburg, spending two nights in Richmond before proceeding south through Petersburg and Emporia to Halifax, North Carolina.

8. The pine with three leaves is probably *Pinus rigida* Mill., pitch pine.

9. *Pinus pungens* Lamb., table mountain pine.

10. Michaux's route from Fredericksburg to Richmond generally follows Virginia Highway 2 to Bowling Green, then US Highway 301 through Hanover into Richmond. Bowling Green was then Bowling Green Ordinary, a tavern located on the main road two miles from the Caroline County courthouse.

11. The pine with two and three needles on the same branch is *Pinus echinata* Mill., shortleaf pine, while the pine with three needles that begins near Bowling Green is *P. taeda* L., loblolly pine.

12. This is the type location of *Ulmus alata* Michx. (IDC 39-10), winged elm (Uttal 1984, 27).

13. "Paterson's Tavern" is likely the Peterson's Inn of Northampton County, North Carolina, shown on the 1808 Price and Strother map between the Virginia state line and the Roanoke River.

14. The pine with three needles and midsize cones is *Pinus taeda* L., loblolly pine; pines with 2 or 3 needles are *P. echinata* Mill., shortleaf pine.

15. Michaux also passed through Enfield on September 4, 1789. His host, whom he refers to as Col. Brandt, was Lt. Col. John Branch (ca. 1750–1806) of the Halifax County militia, a wealthy landowner in the area and one of the most prominent men Michaux visited anywhere in North Carolina. His son John Branch Jr. (1782–1863), a child when Michaux visited, served as governor of both North Carolina and Florida as well as secretary of the US Navy in Andrew Jackson's administration (Haywood 1915).

16. Col. Ethelred (Ethelrid) Phillips of the Edgecombe County militia was one of the most prominent men in that county. Active politically after the war, he served as sheriff of Edgecombe County, and he also served terms in both houses of the legislature (Boddie 1956, 352; Saunders, Clark, and Weeks 1904–7).

17. Tetts Bridge, shown on the 1808 Price and Strother map, spanned the Tar River just upstream from the town of Tarboro, the seat of Edgecombe County after the county had been divided from Halifax County.

18. *Baptisia cinerea* (Raf.) Fernald & B. G. Schub., Carolina wild indigo, retains leaves in

the fall, and the dried plants are conspicuous throughout the winter (Weakley 2011, 421).

19. Town Creek is a tributary that joins the Tar River in southern Edgecombe County, near the modern town of Pinetops.

20. The production of naval stores, of which turpentine was one product, was the principal industry in this region.

21. Peacock's Bridge, near modern Stantonsburg in Wilson County, crossed Contentnea Creek where a Revolutionary War battle occurred and has a North Carolina historical high-way marker (Hill 2001, 215).

22. *Laurus borbonia* is *Persea borbonia* (L.) Spreng., red bay. All the Andromeda varieties that Michaux mentions in this area are now in other genera. See the list in note 26 in this chapter.

23. In his *Flora* Michaux recognized only a single species of *Myrica* south of Pennsylvania, *Myrica cerifera* L., wax myrtle, with three varieties based on height of the plant, leaf shape, and size of fruit (vars. *arborescens*, *media*, and *pumila*). The genus *Myrica* has now been split into *Myrica* and *Morella*. The three species of wax myrtle that he observed in this area are probably those recognized today as *Morella cerifera* (L.) Small, southern bayberry, common wax-myrtle, *M. carolinensis* (Mill.) Small, evergreen bayberry, and *M. pumila* (Michx.) Small, dwarf bayberry, dwarf wax-myrtle.

24. Whitfield's Ferry over the Neuse River was near La Grange in modern Lenoir County. The ferry's location is shown on the 1808 Price and Strother map. Michaux also mentioned crossing there on September 7, 1789.

25. The Duplin Courthouse that Michaux visited was the second location of Duplin County's courthouse; the new site, modern Kenansville, was chosen following the 1784 division of Duplin into Duplin and Sampson Counties. A new Duplin courthouse building was promptly erected in 1785. Col. Alexander Dickson (ca. 1746–1814) was a prominent citizen of this area (McGowan and McGowan 1971, 95–99).

26. *Andromeda axillaris* is *Leucothoe axillaris* (Lam.) D. Don, coastal doghobble; *A. mariana* is *Lyonia mariana* (L.) D. Don, staggerbush; *A. nitida* or *lucida* is *Lyonia lucida* (Lam.) K. Koch, shining fetterbush; *A. paniculata* is *Lyonia ligustrina* (L.) DC. var. *foliosiflora* (Michx.) Fernald, southern maleberry; *A. racemosa* is *Eubotrys racemosus* (L.) Nutt., coastal fetterbush; *A. wilmingtonia* is *Zenobia pulverulenta* (W. Bartram ex Willd.) Pollard, Zenobia.

27. Today the original site of the town of South Washington on the North East Cape Fear River is the Riverview Park Memorial Cemetery in Pender County. Michaux also visited there on September 8, 1789.

28. Perhaps Michaux remembered Duplin Courthouse here but mistakenly gave the distance from South Washington to Wilmington; South Washington was then part of New Hanover County, and the county courts met in rented space in Wilmington.

29. The herbarium specimens "*Atraphaxis?* Wilmington in Carolina" (IDC 124-16) and "*Atraphaxis polygonella dioica*" (IDC 124-17) are both *Polygonella polygama* (Vent.) Engelm. & A. Gray, common October-flower. In his *Flora* (2:240–41), Michaux established the genus *Polygonella* and noted its similarity to the Linnean genus *Atraphaxis*.

30. Uttal (1984, 4) concluded that Warm Spring referred to Warm Spring Road in Frederick County, Virginia.

31. While the route Michaux chose to cross eastern North Carolina was the most direct route from Halifax to Wilmington, traveling through this area was difficult. Travel accounts well into the middle of the nineteenth century (when the Wilmington and Weldon Railroad linked Wilmington to the Halifax area) remark on the difficulty of the roads, the multitude of stream crossings, and the dearth of accommodations for travelers.

32. "The Islands" probably refers to the West Indies and most likely to Saint-Domingue (Haiti). We noted earlier that Michaux had become a true believer in republican France and its ideals.

33. Dr. James LaRoque and his wife, Louisa, were Wilmington residents during this period. James was named in a local court case in May 1793, and he and his wife sold property there in 1808. In 1806 he advertised that "Dr. LaRoque and Son" offered drugs and medicine both wholesale and retail at a store on Market Street formerly occupied by Draper and Metcalf (Walker 1960; Sammons 1996; Fouts 1984).

34. Amasiah Jocelin (Jocelyn, Josselin, Joslin) and his partner Henry DeHerbe were men of property and licensed innkeepers in Wilmington. Jocelin's property included six enslaved persons at the 1790 US Census of North Carolina, and Jocelin posted a bond for one thousand pounds as administrator of DeHerbe's estate in August 1794. Jocelin and DeHerbe's establishment offered accommodations for travelers like Michaux. It was also a boardinghouse for sailors and the site of many public gatherings, including a celebration during George Washington's 1791 visit. The innkeepers also rented rooms in the building for hearings of the Superior Court. In 1810, the building was sold by Samuel R. Jocelin (Joiclyn), an eminent lawyer (Walker 1960; McKoy 1967, 64).

35. As he had in September 1789, Michaux crossed the Cape Fear River from New Hanover County into Brunswick County, continued south toward the South Carolina state line, and again mentioned the tidal river called Lockwood Folly.

36. William Gause (1745–1801) lost a leg as a result of a Revolutionary War wound. He was a wealthy man and a leading citizen of Brunswick County who lived at Gause Landing (adjacent to modern Ocean Isle Beach). He represented Brunswick County in the North Carolina legislature in 1778, and he also hosted both George Washington in 1791 and Bishop Francis Asbury twice between 1791 and 1800 (Berry 1972; Blake 1970).

37. This was likely Alexis Mador Forster (d. 1794), county surveyor, and later sheriff, of Brunswick County, whose name often appears in court records (Thompson 1992; Pruitt 2001; Blake 1970).

38. Both the range and the description of this plant fit *Sageretia minutiflora* (Michx.) C. Mohr, small-flowered buckthorn, rather than *Pisonia inermis*, another opposite-leaved shrub that does not occur north of Florida (S. Hill, pers. comm.).

39. The phrase "kept a tavern" indicates that the local resident routinely offered lodgings to travelers for a fee. The Horry District map in the 1825 *Mills' Atlas* indicates both Vareen and Green families living in South Carolina on the main road just south of the North Carolina state line near Little River, but it is likely that "Wren" and "Wreen" (in the following entry) are misspellings and that Michaux meant Jeremiah Vareen (1748–1808), a Revolutionary War veteran who kept the best inn for travelers in this area. Vareen was mentioned by Schoepf in 1784, and George Washington also stopped at Vareen's on his southern tour in 1791, but Washington was treated as the family's guest and not charged (Daughters of the American Revolution 2003, 3:2801).

40. The stretch of road opposite Long Bay, in Horry County, South Carolina, beginning near Lewis's Swash and continuing south to Murrell's Inlet, ran not inland on high ground, but on the hard-packed sand of the beach's tidal zone for approximately eight miles. It is described as the "sea shore road" in the 1825 *Mills' Atlas*. This road passed through the area that is Myrtle Beach today.

41. William Magill is listed in the US Census of 1790 for South Carolina as a resident of Prince George's Parish in Georgetown District with a household consisting of himself, one adult woman, and fourteen enslaved persons. Rogers (1970) reported that there were also

other Magills present in the area at the time, although these others do not appear in the extant census records (Webber 1925). As he traveled south from Magill's, Michaux entered a region settled largely by French Huguenots a century earlier. These French Protestants came to America as refugees from religious persecution after Louis XIV's revocation of the Edict of Nantes in 1685. Michaux, a Roman Catholic, appears to be sensitive to their story. This was an industrious group, and many Huguenot families accumulated great wealth in South Carolina.

42. This person was certainly a member of the Huguenot Mazyck family.

43. Michaux's route from Murrell's Inlet to Georgetown passed through Waccamaw Neck, a narrow strip of land with the Atlantic shore on the east and the Waccamaw River on the west. Many wealthy rice plantations were established along the Waccamaw, but the houses were set well back from the road, which passed through the poorest land (Salley 1932, 5).

44. Pittcock Ferry does not appear in any records consulted; a fragmentary record does indicate a Paul Pidcock living near Georgetown at this time (Webber 1925).

45. The distance indicates that Michaux traveled directly south to cross the Santee, where at the time there were ferries over the two large branches of the river, which had an island in the middle between them. The ferry over the north branch of the Santee was then named Cook's Ferry (Brevard 1814, 3:393–95). By the 1825 publication of the *Mills' Atlas*, the north Santee ferry had become Collins's Ferry, and the south Santee ferry Mazyck's Ferry. These ferries were not named on the 1780 Faden map. Ferries in South Carolina usually changed names when ownership passed to another family.

46. The Charleston District map in the 1825 *Mills' Atlas* shows the road network that Michaux likely used between the Santee and Charleston and locates "32 Mile House" at approximately where Michaux found the Morell tavern in 1794. The 1780 Faden map indicates three Morel plantations and a Manigault plantation south of Awendaw Creek on the road directly to Charleston, but these locations are not shown on the later map in the 1825 *Mills' Atlas*, and Michaux's directions indicate that he turned inland in the vicinity of Awendaw Creek. The Manigaults had become wealthy and educated; they lived in Charleston. Their plantation in this area was not an elegant place but rather a site where cattle and hogs were allowed to roam the woods as a way of fattening them.

47. John Wigfall (1736–93), who owned three plantations in this area as well as two townhouses in Charleston at the time of his death, was a prominent planter and political leader who served eight times in the legislature. Despite accepting a commission in the Loyalist militia after the fall of Charleston in 1780, Wigfall escaped financial ruin and loss of his property in the shifting tides of the revolution when powerful friends in the Patriot camp stood by him. His heirs appear to have welcomed Michaux (Edgar, Watson, and Bailey 1981, 3:710–11).

48. Clement's Ferry is not indicated on the 1780 Faden map or in the 1825 *Mills' Atlas*, but Simons (1930) indicates the location as five miles upstream on the Cooper River from the city of Charleston. When approaching the city from the north on coastal roads as Michaux did here, crossing at Clement's Ferry offered the most direct route to the garden, which was located ten miles north of Charleston. Clement's Ferry Road retains the old name today.

49. Although they had exchanged letters earlier, this was the first time that the two men had met.

50. André Michaux never directly identified his gardeners in Charleston. One was later identified by François André Michaux as Jacques Renaud, the young man who had originally accompanied the Michauxs from France as a personal servant in 1785, but the identity of the others remains a mystery. There is an indication that while Michaux was away on his journey west to the Mississippi River the following year, the garden was left in the care of the enslaved black men who lived on the property. In spring 1796, shortly after Michaux returned from this

long journey, a visiting Frenchman reported that while he was away "two Negroes took care of his garden and kept it clean of weeds" (La Rochefoucauld-Liancourt 1799, 1:589)

51. Mangourit did not give up the expedition against Florida easily. Fauchet, the new French minister to the United States, issued a proclamation on March 6, 1794, canceling all Genet's commissions and projects. His statements made clear that France was repudiating Genet's actions, and there are some indications that Fauchet unsuccessfully attempted to stop Michaux from delivering Genet's last messages to the consul in Charleston (Kennedy 1976, 628–29). Nonetheless, Mangourit's response to Fauchet's proclamation, which was printed in the Charleston newspapers on March 28, was to meet with the expedition leaders and choose one of them to travel to Philadelphia to plead the expedition's case with Fauchet. When his own replacement, Fontpertius, arrived in Charleston on April 7, Mangourit unsuccessfully tried to stall the new consul's courier long enough that the expedition could begin before orders to cancel it could reach the troops in the St. Marys area (Alderson 2008, 167–71).

52. Royal Swedish Academy of Sciences, Stockholm (RSAS), letters to Nils Collin, L-Ö.

53. Moncks Corner was a day's journey north on the main route into the interior from Charleston. It had several inns for travelers, and Michaux had previously slept there in September 1789.

54. Michaux's *Eryngium aquaticum* L., marsh eryngo (IDC 38-6), was found at Goose Creek Bridge.

55. Revolutionary War sources agree that Manigault's Ferry was about five miles upstream on the Santee from Nelson's (or Neilson's) Ferry.

56. Michaux's earlier reference to *Serratula*, in Kentucky on September 25, 1793, was later placed in the genus *Liatris*; however, *S. fistulosa* (meaning "hollowed-out stem") cannot be identified with any *Liatris*.

57. *Heliotropium indicum* L., turnsole (IDC 31-4), was collected near Eutaw Springs and Camden.

58. Michaux's "*Sida lanceolata?*" (IDC 83-10), now recognized as *Sida acuta* Burm. f., wireweed, was collected on the Santee River.

59. Friedrich Adam Julius von Wangenheim (1749–1800) was a captain in the Hessian cavalry serving with the British army during the American Revolution. A trained forester, he used his free time to study American trees. Michaux demonstrates here that he is aware of Wangenheim's work and compares this unknown oak to Wangenheim's scarlet oak, *Quercus coccinea* Muenchh.

60. This is the type location (Uttal 1984, 34) for *Kalmia cuneata* Michx., white wicky. This very rare species is endemic to the sandhills of North and South Carolina; however, an organized search for it near the Camden battlefield site in 2013 was not successful. The Michaux herbarium specimen is "*Kalmia cuneata* near Camden" (IDC 58-20). He collected live plants at this site on April 7, 1796.

61. "Ninth class" refers to Linnaeus's *Enneandria*, the group with nine stamens.

62. "Wm. Graim" is likely the James Ingram home, where George Washington spent the night when traveling this road in 1791 (Andrews 1998, 137).

63. In 1789 Michaux had visited John Cry (Crye), who lived near the South Carolina state line in the southeastern part of old Mecklenburg County (now in modern Union County), North Carolina. Although Cry had died in 1792, he left his wife a life estate to most of his property, and she was still living on their farm in 1794 (Ferguson 1993).

64. The 1808 Price and Strother map indicates that the Huston Tavern was located near a tributary stream of Twelvemile Creek in modern Union County, North Carolina.

65. Capt. John Springs Jr. (1751–1818); his wife, Sara Shelby Alexander (1762–1842); and

their family lived in southeastern Mecklenburg County. He served in several campaigns of the American Revolution, and his son Adam (1782–1840) was in the first graduating class (1798) of the University of North Carolina (Springs 1965, 10–13, 18).

66. The genus name followed by three dots (*Rhus . . .*) is Michaux's shorthand for a new, unnamed species of *Rhus* that he encountered at this location. In this instance, the location information on his herbarium specimen, "*Rhus pumila*, Carolina mountains" (IDC 42-13), does not agree with the information in his *Flora* (1:182). The analysis of Barden and Matthews (2004) demonstrates that this brief journal entry provides the key to resolving the mystery concerning where Michaux discovered this plant, which we now know as *Rhus michauxii* Sarg., Michaux's sumac. On April 5, 1796, Michaux revisited this site, then in Mecklenburg but now in Union County, and referred to this new species as *Rhus pumilum*, the binomial that he also used in his *Flora*.

67. The herbarium specimen "*Delphinium tridactylom* in the mountains of Carolina and Virginia" (IDC 71-5) is now recognized as *Delphinium exaltatum* Aiton, tall larkspur. This species, which is reported to bloom in July at lower elevations, is not known to occur in or near this area today. The other principal possibility however, *Delphinium tricorne* Michx., dwarf larkspur (IDC 71-6, "*Delphinium tricorne*, very high mountains"), which has been found on McAlpine Creek in Mecklenburg County near the Springs' homesite, is a spring ephemeral, and no part of a plant of *D. tricorne* would be visible above ground in late July (J. Garton, pers. comm.).

68. Benjamin "Bennet" Smith lived near the confluence of Hoyle and Little Hoyle Creeks in old Lincoln County, now Gaston County. He later became a lawyer, sold his farm to his neighbor Peter Smith, and practiced law in Lincolnton and Morganton (Williams 1999, 5–8; (Phifer [1977] 1982, 143).

69. The *Magnolia* described here is *Magnolia macrophylla* Michx., bigleaf magnolia. The new *Stewartia* is *Stewartia ovata* (Cav.) Weath., mountain camellia, a rare plant in North Carolina today and not now known from this county; however, there are indications that it was more widespread in Michaux's time. In their study of John Lyon's journal, Ewan and Ewan (1963) observed that Lyon, another early plant collector, mentioned encountering this species several times during his travels in the Carolinas and Tennessee between 1803 and 1809.

70. Christian Reinhardt (ca. 1735–1817), one of the founders of Lincolnton, was a prosperous landowner and businessman (Williams 1997, 16).

71. This is the same farm on the Henry's Fork of the South Fork River in modern Catawba County previously owned by Heinrich Weidner (1717–92), sometimes referred to as Henry Whitener, who had hosted Michaux in 1789; afterward, his son-in-law Jesse Robinson owned the farm and continued to welcome Michaux.

72. John Rutherford Sr. (1755–1841) was a prominent Burke County planter who sometimes worked with Waightstill Avery on civic improvement projects and who hosted the traveling Methodist bishop Francis Asbury in 1797. Rutherford lived at his plantation Bridgewater, located on the south side of the Catawba and west of Avery's Swan Ponds on the north side of the river (Phifer [1977] 1982). The 1808 Price and Strother map indicates the location of Muddy Creek, a tributary entering the Catawba from the south, as well as both the Avery and Rutherford plantations.

73. Michaux also visited James Ainsworth and obtained guide(s) there in 1789. See chapter 5, note 11.

74. On the east side of the North Fork of the Catawba, several named peaks on Linville Mountain, as well as Dobson's Knob and Bald Knob, rise more than two thousand feet above the elevation of the river near Turkey Cove (USGS, "Little Switzerland Quadrangle, N.C.,"

and "Ashford Quadrangle, N.C.," 7.5 minute series topographic [maps] 1:24,000 [Reston, VA: USGS, 1956, 1960, 1979]).

75. The French foot was slightly larger than the English. The circumference of this tree would be approximately 24½ feet in English measure.

76. *Convallaria umbellata* is the provisional name on Michaux's specimen (IDC 46-16, 17) of *Clintonia umbellulata* (Michx.) Morong, speckled wood-lily, and the berries are typically blue.

77. *Convallaria racemosa* is *Maianthemum racemosum* (L.) Link, false Solomon's-seal, while "*Convallaria multiflora* Canada, Pennsylvania, Carolina mountains" (IDC 47-1) is *Polygonatum biflorum* (Walter) Elliott, Solomon's-seal. *C. majalis* (IDC 46-18 "*Convallaria majalis-americana* mountains of Carolina") is *C. majalis* subsp. *majuscula* (Greene) Gandhi, Reveal, and Zarucchi, American lily-of-the-valley.

78. Michaux's route to, and the specific location of the "Crab Tree" that he visited cannot be pinpointed with the notes in the journal. He was likely near either Crabtree Mountain, less than six miles in a straight line from Turkey Cove, or Crabtree Creek, a tributary of the North Toe River that winds between Sevenmile Ridge and Crabtree Mountain.

79. This is possibly an observation of the plant that Michaux would name *Azalea pilosa* (Rehder 1923, 2). We now recognize the plant as *Rhododendron pilosum* (Michx.) Craven, minniebush. The same descriptive phrase appears again on August 11, 1794, and this plant is likely the *Anonymos azaleoides* mentioned on March 24, 1796.

80. This plant is probably *Vaccinium erythrocarpum* Michx., bearberry (*Flora* 1:227).

81. *Cypripedium calceolaria* is now recognized as *C. parviflorum* Salisb., yellow lady's-slipper, and botanists today recognize two varieties: the larger is known as var. *pubescens*, and the smaller as var. *parviflorum*.

82. This description matches the needle pattern found on branches of *Tsuga caroliniana* Engelm., Carolina hemlock. *T. canadensis* (L.) Carrière, eastern hemlock, has all its needles in one flat plane.

83. Michaux's "*Spiraea tomentosa* 6 miles from Lincoln in Burke County, 3 miles from Charlotte" (IDC 66-5) is *S. tomentosa* L., steeplebush; his "*Spiraea aruncus* Carolina mountains" (IDC 66-15) is *Aruncus dioicus* (Walter) Fernald, goat's-beard.

84. *Podophyllum* with blue berries is *Diphylleia cymosa* Michx., umbrella-leaf (IDC 47-2 "Carolina mountains"; *Flora* 1:203, plates 19 and 20).

85. Michaux's "*Lonicera diervilla* Black Mountain" (IDC 26-18) is *Diervilla lonicera* Mill., northern bush-honeysuckle. Noting the specimens that Michaux collected, Silver (2003, 64) suggests that he climbed to an elevation above four thousand feet on this journey.

86. The *Sedum* is *Hylotelephium telephioides* (Michx.) H. Ohba, Allegheny live-for-ever (IDC 63-1, 63-2; in the *Flora* 1:277 as *Sedum telephioides*).

87. Here in the mountains, *Andromeda axillaris* is *Leucothoe fontanesiana* (Steud.) Sleumer, mountain doghobble; and *A. racemosa* is likely *Eubotrys recurvus* (Buckley) Britton, mountain fetterbush.

88. In 1777 Samuel Bright was one of the five commissioners appointed by the North Carolina legislature to lay off (survey) and mark a road across the mountains to the settlements on the western side. The route then became known as Bright's Trace. Finding desirable land along this route, Bright obtained a land grant in 1780 and soon became the first white settler in the Toe River Valley (Deyton 1947; Phifer [1977] 1982). William Wiseman and his wife, Mary Davenport, then obtained a grant near Bright. Mary's widowed brother Martin Davenport moved to the new neighborhood with his children at the close of the Revolutionary War (Deyton 1947; Vineyard 1997).

89. *Coccinea lutea* (red-yellow) and *flava* (pale yellow) are varying color forms of

*Rhododendron calendulaceum* (Michx.) Torr., flame azalea. Dried flowers persist on these shrubs in the Roan Highlands well into August (J. Donaldson, pers. comm.). Poindexter and Murrell (2008, 307) also reported late-flowering (after mid-July) populations of *R. calendulaceum* on nearby Mount Jefferson in Ashe County, North Carolina.

90. Michaux observed azaleas that were white and pink. Galle (1987, 75) credited Michaux with observing *Rhododendron arborescens* (Pursh) Torr., sweet azalea, in these mountains and noted that the flowers, although white, often have a pink or reddish blush.

91. Michaux's reference for the deciduous azaleas that he encountered in America was Linnaeus's *Azalea nudiflora*, pinxter flower, now recognized as *Rhododendron periclymenoides* (Michx.) Shinners. See Michaux's entry of April 26, 1787.

92. Michaux employed Martin Davenport (ca. 1739–1815) as his guide in these mountains in 1794, 1795, and 1796. Davenport, who sometimes worked for Col. Waightstill Avery, was a prominent local leader in addition to being an experienced woodsman and guide, the qualities that recommended him to Michaux.

93. These plants are in the aster family.

94. In traveling from Roan Mountain to Yellow Mountain (now named Big Yellow Mountain), Michaux followed a route across a series of open mountain summits called "balds" that has since been followed by thousands of hikers on the Appalachian Trail. The views from this ridgeline are spectacular in clear weather.

95. Grandfather Mountain towers dramatically above the nearby terrain, but it is not the highest mountain in the region. While it is 5,946 feet above sea level, several peaks in the vicinity of Roan Mountain and in the Black Mountains are higher.

96. Michaux was interested in all plants, including these small nonflowering plants. Here on Grandfather Mountain, among these "various mosses," he collected four of the twenty lichens, and seven of the fifty-seven mosses, described in his *Flora*, as well as the type specimen of *Bazzania tridenticulata* (Michx.) Trevis, liverwort. The latter, now in the cryptogram collection at the MNHN (MNHN-PC-PC0101824), indicates Grandfather Mountain on the herbarium sheet but not in his *Flora*.

97. On August 9 at the foot of the Black Mountains, Michaux used this binomial with an added question mark for *Tsuga caroliniana* Engelm., Carolina hemlock, while on other occasions he used it for *Picea rubens* Sarg., red spruce.

98. Michaux's climb to the summit of Grandfather Mountain, his subsequent singing of the new French national anthem, "La Marsellaise," and his shouts into the wind calling for the endurance of liberty, of France, and of America have become perhaps the best-remembered event of his career. On the two-hundredth anniversary of Michaux's climb, the late Charles Kuralt, a noted journalist, delivered a stirring speech at the site, honoring Michaux and describing the essential virtues of the French botanist for an audience estimated to be fifteen hundred people.

99. Another adventure not mentioned in the journal probably occurred during a visit with Davenport. François André Michaux reported that his father captured a large salamander about two feet long in the Doe River (F. A. Michaux [1805] 1904, 289). A French colleague at the MNHN named the species *Salamandra alleganiensis* from this specimen (Daudin 1803, 8:231); the modern name is *Cryptobranchus alleganiensis*, eastern hellbender. Harper (1940, 720–21) reported that the specimen had disappeared and that François André Michaux had erred in the name of the river where the salamander was found because he placed Davenport's plantation on the "Doe" River, Tennessee, when it actually was on the North Toe River, North Carolina.

100. Michaux traveled east by a different and more difficult route. Table Rock Mountain

(elevation 3,918 feet) and Hawksbill Mountain (elevation 4,020 feet) overlook the Linville River and are prominent peaks on the ridge that forms the eastern flank of Linville Gorge, an area that remains some of the most rugged country in eastern North America. On his previous journeys Michaux had passed to the south of this formidable landscape on the road from Morganton to Turkey Cove.

101. Thomas Nuttall also visited the same ridgetop in 1816 (he called it the "Catawba Ridge"), and while he also noted *Kalmia buxifolia* (P. J. Bergius) Gift, Kron, & Stevens, sand-myrtle, on the highest summits, he remarked as well on *Pinus pungens* Lamb., table mountain pines, and he made his own remarkable discovery of a new *Hudsonia* that he published as *H. montana* Nutt., mountain-golden heather (Graustein 1967, 109–11).

102. The 1790 US Census of North Carolina indicates that three heads of household named Parks (Benjamin, Larkin, and Thomas) owned property in this area between Table Rock and Morganton, where Michaux was traveling. A case can be made that Michaux might have visited any of these men because all lived within six miles of Table Rock. Recent mapping of the land grants in the area indicates that Larkin and Thomas lived closest to the mountain (Burke County land grant mapping data, North Carolina Room, Burke County Public Library, Morganton, NC).

103. Gen. Charles McDowell (1743–1813) was one of the illustrious fighting McDowells, a family of patriot leaders in the Revolutionary War who were also prominent in political affairs. The 1808 Price and Strother map indicates Gen. McDowell's plantation on the John's River, a tributary that enters the Catawba northeast of Morganton.

104. Dugger (1892) 1934, 263.

105. Each of Michaux's five journeys into or through western North Carolina included a visit with Col. Avery, whose plantation Swan Ponds was on the Catawba northwest of Morganton.

106. Catawba Springs in Lincoln County, named for the mineral springs there, was later a health resort (Hill 2001, 131).

107. Beatties Ford on the Catawba River is named for early settler John Beatty and is shown on the 1808 Price and Strother map. Lake Norman, a hydroelectric project, covers the ford today.

108. This description probably refers to *Trichostema dichotomum* L., common blue curls.

109. Salisbury is located near the historic Yadkin River Trading Ford and was a commercial center and court town founded when Rowan County was created in 1753. An extension of the "Great Wagon Road" south from Philadelphia into Virginia passed through Salisbury, and both the network of roads converging in Salisbury and Michaux's route from Morganton through Lincolnton across the Catawba River at Beatties Ford to Salisbury are indicated on the 1808 Price and Strother map.

110. After crossing the Yadkin River, Michaux traveled through the Uwharrie Mountains in modern Montgomery County, but his exact route has not been determined from his sparse journal account or the road network indicated on the 1808 Price and Strother map.

111. Coker (1943) described the interesting history of *Magnolia acuminata* var. *subcordata* (Spach) Dandy, yellow cucumber-tree, and located Michaux's report of the magnolia at the boundary between Montgomery and Moore Counties.

112. The location of the general store of William Martin (d. 1813) in southern Moore County (juncture of NC Highway 211 and county road 1241 east of Elberton) is indicated on the 1808 Price and Strother map, enabling us to trace Michaux's route toward Fayetteville.

113. Fayetteville is located at the head of navigation of the Cape Fear River. Named for the Marquis de Lafayette, it combined Campbelltown and Cross Creek, two older settlements

founded by Scottish Highlander emigrants. It became an important town and served as the capital of North Carolina from 1789 to 1793.

114. The French revolutionary army known as the "Army of the Sambre-et-Meuse" under Gen. Jean-Baptiste Jourdan scored a decisive victory over the Austrians led by Prince Frederick of Saxe-Coburg at Fleurus, near Charleroi (Belgium), on June 26, 1794.

115. Michaux retraced his route toward the Yadkin to the point where the road to McLaughlan Bridge joined from the south. This road junction is marked "W. R. McKay" on the 1808 Price and Strother map and may have indicated either of the two William McKays of the Fayette District, Richmond County, listed in the 1790 US Census of North Carolina.

116. The 1808 Price and Strother map shows the locations of Drowning Creek, Gum Swamp, and both of the bridges that Michaux mentions (which are only a few miles apart). Botanists continue to validate Michaux's work; in 2004, North Carolina Natural Heritage Program biologist Harry LeGrand found *Paronychia herniaroides* (Michx.) Nutt., Michaux's whitlow-wort, on Green Pond Bay Rim in Scotland County. The plant had not been reported in North Carolina since Michaux's 1794 journey (Weakley 2004).

117. *Comptonia peregrina* (L.) J. M. Coult., sweet-fern, is a rare plant in North Carolina today.

118. Michaux entered South Carolina, traveled across modern Marlboro County, and crossed the Great Pee Dee River into Darlington County. The modern town of Society Hill was then the community of Long Bluff.

119. Col. Lamuel Benton (1754–1818) was one of the most prominent men and largest landowners in the district. He served throughout the Revolutionary War, first as major, then colonel, of the Cheraws militia. After the war he served in a variety of state and local offices as well in the US House of Representatives for three terms. Senator Thomas Hart Benton (1782–1858) of Missouri was his nephew, and pioneering female author Jesse Benton Fremont (1824–1902) was his great niece (Gregg 1867, 386–93; Erwin and Rudisill 1976, 279–85).

120. The route south on September 27 can be followed on roads shown on the Darlington District map of the 1825 *Mills' Atlas*. Michaux traveled the first twelve miles from Long Bluff to Col. Benton's plantation Stony Hill, located near modern Mechanicsville, on what are today secondary roads. From Mechanicsville, he followed the route of SC Highway 35 to cross Black Creek on Williamson's Bridge. Based on his distance of twenty-two miles, Michaux would have encountered Black Swamp in the vicinity of Black Creek. He then continued south across Jeffries Creek, which flows through the modern city of Florence (not founded until the middle of the nineteenth century). Williamson's Bridge over Black Creek predated the American Revolution, and his host Col. Benton could have recommended this familiar route.

121. The roads shown on the Darlington District map in the 1825 *Mills' Atlas* and the distance of forty miles from Society Hill suggest that Michaux followed the general route of modern US Highway 52 south from his crossing of Jeffries Creek (in modern Florence), crossing the Lynches River where it bends north at modern Effingham in Florence County.

122. Michaux continued south along roads shown on the Williamsburg District map of the 1825 *Mills' Atlas* that follow the approximate route of modern US Highway 52, through the future site of Lake City and on to cross the Black River in the vicinity of Kingstree, the seat of Williamsburg County, but he did not mention that town.

123. Murray's Ferry on the Santee, about forty five road miles upstream from the river's mouth, and the road network that this ferry served are indicated on several period maps. Michaux followed a road on the north side of the Santee to reach Lenud's Ferry downstream.

124. The route via Strawberry Ferry over the western branch of the Cooper River in modern Berkley County provided the shortest and most direct route to Michaux's garden from

Lenud's Ferry on the Santee. Michaux also mentions this ferry on November 4–5, 1787, and on January 16, 1791.

125. Michaux's garden was located near the inn owned by Thomas Tims referred to as Ten Mile House.

126. RSAS, letters to Nils Collin, L-Ö.

127. "Mr. Lebours" is probably Alexandre Le Rebours (b. ca. 1763), a French intellectual then living in Philadelphia who may have been a tutor in the Bartram household. He was an active member of the APS and corresponded with Jefferson (Hallock, Hoffman, and Fry 2010, 346–47).

## Chapter Twelve

1. Drayton Hall plantation, although near Michaux's garden outside Charleston, was on the opposite side of the Ashley River, and visiting it required a ferry crossing or the use of other watercraft. Today the imposing eighteenth-century plantation house is a beautifully maintained site of the National Trust for Historic Preservation.

2. Orsolits 2003.

3. See note 5 in Chapt. 10.

4. Letouzey 1989; Spary 2000.

5. RSAS, letters to Nils Collin, L-Ö.

6. Thouin had left Paris for Belgium and Holland (from September 1794 to July 1795) because he had been appointed as one of four commissioners to follow the French army as it occupied these countries. He was to report on agriculture, education, art, manufacturing, science, and health. He was also to select and send back to France valuable paintings and books as well as natural history collections (Letouzey 1989, 393–448).

7. MNIIN, Archives Thouin, transcription by Régis Pluchet. Michaux now often gives the date according to the new republican calendar, which was in use from late 1792 to 1805. The twelve months, whose names were based on the weather and agriculture, were each divided into three ten-day weeks, with a few extra days at the end of the year. It was an attempt to de-Christianize the calendar.

8. Letter was received by André Thouin, 12 germinal year 3 (October 1, 1795). MNHN, Archives Thouin, transcription by Régis Pluchet. An interesting facet of André Michaux's letter to Thouin that is not apparent in English translation is Michaux's use of the familiar form of *you* in the original French. Michaux is not known to have used this form in his earlier letters to Thouin.

9. Michaux was in Philadelphia at the time of Genet's visit.

10. Michaux's original instructions from Count d'Angiviller in 1785 directed that more than 90 percent of his shipments be sent to the royal nurseries at Rambouillet. The Jardin du Roi in Paris received only a small fraction of the plant material that Michaux shipped from 1785 to 1791.

11. Michaux writes of concrete observations. Bridges were often lacking in America and virtually absent west of the Appalachian Mountains. In his letter to Thouin dated April 13, 1796, Michaux remarked that sometimes he had to build rafts for himself and swim his horses across rivers.

12. This plant was probably *Heuchera americana* L., American alumroot (IDC 39-7 "Illinois, Carolina").

13. The Charleston District map in the 1825 *Mills' Atlas* indicates that Forty-Five Mile House Tavern was located about halfway between Moncks Corner and Nelson's Ferry on the Santee River. Michaux followed the route north through Moncks Corner that he had followed

in 1789 and 1794. The area around Eutaw Springs and the ferry is now submerged beneath the waters of Lake Marion, a hydroelectric lake.

14. The new tree from the Santee is *Planera aquatica* J. F. Gmel., planer-tree, water-elm; six months later on the banks of the Tennessee River, Michaux called the tree *Anonymos ulmoides*.

15. Michaux's herbarium offers some possibilities for identifying these plants. His *Phlox subulata* "High Hills of Santee" (IDC 34-9) appears to be *Phlox nivalis* Lodd. ex Sweet, pine-land phlox. His *Phlox aristata* "River Santee" (IDC 34-3) also has "*Phlox subulata* Walt." written on the sheet; it is likely *Phlox pilosa* L., downy phlox, and is not limited to the sandhills.

16. Michaux's route from Nelson's Ferry to Stateburg can be followed on the Sumter District map in the 1825 *Mills' Atlas*, which also indicates the relief for the High Hills of Santee.

17. This is a reference to *Kalmia cuneata* Michx., white wicky, previously noted on July 18, 1794. See chapter 11, note 60.

18. The main road north, described on the Kershaw District map in the 1825 *Mills' Atlas* as the Salisbury Road, ran past Hanging Rock and Hanging Rock Creek, where the road forked, providing three different routes north through the Lancaster District. The Flat Rock here has sometimes been mistaken for the town of the same name in the Blue Ridge Mountains of North Carolina. Uttal (1984) notes that this Kershaw County, South Carolina, Flat Rock is the type location of *Sedum pusillum* Michx., Puck's orphine, granite stonecrop (IDC 63-4; *Flora* 1:276).

19. This is John May, a merchant "lately of Charleston" who purchased 150 acres of land on Cane Creek, Lancaster County, in 1792 (Holcomb 1981, 48). In all likelihood this was not a visit with a stranger. John May is listed in the 1790 US Census of South Carolina as a resident of St. James Goose Creek Parish in the Charleston District, and Mr. May is mentioned in Michaux's letter to Consul Mangourit of January 15, 1793.

20. This is Joseph Lee (1742–1814), a local man of substance who was justice of the peace of the Lancaster District. Either Lee's eighteen-year-old son Thomas, who would be sheriff of the Lancaster District from 1804 to 1809, or his sixteen-year-old son William, helped Michaux search for the lost horse (Holcomb 1981; L. Pettus, pers. comm.).

21. This crossing of the Catawba River between Lancaster and Chester Counties is mentioned in several accounts of the Revolutionary War. It was later the site of the Landsford Canal, an early nineteenth-century project to bypass the shoals located there. Today, it is the site of Landsford Canal State Park.

22. The South Carolina legislature authorized McCleneghan's Ferry, located a few miles upstream from Land's Ford (which later became Landsford), in 1795 (McCord 1841).

23. Col. William Hill (1741–1816) served in the Revolutionary War and was a partner in the Aera Ironworks, a gristmill, and a sawmill, all located on Alison Creek in York County, South Carolina. This complex of businesses was the largest and most important industrial enterprise in the Carolina Piedmont in the late eighteenth century. During the Revolutionary War, the ironworks had been destroyed by British and Loyalist regulars, but it was subsequently rebuilt and was functioning again at the time of Michaux's visit (Hill and Salley 1921; Cowan 1987; Buchanan 1997, 112–15).

24. Armstrong Ford, a crossing of the South Fork of the Catawba in modern Gaston County, North Carolina, is just south of the modern town of Belmont. Between Hill's Ironworks and Armstrong Ford, Michaux passed the future site of the Daniel Stowe Botanical Garden.

25. When Michaux visited, Bennet (Benjamin) Smith owned property near the confluence of Hoyle and Little Hoyle Creeks in what is now Gaston County. Bennet Smith later became a lawyer, sold this property to his neighbor Peter Smith (whom Michaux had visited in 1789),

and moved first to Lincolnton and later to Morganton, where he took an active part in public affairs and served as state's sttorney in Burke County from about 1798 until 1805 (Phifer [1977] 1982, 143). The magnolia at Smith's was *Magnolia macrophylla* Michx., bigleaf magnolia. The botanist first encountered this species in this vicinity in 1789 and would collect live plants here in 1796.

26. This is the same route that Michaux followed through this area in 1789 and 1794. Each journey goes through Lincolnton (Lincoln Courthouse). Observing Michaux's distances and direction to find the correct location of the "good man Wilson," we note that he is most likely to have been James Wilson Sr., who is listed in the 1790 US Census of North Carolina in the Morgan District of Lincoln County as the head of a large family, or possibly his son James Jr., also a prosperous farmer. Old deed books indicate that father and son had extensive landholdings near present Maiden in modern Catawba County (Pruitt 1988). This Wilson was not Zaccheus Wilson, as suggested by Thwaites (Michaux 1904). Zaccheus Wilson lived in the part of old Mecklenburg that is now Cabarrus County until he moved to Tennessee in 1796.

27. Morganton was also called Burke Courthouse.

28. John Wagely owned land on both sides of the Linville River in Burke County, near where the road leading west into the mountains crossed this river, and his home was within one-half mile of the river. The 1790 US Census of North Carolina reported that his household consisted of two men over the age of sixteen and four women. Wagely lived in a majestic landscape near the base of Shortoff Mountain, which rises precipitously eighteen hundred feet above the Linville River. This is Michaux's first journal reference to the mountain magnolia as *Magnolia auriculata*, William Bartram's name for this tree. At other times Michaux used the names *M. cordata* and *M. hastata* for this tree, which we recognize today as *M. fraseri* Walter, Fraser magnolia, mountain magnolia (Phifer [1977] 1982; Newsome 1934; USGS, "Ashford Quadrangle, NC" 7.5 minute series topographic [map] 1:24,000 [Reston, VA: USGS, 1956]).

29. Thomas Young (1747–1829) and his wife, Naomi, lived in North Cove at the foot of the Blue Ridge on the North Fork of the Catawba River. Young was not part of the Turkey Cove settlement; he lived nearby in the more open valley slightly to the north of Turkey Cove and beyond the point where Pepper Creek enters the North Fork of the Catawba (Young 2003). From Young's it was possible to ascend the Blue Ridge either by going south through Turkey Cove or by following another steep trail uphill along Pepper Creek. In 2006 Young's descendants erected a monument honoring their ancestors Thomas and Naomi in North Cove. The inscription notes that the Youngs hosted both André Michaux and Bishop Francis Asbury (P. D. Young, pers. comm.).

30. This is a note about routes and distances between settlers that Michaux visited more than once. The botanist knew this area by 1795, having visited twice in 1789 and having extensively explored the area with Martin Davenport in 1794.

31. The stream that runs between these mountains is Roaring Creek, not the North Toe River.

32. Here he refers to several species not individually named; they are likely some of the same species that he had seen in fruit and recorded in his journal for August 1 and 2, 1794.

33. *Podophyllum diphyllum* was Linnaeus's name for twinleaf, now known as *Jeffersonia diphylla* (L.) Pers. It is not found in this area today, perhaps due to the soil's high acidity. In Virginia on July 3, 1789, Michaux collected live plants of the species, which he shared with the Bartrams (see chapter 5, note 21). Note that in the list of plants for May 25, 1795 (in the present chapter), Michaux uses the new genus name for the first time. Michaux's *Podophyllum umbellatum* is *P. peltatum* L., may-apple, and is commonly found in this area today.

34. Humpback Mountain is on the northern McDowell–eastern Avery County line, on the crest of the Blue Ridge.

35. *Azalea lutea* is *Rhododendron calendulaceum* (Michx.) Torr., flame azalea. The azalea found along rivers with brilliant red flowers is *R. flammeum* (Michx.) Sarg., Oconee azalea, and the one with white flowers is *R. arborescens* (Pursh) Torr., sweet azalea.

36. John Miller lived on the upper reaches of the Doe River in the vicinity of modern Roan Mountain, Tennessee. He was the blacksmith who shod horses of the Overmountain Men when they stopped at Shelving Rock (sometimes called Resting Place) the first night after they left Sycamore Shoals en route to the Battle of King's Mountain in 1780 (Alderman 1986, 85).

37. Tennessee's Doe River, not to be confused with the nearby North Toe River in North Carolina, empties into the Watauga River near Landon Carter's house in present Elizabethton (see chapter 5, note 16). In January 1795, land agent John Brown followed the route along the Doe in the opposite direction and reported an experience similar to that reported by Michaux (Newsome 1934, 289).

38. Col. John Tipton (1730–1813) lived on Sinking Creek, now in Johnson City, Tennessee. A strong-minded and politically active natural leader, Tipton served in the legislatures of Virginia, North Carolina, and Tennessee. He was in the thick of the conflict that developed around the attempt to form the state of Franklin from North Carolina's western lands in the 1780s. His chief adversary was Col. John Sevier (1745–1815), one of the heroes of the Battle of King's Mountain and the struggle against the Cherokee, who led the state of Franklin proponents. Tipton headed the group that sought to enforce the laws of North Carolina after the state reneged on its early plan to cede the region and opposed the state of Franklin movement. The dispute eventually led to bloodshed before the crisis dissipated with the ratification of the US Constitution and North Carolina's final cession of her western lands (Tennessee) to the federal government. The region that had attempted to form the state of Franklin then became part of the awkwardly named "Territory of the United States south of the river Ohio," with its own governor, William Blount (1749–1800). Blount united the widely separated settlements in present eastern and middle Tennessee, and the area achieved statehood in 1796. The Tipton homesite, today Tipton-Haynes State Historic Site, has a Tennessee historical marker and a museum display remembering Michaux's visits with Tipton (*DAB*, s.v. "Tipton, John"; Williams [1924] 1993).

39. Jonesborough, the oldest town in the state, was established in 1779 as the seat of Washington County, Tennessee's first county, which was organized when the area was still a part of North Carolina (Forester 1996, 105).

40. Anthony Moore was an early settler in this region. He appears on the first tax list for Greene County in 1783, and he signed the state of Franklin petition to the North Carolina legislature in 1787 (Alderman 1986, 242, 248).

41. "Earnest" probably refers to Henry Earnest (ca. 1732–1809), originally from Switzerland, who settled on the Nolichucky River in 1777 in what would become Greene County. His farm, now known as Elmwood Farm, is the oldest Tennessee Century Farm and is listed on the National Register of Historic Places; Andrew Fox's farm was nearby on the Nolichucky (Griffey 2000, 179–80, 192; West 1998, 385–86).

42. Greeneville, the seat of Greene County, was founded in 1783 and named for Gen. Nathanael Greene. It was later the home of US president Andrew Johnson (1808–75). Michaux followed the route of present-day US Highway 11E from Jonesborough to Greeneville.

43. With a choice of routes to Knoxville at Greeneville, Michaux chose the right fork in order to approach Knoxville through the Holston River Valley and take advantage of McBee's Ferry over the Holston.

44. Bull's Gap is a gap in Bay's Mountain, northwest of Greeneville. Today, US Highway 11E goes west from Greenville through Bull's Gap before turning southwest toward Knoxville.

45. Christopher Haynes was operating the ironworks at Mossy Creek (now Jefferson City) in 1795 (Jefferson County Genealogical Society 1996).

46. It seems likely that what Michaux found was hematite in its red form. Commonly referred to as red ochre, this substance has been widely used since ancient times for facial paint and other such products.

47. Isaac Newman (1755–1832), a Revolutionary War veteran from Guilford County, North Carolina, was residing at Mossy Creek (modern Jefferson City) in the early 1790s. His early land grants were slow to be issued by North Carolina, and the documents indicate Hawkins, the original county, and not Jefferson, which was formed from Hawkins and Greene in 1792 (Griffey 2000, 314). One of Isaac Newman's descendants was a founder of Carson-Newman University in Jefferson City.

48. Michaux was correct that there was a mineral deposit. A substantial deposit of zinc was later mined near the old mill at Mossy Creek (Perdue 1912).

49. Revolutionary War veteran Col. James King (1752–1825) was an important early East Tennessee pioneer businessman and a major landowner with several plantations, who once purchased the land that is now the city of Bristol. He was the leading partner in the pioneer ironworks located near Bristol, and he also had other manufacturing and mercantile interests. In 1794 Col. King purchased the land at Strawberry Plains on the Holston River across from McBee's Ferry, and he lived there before selling it to his brothers-in-law in 1797. He was an active community leader in early Knoxville, and in 1807 he became a charter trustee of East Tennessee College, now the University of Tennessee (Phillips 1999, 6–8, 247; University of Tennessee 1898, 247).

50. McBee's Ferry over the Holston River was established by William McBee in the 1790s and is mentioned in the accounts of several early travelers. The site is near the railroad bridge at Strawberry Plains in Jefferson County (Forester 1996, 116).

51. Knoxville was founded in 1786 by Capt. James White, a Revolutionary War veteran, and first called White's Fort. It was soon renamed for Gen. Henry Knox, Washington's artillery commander during the Revolutionary War and later secretary of war. The town was located four miles downstream from the junction where the Holston and French Broad Rivers come together to form the Tennessee River. Territorial governor William Blount resided in Knoxville, and Michaux met the governor at this time, but he failed to mention their meeting in his journal. Governor Blount recommended that Michaux visit Col. Robert Hays when in Nashville; see Michaux's journal entry for February 16, 1796 (in chapter 13). Blount's Knoxville home is a Tennessee historic site today.

52. Both here in Knoxville and again on June 15–17 in the Nashville area and elsewhere, Michaux's plant lists include several oaks. Although Michaux later published a monograph on North American oaks and significantly advanced botanical understanding of the genus, some of the species names used in his journal are not repeated in the monograph.

53. *Quercus alba pinnatifida* is one of the two varieties of *Q. alba* L. that Michaux (1801) described in his monograph *Histoire des chênes de l'Amérique*.

54. *Nyssa aquatica* L., water tupelo, is not found near Knoxville today, but it was a species that Michaux had observed regularly and examined closely in South Carolina, and he could be expected to identify it correctly. Michaux's plant list at the end of his letter to Count d'Angiviller of December 26, 1786, includes a comparison of *N. sylvatica* Marshall and *N. aquatica* L. (APS MSS film 330).

55. At this time most of the land between Knoxville and Nashville belonged to the Cherokee

and was known as the Wilderness. No white settlements were permitted, and the Cherokee also objected to the road through their land and frequently attacked passing travelers.

56. In 1787 Col. David Campbell (1753–1832) established Campbell's Station, a fortified outpost on the main road between Knoxville and the Clinch River. He had been a captain of the Virginia militia at the Battle of King's Mountain, and he later served in the legislatures of North Carolina and Tennessee and was a partner in the mercantile business with Charles McClung of Knoxville (Forester 1996, 18; Williams 1928, 316; West 1998, 118).

57. This spot is now the town of Kingston, located at the junction of the Clinch and Tennessee Rivers. The militia-garrisoned fortification that Michaux visited, the Southwest Point Blockhouse, was soon upgraded to a fort known as Fort Southwest Point, which was the base for the regular army in the region for the next decade. After extensive archaeology, the fort has been reconstructed as part of a city park (West 1998, 330–31).

58. Although any specific route from the Clinch to the Cumberland is speculative, Michaux's path between June 6 and June 10 likely followed the Avery Trace. This route is approximated today by US Highway 70 from Kingston to the vicinity of Crossville. There it follows US Highway 70N to the vicinity of Monterey and Cookeville, where it swings north to cross the Cumberland River in the vicinity of modern Gainesboro. His notes on June 9 suggest that he may have passed through the mountainous area northeast of modern Cookeville. Once across the Cumberland River, he followed trails westward approximated by modern Routes 85 and 25. This analysis conforms to the route indicated on a period map by a dotted line leading from near the confluence of the Clinch and Tennessee Rivers northwest to the mouth of Fleen's Creek at the Cumberland River then on the north side of the Cumberland to the vicinity of Nashville; see Gen. Daniel Smith's "Map of the Tenassee Government formerly Part of North Carolina," published in Philadelphia in 1794 by Matthew Carey (Wells 1976). For the modern approximation of the route, see the Avery Trace Association's brochure and map of the Avery Trace, titled "Trace 200 Years of Tennessee History in 200 Miles . . . Historic Avery Trace, Blazed 1788."

59. This is an observation of *Magnolia macrophylla* Michx., bigleaf magnolia, in flower. Although Michaux was familiar with this new species from earlier observations in North Carolina, the type specimen was likely collected here.

60. Bledsoe's Station, now known as Castalian Springs, was established by Anthony Bledsoe (1753–88), an important pioneer of Middle Tennessee. He and several members of his family were casualties of the war between the settlers and the Native Americans (Forester 1996, 258–59; West 1998, 71).

61. Col. James Winchester (1752–1826) was a Revolutionary War veteran, later a general, from Maryland who settled on Bledsoe's Creek in 1785. He was both a military leader and a successful businessman. Later, in partnership with Andrew Jackson and John Overton, he developed the tract on the Mississippi River that became Memphis. When André Michaux visited, Winchester was living in a log home, but he was soon to build a mansion. Visiting six years later, François André Michaux commented on the grandeur of "Cragfont." This then-uncompleted stone house in the wilderness is now a Tennessee historic site, restored and open to the public (*DAB*, s.v. "Winchester, James").

62. Andrew Jackson (1767–1845), later seventh president of the United States, at this time owned a plantation called Poplar Grove on a bend in the Cumberland River east of Nashville. The site, in an industrial area today, does not have a historical marker and is the most obscure of Jackson's Tennessee residences. Jackson's correspondence indicates that he—who often traveled swiftly without escort through the Wilderness—was also traveling west at this time, two to three days behind Michaux. It is not known whether Jackson, traveling more quickly,

had reached Poplar Grove when Michaux visited on June 15, 1795. They did meet, though if not on this occasion, then at a later time, since "Mr. Jackson, attorney" appears on Michaux's 1796 list of persons known in Nashville, found with his journal entries for January 19, 1796, but omitted from the 1889 transcription of his journal (Jackson, Smith, Owsley, and Moser 1980, 1:59–62; Horn 1961).

63. This is a reference to *Quercus macrocarpa* Michx., bur oak, mossycup oak (IDC 117-21 "overcup white oak, *chene frise*, Illinois, Cumberland, Kentucky").

64. Nashville, first called French Lick and named for Revolutionary War general Francis Nash, was founded in 1779–80 by settlers from the Watauga settlements. Located far from other white settlements that might have regularly given them aid, the Cumberland settlers waged a long, bloody, but successful war to wrest the area from the Native Americans. By fall 1795 when the first census of the area was taken, the population of Nashville and surrounding Davidson County had grown to more than thirty-six hundred (Albright 1909).

65. One of the most important men in early Tennessee, Virginia native and veteran of the wars with the Native Americans and the Revolutionary War, Daniel Smith (1745–1818) moved to the Cumberland settlements in 1783. There, he continued the remarkable career in military and public service he had begun in East Tennessee. He became brigadier general of militia and a delegate to the North Carolina convention that ratified the US Constitution. At the time that Michaux visited in 1795, Smith was serving as territorial secretary and acting governor during Governor Blount's absences from the territory. Smith later helped draw up the state constitution of Tennessee and served twice as US senator. Smith County, Tennessee, was named in his honor in 1799. When Michaux and Smith met, the former was likely aware that the latter had published a description and a map of the territory of Tennessee. In any case, François André Michaux, who visited Smith in 1802, definitely discussed Smith's map with him and observed that Smith was also an experimental chemist who owned English translations of the chemistry books of Lavoisier and Fourcroy (Wells 1976, 123–24; F. A. Michaux [1805] 1904, 255–56; *DAB*, s.v. "Smith, Daniel").

66. James Robertson (1742–1814) was an early leader in both the Watauga (East Tennessee) and Cumberland settlements (Middle Tennessee); he led the first group of permanent settlers to what is now Nashville from the Watauga settlement in 1779. There he was the able and vigorous leader of the new settlement during its difficult early years. He served in the North Carolina legislature before the state ceded its western lands, then he worked closely with Governor Blount and later with Governor Sevier negotiating treaties with the Native Americans, and he served as one of the territory's brigadier generals (*DAB*, s.v. "Robertson, James").

67. Capt. John Gordon (1763–1819) was a noted frontier fighter who helped defend the Nashville settlement during its early days. He was made captain of the mounted infantry of the Davidson County Regiment in 1793, and in 1795 he was justice of the peace of Davidson County. He permanently settled south of Nashville, where he continued to serve in the militia and is especially remembered for his service leading a company of scouts in the Creek War (Ramsey [1853] 1967, 604–7, 612; Carter 1936, 4:397, 4:468).

68. This is probably George M. Deaderick (ca. 1756–1816), a wealthy merchant, banker, and real estate dealer who had arrived in Nashville from Virginia about 1788. He was a close friend and associate of Andrew Jackson, and he was said to be the wealthiest man in Nashville (West 1998, 238).

69. Dr. James White (1749–1809) was trained in divinity, law, and medicine, an unusual background for the Tennessee frontier. He was a native of Pennsylvania, but he received his early education in France. In Tennessee he was active in political affairs and practiced both law and medicine, before moving to Louisiana, where he became a judge. His son became a

five-term US congressman, and his grandson chief justice of the US Supreme Court. He is sometimes confused with Gen. James White, founder of Knoxville, but no connection between them is known (Bloom 1995).

70. Rev. Thomas Craighead (ca. 1750–1825), Nashville's first minister, was a graduate of Princeton and the son of the fiery Presbyterian minister Alexander Craighead (d. 1766). In 1780 Rev. Craighead married Elizabeth Brown, sister of the influential John Brown, later US senator from Kentucky, and James Brown, secretary of state of Kentucky. When Michaux traveled to Kentucky in 1793, he carried a letter of introduction to Craighead. The Craigheads had moved to the vicinity of Nashville about 1785. In Nashville, Craighead established a Presbyterian church called Spring Hill as well as a school (West 1998, 216).

71. This plant is possibly *Carya cordiformis* (Wangenh.) K. Koch, bitternut hickory.

72. *Rhamnus alaternus latifolius* is probably *Frangula caroliniana* (Walter) A. Gray, Carolina buckthorn.

73. Small, short-tailed flycatchers with eye rings and wing bars are now in the genus *Empidonax*.

74. Many European, Asiatic, and African woodpeckers are classified in the genus *Picus*. There is no species of American woodpecker classified in the genus *Picus* today, although there were some in Michaux's era.

75. Anthony Sharpe (d. 1812), an officer of the Continental Line in the Revolutionary War, settled about forty miles northeast of Nashville on the Red River, where he lived until 1809, when he moved near the town of Franklin, south of Nashville. He was the local commander in charge of all Sumner County's militia units, forts, and stations during the desperate fighting of 1792 (Pension Application of Anthony Sharpe, dated September 2, 1839, available at persi. heritagequestonline.com; Carter 1936, 358; Ramsey [1853] 1967, 565). A few months after André Michaux's visit, Louis-Philippe, the future king of France (1830–48), then evading the French Revolution with his brothers, also reported staying at Maj. Sharpe's, which he said was a good inn (Louis-Philippe 1977, 110). François André Michaux also stayed there in 1802, remarking that he found "every accommodation" at this inn (F. A. Michaux [1805] 1904, 217).

76. Andrew McFaddin's (McFadden) Station and Ferry at the site of modern Bowling Green, Kentucky, are believed to have been established about 1785. His station was located about one-half mile from the confluence of Drake's Creek and the Big Barren River on a bluff near a spring on the north side of the river (Wells 2002, 1665).

77. Although Michaux's account lacks details, his route from the Big Barren River to the Little Barren River probably followed the principal route of the Cumberland Trace northeast through sections of four modern Kentucky counties: Warren, Edmonson, Barren, and Hart (Settle 1963, 20–21). He traveled quickly and surely through this dry, relatively level landscape in only two days to cross into Green County at the Little Barren River and continue across the Green River near modern Greensburg (a distance of more than fifty miles on the route of the trace). Seven years later, his son also passed this way, and he described becoming lost crossing the barrens (F. A. Michaux [1805] 1904, 214–15).

78. Continuing along the route of the Cumberland Trace, Michaux traveled northeast across Green and Taylor Counties, then down Muldraugh Hill into Marion County, where he crossed the Rolling Fork at present-day New Market, and then, at modern Lebanon, he turned almost due east toward Danville (Settle 1963, 20–21).

79. When Michaux wrote that it was 117 miles from Nashville to Danville, he meant the distance from McFaddin's Station on the Big Barren River (Bowling Green) to Danville. His mileages between those two points total 114 miles, and when his "3 miles from the creek . . ." is added, the total is 117 miles. When all his mileage entries from Nashville to Danville are totaled, the result is 173 miles.

80. These animals would have been *Tamias striatus*, the eastern chipmunk; these Michaux specimens are now lost, but the animal was described as *Scurpus striatus* by E. Geoffroy Saint-Hilaire in his 1803 *Catalogue des mammifères du Muséum national d'histoire naturelle*.

81. Michaux remained in Danville from June 27 until July 15, but he made only one entry in his journal between July 5 and July 16. He placed his note about dining with Governor Isaac Shelby, the only person mentioned by name, at the bottom of a page that is otherwise blank, and he also left the top third of the following page blank. It seems that his political activities on behalf of Genet two years earlier did not make him unwelcome in Kentucky.

82. This entry possibly refers to the James Standeford (Standiford) who, with Lewis Fields and Isaac Hornbeck, trustees of the town of Newtown in Jefferson County (located near Mann's Lick), offered lots for sale in Newtown on February 7, 1795 (Green 1983, 117).

83. Michaux never explained how he shipped his baggage to Kentucky in 1795; perhaps most of it was already stored there after his 1793 journey. Ten days before leaving on his mission for Genet in July 1793, he shipped baggage weighing 560 pounds from Philadelphia to Pittsburgh (see chapter 10, note 6). When he returned to Philadelphia by way of the dangerous Wilderness Road in November 1793 to confer with Genet, he could not carry this heavy baggage with him and left most of it behind. He expected to be returning to Kentucky soon.

84. Salt was a critical and difficult item to obtain in Kentucky, and salt production at Mann's Lick was an important enterprise (Threlkel 1927). Concerning the fossils, see also chapter 10, note 47.

85. *Quercus cerroides* is a provisional name for *Q. macrocarpa* Michx., bur oak, mossycup oak, not (as some botanists have believed) a form of *Q. alba* L., white oak. On October 9, 1795, also using the binomial *Q. cerroides*, Michaux observed that it is "known by the French as curly oak [*chêne frisé*], by the Americans [as] overcup white oak." These are vernacular names that he provides for *Q. macrocarpa* in *Histoire des chênes de l'Amérique* (1801) and on his herbarium specimen. The French description *frisé* [curly] refers to the characteristic fringed acorn cup of the species.

86. This plant is probably *Hibiscus laevis* All., smooth rose-mallow.

87. See Fisk and Krings 2009, 739–40.

88. Michael Lacassagne (ca. 1750–97) was a wealthy, well-educated businessman of French origin living in Louisville who had commercial contacts in New Orleans and Philadelphia. He served as a delegate to the Kentucky convention of 1787 and as the city's first postmaster. Lacassagne's home, a French-style cottage with a veranda on three sides featuring a beautiful garden, became a popular rest stop for visitors to Louisville (Burnett 1976, 7; Dupre 1941a, 86; Kleber 2001, 496). Linklater (2009, 118–21) observed that Lacassagne could be absolutely ruthless in his business dealings and proposed that it was his relentless pressure on Gen. James Wilkinson for repayment of debts that helped to nudge Wilkinson from his early role as a political friend of Spain in Kentucky into one of outright treason by becoming a well-paid Spanish spy at the time that he was brigadier general in the US Army. One of the letters that Michaux carried on his mission for Genet in 1793 was addressed to Wilkinson. Michaux's journal does not indicate whether he delivered the letter.

89. This plant is now known as *Acer saccharinum* L., silver maple.

90. The herbarium specimens labeled as *Juglans alba* (IDC 114-6 and IDC 115-2) have five leaflets and appear to be *Carya ovata* (Mill.) K. Koch, common shagbark hickory.

91. Clarksville, Indiana, was a small (forty residents in 1793) community on a land grant given to George Rogers Clark (Madison 1986, 29). Michaux traveled across southern Indiana from Clarksville to Vincennes along the route called the Buffalo Trace or sometimes the Vincennes Trace (or other local names); it was the only direct land route of this period. The trace ran west from Clarksville through sections of Clark, Floyd, Harrison, Crawford,

Washington, Orange, Dubois, Pike, and Knox Counties to Vincennes, where it crossed the Wabash River and continued into Illinois. Unlike the later stagecoach route, the old trace traversed the higher land between the East Fork of the White River and the Patoka River, approaching Vincennes from the southeast (Wilson and Thornburgh 1946). Like the wilderness areas of Tennessee and Kentucky, the Buffalo Trace in southern Indiana was a particularly dangerous area for travelers, and Michaux traveled it in company with other travelers who are not named.

92. The river that Michaux crossed twenty miles before Vincennes, Indiana, was the White River, a major tributary of the Wabash. Michaux's likely crossing point was the ford located approximately six miles west of the junction of the two forks of the White River near modern Petersburg, Pike County. A state historical marker for the Buffalo Trace and the ford is located near where IN Highway 61 crosses the White River today. In 1796, Vincennes was a compact village of about fifty houses located on the east bank of the Wabash River and surrounded by agricultural fields on both sides of the river. The settlement had begun in the early 1730s as a fortified trading post for the fur trade, and it evolved into an agricultural settlement (Ekberg 1998, 82–85). Vincennes had been captured, lost, and dramatically recaptured by George Rogers Clark in 1778–79. A significant part of Clark's military reputation rested on his brilliantly conducted campaign to recapture Vincennes from the British.

93. Michaux's "*Verbena rigens* Illinois, Post Vincennes, Kaskaskia" (IDC 77-10) is *V. hastata* L., common vervain, blue vervain.

94. Michaux's "*Verbena bracteosa* Nashville, Vincennes, Kaskaskia" (IDC 77-17) is recognized today as *V. bracteata* Lag. & Rodr., prostrate vervain, big-bracted vervain.

95. *Silphium connatum* L. is not found in this region. This plant is probably *S. integrifolium* Michx., prairie rosinweed.

96. *Holcus*? is possibly *Phalaris arundinacea* L., reed canary-grass.

97. *Quercus latifolia*, shingle oak or ram's oak, is *Q. imbricaria* Michx., shingle oak. The range given in Michaux's *Histoire des chênes* (1801) specifically mentions that the species occurs along the Wabash River.

98. *Trifolium*? *pentandrum* "larger" is probably *Orobexilum onobrychis* (Nutt.) Rydb., lance-leaf scurfpea; *Trifolium*? *pentandrum* "flowers purple" is probably *Orobexilum pedunculatum* (Mill.) Rydb., Sampson's snakeroot.

99. C. S. Sargent (1889), when annotating the 1889 transcription of the Michaux journals, believed that Michaux intended to write *Spigelia marilandica* instead of *Sanicula marilandica*. The roots of both these herbs were used by Native Americans to make medicinal teas (Foster and Duke 1990, 62, 148). *Sanicula* is the more widespread and was used as a treatment for veneral disease.

100. Michaux's route from Vincennes to Kaskaskia is indicated on Thomas Hutchins and T. Cheevers's *A New Map of the Western Parts of Virginia, Pennsylvania, Maryland and North Carolina; Comprehending the River Ohio, and All the Rivers, Which Fall into It* . . . (London: 1778). This route did not later develop into a road. The small river where Michaux camped is the Embarras River in modern Lawrence County, Illinois; it is identified as the Troublesome River on Hutchins and Cheevers's map.

101. *Gerardia auriculata* Michx. (IDC 78-16, "prairies of Illinois, along the Mississippi") is recognized today as *Agalinis auriculata* (Michx.) S. F. Blake, earleaf foxglove. This is the type location of this species (Uttal 1984, 42).

102. The town of Kaskaskia and the river take their name from the Kaskaskia, one of the tribes of the Illinois Confederacy living in the area when French explorers first visited in the late seventeenth century. The settlement began to develop around a French Jesuit mission

established in 1703. Eventually Kaskaskia became the territorial capital and then the first state capital of Illinois. A change in the course of the Mississippi River later destroyed the town (Belting 1948, 8, 10).

103. Michaux's poor first impression of the French of Illinois echoes the observations of the area's residents by the Spanish official Don Pedro Piernas a quarter of a century earlier (see Houck 1909, 1:71). While recognizing the problems during the time of Michaux's visit and earlier, modern historians have a much more positive view of the French residents of the area, see notes 105 and 106 in this chapter.

104. St. Philippe was founded in 1723 by Philippe Renaut (Renault), who mined lead in the rich deposits on the west side of the Mississippi. The settlement of Prairie du Rocher followed a few miles to the southeast in the 1730s (Ekberg 1998, 35).

105. The villages mentioned here and the road linking them are shown on Thomas Hutchins's (1730–89) inset map "Plan of the Several Villages in the Illinois Country" in his book *A Topographical Description of Virginia, Pennsylvania, Maryland, and North Carolina*, published in London in 1778. The compact area of only a few hundred square miles encompassing these villages lay along the eastern side of the Mississippi from north of Cahokia south to the Kaskaskia River and came to be called the American Bottom. It was one of the most fertile strips in the whole of the Mississippi Valley. With wheat as its main crop, it was the breadbasket of the French colonies in America, and its produce was shipped to other French colonies down the Mississippi and to Canada and the Caribbean (Belting 1948, 10–11). This rich land and the long distance from colonial administrative centers enabled the French settlers to enjoy a high standard of living; they behaved more like New Englanders than like the other French colonials in North America. There were few vestiges of feudalism so common elsewhere, land ownership was widespread, and there was considerable social mobility. The settlers elected clergymen and militia officers and had a sense of self-worth and assertiveness not found among other French colonials (Davis 1998, 48–50).

106. Cahokia, across the Mississippi from Saint Louis, was the oldest and most stable French settlement in the Illinois country (Ekberg 1998, 33). Had Michaux arrived at Cahokia before he reached Kaskaskia, his first impression of the inhabitants might have been different. The governmental changes from French to British and then to American control led to a period of lawlessness after the American Revolution that was more pronounced in Kaskaskia than in Cahokia, where the courts continued to function. The British never replaced the French civil administration in the Illinois country with their own, keeping the area under a military government, and US control came slowly. No ordered civil government was set up in Kaskaskia until the governor of the Northwest Territory, Arthur St. Clair, arrived there in March 1790. For part of the decade after Clark's troops left in 1780, the Indian agent John Dodge became a renegade, seized control of the area, and terrorized Kaskaskia's residents (Davis 1998, 87–88, 98–100).

107. Michaux provides population figures only for the settlements in US territory. Bellefontaine is Waterloo in Monroe County, Illinois. It appears on Hutchings's (1778) "Plan of the Several Villages in the Illinois Country" (Ekberg 1998, 29).

108. The word *Florissant* in Michaux's account, mistakenly interpreted by some writers as the French adjective *flourishing* and thought, therefore, to describe Saint Louis, was instead the name of a then small settlement located four leagues (twelve miles) north-northwest of Saint Louis on Cold Water Creek, a stream that flows into the Missouri River (Finiels, Ekberg, and Foley 1989, 70). The earlier reports that suggested that Michaux crossed the Mississippi and visited Saint Louis were due to errors in interpreting his journal. These errors were noted by McKelvey (1955, 107–8).

109. Here, Michaux probably abbreviated Petites Côtes. It is the original name of Saint Charles, Missouri (Finiels, Ekberg, and Foley 1989, 32).

110. Vide Poche 'Empty Pocket' was an early name for Carondolet, now part of Saint Louis. The name was coined because of the poverty of its small population (Finiels, Ekberg, and Foley 1989, 32).

111. Ste. Geneviève is located on the west bank of the Mississippi in Missouri, while Kaskaskia was on the opposite bank in Illinois. Ste. Geneviève is shown on Thomas Hutchins's 1778 map, and it also appears on Hutchins's "Plan of the Several Villages in the Illinois Country" from his book *A Topographical Description of Virginia, Pennsylvania, Maryland, and North Carolina*, published at the same time (Ekberg 1998, 29).

112. Rivière des Arcs is possibly Rivière d'Ardennes, now Dardenne Creek in St. Charles County, Missouri. The creek empties into the Mississippi just north of the great waterway's junction with the Illinois River (Finiels, Ekberg, and Foley 1989, 32).

113. The intriguing place-names at the end of this list were not included in the 1889 transcription of Michaux's journals. Michaux did not travel to Mobile in Alabama, Natchez in Mississippi, or to New Orleans in Louisiana on this journey, but these notes are evidence that he was engaged in gathering information for future journeys. He planned another exploration west of the Mississippi that is mentioned in his letter to André Thouin of April 1, 1795, and he made a note in his journal on February 20, 1796, concerning the expense of hiring guides for such an expedition.

114. The French first established Fort Massac in 1761, approximately opposite the point where the Tennessee River enters the Ohio. Abandoned about 1764, the site was refortified by Gen. Anthony Wayne in 1794 in connection with the US campaign against a confederation of Native American tribes north of the Ohio River. A trail from Fort Massac to Kaskaskia was the route taken by George Rogers Clark and his Virginia militia in their campaign of 1778–79. Both Fort Massac and a trail connecting it to Kaskaskia appear on Thomas Hutchins's 1778 map. Today, Fort Massac State Park occupies the site, which is located near the modern city of Metropolis in Massac County, Illinois (Caldwell 1950).

115. Michaux meant the Ohio rather than the Mississippi River at this point in his narrative, since Fort Massac is on the Ohio River and over thirty miles from the Mississippi. He corrected his error when he returned to Fort Massac on October 20.

116. *Quercus imbricaria* Michx., the shingle oak, described here, is not known to occur in South Carolina.

117. *Brunnichia ovata* (Walter) Shinners, American buckwheat vine (IDC 62-1) is probably the species indicated here and again on Oct. 15 ("*Rajania* . . . 8 stamens").

118. This species is possibly *Forestiera acuminata* (Michx.) Poir., swamp-privet. In his review of type specimens from Illinois, Jones (1952) noted that the descriptions and ranges of *Adelia acuminata* Michx., swamp-privet, and *A. ligustrina* Michx., southern-privet (now recognized as *Forestiera acuminata* [Michx.] Poir. and *F. ligustrina* [Michx.] Poir., respectively) were transposed in Michaux's *Flora* (2:224).

119. Earlier in his travels, Michaux had mistaken this species, *Betula nigra* L., river birch, for *B. papyrifera* Marshall, paper birch.

120. *Smilax hispida* Raf., bristly greenbrier, is the most common *Smilax* in southern Illinois and nearby. *S. pseudochina* L. is an endemic species of the Atlantic coastal plain, reported only from New Jersey south to Georgia.

121. Sargent (1889) changed Michaux's first entry dated October 15 to October 16. Therefore the 1889 transcription has two entries dated October 16 but only one dated October 15, the reverse of what is found in the manuscript. Also in the manuscript, in the second paragraph

of Michaux's first entry for October 15, the phrase "Suite des plantes" precedes the phrase "Rives de la rivièrre Cheroquis (Tenasee)," but the initial phrase is missing from the 1889 transcription.

122. The Philadelphia newspaper *Gazette of the United States*, in its June 16, 1795, issue, reported the arrival on June 9 of the new French diplomatic delegation at Newport, Rhode Island, aboard the frigate *Medusa*. Unlike Citizen Genet two years earlier, these French diplomats made a speedy journey to Philadelphia and promptly presented their credentials to the US government.

123. *Liquidambar styraciflua* is sometimes called *Liquidambar resineux* for the resin derived from this tree.

124. This plant is recognized today as *Veronicastrum virginicum* (L.) Farw., Culver's-root (IDC 1-15 *Veronica virginica* "Illinois").

125. While Michaux's journal is silent about political matters, his presence in the area had been noted, and his motives for being there questioned, because his earlier involvement with Genet and Clark was widely known. He seemed unaware that his presence in Illinois was causing anxiety to Spanish officials and that he was subject to arrest if he entered Spanish territory. Coincidentally, the Spanish governor of Upper Louisiana, Manuel Gayoso de Lemos, was on a tour of Spanish settlements on the upper Mississippi. While visiting New Madrid, Missouri, in October, the governor was informed of the arrival in Illinois of Michaux and another Frenchman. He was profoundly suspicious of Frenchmen and ordered the arrest of both men (Nasatir 1968, 124). The following is an excerpt from a letter from Don Zenon Trudeau, commandant at Saint Louis, to Francisco Luis Hector Carondelet, governor general of Louisiana in New Orleans, dated October 26, 1795 (Nasatir 1990, 1:366):

> At the time, he [Arthur St. Clair, governor of the Northwest Territory of the United States] informed me that there had just arrived at their place [Illinois] a man named Micheaux, already known for having been in the intrigue of Genet in his former project against our colony; also this man might have some friends, but if that happened, he [St. Clair] could assure that he [Michaux] would not be successful in his district. That was an enigma for me, in that I cannot conceive the projects that he is planning which makes me (after this advice), keep on my guard. I learned that he [is] mistrusted [in] our district [Upper Louisiana] where, in effect, he would be arrested, if he appeared there.

126. The land rises abruptly and sometimes spectacularly with rock outcrops and cliffs sharply defining the limits of the wide flat plain of the American Bottom.

127. The French built three discrete versions of Fort de Chartres in the same general vicinity. The first, begun in 1719, and the second, which replaced it, were wooden palisaded structures. The third, which Michaux found in ruins, had been an imposing stone fortress built at great expense (Ekberg 1998). Today a partial reconstruction of the stone structure is an Illinois State Historic Site open to the public.

128. Pierre Richard, a resident of Kaskaskia, made voyages on the rivers (Michaux 1904, 78; Nasatir 1968, 320). Richard had supplied Col. George Rogers Clark's Virginia militia with thirty-eight pounds of gunpowder on credit in 1779. Richard's heirs and assigns were still attempting to collect this debt in 1834 (Virginia General Assembly House of Delegates 1835, 72).

129. This note refers to the Spanish-sponsored expedition up the Missouri River in 1795, led by James MacKay (1761–1822) and John Evans (1770–99), who reached the Mandan villages on the upper Missouri. Jacques Clamorgan (or Clanmorgan) was a prominent businessman and trader who came from the Caribbean to settle in Saint Louis. He prospered during the Spanish and early US periods of the city (Wood 2003).

130. Quoted by Ekberg (2011, 130).

131. Translated, from the Delassus Collection, Missouri Historical Society, Saint Louis.

## Chapter Thirteen

1. These boats were substantial watercraft capable of carrying tons of cargo and passengers. Equipped with a cabin, twelve to fourteen oars, a mast with a sail, and a rudder, these boats linked the distant settlements along the network of navigable rivers. When going upstream against river currents, these boats were propelled by oarsmen (Baldwin 1941, 42–49; Nasatir 1968, 49–51).

2. The salt spring was located approximately three miles south of the town of Ste. Geneviève on the Missouri side of the Mississippi. The saltwater creek there was called La Saline (Stepenhoff 2006, 15, 23, 35; Houck 1909 1:71–72).

3. The Treaty of Basel, which ended Spain's participation in the war against France, was signed on August 3 and celebrated in Madrid on August 7, 1795.

4. *Prêle* is an abbreviation of *aprêle*, from the Latin *asper*, meaning "rough." The *Equisetum* is *Equisetum hyemale* L., tall scouring-rush.

5. Cape Cinque Hommes is shown on Thomas Hutchins's 1778 map, but it was indicated as Cape St. Cosme on some other early maps. It is located on the western bank of the Mississippi River in Perry County, Missouri.

6. The community of Cape Girardeau, Missouri, was founded in 1793 by Louis Lorimer, the recipient of a Spanish land grant.

7. Manuel Gayoso de Lemos (1747–99), based in Natchez, was the highest ranking Spanish official in Upper Louisiana. His close contacts with the treacherous US general James Wilkinson had helped thwart George Rogers Clark's projected attack on Louisiana; he was aware that Michaux had assisted Clark. In the fall of 1795, Gayoso de Lemos was traveling up the Mississippi River with Spain's fleet of rowed river gunboats. On December 10, while he was docked at New Madrid, Missouri, he received a copy of Michaux's letter to de Luzières. The warning that Michaux had given de Luzières concerning a possible new French attack on Louisiana disturbed him. After one of his subordinates found Michaux on a boat that he was inspecting, Gayoso de Lemos promptly crossed the Mississippi to question the Frenchman himself. His diary provides a different viewpoint about their meeting as well as new information that is not found in Michaux's notes (quoted in Nasatir 1968, 319–20):

> Thursday December 16, 1795, at four o'clock in the afternoon a bercha was sighted which was descending the Mississippi following close around the point of the Ohio. I dispatched Cruzat in the bercha to see who it was, and in spite of the attempts that he made, he could not reach it before it entered the Ohio; but having landed in the American district, Cruzat went on board with the purpose of seeing the new state of things that might have occurred. At six he returned to the naval station and with him Mr. Richard, owner of the bercha that was going to Cumberland, who told me he had on board the man named Michaux. Immediately I embarked in order to go and see him to investigate, if I could, on what he based the notice he had given to Mr. [de] Luzière [regarding a possible French attack on Louisiana]. I arrived at the place they were encamped and Mr. Micheaux received me with many signs of friendship. He did not take long to enter into a political conversation. As soon as we were through, I took leave of him and left for the naval station [New Madrid].

After their meeting, Gayoso de Lemos rescinded his arrest order for Michaux (Nasatir 1968, 124, 318).

8. Michaux's location here is modern Eddyville in Lyon County, Kentucky. The town was given that name because of two large eddies in the river, one just above, and the other two miles below the town. The place became famous for its boat-building industry (Bedford 1919).

9. The route upstream on the Cumberland River that Michaux followed after December 21 and his landmarks are now submerged beneath the waters of Lake Barkley. The Little River enters the lake in Trigg County, Kentucky.

10. The small green parrot was the now extinct Carolina parakeet, *Conuropsis carolinensis*.

11. The Cumberland settlements soon became part of the new state of Tennessee.

12. The *Anonymos ligustroides* was probably *Forestiera acuminata* (Michx.) Poir., swamp-privet.

13. Both Yellow Creek and Blooming Grove Creek join the Cumberland River in modern Montgomery County, Tennessee, downstream from Clarksville.

14. Red Paint Hill, a rock bluff at the confluence of the Cumberland and Red Rivers, was a noted early landmark shown on the Hutchins map of 1778. Clarksville, named for Gen. George Rogers Clark, was founded in 1784, when the area was still part of North Carolina. It is now the seat of Montgomery County (Beach 1964).

15. A Revolutionary War veteran, Dr. Morgan Brown (1758–1840) was a prominent planter, physician, and state legislator in the Cheraws region of South Carolina before he moved his family to Tennessee in 1795. In addition to being a practicing physician, Dr. Brown was a successful businessman with many ventures, including an early ironworks. In Tennessee he established the town then known as Blountsboro (now Palmyra), located about ten miles downstream from Clarksville on the Cumberland (now Lake Barkley). Dr. Brown succeeded in having the town designated the first port of entry for goods arriving in Tennessee from the Ohio and Mississippi Rivers, thus increasing its early commercial activity (Edgar, Watson, and Bailey 1981, 3:98–99).

16. Clarksville is meant here because adding Michaux's reported distances from Ebenston mill to Nashville is a total of forty-three miles, not ten. Thwaites (Michaux 1904) and Williams (1928) both wrote that Ebenston was the John Edmiston who was appointed captain of the Davidson County militia in 1792, but this is questionable. Although the name was probably spelled either Edmiston or Edmonson (both appear in early Tennessee land records), the man Michaux met was most likely a resident of the county then called Tennessee (now Montgomery), not Davidson.

17. Frederick Stump Sr. (1724–1822), one of the earliest settlers of the Nashville region, was a signer of the Cumberland Compact of 1780. He was a stout fighter against the Cherokee in the difficult early years of the settlement, and he became a wealthy businessman and planter. His two-story cabin of red cedar logs (built in 1797 or earlier) survives and is listed on the National Register of Historic Places (Guice 1991, 65–71; Williams 1928, 508–9; Public Library of Nashville, Special Collections, "Stump, Frederick" [file]).

18. Timothy DeMonbreun (1747–1826) was a pioneer who arrived at the future site of Nashville from Canada by way of Illinois to trade with the Native Americans around 1769. At first he divided his time between Kaskaskia and Nashville, but after 1790 he resided full-time in Nashville, where he built a substantial fortune as a merchant. A major Nashville thoroughfare is named for him today (West 1998, 243).

19. Arnow (1961, 153) reported that John Deaderick was a bachelor who liked to live well. In 1795, he was a justice of the peace. His brother George was a prominent merchant, pioneer banker, and real estate dealer (see also chapter 12, note 68).

20. "Tetts" is likely William Taitt, an associate of Andrew Jackson in his Nashville mercantile business (Jackson, Smith, Owsley, and Moser 1980, 1:81–82, 1:105–6, 1:333–34).

21. Andrew Jackson encouraged Seth Lewis (1766–1848) to study law. Lewis became a representative from Davidson County in the first Tennessee legislature after statehood and went on to become a US attorney in Mississippi and a judge in Louisiana (McBride and Robison 1975, 1:447–48).

22. John Overton (1766–1833), a lawyer, businessman, and judge, was a close and trusted personal friend of Andrew Jackson. He is best remembered for his partnership with Jackson and James Winchester (1752–1826) in founding the city of Memphis (*DAB*, s.v. "Overton, John").

23. This is almost certainly Capt. John Gordon; see chapter 12, note 67.

24. Hall and Hall (1975, 1–2) identify this physician as Dr. Mark Brown Sappington (1742–1805), who moved to the Cumberland settlements from Maryland in 1785. He was the first trained physician in the area to earn his livelihood with a medical practice. His sons, of whom John (1776–1856) became the most famous, trained with their father to become physicians. The Sappingtons were strong supporters of Andrew Jackson, but they were not his physicians.

25. No "Dr. Holland" has been identified in contemporary records.

26. James Maxel and William Betts were both licensed tavern keepers in Nashville as early as 1793 (Wells 1990–91, 2:7).

27. A Continental Army officer in the Revolutionary War, Col. Robert Hays (1758–1819) was Andrew Jackson's brother-in-law and one of his closest and most trusted associates. After moving to the Cumberland settlements about 1784, Hays, who was known for his good humor and hospitality, became prominent in the community and active in the struggle against the Cherokees. Michaux visited with Hays again after he returned from Kentucky in February and purchased a horse from him (Robison 1967).

28. The last name on this list, "Mr. Jackson, attorney," is Andrew Jackson (1767–1845), future president of the United States, then a Nashville attorney and partner in a mercantile business with his brother-in-law Samuel Donelson. Several of the other twelve people on this list were Jackson's close friends and business associates.

29. Michaux headed north to retrieve the collection that he had stored in Louisville the previous summer and spent the first night at Anthony Sharpe's inn, where he had also stayed on June 22, 1795.

30. The lodging that Chapman offered his guests was the floor space between the two beds in the common room of his cabin. We know this from Louis-Philippe, future king of France, who offered a colorful description of a night spent with the Chapmans a few months after Michaux's visit. Capt. Chapman and his wife occupied one bed. They were busy talking about the day's activities. In the other bed was an unmarried daughter. She was joined by a newly arrived young man, who undressed, and soon, there were indications of love-making between the two. This activity did not seem to disturb the family; soon, another sister blew out the candle and crept into the bed next to the young man (Louis-Philippe 1977, 111–14).

31. Michaux also visited Andrew McFaddin's Station and Ferry at the site of modern Bowling Green, Kentucky, on June 24, 1795.

32. Thwaites (Michaux 1904, 87) thought that "Mr. Maddisson" was George Madison (1763–1816), youngest son of John and Agatha Strother Madison of Augusta County, Virginia, and the fifth governor of Kentucky; however, we believe, based on the evidence that follows, that "Mr. Maddisson" was George's brother Rowland.

The evidence is clear that George Madison owned property in the area of Warren County where Michaux was traveling (Murray 1985, 1). However, the date on the marriage record of George Madison and Jane Smith in Frankfort, Kentucky (more than two hundred miles from Warren County), is reported as February 11, 1796 (Kentucky Historical Society 1983, 205). This date is two weeks after Michaux's visit with the Madisons on January 28–29. On March 7,

1796, George Madison was appointed state auditor by Governor Isaac Shelby, and George and Jane Smith Madison resided in Frankfort thereafter (Harrison 1985, 18).

Two of George Madison's brothers, Gabriel and Rowland (Roland), also lived in Kentucky in 1796. Gabriel is not known to have lived in this area, but evidence indicates that Rowland did. George and Rowland were both in the party that laid out the town of Bowling Green, which soon became the seat of newly formed Warren County. Rowland died in Warren County soon afterward (Strange 1889, 120–22). Additional evidence that Rowland Madison lived along Michaux's route through Warren County includes an 1801 court order for Robert Wallace indicating that his neighbor Rowland Madison had a bounty land grant north of the Big Barren River for his Revolutionary War service. Rowland Madison's estate was inventoried in Warren County in 1802 (King 1969, 157, 228).

Like George's wife, Rowland's wife, Anne Lewis, the daughter of Gen. Andrew Lewis (1720–81), came from a distinguished Virginia family, and she also fits Michaux's description of a cultured hostess (Klotter and Dawson 1981, 374–75). This information leads us to suggest that Michaux visited the home of Rowland and Anne Lewis Madison rather than that of George and Jane Smith Madison, who were not married on the date of his visit, and that it was Anne Lewis Madison who made the valuable suggestion that would protect Michaux's feet from frostbite.

33. Michaux's reference here is to James Madison (1751–1836), later the fourth president of the United States.

34. On this journey across Kentucky, Michaux first traveled northeast through modern Warren, Edmonson, Barren, and Hart Counties. Turning north in Hart County, he crossed the Green River and Bacon Creek, then he entered modern Larue County, where he passed through Hodgenville. Continuing almost due north, he crossed the Rolling Fork into Nelson County and crossed Beech Fork soon afterward. He did not go through Bardstown as he had on his previous journeys but rather continued on a new road directly north across Long Lick Creek and on through modern Shepherdsville in Bullitt County, before entering Jefferson County and reaching his destination, Louisville.

35. This person was probably George Close Jr., who is listed in the tax records for old Hardin County (Clift [1954] 1993, 56).

36. Robert Hodgen (d. 1810) built a gristmill in 1789 on the North Fork of the Nolin River, at the site of modern Hodgenville, the seat of Larue County and later the birthplace of Abraham Lincoln (Ingmire 1982, 10–11).

37. "Mr. Scoth" was likely one of several settlers with the surname Scott living in this area.

38. Michaux had lodged with a Standeford on July 18, 1795; see chapter 12, note 82.

39. "Prince" is probably the Sylvanus Prince who is on the Jefferson County tax lists for both 1790 and 1800; he is the only person with the surname Prince on the 1790 list (Heinemann [1940] 1971, 77; Clift [1954] 1993, 243).

40. Michael Lacassagne was the Louisville merchant who had stored Michaux's collections while he had traveled to the Illinois country in 1795. Michaux's notes about Lacassagne's agents in Philadelphia and New Orleans could mean that he was examining alternate ways to ship his collections to Charleston. See also chapter 12, note 88.

41. John Clark (1725–99) was George Rogers Clark's father (*Early Kentucky Settlers* 1988, 230).

42. Samuel Fulton was George Rogers Clark's agent in his attempt to recover from the French government the expenses that he had incurred in Genet's scheme. Fulton traveled to Philadelphia and then to Paris in 1795 on this mission. After Fulton met in Philadelphia with Pierre Adet, French minister to the United States, Adet wrote in March 1796, advising Fulton

that Clark needed affidavits from Michaux in order to pursue his effort to be reimbursed for his expenses of preparing for the invasion of Louisiana (American Historical Association 1897, 1097). Michaux did meet with Fulton in Paris on December 26, 1796, soon after his own return to France.

43. Michaux refers here to Kentucky's oldest commercial salt-making operation, the Bullitt's Lick Salt Works near Shepherdsville, in operation since 1779 (Kleber 2001, 782–83).

44. The *Pinus* with two needles is *P. virginiana* Mill., Virginia pine.

45. Typically, tulip-tree sapwood is white, and heartwood is yellowish. Heartwood, the hard inactive central wood of a tree, is usually darker than the outer sapwood that surrounds it.

46. This orchid is *Aplectrum hyemale* (Muhl. ex Willd.) Torr., Adam and Eve.

47. "Huggins" refers to the same Robert Hodgen whom Michaux visited on January 31, 1796.

48. In this locale, *Quercus nigra* is *Q. marilandica* Münchh., blackjack oak.

49. This person is Robert Hays, Andrew Jackson's brother-in-law. His name was on Michaux's list of people known in Nashville (see note 27 in this chapter).

50. Michaux's strong desire to explore west of the Mississippi continues. Here he attempts to determine how much he would have to pay a guide on such an expedition.

51. Kaspar (Kasper) Mansker (ca. 1750–1820), said to have been born at sea of German immigrant parents, was one of the earliest arrivals and most tenacious fighting men of the Cumberland settlements. We have no account from Mansker concerning the disagreement with Michaux, but nothing known about him suggests that he was ever a partisan of royalty. He was one of the surviving members of the small group of white settlers who wrested the Cumberland region from the Cherokees during a period of border warfare lasting more than a decade, taking the lives of many of his contemporaries (Durham 1971). Whatever the actual cause of their dispute, Michaux used the episode to assert that he was a good French Republican who chose his country's honor above all else. Today, a modern reconstruction of Mansker's Fort in present Goodlettsville is a center for historical interpretation of eighteenth-century life in the Cumberland region.

52. Fort Blount was a frontier post located on the western side of the Cumberland River near the mouth of Flynn's Creek and was initially garrisoned by Tennessee militia. The site has been the subject of extensive historical and archaeological research (Smith and Nance 2000).

53. *Sophora* is a worldwide genus in *Fabaceae*, the pea family. The new species was the yellow-wood, recognized today as *Cladrastis kentukea* (Dum. Cours.) Rudd. It does not appear in Michaux's *Flora* as he did not have access to flowering material. Either André Michaux or his son (who collected seeds in the same area in 1802) must have given viable yellow-wood seeds to the horticulturist Jacques-Philippe Martin Cels or to André Thouin, because when the horticulturist Dumont de Courset (1746–1824) wrote the second edition of his *Botaniste cultivateur* (1811–14), he described and named it *Sophora kentukea*. François André Michaux also included it in his *Histoire des arbres forestiers de l'Amérique septentrionale* (1813) as *Virgilia lutea*. A very large yellow-wood, believed to have been planted by William Bartram from seed given by one of the Michauxs, grows in Bartram's Garden in Philadelphia, where it is a living link to these botanists. Fortunately, this old tree survived serious storm damage in 2010 (J. Fry, pers. comm.).

54. Capt. Sampson Williams (1762–1841) was identified by both Thwaites (Michaux 1904) and S. C. Williams (1928) as the soldier at Fort Blount who cut down the tree for Michaux. Sampson Williams, a close associate and admirer of Andrew Jackson, was an active campaigner in the struggle against the Cherokee who rose to prominence and held a variety of offices in the Cumberland settlements before he became the founder of Jackson County. In their recent study of Fort Blount, Smith and Nance (2000, 54) clarify that it was actually Sampson

Williams's younger brother Oliver who assisted Michaux in collecting seeds while Sampson, who operated an inn at the site, provided hospitality to the traveler.

55. The yellow-wood tree is probably Michaux's most widely reported plant discovery. Michaux's letter to Governor Blount and the governor's reply were soon published in the *Knoxville Gazette*, although the March 15, 1796, date of publication previously reported by François André Michaux and then repeated by Thwaites (Michaux 1904) and S. C. Williams (1928) is incorrect. Nonetheless, the account published in the Knoxville newspaper was reprinted in the Charleston, South Carolina, *Columbian Herald* on May 15, 1796:

> [I] arrived upon the territory of your government. I flatter myself of having the satisfaction to present, in a few days, my respects to your excellency, and my sincere gratitude for your kind reception last year.
>
> Knowing the warm interest you take in all researches relating to the public good—I have the pleasure to announce to you an useful discovery. It is a small tree, native in the neighborhood of Fort Blount, upon the Cumberland river, and about the head of Flint [Flynns] Creek. That tree is of the class or rather the genus of the sophora. The fresh roots give out an yellow-orange color, very light and beautiful. It appears to me, that it has much affinity with the sophers of China and Japan, employed in those countries for dying [dyeing] and lately introduced into the botanical gardens of Europe. I observed that tree in June last; it was not then in blossom; therefore, I have not been able to determine the essential characters of its genus; but by the accessories of the foliation[,] fructification and some other circumstances, I may assert that that tree is of the genus sophora.
>
> Having stopped at fort Blount in order to obtain some of the young plants, I have been prevented by the snow now on the ground, to get them. I am indebted to Captain Williams, jun. who accompanied me to gather a few seeds remaining on the trees. It is to be observed, those seeds ought to be gathered in the fall, because those remaining on the trees are, the greater part, unsound.

A shorter announcement reprinting a notice of the discovery from a Winchester, Virginia, paper also appeared in the *Argus, Or Greenleaf's New Daily Advertiser* in New York on May 10, 1796.

56. The little bulbous umbellifer is probably *Erigenia bulbosa* (Michx.) Nutt., harbinger-of-spring.

57. This location is Fort Southwest Point.

58. Michaux had met Gov. Blount when he visited Knoxville between May 25 and June 4, 1795. Blount's house in Knoxville is a Tennessee historic site today, and a yellow-wood is a prominent part of the landscaping.

59. The innkeeper was possibly Capt. Moses Looney, who owned property on the Holston River in Knox County, and whose name appears often in the early Knox County court records (Griffey 2000, 272; Tennessee County Court and Tennessee Historical Records Survey 2010).

60. Here Michaux meant not the Cumberland but the Holston River (now the Tennessee River).

61. The saxifrage is probably *Tiarella cordifolia* L., foamflower.

62. Capt. William Rickard was a US Army officer who often assisted Gov. Blount by certifying militia rosters and payrolls (Clark 1990).

63. This is Michaux's second use of the new genus name *Jeffersonia* in his journal. See his list of plants found in the vicinity of Knoxville, May 25, 1795. His herbarium specimen is *Podophyllum diphyllum* L. (IDC 53-10). Also written on the sheet below the name is: "*Biliaria tinctoria*, *Jeffersonia*. Barton"; no locality is given. The species is not included in his *Flora*.

64. Governor Blount's response to Michaux's letter follows (reprinted from the *Knoxville Gazette* [date uncertain] in the *Columbian Herald*, Charleston, South Carolina, May 15, 1796):

> If proofs were wanting of the disposition of the French republic, to promote the general happiness of the whole human family, the researches in which you are engaged, under their authority, could be adduced as one—and to contribute to your success, in any degree, would, to me, be a high gratification.
>
> I beg you, Sir, to believe, that I shall be happy in your return to this place . . .

65. Attorney James Reece came to Tennessee from North Carolina about 1784 and was a supporter of the state of Franklin movement. His son William B. Reece was later a justice of the Tennessee Supreme Court and president of East Tennessee College, now the University of Tennessee (Williams [1924] 1993, 325).

66. Lick Creek, which enters the Nolichucky River near the boundary of modern Green, Cocke, and Hamblen Counties, was the next physical feature that Michaux noted after crossing from the watershed of the Holston River through Bays Mountain at Bull Gap, following the route of modern highways 11E and 34 into Greeneville.

67. Maj. Carter is Landon Carter (1760–1800), whom Michaux had visited on June 28–29, 1789.

68. This is a return visit with Col. John Tipton, whose home is now the Tipton-Haynes Historic Site in Johnson City. Michaux had visited him on May 14–15, 1795.

69. Charles Colyar (Colyer or Collier, 1757–ca. 1840) had a land grant of 640 acres in Limestone Cove, then Washington (now Unicoi) County, Tennessee. A Collier family member briefly guided the 1799 North Carolina–Tennessee boundary survey party in the Iron Mountains, memorably providing the insect-plagued surveyors with two gallons of good whiskey that eased their suffering (or made them oblivious to it). Charles Colyar and other family members later moved to Kentucky (Griffey 2000, 144; Williams 1920, 46–57; [Revolutionary War] Pension Application of Charles Colyer, dated August 17, 1834, available at persi.heritagequestonline.com).

70. François André Michaux followed this little-used route across the Iron Mountains in September 1802 and remarked on its difficulty and danger (F. A. Michaux [1805] 1904, 283–84). "Becker" is David Baker (1749–1838), a Revolutionary War veteran and the earliest settler of modern Bakersville, now the seat of Mitchell County, North Carolina. He married twice, first to Mary Webb and then to Dorothy Wiseman (1765–1855). Baker's Revolutionary War service included participation in the battles of Princeton and Trenton as well as wintering at Valley Forge ([Revolutionary War] Pension Application of David Baker, dated September 26, 1832, available at persi.heritagequestonline.com).

71. No settler named Rider is mentioned in this area in local records, so this man was likely one of the four Griders listed in the 1790 US Census of North Carolina along with "Saml Ramsy." There were no settlers named Nigh, but the two heads of household named Night listed in the 1790 US Census of North Carolina do not appear in the 1800 enumeration, so a widow with that surname in 1796 is plausible (United States Bureau of the Census 1908a, 11:109).

72. Michaux indicates that he traveled from Baker's house to Davenport's house, a distance of twenty-three miles. Although at first glance this entry appears to give a list of distances from one settler's cabin to another along the route that he followed between these two points, a close examination indicates that it is not. His mileages add up to twenty-eight miles, not twenty-three miles. Moreover a journey by way of all the settlers' cabins listed here would have been a distance of thirty-three miles because Cox and Young did not live between Baker

and Davenport. They lived more than 1,500 feet downhill at the foot of the Blue Ridge and ten additional miles of difficult travel (five miles down the mountain, then five miles back up) would have been required to arrive at Davenport's house had Michaux first visited Cox and Young on March 23. Since he reported his total distance traveled as only twenty-three miles, he did not visit Cox and Young.

73. *Azalea lutea* and *A. fulva* (golden yellow and tawny yellow azaleas) are descriptive references to *Rhododendron calendulaceum* (Michx.) Torr., flame azalea. *Anonymos azaleoides* is possibly an observation of the plant that Michaux would name *Azalea pilosa* (Rehder 1923, 2). We now recognize the plant as *Rhododendron pilosum* (Michx.) Craven, minniebush.

74. Michaux had previously mentioned Thomas Young (1747–1829) in his journal entries for September 7, 1794, and May 3–4, 1795 (see chapter 12, note 29).

75. Here, although he was traveling south, Michaux used the indication of left and right as if going from south to north. There are two other examples of this practice in his journal during this same week: on April 4 at the Tuckasege Ford on the Catawba River, and on April 6 describing a route to bypass Charlotte.

76. This is Michaux's fifth recorded visit with Col. Avery. Each of his visits to the high mountains of northwestern North Carolina included a visit with Avery (June 13, 1789; November 19, 1789; September 8, 1794, May 2, 1795; and here on March 31, 1796). See also chapter 5, note 7.

77. This was Michaux's fourth visit to the home of Heinrich Weidner (d. 1792), now owned by his daughter Molliana and son-in-law Jesse Robinson. It was located in present-day Catawba County, a few miles south of modern Hickory, North Carolina. See also chapter 5, note 88.

78. Michaux had also visited Christian Reinhardt (1735–1817) on July 24 and September 11, 1794.

79. Michaux also visited Ben (Benjamin) Smith on July 23, 1794, and April 28–29, 1795. Ben Smith's farm, at the confluence of Hoyle and Little Hoyle Creeks near modern Stanley in Gaston County, was Michaux's collection site for live plants of *Magnolia macrophylla*. It is now protected as the Jack C. Moore Preserve of the Catawba Lands Conservancy.

80. Robert J. Alexander (1749–1813) was a Revolutionary War veteran and prosperous local resident who had served in the North Carolina legislature several times in the 1780s. His plantation house was visible from the Tuckasege Ford, and Michaux likely made his acquaintance at an earlier crossing of this ford (Ray [1950] 2002, 349).

81. *Rhus pumilum* is *Rhus michauxii* Sarg., Michaux's sumac, and was collected on a side trip to revisit the site between the Springs' home and the Huston Tavern where Michaux had first observed this rare plant on July 21, 1794 (Barden and Matthews 2004).

82. Maj. (not Col.) Robert Crawford (ca. 1728–1801) served in the Revolutionary War and was at the Battle of Hanging Rock, where his young kinsman thirteen-year-old Andrew Jackson received his own initial combat experience. After the war Jackson lived briefly in Crawford's home before he began the study of law in Salisbury, North Carolina. It is probably not a coincidence that Michaux visited Crawford after meeting Jackson in Tennessee (Andrews 1998, 137–38).

83. This was the last of Michaux's five visits to Camden. Manchester was once a thriving town, but it has disappeared today. A historical marker on SC Highway 261 located 8.5 miles south of Stateburg indicates the site of the town (Andrews 1998, 220).

84. The Manigault Ferry was a few miles upstream from the better-known Nelson's Ferry and was later named Vance's Ferry. Michaux had crossed the Santee by the Manigault Ferry on July 16, 1794, when the high waters on the Santee made Nelson's Ferry unsafe.

85. Gaillard Road, shown on the Orangeburg District map in the 1825 *Mills' Atlas*, connected with other roads into Charleston from the Orangeburg District. This route from the Santee did not pass through Moncks Corner.

## Chapter Fourteen

1. MNHN, Archive Thouin, transcription by Régis Pluchet.

2. The settlements in eastern Tennessee are meant here.

3. Michaux's note on October 9, 1795, about the large-seeded *Guilandia dioica*, now recognized as *Gymnocladus dioicus* (L.) K. Koch, Kentucky coffee-tree, to be found along the Illinois River, indicates that he botanized along this river that enters the Mississippi north of Cahokia, but he did not specifically mention it in his journal.

4. Victor Marie DuPont (1767–1827), older brother of E. I. DuPont (founder of the enterprise that became the DuPont chemical company), served as French consul in Charleston from 1795 to 1797.

5. Prairie du Chien (Wisconsin) was an old French fur-trading post located north of Saint Louis, where the Wisconsin River enters the Mississippi.

6. Michaux's information was correct. French physician and botanist Joseph Dombey, who had collected extensively on the west coast of South America, died on the Caribbean island of Montserrat in 1794.

7. Dr. Jean Rouelle had come to America after being chosen by an advisory committee in Paris for a teaching position at Quesnay de Beaurepaire's proposed academy in Richmond, Virginia. Described as a physican and a traveler, he was well connected in the French natural history circle; both his father and his uncle had been noted chemists at the Jardin du Roi. Rouelle was elected to membership in the APS in 1792 at the same time as his friend Palisot de Beauvois. Although the academy at Richmond failed before admitting its first student, Rouelle proved to be a practical scientist. He studied and reported on the mineral springs of Virginia (Library Board of the Virginia State Library 1922, 49–50; Forsyth 1932, 28–29).

8. Palisot de Beauvois was a French botanist whom Michaux had met in Philadelphia in 1792. His reason for visiting Charleston is discussed later in this chapter.

9. Hippolyte Nectoux (1759–1836) was director of the botanical garden on Saint-Domingue (Haiti). He came to the United States during the revolution on the island. After returning to France, Nectoux was one of the scientists who went to Egypt with Napoleon (McClellan and Regourd 2011, 541).

10. This plant is *Fraxinus quadrangulata* Michx., blue ash (IDC 128-6).

11. This plant is *Gymnocladus dioicus* (L.) K. Koch, Kentucky coffee-tree (IDC 124-19; it is in the *Flora* 2:241 as *G. canadense* with an illustration by P. J. Redouté).

12. This plant is *Panax quinquefolius* L., American ginseng (IDC 128-7).

13. This plant is *Panax trifolius* L., dwarf ginseng (IDC 128-8).

14. This plant is *Hypericum spherocarpum* Michx., roundseed St. John's-wort (IDC 91-15).

15. This plant is probably *Silphium terebinthinaceum* Jacq., prairie-dock.

16. This plant is *Phryma leptostachya* L., American lopseed (IDC 77-3).

17. *Dracocephalum virginiana* is *Physostegia virginiana* (L.) Benth., obedient-plant (IDC 76-10).

18. This grass is *Bouteloua curtipendula* (Michx.) Torr., side oats grama (IDC 14-9, 14-10).

19. The blue *Sophora* is *Baptisia australis* (L.) R. Br., blue wild indigo (IDC 60-15). It appears in the *Flora* (1:264) as *Podalyria coerulea*, and it grows in sandy places along the Ohio that are flooded part of the year.

20. This plant is *Campanula americana* L., tall bellflower (IDC 26-21).

21. The genus *Serratula* has been split into three modern genera: *Liatris*, *Carphephorus*, and *Vernonia*. The *Serratula* with plumose pappus could be *Liatris squarrosa* (L.) Michx., scaly blazing star (IDC 93-20).

22. This plant is *Tragia cordata* Michx., heartleaf noseburn (IDC 111-4 "*Tragia volubilis?*"). See chapter 10, note 57.

23. *Leontice thalictroides* is *Caulophyllum thalictroides* (L.) Michx., common blue cohosh (IDC 47-4).

24. *Illecebrum divaricatum* is *Paronychia dichotoma* (Michx.) A. Nelson, a synonym for *Paronychia canadensis* (L.) Wood, Canada whitlow-wort (IDC 28-1, 28-2; in the *Flora* 2:113 as *Anycha dichotoma*).

25. The plant similar to *Rhinanthus*, four feet high, is possibly an *Aureolaria*, oak-leach, false-foxglove.

26. The new species is *Viburnum molle* Michx., soft arrow-wood (IDC 40-13 "Kentucky near Danville"). This is the type location for this species (Uttal 1984, 28).

27. *Polymnia uvedalia* is *Smallanthus uvedalia* (L.) Mack. ex Small, bearsfoot (IDC 106-1, 106-2 "Ohio River"; *Flora* 2:147).

28. *Nepeta altissima* is *Agastache nepetoides* (L.) Kuntze, yellow giant-hyssop (IDC 74-10, 74-11). It is in the *Flora* (2:3) as *Hyssopus nepetoides* L.

29. *Aster grandiflorus* is *Symphyotrichum novae-angliae* (L.) G. L. Nesom, New England aster (IDC 97-15 "Warm Spring, Kentucky"). It is in the *Flora* (2:111) with location as the mountains of Virginia and North Carolina.

30. *Urtica divaricata* is *Laportea canadensis* (L.) Wedd., wood nettle. The other species would be *U. gracilis* Aiton, American stinging nettle, or *U. chamaedryoides* Pursh, dwarf stinging nettle.

31. *Sarothra gentianoides* is *Hypericum gentianoides* (L.) BSP., pineweed, orange-grass (IDC 92-2). It is in the *Flora* (2:79) as *H. sarothra*.

32. *Mimosa mississippi* (IDC 127-14 "islands in Tennessee and Mississippi rivers") is *Desmanthus illinoensis* (Michx.) MacMill. ex B. L. Rob. & Fernald, bundleflower, prairie mimosa. It is the type collection of the species (Uttal 1984, 63). It is in the *Flora* (2:254) as *Mimosa glandulosa*.

33. See Gillispie 1992.

34. Philippe André Joseph Létombe was accredited consul general of France on June 22, 1795, and became minister in 1797 (Nasatir and Monell 1967, 563). The following year President Adams revoked all French consular credentials during the naval quasi war with France.

35. Furstenberg 2014, 345.

36. Palisot de Beauvois to Caspar Wistar, May 20, 1796, APS MSS B. W76.

37. "Rapport sur le jardin de la République situé a deux milles de Charleston dans la Caroline du Sud et sous la direction du citoyen Michaux botaniste." AAE, Correspondence des Consuls, Charleston, SC, 1793–1799.

38. Daubenton was Count de Buffon's collaborator on the monumental *Histoire naturelle* (1749–1804). In 1793 he became the first professor of mineralogy at the MNHN.

39. *Castor* L. is the genus of the beaver not the raccoon.

40. Palisot de Beauvois ignored the fact that Michaux had already established a northern garden in New Jersey and had assigned Saunier, his gardener trained by Thouin, to live there and manage the property. Saunier, with the assistance of the consuls in New York, had made shipments from this garden to France until at least 1791. The gardener was still there in 1796 and would remain on this property until his death in 1817. It is difficult to imagine how Palisot de Beauvois, who had met Michaux in Philadelphia in 1792, could have been unaware of this

northern garden. We interpret his lapse as a knowing concealment of the facts and suggest that he was party to a deliberate scheme to divert any government support going to Michaux, toward himself and Dr. Jean Rouelle, who would be the two "new directors" of the northern garden that he proposed in this document.

41. In Michaux's account of this episode (found in his letter to Bosc of August 22, 1798, which follows this one), Palisot de Beauvois did not personally speak with him but rather visited the garden when Michaux was absent.

42. VERS, dossier Bosc, transcription by Régis Pluchet.

43. Pierre Bénézech was minister of the interior from 1795 until 1797. Michaux obtained financial support from him after he returned to France.

44. Palisot de Beauvois was a member of the aristocracy in the Old Regime. Like many others with royalist connections, he had been prohibited from returning to France after the revolution.

45. As noted earlier, Dr. Jean Rouelle was well connected in the French natural history circle. Perhaps these memoranda concerning the Charleston garden would have been viewed differently had Michaux's friend and correspondent André Thouin not been away in Italy at this time.

46. Louis-Marie la Révellière-Lépeaux (1753–1824) was a member of the five-man Directory, which ruled France from 1795 until 1799. Bosc had hidden him and helped him escape the Reign of Terror.

47. Michaux has used the phrase "true Republican" to describe individuals earlier, and it suggests a purity of belief in the ideals of the revolution. He characterized as "true Republicans" people who were egalitarian in their thinking. He experienced problems after he returned to France both because he lacked money and because, it appears, some of the scholars in the natural history community did not see him as an equal. Although many of these men were commoners like Michaux rather than aristocrats, most of them, unlike Michaux, had more formal schooling, or in some cases their families had been connected with the Jardin du Roi or had been gardeners on royal estates for generations. Michaux was a first-generation scientist and apparently was seen as something of an upstart. He, of course, also had a strong independent streak.

48. The Jussieu herbarium contains more than five hundred plant specimens collected by André Michaux in North America. Most are duplicates, but some are the only surviving specimens of a particular species collected by Michaux. For example, the only surviving complete specimen of Michaux's *Shortia galacifolia* is in the Jussieu herbarium; it includes a note indicating that Michaux collected it in 1787. The specimen in the Michaux herbarium studied by Asa Gray in 1839 and used to describe the species has been lost. The only portion of this lost specimen that survives is the fragment that Gray took with him to Harvard.

49. HU, HM 71884.

50. A reference to John Fraser's contributions to Thomas Walter's *Flora Caroliniana*, 1788.

51. Victor Marie DuPont to Pierre Auguste Adet, 12 thermidor year 4 (July 30, 1796). Hagley Library and Archives, Wilmington, DE (HA), Winterthur Manuscripts Group 3, Victor M. DuPont Papers, Consular correspondence, v. 1, pp. 151–52.

52. Savage and Savage 1986, 160.

53. Adet, a progressive chemist and member of the scientific establishment of France, served in the United States from June 1795 until May 1797. His clumsy attempts to influence the US presidential election of 1796 in favor of Jefferson only added to the deterioration of the United States' relations with France (Malone 1962, 284–89; Conlin 2000).

54. Turner 1903, 959–60.

55. Michaux's transit was indeed not a happy one. His ship was wrecked along the coast of Holland on October 10, 1796. Adet, of course, could not foresee the future, but misadventures at sea were very frequent in those days.

56. Palisot de Beauvois's trip was cut short because of ill health.

## Chapter Fifteen

1. The Isles of Scilly are an archipelago of more than 150 islands located in the Atlantic Ocean off the southwest tip of England. Ushant is an island off the Brittany coast and is the westernmost point of France.

2. The *Ophir* had been at sea for fifty-six days when it arrived at the coast of Holland. For the past eleven years, Michaux had experienced diseases, perilous conditions from both the land and its inhabitants, and severe weather. He regarded his own body as nearly indestructible, but now in the midst of a shipwreck, Michaux was reconciled to the fact that his life might soon end.

3. The accounts of Michaux's rescue found in some secondary sources such as Coulter (1883) and Eifert (1965) are exaggerated when compared to the actual narrative in Michaux's journal. Michaux did not tie himself to a plank, nor was his herbarium lashed securely to his body.

4. Pinson d'Ardenne are chaffinches of the Ardennes. The Ardennes is a region of wooded plateaus and hills in northern France, southeast Belgium, and northern Luxembourg.

5. MNHN, Paris MS-Cry-505/1138, transcription by Régis Pluchet.

6. Duprat 1957, 248.

7. Michaux remained in Egmond for over five weeks, during which time he regained his health and restored order to his collections.

8. Bosc was in Charleston, but he did not take over the care of the Carolina nursery as stated by Savage and Savage (1986). Bosc stayed at the Michaux house and did study and collect many natural history specimens in the Charleston area. He brought back seeds from plants in the Michaux garden and gave some of these to Cels. Michaux had been a guest in the Charleston home of Gen. Charles Cotesworth Pinckney. The botanist had written him via New York before the general left for France; the latter was in Paris from December 1796 to February 1797.

9. VERS, dossier Bosc, transcription by Régis Pluchet.

10. Although Michaux did not mention so earlier, this entry indicates that Michaux may have shared the house near Charleston with other persons who are not named.

11. Sebald Justinus Brugmans (1763–1829) was professor of natural history at Leiden. Michaux had lost his books in the shipwreck.

12. This account and others in the journal clearly indicate that Michaux's interest in birds had increased since his earlier days in America.

13. These two towns are just south of Lille in northern France.

14. The two species of ducks are known today as the wood duck (*Aix sponsa*) and the mandarin duck (*Aix galericulata*). Males of both species are highly colored. Michaux was especially fond of the wood duck—or summer duck, as he called the species—and a number of living specimens were shipped to France. Michaux visited Louis Jean Marie Daubenton (1716–99), French naturalist, anatomist, and collaborator of Count de Buffon (1707–88) in his encyclopedic *Histoire naturelle* (1749–1804). Louis Claude Marie Richard (1754–1821) was later to help in writing Michaux's *Flora*. René Louiche Desfontaines (1750–1833) was a prominent professor of botany who was associated with the conversion of the Jardin du Roi into the Jardin des Plantes and later became its director. The Thouin mentioned here is either Jean Thouin, brother to André Thouin (1747–1824), or the Thouin household. André Thouin himself had been named

commissioner of the French Republic attached to the group following Gen. Bonaparte to Italy in May 1796 to access cultural material for France. He did not return to France until April 1798 (Letouzey 1989).

15. Jacques-Phillippe Martin Cels was an old friend from Versailles, a horticulturist who had a well-known nursery in Montrouge (now part of Paris), to whom Michaux had been sending seeds from America. Tessier is Alexandre Henri Tessier (1741–1837), who had been codirector (with Abbé Nolin) of the plant nursery at Rambouillet, where most of Michaux's shipments of plants and seeds were sent before the revolution.

16. Charles Louis l'Héritier de Brutelle was a famous French naturalist, former prosecutor for the king, and supervisor of waters and forests of Paris. He worked on *Cornus* and *Geranium*, and he described *Michauxia*, an attractive genus of the *Campanulaceae* that Michaux collected during his Persian travels.

17. AN F10/392, dossier 1, #49, transcription by Régis Pluchet.

18. Mangourit was consul in Charleston when the Girondins were in power during the French Republic (1792–94), and he and Michaux became good friends. Although he was recalled to France by the Jacobins, Mangourit was later named as chargé d'affaires to the United States when the Girondins returned to power, but he was not approved by the United States and never returned.

19. Col. Robert Fulton was incorrectly identified as the famous engineer who developed the steamboat (Savage and Savage 1986; Taylor and Norman 2002). The Col. Fulton to whom Michaux refers here was Samuel Fulton, the associate of Gen. Clark, who was in Paris from 1795 to 1798 trying to resurrect French action along the Mississippi and to obtain reimbursement for Clark's services from the French government.

20. AN F10/392, dossier 1, #32, transcription by Régis Pluchet.

21. AN F10/392, dossier 1, #17, transcription by Régis Pluchet.

22. AN F10/392, dossier 1, #28, transcription by Régis Pluchet.

23. Michaux returned to his birthplace of Satory at Versailles, whose land he had cultivated for many years and where his brother André-François lived.

24. Louis-Guillaume Lemonnier (1717–99) was Michaux's mentor and former principal professor of botany at Jardin du Roi.

25. Jean Thouin, younger brother to André Thouin, became head gardener at the Jardin des Plantes when the latter was named professor of horticulture in 1793. Madame Gilbert is Marie-Jeanne Guillebert, oldest sister of André Thouin and head of the Thouin household. Madame Le Clerc was another sister of André Thouin. Madame Trouvé was André Thouin's natural daughter, by Buffon's son's nursemaid, who was adopted by the Thouin household at the age of two. She was known as Gorelli before her marriage to Claude Joseph Trouvé, editor-in-chief of the newspaper *Moniteur universel* (Letouzey 1989).

26. Jean-Anthelme Barquet (ca. 1760–1809), a former Catholic priest, had married Michaux's sister Gabrielle (1752–1827) while the botanist was in America.

27. M. du Pont was most probably Pierre Samuel du Pont de Nemours (1739–1817), the well-known economist, whose oldest son, Victor, was French consul in Charleston in 1796, and whose youngest son, Eleuthère Irénée (1771–1834), was soon to begin an industrial empire in Delaware. In the early 1800s, E. I. du Pont and his father attempted for several years to lobby the French government to keep the French gardens in America, especially the one in New Jersey, from being sold (DuPont 1924). E. I. du Pont was genuinely interested in plants (Wilkinson 1972); he had studied with de Jussieu in Paris as a young man. In the Hagley Library in Wilmington, there is a letter of December 10, 1801 (W4-606), from François André

answering E. I. du Pont's request for a copy of Michaux's *Histoire des chênes*. This botanical interest was continued by his descendants, especially his great-grandson Pierre Dupont (1870–1954), who established the world-famous Longwood Gardens, near Philadelphia. Louis-Marie la Révellière-Lepéaux (1753–1824) was an amateur botanist. He was influenced by J. J. Rousseau, and he became a member of the Revolutionary Consul and the Directory and head of the Constitutional Convention. La Révellière-Lépeau was a great friend of Thouin. Michaux's friend Bosc had saved him along with several other Girondin leaders during the Reign of Terror.

28. Michaux dined at the home of the brother of his late wife, Anne-Cécile. Remi Claye was a dry goods merchant in Paris.

29. Pierre-Joseph Redouté (1759–1840) and his brother H. J. Redouté were painters at the Louvre. The former did the illustrations for André Michaux's *Histoire des chênes de l'Amérique* (Paris: 1801) and *Flora Boreali-Americana* (Paris: 1803), and both contributed to François André's *North American Sylva* (Paris: 1817–19). No doubt there was already conversation with the Redoutés regarding the illustrations for Michaux's future works.

30. Etienne Pierre Ventenat (1757–1808) was first trained as a monk, and he later became head librarian of the original Ste. Geneviève Church with its well-known collection of ancient books. His interest turned to science upon his visit to English gardens. Later he was appointed librarian at the Panthéon and became member of the Institut. He described several new species from Michaux's collections from North America in his book *Descriptions des plantes nouvelles et peu connues cultivées dans le Jardin de J. M Cels* (1803).

31. The Panthéon, now a famous monument in Paris, was initially built (1754–89) as a church, known as Église Ste Geneviève. After the revolution, the monument became a repository for the ashes of famous men and women.

32. Jean Baptiste Dubois (1753–1808) was a naturalist and an economist who was head of the agriculture division in the Ministry of the Interior. He supported Michaux's demands for indemnity and tried to find him a position (R. Pluchet, pers. comm.).

33. Claude Jean Veau Delaunay (1755–1826) was a student of Louis Jean Marie Daubenton.

34. Jean-Baptiste Louvet was a journalist, a deputy during the Girondin regime, and a good friend of Bosc and Révellière-Lépeaux.

35. Michaux wrote "Bernard de Ste Afrique," not "Bernardde Ste Afrique" as given in Sargent (1889). This person may have been Jacques Henri Bernadin de Saint-Pierre (1737–1814), writer, naturalist, and intendant of the Jardin des Plantes in 1792–93, but more likely, as pointed out by R. Pluchet (pers. comm.), he was Louis Bernard de Saint-Affrique, member of the legislative body under the Directory and head of the Council of Five Hundred, the subsequent legislative body in power from 1795 to 1799. He was from Ste. Affrique in the Gard Region of southern France.

36. Capt. "Baas" is likely the Capt. Bass whose ship made voyages between Charleston and Bordeaux. Michaux mentioned him on March 8, 1790, in connection with a shipment sent to Monsieur, brother of the king, by way of Bordeaux. Citizen Droict-Bussy was chancellor of the consulate in New York. P. G. Chion, of French origin, was a Charleston merchant and friend of Michaux's. Saulnier is Pierre-Paul Saunier, now on his own and in charge of the New Jersey nursery. Usually Michaux used the latter spelling of Saunier's name.

37. Jean Jacques Himely, native of Switzerland, was Michaux's friend and patron while in Charleston. Himely purchased the Carolina nursery, which Michaux sold for fifty-three guineas at a public auction on March 27, 1792. Himely ceded the property back to Michaux (Savage and Savage 1986). Michaux entrusted the Charleston garden to Himely in 1796, pending his return to the United States. Himely kept the Charleston garden from going to ruins.

## Epilogue

1. Dorr 2004; Pluchet 2010.

2. AM to Samuel du Pont, 1799. HA W2-2340.

3. AM to Per Afzelius, February 18, 1797. UP, Thunberg archives, signature G 300s.

4. AM to Carl Thunberg, November 16, 1797; May 28, 1798; May 4, 1799. UP, Thunberg archives, signature G 300s.

5. AM to Rev. Nicholas Collin, 1796. RSAS, archives of Nils Collin, L-Ö.

6. Alexander von Humboldt to AM, circa 1798. APS B M 58.

7. Grégoire 1810, 159.

8. Helferich 2004, 19.

9. Jangoux 2004, 56; Pluchet 2010, 206; AM letters to officials, AN F10-392, dossier 2, and AJ/15/569.

10. Jangoux 2013, 405.

11. AM, "Aperçu du plan de mon voyage," circa 1800. AN AJ/15/569.

12. Grégoire 1810, 159; Wadström 1798, 88; Etienne Geoffroy Saint-Hilaire, Lettre aux professeurs, séance du 28 germinal an xii (April 18, 1804), AN AJ/15/592; C. Jouanin., pers. comm.

13. Antoine-Laurent de Jussieu to Jan Hendrik van Swinden, cited by Rousseau 1964, 228–30.

14. AM to Antonio José Cavanilles, August 8, 1800. Archives Real Jardin Botánico, Madrid (MABG), corresp. científica Cavanilles, cat. 194 # 43.

15. Oppert 1895; "Lucien Bonaparte authorizes the purchase of the Caillou," 24 vendémiaire an ix (October 16, 1800). AN AJ/15/569, Dossier Baudin, folios 362–65, transcribed by Régis Pluchet.

16. "Assemblée des professeurs du Muséum," séance 28 germinal an xii (April 18, 1804). AN AJ/15/592.

17. Pluchet 2010, 207.

18. Pluchet 2010, 207; "Procuration à François-André Michaux," 4 vendémiaire an ix (September 26, 1800). AN Minutier central des notaires de Paris, ETIX-839.

19. AM to Antoine-Laurent de Jussieu, 12 vendémiaire an ix (October 4, 1800). Archives de l'Académie des sciences, Paris (AAS), dossier Michaux.

20. Pluchet 2010, 207.

21. AM to Antoine-Laurent de Jussieu, 25 germinal an x (April 15, 1802). AAS, dossier Michaux.

22. Jouanin 2004.

23. Pluchet 2010, 205.

24. Deleuze 1805, 349.

25. Dorr 2004.

26. Michaud and Michaud 1826; R. Pluchet, pers. comm.

27. Pluchet 2010, 208; AM au citoyen Michaux fils, 19 vendémiaire an x (October 11, 1801), APS B58.

28. Pluchet 2010, 209.

29. Pluchet 2010, 209.

30. Dorr 2004, 13; Ly-Tio Fane 1996.

31. Pluchet 2004, 228, 231.

32. Pluchet 2010, 209; [Assemblée des professeurs du Muséum, séance du 15 thermidor an x (Aug. 3, 1803). AN AJ/15/5990].

33. "Rougey de la Réunion" is perhaps *Syzium cymosum*, known today as *bois de pomme* because of its small, red, apple-like fruits. This species is endemic to Réunion and Mauritius.

34. Pluchet 2010, 210.

35. E. Geoffroy Saint-Hilaire, Lettre lue a l'assemblée des professeurs du Muséum, séance du 28 germinal an xii (April 18, 1804). AN AJ/15/592, transcribed by Christian Jouanin.

36. Grégoire 1810, 77.

37. Leschenault 1810; R. Pluchet, pers. comm.

38. Pluchet 2010, 210; F. A. Michaux, Procuration à Paul Martin de Moncamp, 14 messidor an xii (July 3, 1804). AN, Minutier central des notaires de Paris, ETIX-881.

39. Pluchet 2010, 210; Archives du Muséum d'histoire naturelle, Le Havre, Ms. 15034. Jangoux (2013, 222–23) further reported that the marine office sold Michaux's possessions.

40. É. Geoffroy Saint-Hilaire, Lettre lue a l'assemblée des professeurs du Muséum, séance du 28 germinal an xii (April 18, 1804). AN AJ/15/592, transcribed by Christian Jouanin.

41. Pluchet 2010, 211; François 1940.

42. Pluchet 2004, 228–32.

43. Robbins and Howson 1958.

44. Robbins and Howson, 367.

45. Wilkinson 1972.

46. Photo and letter of P. Saunier (descendant of Pierre-Paul) to William J. Robbins, 1957. New York Botanical Garden Library and Archives, New York (NYBG), Robbins papers.

47. Thouin 1802.

48. Coker 1911.

49. FAM in Paris to John Simon, collector, in Charleston, September 5, 1803; and FAM in Paris to Philippe Noisette in Charleston, June 10, 1804. MNHN, transcriptions by Régis Pluchet.

50. Cothran 1995.

51. Joyce 1988, 2009.

52. *Charleston Post and Courier*, June 5, 2016.

53. Coker 1911.

54. Hunt 1947.

55. MacPhail 1981.

56. Swartz 1986.

# Bibliography

Albright, Edward. 1909. *Early History of Middle Tennessee*. Nashville, TN: Brandon.

Alderman, Pat. 1986. *The Overmountain Men*. Johnson City, TN: Overmountain Press.

Alderson, Robert J. 2008. *This Bright Era of Happy Revolutions: French Consul Michel-Ange-Bernard Mangourit and International Republicanism in Charleston, 1792–1794*. Columbia: University of South Carolina Press.

Allen, Gay Wilson, and Roger Asselineau. 1987. *Saint John de Crevecoeur: The Life of an American Farmer*. New York: Viking Press.

Al-Zein, Mohammad S., and Lytton Musselman. 2004. "*Michauxia* (Campanulaceae): A Western Asian Genus Honoring a North American Pioneer Botanist." In Baranski 2004, 200–205.

American Historical Association. 1897. "Correspondence of Clark and Genet." In *American Historical Association Annual Report 1896*, 1:930–1107. Washington, DC: Government Printing Office.

Ammon, Harry. 1973. *The Genet Mission*. New York: W. W. Norton.

Andrew, Rod, Jr. 2017. *The Life and Times of General Andrew Pickens: Revolutionary War Hero, American Founder*. Chapel Hill: University of North Carolina Press.

Andrews, Judith M. 1998. *South Carolina Highway Historical Marker Guide*. Columbia: South Carolina Department of Archives and History.

Arnold, Scott David. 2007. *Guide to Virginia's Historical Markers*. 3rd ed. Charlottesville: University of Virginia Press.

Arnow, Harriette Simpson. 1961. "Education and the Professions in the Cumberland Region." *Tennessee Historical Quarterly* 20:120–58.

Asbury, Francis, and Grady L. E. Carroll. 1964. *Francis Asbury in North Carolina*. Nashville, TN: Parthenon Press.

Asselin, Alain, Jacques Cayouette, and Jacques Mathieu. 2017. *Curieuses histoires de plantes du Canada*. Vol. 3, *1760–1867*. Québec: Les éditions du Septentrion.

Avery, Issac Thomas, Jr. 1979. "Waightstill Avery." In *Dictionary of North Carolina Biography*, edited by William S. Powell, 1:70–71. Chapel Hill: University of North Carolina Press.

Baird, Eleanora Gordon. 2003. "Moses Bartram's Account Book 1778–1788: Notes Made by a Philadelphia Apothecary." *Bartram Broadside* 2003 (Spring): 1–8.

Baldwin, Leland Dewitt. 1941. *The Keelboat Age on Western Waters*. Pittsburgh, PA: University of Pittsburgh Press.

Baranski, Michael J., ed. 2004. *The Proceedings of the André Michaux International Symposium*. *Castanea*: Occasional Papers in Eastern Botany 2. Norris, TN: Southern Appalachian Botanical Society.

Barden, Lawrence S., and James F. Matthews. 2004. "André Michaux's Sumac—*Rhus michauxii* Sargent: Why Did Sargent Rename It and Where Did Michaux Find It?" *Castanea* 69 (2): 109–15.

Barton, Benjamin Smith. 1793. "A Botanical Description of the *Podophyllum Diphyllum* of Linnaeus, in a Letter to Charles Peter Thunberg, M.D." *Transactions of the American Philosophical Society* 41:334–48.

Bartram, William. 1958. *Travels of William Bartram*. Naturalist's Edition. Edited by Francis Harper. New Haven, CT: Yale University Press.

Beach, Ursula Smith. 1964. *Along the Warioto, or A History of Montgomery County, Tennessee*. Nashville: Clarksville Kiwanis Club and the Tennessee Historical Commission.

Beckner, Lucien. 1948. "John D. Shane's Copy of Needham Parry's Diary of Trip Westward." *Filson Club History Quarterly* 22 (4): 232–53.

Bedford, John R. 1919. "A Tour in 1807 Down the Cumberland, Ohio, and Mississippi Rivers to New Orleans." *Tennessee Historical Magazine* 5 (1): 40–69.

Belting, Natalia Maree. 1948. *Kaskaskia under the French Regime*. Urbana: University of Illinois Press.

Bergh, Albert Ellery, ed. 1907. *The Writings of Thomas Jefferson*. 20 vols. Washington, DC: Thomas Jefferson Memorial Association of the United States.

Berkeley, Edmund, and Dorothy Smith Berkeley. 1963. *John Clayton: Pioneer of American Botany*. Chapel Hill: University of North Carolina Press.

———. 1982. *The Life and Travels of John Bartram: From Lake Ontario to the River St. John*. Tallahassee: University Presses of Florida.

Berry, Connelly Burgin. 1972. Gause Family Data Collected by Connelly Burgin Berry. New Hanover County Public Library, Wilmington, NC.

Best, Edna Hunter. 1936. *Sketches of Washington, Mason County, Kentucky*. Washington, KY: n.p.

Beyer, George R. 2000. *Guide to the State Historical Markers of Pennsylvania*. Harrisburg: Pennsylvania Historical and Museum Commission.

Blake, Lucille Dresser. 1970. *Brunswick County Court Minutes 1782–1786, 1789, 1801*. Wilmington, NC: North Carolina Room, New Hanover County Public Library.

Bloom, Jo Tice. 1995. "Establishing Precedents, Dr. James White and the Southwest Territory." *Tennessee Historical Quarterly* 54 (4): 325–35.

Blythe, LeGette. 1961. *The Hornet's Nest*. Charlotte, NC: Public Library of Charlotte and Mecklenburg County.

Boddie, John Bennett. 1956. *Southside Virginia Families*. Redwood City, CA: Pacific Coast.

Boewe, Charles. 2011. *The Life of C. S. Rafinesque, a Man of Uncommon Zeal*. Philadelphia: American Philosophical Society.

Boiteau, P. 1975. "La dynastie des Richard, jardiniers-botanistes." *Actes du 100ième Congres des sociétés savantes, Paris* 3:13–29.

Bory de Saint Vincent, J. B. 1804. *Voyage dans les quatres principales iles des mers d'Afrique*. Paris: Buisson.

Brasher, J. Lawrence. 2004. "Bedazzled and Bedeviled: The Religious Sensibilities of André Michaux." In Baranski 2004, 140–46.

Brevard, Joseph. 1814. *An Alphabetical Digest of Public Statute Law of South Carolina*. Vol. 3, Title 19, *Roads, Bridges, and Ferries*. Charleston, SC: John Hoff.

Brissot de Warville, Jacques-Pierre. 1964. *New Travels in the United States of America, 1788*. Edited by Durand Echeverria. Cambridge, MA: Belknap Press of Harvard University Press.

Brock, R. A. (1888) 1973. "James McDowell." In *Virginia and Virginians*, 204–8. Spartanburg, SC: Reprint.

Brown, Douglas Summers. 1968. *Sketches of Greensville County Virginia, 1650–1967*. Emporia, VA: Riparian Woman's Club.

Brugger, Robert J. 1988. *Maryland, a Middle Temperament*. Baltimore: Johns Hopkins University Press and the Maryland Historical Society.

Brunet, Abbé Ovide. 1864. "Michaux and His Journey in Canada." *Canadian Naturalist and Geologist* 1:331–43.

Buchanan, John. 1997. *The Road to Guilford Courthouse: The American Revolution in the Carolinas*. New York: John Wiley and Sons.

Bullard, Mary R. 1995. *Robert Stafford of Cumberland Island: Growth of a Planter*. Athens: University of Georgia Press.

Burnett, Robert A. 1976. "Louisville's French Past." *Filson Club History Quarterly* 50 (2): 5–27.

Caldwell, Norman W. 1950. "Fort Massac: The American Frontier Post 1778–1805." *Journal of the Illinois State Historical Society* 43 (4): 265–80.

Carey, Matthew. 1796. *American Pocket Atlas*. Philadelphia: Lang and Ustick.

———. 1814. *Carey's General Atlas, Improved and Enlarged*. Philadelphia: printed by the author.

Carter, Clarence Edwin. 1936. *Territorial Papers of the United States*. Vol. 4, *The Territory South of the River Ohio*. Washington, DC: Government Printing Office.

Castiglioni, Luigi. 1983. *Luigi Castiglioni's Viaggio: Travels in the United States of North America 1785–87*. Translated and edited by Antonio Pace. Syracuse, NY: Syracuse University Press.

Catanzariti, John. 1990. *The Papers of Thomas Jefferson*. Vol. 24, *1 June to 31 December 1792*. Princeton, NJ: Princeton University Press.

———. 1992. *The Papers of Thomas Jefferson*. Vol. 25, *1 January to 10 May 1793*. Princeton, NJ: Princeton University Press.

———. 1995. *The Papers of Thomas Jefferson*, Vol. 26, *11 May to 31 August 1793*. Princeton, NJ: Princeton University Press.

Catesby, Mark. 1771. *The Natural History of Carolina, Florida and the Bahama Islands*. 2 vols. London: B. White.

Cayouette, Jacques. 2014. *A la découverte du Nord: Deux siècles et demi d'exploration de la flore nordique du Québec et du Labrador*. Québec: Éditions Multimondes.

Centre national de la recherche scientifique. 1957. *Les botanistes français en Amérique du Nord*. Paris: Centre national de la recherche scientifique.

Chapelier, L. A. 1905. "Lettres de Chapelier." *Bulletin de l'Académie Malgache* 4:4.

Chastellux, François Jean. 1963. *Travels in North America*. Chapel Hill: University of North Carolina Press.

Chinard, Gilbert. 1943. "Jefferson and the American Philosophical Society." *Proceedings of the American Philosophical Society* 87:263–76.

———. 1957. "André and François-André Michaux and Their Predecessors: An Essay on Early Botanical Exchanges between America and France." *Proceedings of the American Philosophical Society* 101:344–61.

Christian, John H. 1990. *Founders of St. Marys*. N.p.: n.p.

———. 1995. *Reconstructed 1800 Census Index, Camden County, GA*. Woodbine, GA: Bryan-Lang Library.

Clark, Murtie June. 1990. *American Militia in the Frontier Wars, 1790–1796*. Baltimore, MD: Genealogical.

Clift, G. Glenn. 1936. *History of Maysville and Macon County*. Lexington, KY: Transylvania Printing.

———. (1954) 1993. *"Second Census" of Kentucky 1800*. Baltimore, MD: Genealogical.

Coker, William Chambers. 1911. "The Garden of André Michaux." *Journal of the Elisha Mitchell Scientific Society* 27 (2): 65–73.

———. 1943. "*Magnolia cordata* Michaux." *Journal of the Elisha Mitchell Scientific Society* 59 (1): 81–88.

Conlin, Michael F. 2000. "The American Mission of Citizen Pierre-Auguste Adet: Revolutionary Chemistry and Diplomacy in the Early Republic." *Pennsylvania Magazine of History and Biography* 124 (4): 489–520.

Cooke, J. W. 1963. "Governor Shelby and Genet's Agents." *Filson Club History Quarterly* 32 (2): 162–70.

Corbitt, David Leroy. 1950. *The Formation of the North Carolina Counties, 1663–1943*. Raleigh: North Carolina Division of Archives and History.

Cothran, James. 1995. *Gardens of Historic Charleston*. Columbia: University of South Carolina Press.

———. 2004. "Treasured Ornamentals of Southern Gardens—Michaux's Lasting Legacy." In Baranski 2004, 149–57.

Coulter, E. Merton. 1960. *A Short History of Georgia*. Chapel Hill: University of North Carolina Press.

———. 1965. *Old Petersburg and the Broad River Valley of Georgia*. Athens: University of Georgia Press.

Coulter, J. M. 1883. "Some North American Botanists. III. André Michaux." *Botanical Gazette* 8 (3): 181–83.

Cowan, Thomas. 1987. "William Hill and the Aera Ironworks." *Journal of Early Southern Decorative Arts* 13 (2): 1–31.

Craig, Neville B. 1854. *A Sketch of the Life and Services of Isaac Craig*. Pittsburgh, PA: J. S. Davidson.

Culler, Daniel M. 1995. *Orangeburgh District, 1768–1868: History and Records*. Spartanburg, SC: Reprint.

Daudin, F. M. 1803. *Histoire naturelle, générale et particulière des rèptiles*. 8 vols. Paris: Dufart.

Daughters of the American Revolution. 2003. *DAR Patriot Index*. 3 Vols. Baltimore, MD: Gateway Press.

Davidson, Grace Gillam. 1932. *Early Records of Georgia*. Vol. 1, *Wilkes County*. Macon, GA: Daughters of the American Revolution.

Davies, P. A. 1956. "Type Location of *Shortia galacifolia*." *Castanea* 21 (3): 107–13.

Davis, James E. 1998. *Frontier Illinois*. Bloomington: Indiana University Press.

Deleuze, J. P. F. 1805. "Memoir of the Life and Botanical Travels of André Michaux." Translated by Charles König and John Sims. *Annals of Botany* 1:321–55. Originally published in 1804 as "Notice historique sur André Michaux." *Annales du Museum d'histoire naturelle* 3:191–227.

———. (1805) 2011. *Annotated Memoirs of the Life and Botanical Travels of André Michaux*. Edited by Charlie Williams. Athens, GA: Fevertree Press.

De Sacy, Jacques Silvestre. 1953. *Le comte d'Angiviller: Dernier directeur général des bâtiments du roi*. Paris: Plon.

Deyton, Jason Basil. 1947. "The Toe River Valley to 1865." *North Carolina Historical Review* 24 (4): 423–66.

Dorr, Laurence J. 2004. "André Michaux in Madagascar." Revision of unpublished paper presented at the André Michaux International Symposium, Belmont Abbey College, May 2002.

Doyle, Barbara, Mary Edna Sullivan, and Tracey Todd. 2008. *Beyond the Fields: Slavery at Middleton Place*. Charleston, SC: Middleton Place Foundation.

Draper, Lyman C. (1881) 1996. *King's Mountain and Its Heroes: History of the Battle of King's Mountain, October 7th, 1780*. Johnson City, TN: Overmountain Press.

Drayton, John. 1943. *The Carolinian Florist of Governor John Drayton of South Carolina*. Edited by Margaret Babcock Meriwether. Columbia: South Caroliniana Library of the University of South Carolina.

Dugger, Shepherd. (1892) 1934. *The Balsam Groves of the Grandfather Mountain*. Banner Elk, NC: S. M. Dugger.

Dumont de Courset, Georges Louis Marie. 1811–14. *Le botaniste cultivateur*. 2nd ed. 7 vols. Paris: Deterville.

Du Pont, Bessie Gardner. 1923. *Life of Eleuthère Irénée du Pont from Contemporary Correspondence*. Newark: University of Delaware Press.

Du Pont, Eleuthère Irénée. 1924. *Life of Eleuthère Irénée du Pont from Contemporary Correspondence*. Vol. 5, *1799–1802*. Translated by Bessie Gardner Du Pont. Wilmington: University of Delaware Press.

Duprat, Gabrielle. 1957. "Essai sur les sources manuscrites conservées au Muséum national d'histoire naturelle." In Centre national de la recherche scientifique 1957, 241–43.

Dupre, Huntley. 1941a. "The French in Early Kentucky." *Filson Club History Quarterly* 15 (2): 78–104.

———. 1941b. "The Political Ideas of George Nicholas." *Register of the Kentucky State Historical Society* 39:201–23.

Durham, Walter T. 1971. "Kasper Mansker: Cumberland Frontiersman." *Tennessee Historical Quarterly* 30 (2): 154–71.

Dutilly, Arthème, and Ernest Lepage. 1945. "Retracing the Route of Michaux's Hudson Bay Journey of 1792." *Revue de l'Université d'Ottawa* 1 (15): 89–102.

Dwelle, Mary Myers. 1968. *A Sketch of the Life of Queen Charlotte*. Charlotte, NC: Heritage Printers.

*Early Kentucky Settlers: The Records of Jefferson County, Kentucky, from the Filson History Quarterly*. 1988. Baltimore, MD: Genealogical.

Edgar, Walter. 1998. *South Carolina: A History*. Columbia: University of South Carolina Press.

———. 2006. *South Carolina Encyclopedia*. Columbia: University of South Carolina Press.

Edgar, Walter, Inez Watson, and N. Louise Bailey. 1981. *Biographical Directory of the South Carolina House of Representatives*. Vol. 3, *1775–1790*. Columbia: University of South Carolina Press.

Eifert, Virginia. 1965. *Tall Trees and Far Horizons: Adventures and Discoveries of Early Botanists in America*. New York: Dodd, Mead.

Ekberg, Carl J. 1998. *French Roots in the Illinois Country: The Mississippi Frontier in Colonial Times*. Urbana: University of Illinois Press.

———. 2011. *A French Aristocrat in the American West: The Shattered Dreams of De Lassus de Luzières*. Columbia: University of Missouri Press.

Erwin, Eliza Cowan, and Horace Fraser Rudisill. 1976. *Darlingtoniana: A History of People, Places and Events in Darlington County, South Carolina*. Spartanburg, SC: Reprint.

Ewan, Joseph, and Nesta Ewan. 1963. "John Lyon, Nurseryman and Plant Hunter, and His

Journal, 1799–1814." *Transactions of the American Philosophical Society* 53 (2): 1–69.

———. 2007. *Benjamin Smith Barton, Naturalist and Physician in Jeffersonian America.* Saint Louis: Missouri Botanical Garden Press.

Faragher, John Mack. 1990. *The Encyclopedia of Colonial and Revolutionary America.* New York: Facts on File.

Federal Writers' Project. (1939) 1988. *North Carolina: The WPA Guide to the Old North State.* Columbia: University of South Carolina Press.

Ferguson, Herman W. 1993. *Mecklenburg County, North Carolina: Will Abstracts, 1791–1868, Books A–J.* Rocky Mount, NC: H. W. Ferguson.

———. 1995. *Mecklenburg County, North Carolina: Minutes of the Court of Common Pleas and Quarter Sessions.* Vol. 1, *1780–1800.* Rocky Mount, NC: H. W. Ferguson.

Fernald, Merritt L. 1950. "Betula michauxii, a Brief Symposium." *Rhodora* 52:26–33.

Filson, John. 1784. *The Discovery, Settlement and Present State of Kentucke.* Wilmington, DE: James Adams.

Finiels, Nicolas de, Carl J. Ekberg, and William E. Foley. 1989. *An Account of Upper Louisiana.* Columbia: University of Missouri Press.

Fischer, David Hackett, and James C. Kelly. 2000. *Bound Away, Virginia and the Westward Movement.* Charlottesville: University of Virginia Press.

Fischer, Marjorie Hood. 1996. *Tennesseans before 1800, Washington County.* Galveston, TX: Frontier Press.

Fisk, Connie, and Alexander Krings. 2009. "Clarification of the Typification of Michaux Names in Eastern North American *Vitis* (Vitaceae)." *Journal of the Botanical Research Institute of Texas* 3 (2): 739–40.

Forbes, James Grant. (1821) 1964. *Sketches, Historical and Topographical, of the Floridas; More Particularly of East Florida.* Gainesville: University of Florida Press.

Forester, Cathy Tudor. 1996. *Tennessee Historical Markers.* 8th ed. Nashville: Tennessee Historical Commission.

Forsyth, King Logan. 1932. "Quesnay de Beaurepaire's Project for an Academy of Science and Fine Arts in Richmond." Master's thesis, University of Virginia.

Foster, Stephen, and James A. Duke. 1990. *A Field Guide to Medicinal Plants: Eastern and Central North America.* Boston: Houghton Mifflin.

Fouts, Raymond Parker. 1984. *Abstracts of Newspapers from Wilmington, N.C.* Vol. 3, *1801–1803.* Cocoa Beach, FL: Gen Rec Books.

François, E. 1940. "Notes sur les tombes de Michaux et Chapelier." *Collection de documents concernant Madagascar* 2:169.

Frappaz, T. 1939. *Les voyages du lieutenant de vaisseau Frappaz dans les mers des Indes.* Tananarive: Pitot de la Beaujardiere.

Fraser, John. 1789. *A Short History of the* Agrostis cornucopiae: *Or the New American Grass.* London: printed for the author.

Freeze, Gary R. 1995. *The Catawbans: Crafters of a North Carolina County, 1746–1900.* Newton, NC: Catawba County Historical Association.

Fressange, J. B. 1808. "Voyage à Madagascar en 1802–1803." In *Annales des voyages de la géographie et de l'histoire,* edited by C. Maltebrun, 2:39. Paris: Buson.

Fry, Joel T. 2000. "*Franklinia alatamaha*, A History of That 'Very Curious' Shrub, Part 1: Discovery and Naming of the Franklinia." *Bartram Broadside* 2000 (Spring): 1–24.

Fry, Joshua, and Peter Jefferson. (1755) 2000. "A Map of the Most Inhabited Part of Virginia . . ." In Stephenson and McKee 2000, 83–88.

Fulcher, Bob. 1998. "Muir, Michaux, and Gray on the Roan." *Tennessee Conservationist* 64 (5): 14–20.

Furr, Mary Lou Avery. 2007. "Is There a 'Southern Hive' of Averys?" *Avery Newsletter* 36 (2): n.p. Avery Memorial Association, Groton, CT.

Furstenberg, François. 2014. *When the United States Spoke French: Five Refugees Who Shaped a Nation*. New York: Penguin Press.

Galle, Fred C. 1987. *Azaleas*. Portland, OR: Timber Press.

Gandhi, K. N., J. L. Reveal, and J. L. Zarucchi. 2012. "Nomenclatural and Taxonomic Analysis of *Convallaria majalis*, *C. majuscula* and *C. montana* (Ruscaceae/Liliaceae)." *Phytoneuron* 17:1–4.

Gerard, W. R. 1885. "The Word Savoyanne." *Bulletin of the Torrey Botanical Club* 12:72–73.

Gibbons, Whit, and Michael Dorcas. 2005. *Snakes of the Southeast*. Athens: University of Georgia Press.

Gillispie, Charles C. 1992. "Palisot de Beauvois and the Americans." *Proceedings of the American Philosophical Society* 136 (1): 33–50.

Gilman, Arthur V. 2010. "Checklist of the Flora of Vermont." *Vermont Botanist* 2:2–37.

———. 2015. *New Flora of Vermont*. New York: New York Botanical Garden Press.

Gilmore, James R. (1887) 1997. *John Sevier as Commonwealth Builder*. Johnson City, TN: Overmountain Press.

Gleason, Henry A., and Arthur Cronquist. 1991. *Manual of Vascular Plants of Northeastern United States and Adjacent Canada*. New York: New York Botanical Garden Press.

Gottschalk, Louis. 1976. *The Letters of Lafayette to Washington, 1777–1799*. Philadelphia: American Philosophical Society.

Graustein, Jeannette E. 1967. *Thomas Nuttall, Naturalist: Explorations in America, 1808–1841*. Cambridge, MA: Harvard University Press.

Green, Karen Mauer. 1983. *The Kentucky Gazette, 1787–1800: Genealogical and Historical Abstracts*. Galveston, TX: Frontier Press.

Gregg, Alexander. 1867. *History of the Old Cheraws*. New York: Richardson.

Grégoire, Henri-Baptiste. 1810. *Enquiry Concerning the Intellectual and Moral Faculties, and Literature of Negroes*. Translated by D. B. Warren. Brooklyn, NY: Thomas Kirk.

Griffey, Irene M. 2000. *Earliest Tennessee Land Records and Earliest Tennessee Land History*. Baltimore, MD: Genealogical.

Guice, Julia Cook. 1991. *Frederick Stump: The Rest of the Story*. Biloxi, MS: J. C. Guice.

Hairr, John. 2007. *North Carolina Rivers: Facts, Legends and Lore*. Charleston, SC: History Press.

Hall, Thomas B., Jr., and Thomas B. Hall III. 1975. *Dr. John Sappington of Saline County, Missouri*. Arrow Rock, MO: Friends of Arrow Rock.

Hallock, Thomas, Nancy Hoffman, and Joel Fry. 2010. *William Bartram, the Search for Nature's Design: Selected Art, Letters, and Unpublished Writings*. Athens: University of Georgia Press.

Hammond, Neal O. 2002. "Kentucky Pioneer Forts and Stations." *Filson History Quarterly* 76 (4): 523–86.

Harper, Francis. 1940. "Some Works of Bartram, Daudin, Latreille, and Sonnini, and Their Bearing upon North American Herpetological Nomenclature." *American Midland Naturalist* 23 (3): 692–723.

Harrison, Lowell H. 1985. *Kentucky Governors, 1792–1985*. Lexington: University Press of Kentucky.

Haynes, Robert R. 1986. "Typification of Linnaean Species of *Potamogeton* (Potamogetonaceae)." *Taxon* 35 (3): 563–73.

Haywood, Marshall DeLancey. 1915. *John Branch, 1782–1863, Governor of North Carolina, United States Senator, Secretary of the Navy, Member of Congress, Governor of Florida, etc.* Raleigh, NC: Commercial.

Heinemann, Charles Brunk. (1940) 1971. *First Census of Kentucky, 1790*. Baltimore, MD: Genealogical.

Heitman, Francis B. (1914) 1982. *Historical Register of Officers of the Continental Army during the War of the Revolution, April 1775 to December 1783*. Baltimore, MD: Genealogical.

Heitzler, Michael J. 2005. *Goose Creek: A Definitive History, 1670–2003*. Vol. 1, *Planters, Politicians and Patriots*. Charleston, SC: History Press.

Helferich, Gerard. 2004. *Humboldt's Cosmos: Alexander von Humboldt and the Latin American Journey That Changed the Way We See the World*. New York: Gotham Books.

Henderson, Archibald. 1920. "Isaac Shelby and the Genet Mission." *Mississippi Valley Historical Review* 6 (4): 451–69.

Hendricks, Christopher E. 2006. *The Backcountry Towns of Colonial Virginia*. Knoxville: University of Tennessee Press.

Hill, Michael. 2001. *Guide to North Carolina Historical Highway Markers*. 9th ed. Raleigh: North Carolina Division of Archives and History.

Hill, William, and A. S. Salley. 1921. *Colonel William Hill's Memoirs of the Revolution*. Columbia, SC: Historical Commission of South Carolina.

Hirsch, Arthur H. (1928) 1999. *The Huguenots of Colonial South Carolina*. Columbia: University of South Carolina Press.

Hochreutiner, B. P. G. 1934. "Validity of the Name Lespedeza." *Rhodora* 36:390–92.

Hogeland, William. 2006. *The Whiskey Rebellion*. New York: Scribner.

Holcomb, Brent. 1978. *Winton (Barnwell) County, South Carolina, Minutes of County Court and Will Book I, 1785–1791*. Easley, SC: Southern Historical Press.

———. 1981. *Lancaster County, South Carolina, Deed Abstracts, 1787–1811*. Easley, SC: Southern Historical Press.

Holley, Horace. 1823. *A Discourse Occasioned by the Death of Col. James Morrison: Delivered in the Episcopal Church, Lexington, Kentucky, May 19th, 1823*. Lexington, KY: J. Bradford.

Horn, Stanley F. 1961. "The Hermitage, Home of Andrew Jackson." *Tennessee Historical Quarterly* 20 (2): 4–5.

Houck, Louis. 1909. *The Spanish Regime in Missouri*. 2 vols. Chicago: R. R. Donnelly.

Hudson, Frank Parker. 1988. *A 1790 Census for Wilkes County, Georgia: Prepared from Tax Returns*. Spartanburg, SC: Reprint.

Hunt, Kenneth W. 1947. "Charleston Woody Flora." *American Midland Naturalist* 37:670–786.

Hutchins, Thomas. 1778. *A Topographical Description of Virginia, Pennsylvania, Maryland, and North Carolina*. London: printed by the author.

Hutchins, Thomas, and T. Cheevers. (1778) 2000. "A New Map of the Western Parts of Virginia, Pennsylvania, Maryland and North Carolina; Comprehending the River Ohio, and All the Rivers, Which Fall into It . . ." In Stephenson and McKee 2000, 97–101.

Hyde, Elizabeth. 2013. "André Michaux and French Botanical Diplomacy in the Cultural Construction of Natural History in the Atlantic World." In *Elephants and Roses: French Natural History 1790–1830*, edited by Sue Ann Prince, 88–100. Philadelphia: American Philosophical Society.

IDC (Inter Documentation Company). 1968. Microfiche set of André Michaux's Herbarium. IDC 6211, 145 microfiches. Zug, Switzerland: Inter Documentation.

Imlay, Gilbert. 1793. *Description of the Western Territory of North America*. Dublin: William Jones.

Ingmire, Frances Terry. 1982. *Hardin County, Kentucky, Will Book "B," 1810–1816*. Signal Mountain, TN: Mountain Press.

Jackson, Andrew, Sam B. Smith, Harriett F. C. Owsley, and Harold D. Moser. 1980. *The Papers of Andrew Jackson*. Vol. 1, *1770–1803*. Knoxville: University of Tennessee Press.

Jackson, Donald, and Dorothy Twohig. 1979. *Diaries of George Washington*. Vol. 6, *January 1790–December 1799*. Charlottesville: University of Virginia Press.

James, James Alton. 1928. *The Life of George Rogers Clark*. Chicago: University of Chicago Press.

Jandin, S. 1995. "L'itinéraire d'un naturaliste Louis-Claude Richard (1754–1821)." Master's thesis, Université de Paris.

Jangoux, Michel. 2004. "Les zoologistes et botanistes qui accompagnèrent le capitaine Baudin aux terres australes." *Australian Journal of French Studies* 41:55–78.

———. 2013. *Le voyage aux terres australes du commandant Nicholas Baudin*. Paris: University Presses.

Jefferson County Genealogical Society. 1996. *Jefferson County, Tennessee: Families and History, 1792–1996*. Waynesville, NC: Don Mills and the Jefferson County Genealogical Society.

Johnson, Amandus. 1936. *Journal and Biography of Nicholas Collin*. Philadelphia: New Jersey Society of Pennsylvania.

Johnson, Randy. 2016. *Grandfather Mountain: The History and Guide to an Appalachian Icon*. Chapel Hill: University of North Carolina Press.

Jones, Charles C. (1878) 1997. *Dead Towns of Georgia*. Savannah, GA: Oglethorpe Press.

Jones, George Neville. 1952. "Type Localities of Vascular Plants First Described from Illinois." *American Midland Naturalist* 47 (2): 487–507.

Jones, Ronald L. 2005. *Plant Life of Kentucky: An Illustrated Guide to the Vascular Flora*. Lexington: University Press of Kentucky.

Joslin, Mary Coker. 2004. "A Genealogical Discovery: Contact with a Member of the Michaux Family in France." In Baranski 2004, 223–37.

Jouanin, Christian. 2004. "Nicolas Baudin chargé de réunir une collection pour la future Impératrice Joséphine." *Australian Journal of French Studies* 41:43–54.

Joyce, Dee Dee. 1988. *Preliminary Report on the Archaeological Investigation of the André Michaux Site 38CH1022*. Charleston, SC: College of Charleston.

———. 2009. *Final Report of Archaeological Testing of the André Michaux Garden Site (38CH1022), Spring 2009*. Charleston, SC: College of Charleston.

Kammen, Michael. 1975. *Colonial New York: A History*. New York: Scribners.

Kegley, Mary B. 1989. *Wythe County Virginia: A Bicentennial History*. Wytheville, VA: Wythe County Board of Supervisors.

Kennedy, Michael. 1976. "La Société française des amis de la liberté et de l'égalité de Philadelphie (1793–1794)." *Annales Historiques de la Révolution Française* 226:614–36.

Kentucky Historical Society. 1983. *Kentucky Marriage Records: From the Register of the Kentucky Historical Society*. Baltimore, MD: Genealogical.

Kincaid, Robert. 1947. *The Wilderness Road*. Indianapolis, IN: Bobbs-Merrill.

King, J. Estelle Stewart. 1969. *Abstract of Early Kentucky Wills and Inventories*. Baltimore, MD: Genealogical.

Kirkland, Thomas J., and Robert M. Kennedy. 1905. *Historic Camden*. Vol. 1, *Colonial and Revolutionary*. Columbia, SC: State.

Kleber, John E. 2001. *The Encyclopedia of Louisville*. Lexington: University Press of Kentucky.

Klotter, James C., and Nelson L. Dawson. 1981. *Genealogies of Kentucky Families: From the Register of the Kentucky Historical Society, O–Y*. Baltimore, MD: Genealogical.

Knight, Lucian Lamar. 1914. *Georgia's Landmarks, Memorials and Legends*. 2 vols. Atlanta, GA: Byrd.

———. (1920) 1967. *Georgia's Roster of the Revolution*. Baltimore, MD: Genealogical.

Lamaute, Marie Florence. 1981. "André Michaux et son exploration en Amérique du Nord (1785–1796), d'apres les sources manuscrites." Master's thesis, Universitié de Montreal.

Lamy, G. 2005. "L'éducation d'un jardinier royal au Petit Trianon: Antoine Richard (1734–1807)." *Polia: Revue de l'art des jardins* 4:57–73.

La Rochefoucauld-Liancourt, François-Alexandre-Frédéric de. 1799. *Travels through the United States of North America: The Country of the Iroquois, and Upper Canada, in the Years 1795, 1796, and 1797; with an Authentic Account of Lower Canada*. 4 vols. London: R. Phillips.

Leavell, B. S. 1977. "Thomas Jefferson and Smallpox Vaccination." *Transactions of the American Clinical and Climatological Association* 88:119–27.

Lee, Lucy C. 1903. "General Henry Lee." *Register of the Kentucky Historical Society* 1 (3): 82–88.

Leschenault, M. 1810. "Mémoire sur le *Strychnos tieute* et l'*Antiaris toxicaria*, plantes vénéneuses de Java avec le suc desquelles les indigènes empoisonnent leurs flêches." *Annales du Muséum d'histoire naturelle* 16:459–82.

Letouzey, Yvonne. 1989. *Le Jardin des plantes à la croisée des chemins avec André Thouin (1747–1824)*. Paris: Muséum national d'histoire naturelle.

Library Board of the Virginia State Library. 1922. *Eighteenth Annual Report, 1920–1921*. Richmond: Davis Bottom, Superintendent of Public Printing.

Lindsay, Ann. 2005. *Seeds of Blood and Beauty: Scottish Plant Explorers*. Edinburgh, UK: Birlinn.

Linklater, Andro. 2009. *An Artist in Treason: The Extraordinary Double Life of General James Wilkinson*. New York: Walker.

Linnaeus, Carl. (1753) 1957–59. *Species Plantarum: A Facsimile of the first edition*. 2 vols. Introduction by W. T. Stearn. Appendix by J. L. Heller and W. T. Stearn. London: Ray Society.

Lipscomb, A. A., and A. E. Bergh. 1903–4. *The Writings of Thomas Jefferson*. 20 vols. Washington, DC: Thomas Jefferson Memorial Association of the United States.

Louis-Philippe. 1977. *Diary of My Travels in America*. Translated by Stephen Becker. New York: Delacorte Press.

Lowitt, Richard. 1948. "Activities of Citizen Genet in Kentucky, 1793–1794." *Filson Club History Quarterly* 22:252–67.

Ly-Tio Fane, Madeleine. 1996. "Botanic Gardens: Connecting Links in Plant Transfer between Indo-Pacific and Caribbean Regions." *Harvard Papers in Botany* 1 (8): 7–14.

MacPhail, Ian. 1981. *André and François André Michaux: The Sterling Morton Library Bibliographies in Botany and Horticulture*. Lisle, IL: Morton Arboretum.

Madison, James, William Prentis, and William Davis. (1807) 2000. "A Map of Virginia Formed from Actual Surveys." In Stephenson and McKee 2000, 139–45.

Madison, James H. 1986. *The Indiana Way: A State History*. Bloomington: Indiana University Press.

Malone, Dumas. 1962. *Jefferson and the Ordeal of Liberty*. Boston: Little, Brown.

Maroteaux, Vincent. 2000. *Versailles, le Roi et son Domaine*. Paris: Editions Picard.

McAllister, Ann Williams, and Kathy Gunter Sullivan. 1988. *Henry Weidner Memorial Booklet*. Hickory, NC: printed by the authors.

McBride, Robert M., and Dan M. Robison. 1975. *Biographical Directory of the Tennessee General Assembly*. Vol. 1, *1796–1861*. Nashville: Tennessee State Library and Archives and Tennessee Historical Commission.

McCall, Ettie Tidwell. (1941) 1968 69. *Roster of Revolutionary Soldiers in Georgia* 3 vols. Baltimore, MD: Genealogical.

McClellan James E., III, and François Regourd. 2011. *The Colonial Machine: French Science and Overseas Expansion in the Old Regime*. Turnhout, Belgium: Brepols.

McCord, David J. 1841. *Statutes at Large of South Carolina*. Vol. 9, *Acts Relating to Roads, Bridges and Ferries*. Columbia, SC: A. S. Johnson.

McGowan, Faison, and Pearl McGowan. 1971. *Flashes of Duplin's History and Government*. Raleigh, NC: Edwards and Broughton.

McKelvey, Susan Delano. 1955. *Botanical Exploration of the Trans-Mississippi West, 1790–1850*. Jamaica Plain, MA: Arnold Arboretum of Harvard University.

McKoy, Elizabeth Francenia. 1967. *Early Wilmington Block by Block*. Wilmington, NC: Edwards and Broughton.

McMillan, Patrick D. 2007. "André Michaux Botanist-Explorer of the South Carolina Upstate: Who Was this Man?" In Sanders 2007, 5–11.

Michaud, L. G., and J. F. Michaud. 1826. "Du Petit-Thouars, L. M. A." In *Biographie universelle et moderne*, 44:226–27. Paris: Michaud.

Michaux, André. 1781. "Modèle d'une caisse propre à transporter des plantes délicates dans des voyages de longs temps, 1 Sept. 1781." Table des rapports de l' Académie des sciences. Procès verbaux 110, Rapports 1699–1793. Procès verbaux de l'Acadèmie royale des sciences, Tome 100, 10 janvier 1781–22 décembre 1781. Institut de France, Académie des sciences, Services des archives, Paris.

———. 1801. *Histoire des chênes de l'Amérique*. Paris: Leverault frères.

———. 1803. *Flora Boreali-Americana*. 2 vols. Paris: C. d'Hautel.

———. 1904. "Journal of André Michaux, 1793–96." In Thwaites 1904, 27–104.

———. (1803) 1974. *Flora Boreali-Americana*. 2 vols. Facsimile edition, with an introduction by Joseph Ewan. New York: Hafner Press.

———. 1998. "Michaux Expenses Journal November 1785–April 1787." Unpublished manuscript. Translated by Marie-Eve Berton Tomlin. Archives of the 2002 André Michaux International Symposium. Abbot Vincent Taylor Library, Belmont Abbey College, Belmont, NC.

Michaux, André, and Francis Boott. 1826. "Memorial of the Botanical Labours of André Michaux." *Edinburgh Journal of Medical Science* 1:126–32.

Michaux, François André. 1817–19. *North American Sylva*. 3 vols. Paris: C. d'Hautel.

———. (1805) 1904. "'Travels to the Westward of the Alleghany Mountains,' by François André Michaux." In Thwaites 1904, 105–306.

Mills, Robert. (1825) 1979. *Mills' Atlas of South Carolina: An Atlas of the Districts of South Carolina in 1825*. Lexington, SC: Sandlapper Store.

Morison, Samuel Eliot. 1972. *Samuel de Champlain: Father of New France*. Boston: Little, Brown.

Morrill, Dan L. 1993. *Southern Campaigns of the American Revolution*. Charleston, SC: Nautical and Aviation.

Morton, Conrad V. 1967. "The Fern Herbarium of André Michaux." *American Fern Journal* 57:166–82.

Moss, Bobby G. 1983. *Roster of South Carolina Patriots in the American Revolution*. Baltimore, MD: Genealogical.

Motte, Jacob Rhett. 1953. *Journey into Wilderness: An Army Surgeon's Account of Life in Camp and Field during the Creek and Seminole Wars, 1836–1838*. Gainesville: University of Florida Press.

Mowat, Charles Loch. (1943) 1964. *East Florida as a British Province, 1763–1784*. Gainesville: University of Florida Press.

Murray, Joyce Martin. 1985. *Deed Abstracts of Warren County, Kentucky, 1797–1812*. Dallas, TX: J. M. Murray.

Nasatir, Abraham P. 1968. *Spanish War Vessels on the Mississippi, 1792–1796*. New Haven, CT: Yale University Press.

———. 1990. *Before Lewis and Clark*. 2 vols. Lincoln: University of Nebraska Press.

Nasatir, Abraham P., and Gary Elwyn Monell. 1967. *French Consuls in the United States: A Calendar of Their Correspondence in the Archives Nationales*. Washington, DC: Library of Congress.

Newsome, A. R. 1934. "John Brown's Journal of Travel in Western North Carolina in 1795." *North Carolina Historical Review* 11 (4): 284–313.

Nicholson, Edgar P. 1914. "James McDowell." *John P. Branch Historical Papers of Randolph-Macon College* 4 (2): 5–33.

Norman, Eliane M. 1976. "An Analysis of the Vegetation at Turtle Mound." *Florida Scientist* 39 (1): 19–31.

Norman, Eliane M., and Susan S. Hawley. 1995. "An Analysis of the Vegetation at Turtle Mound, Volusia County, Florida: Twenty Years Later." *Florida Scientist* 58:258–69.

Ogden, Scott. 2007. *Garden Bulbs for the South*. 2nd ed. Portland, OR: Timber Press.

Oppert, Jules. 1895. "Le caillou de Michaux." *Comptes rendus des séances de l'Académie des inscriptions et belles-lettres* 39:108–13.

Orsolits, Barbara. 2003. "Drayton Hall and the Michaux Connection." *Magnolia, Bulletin of the Southern Garden History Society* 18 (2): 1–6.

Padgett, Donald J., Donald H. Les, and Garrett E. Crow. 1999. "Phylogenetic Relationships in *Nuphar* (Nymphaeaceae): Evidence from Morphology, Chloroplast DNA, and Nuclear Ribosomal DNA." *American Journal of Botany* 86 (9): 1316–24.

Perdue, A. H. 1912. *The Zinc Deposits of Northeastern Tennessee*. State of Tennessee Geological Survey, Bulletin 14. Nashville: TN: Brandon.

Phifer, Edward W., Jr. (1977) 1982. *Burke: The History of a North Carolina County, 1777–1920*. Morganton, NC: Burke County Historical Society.

Phillips, V. N. Bud. 1999. *The Book of Kings: The King Family's Contribution to the History of Bristol, Tennessee/Virginia*. Johnson City, TN: Overmountain Press.

Pluchet, Régis. 2004. "Michaux Mysteries Clarified." In Baranski 2004, 228–32.

———. 2010. "En marge de l'expédition vers les Terres australes: Un portrait du botaniste André Michaux." In *Portés par l'air du Temps: Les voyages du capitaine Baudin*, edited by Michel Jangoux, 205–11. Brussels: Éditions de l'Université de Bruxelles.

———. 2014. *L'extraordinaire voyage d'un botaniste en Perse*. Toulouse: Privat.

Poindexter, Derick, and Zack E. Murrell. 2008. "Vascular Flora of Mount Jefferson State Natural Area and Environs, Ashe County, North Carolina." *Castanea* 73 (4): 283–327.

Powell, William S., and Michael Hill. 2010. *The North Carolina Gazetteer: A Dictionary of Tar Heel Places and Their History*. Chapel Hill: University of North Carolina Press.

Preslar, Charles J. 1954. *A History of Catawba County*. Salisbury, NC: Catawba County Historical Association.

Prewitt, Richard A. [1976?]. "Isham Prewitt of Virginia." Unpublished genealogical study. Filson Club Library, Louisville, KY.

Price, Henry R. 1996. *Hawkins County Tennessee: A Pictorial History*. Virginia Beach, VA: Donning.

Price, William Jennings. 1940. "Danville Was the First Post Office Established in Kentucky and in the Territory beyond the Alleghenies." *Filson Club History Quarterly* 14 (4): 191–203.

Pruitt, A. B. 1988. *Abstract of Deeds: Lincoln County, N.C., 1793–1800*. N.p.: printed by the author.

———. 2001. *Abstracts of Land Warrants, Brunswick County, N.C., 1778–1800*. N.p.: printed by the author.

Ramsey, J. G. M. (1853) 1967. *The Annals of Tennessee to the End of the Eighteenth Century*. Kingsport, TN: Kingsport Press.

Raven, James. 2002. *London Booksellers and American Customers*. Columbia: University of South Carolina Press.

Ray, Worth S. (1950) 2002. *Tennessee Cousins: A History of Tennessee People*. Baltimore, MD: Genealogical.

Rehder, Alfred. 1923. "Michaux's Earliest Notes on American Plants." *Journal of the Arnold Arboretum* 4:1–8.

Rembert, David H., Jr. 1980. *Thomas Walter, Carolina Botanist*. Columbia: South Carolina Museum Commission.

———. 1984. "The Type Specimen for *Aesculus parviflora* Walter." *American Journal of Botany* 71 (5, part 2): 107.

———. 2004. "André Michaux's Travels and Plant Discoveries in the Carolinas." In Baranski 2004, 107–18.

Reveal, James L. 2004. "No Man Is an Island: The Life and Times of André Michaux." In Baranski 2004, 22–68.

Reveal, James L., and Margaret J. Seldin. 1976. "On the Identity of *Halesia carolina* L. (Styracaceae)." *Taxon* 25 (1): 123–40.

Rice, Howard C. 1936–37. "French Consular Agents in the United States, 1778–1791." *Franco-American Review* 1:369–70.

———. 1938. *Barthélemi Tardiveau: A French Trader in the West*. Baltimore, MD: Johns Hopkins University Press.

Ricker, P. L. 1934. "The Origin of the Name Lespedeza." *Rhodora* 36:130–32.

Robbins, Paula Ivanska. 2007. *The Travels of Peter Kalm: Finnish-Swedish Naturalist through Colonial North America, 1748–1751*. Fleischmanns, NY: Purple Mountain Press.

Robbins, William J. 1958. "Louis-Guillaume Lemonnier, Physicist, Botanist, Physician, Courtier, 1717–1799." *Garden Journal of the New York Botanical Garden* 8 (3): 80–81, 93.

Robbins, William J., and Mary Christine Howson. 1958. "André Michaux's New Jersey Garden and Pierre Paul Saunier, Journeyman Gardener." *Proceedings of the American Philosophical Society* 102 (4): 351–70.

Robertson, James Rood. 1914. *Petitions of the Early Inhabitants of Kentucky to the General Assembly of Virginia, 1769–1792*. Filson Club Publication 27. Louisville, KY: John P. Morton.

Robison, Dan E. 1967. "Robert Hays, Unsung Pioneer of the Cumberland Country." *Tennessee Historical Quarterly* 26 (3): 263–78.

Rogers, George C. 1970. *History of Georgetown County, South Carolina*. Columbia: University of South Carolina Press.

Romans, Bernard. (1775) 1999. *A Concise Natural History of East and West Florida*. Edited by Kathryn E. Holland Braund. Tuscaloosa: University of Alabama Press.

Roselli, Bruno. 1933. *Vigo: A Forgotten Builder of the American Republic*. Boston: Stratford.

Rousseau, Jacques. 1931. "Lettres du Dr. J. M. Nooth à Sir Joseph Banks." *Naturaliste Canadien* 58:139–47, 170–77.

———. 1948. "Le voyage d'André Michaux au lac Mistassini en 1792." *Revue d'historie de l'Amérique française* 3 (2): 390–423.

———. 1964. "De la forêt Hudsonienne à Madagascar avec le citoyen Michaux." *Cahier des dix* 29:223–45.

Rusby, H. H. 1884. "Michaux's New Jersey Garden." *Bulletin of the Torrey Botanical Club* 8 (2): 88–90.

Rush, Benjamin, and George W. Corner. 1948. *Autobiography of Benjamin Rush: His "Travels through Life" Together with His Commonplace Book for 1789–1813*. Philadelphia: American Philosophical Society.

Salley, A. S. 1932. *President Washington's Tour through South Carolina in 1791*. Bulletin of the Historical Commission of South Carolina 12. Columbia, SC: State.

Sammons, Helen Moore. 1996. *New Hanover County Minutes of the Court of Pleas and Quarter Sessions, 1805–1808*. Wilmington, NC: North Carolina Room of the New Hanover County Public Library and Old New Hanover Genealogical Society.

Sanders, Brad. 2002. *Guide to William Bartram's Travels*. Athens, GA: Fevertree Press.

———. 2007. *[Proceedings of] An Oconee Bell Celebration, March 16–18, 2007, Clemson University*. Athens, GA: Fevertree Press.

Sargent, Charles S. 1886. "Some Remarks upon the Journey of André Michaux to the High Mountains of Carolina, in December 1788, in a Letter Addressed to Professor Asa Gray." *American Journal of Science*, 3rd series, 32:466–73.

———. 1889. "Portions of the Journal of André Michaux, Botanist, Written During His Travels in the United States and Canada, 1785 to 1796. With an Introduction and Explanatory Notes by C. S. Sargent." *Proceedings of the American Philosophical Society* 26 (129): 1–145.

———. 1915. "Washington and Michaux." *Rhodora* 17:49–50.

Sarudy, Barbara Wells. 1992. "South Carolina Seed Merchants and Nurserymen before 1820." *Magnolia: Bulletin of the Southern Garden History Society* 8 (3): 6–10.

Saunders, W. L., Walter Clark, and Steven Weeks. 1904–7. *State Records of North Carolina*. 26 vols. Goldsboro, NC: Nash Brothers.

Savage, Henry, Jr. 1956. *River of the Carolinas: The Santee*. Chapel Hill: University of North Carolina Press.

Savage, Henry, Jr., and Elizabeth J. Savage. 1986. *André and François André Michaux*. Charlottesville: University of Virginia Press.

Schöpf, David. (1911) 1968. *Travels in the Confederation, 1783–1784: From the German of Johann David Schoepf*. 2 vols. Translated and edited by Alfred J. Morrison. New York: Bergman.

Schwarzkopf, S. Kent. 1985. *A History of Mt. Mitchell and the Black Mountains*. Raleigh: North Carolina Division of Archives and History.

Seaborn, Margaret Mills. 1976. *André Michaux's Journeys in Oconee County, South Carolina, in 1787 and 1788*. Walhalla, SC: Oconee County Library.

Settle, Simon Dewitt. 1963. "The Barrens." Unpublished manuscript. Kentucky Library, Western Kentucky University, Bowling Green.

Sharpe, Bill. 1954–65. *A New Geography of North Carolina*. 4 vols. Raleigh, NC: Sharpe.

Siebert, Wilbur Henry. 1929. *Loyalists in East Florida, 1774–1785*. 2 vols. Deland: Florida State Historical Society.

Silver, Timothy. 2003. *Mt. Mitchell and the Black Mountains: An Environmental History of the Highest Peaks in Eastern America*. Chapel Hill: University of North Carolina Press.

Simons, Katherine Drayton. 1930. *Stories of Charleston Harbor*. Columbia, SC: State.

Simpson, Marcus B., Jr., Stephen Moran, and Sallie W. Simpson. 1997. "Biographical Notes on John Fraser (1750–1811): Plant Nurseryman, Explorer, and Royal Botanical Collector to the Czar of Russia." *Archives of Natural History* 24 (1): 1–18.

Smith, Henry A. M. 1913. "Some Forgotten Towns in Lower South Carolina." *South Carolina Historical and Genealogical Magazine* 14 (4): 198–208.

———. 1919. "The Ashley River: Its Seats and Settlements." *South Carolina Historical and Genealogical Magazine* 20 (1): 3–51.

———. 1928. "Goose Creek." *South Carolina Historical and Genealogical Magazine* 29 (1): 1–25.

Smith, Samuel D., and Benjamin C. Nance. 2000. *An Archaeological Interpretation of the Site of Fort Blount, a 1790s Territorial Militia and Federal Military Post, Jackson County, Tennessee*. Nashville: Tennessee Department of Environment and Conservation, Division of Archaeology.

Spary, Emma C. 2000. *Utopia's Garden: French Natural History from Old Regime to Revolution*. Chicago: University of Chicago Press.

Speed, Thomas. 1886. *Wilderness Road, A Description of the Routes of Travel by Which the Pioneers and Early Settlers First Came to Kentucky*. Filson Club Publication 2. Louisville, KY: John P. Morton.

——. 1894. *The Political Club, Danville, Kentucky, 1786–1790*. Filson Club Publication 9. Louisville, KY: John P. Morton.

Spoden, Muriel M. C. 1991. *Kingsport Heritage: The Early Years, 1700 to 1900*. Johnson City, TN: Overmountain Press.

Spongberg, Stephen. 1990. *A Reunion of Trees: The Discovery of Exotic Plants and Their Introduction into North American and European Landscapes*. Cambridge, MA: Harvard University Press.

Sprague, Stuart Seely. 1973. "Kentucky and the Navigation of the Mississippi: The Climactic Years, 1793–1795." *Register of the Kentucky Historical Society* 71:364–92.

Springs, Katherine W. 1965. *The Squires of Springfield*. Charlotte, NC: W. Loftin.

Sprunt, James. (1914) 2005. *Chronicles of Cape Fear River, 1660–1916*. Wilmington, NC: Dram Tree.

Stadelman, Bonnie S. 1971. "Flora and Fauna versus Mice and Mold." *William and Mary Quarterly*, 3rd series, 28 (4): 595–606.

Stepenoff, Bonnie. 2006. *From French Community to Missouri Town: Ste. Genevieve in the Nineteenth Century*. Columbia: University of Missouri Press.

Stephenson, Richard W., and Marianne M. McKee. 2000. *Virginia in Maps: Four Centuries of Settlement, Growth and Development*. Richmond: Library of Virginia.

Stevens, Peter F. 1994. *The Development of Biological Systematics: Antoine-Laurent de Jussieu, Nature, and the Natural System*. New York: Columbia University Press.

Stewart, Robert Armistead. 1926. "Stubblefield of Spotsylvania." *Researcher: A Magazine of History and Genealogical Exchange* 1:55–56.

Strange, Agatha Rochester. 1889. *The House of Rochester in Kentucky*. Harrodsburg, KY: Democrat.

Swartz, Kenneth D. 1986. "History, Michaux State Forest District No. 1." Staff working paper, Michaux State Forest District Office, Chambersburg, PA.

Swetnam, George, and Helene Smith. 1976. *A Guidebook to Historic Western Pennsylvania*. Pittsburgh, PA: University of Pittsburgh Press.

Talbert, Charles Gano. 1962. *Benjamin Logan, Kentucky Frontiersman*. Lexington: University of Kentucky Press.

Taylor, Walter Kingsley, and Eliane M. Norman. 2002. *André Michaux in Florida: An Eighteenth Century Botanical Journey*. Gainesville: University Press of Florida.

Tennessee County Court (Knox County), and Tennessee Historical Records Survey. 2010. *Knox County, Tennessee Minutes of the County Court, 1792–1795*. Greenville, SC: Southern Historical Press.

Thompson, Doris Lancaster, ed. 1992. *Brunswick County, North Carolina: Minutes of Court of Pleas and Quarter Sessions*. Vol. 2, *1792–1797*. Wilmington, NC: North Carolina Room, New Hanover County Public Library.

Thompson, Joseph J. 1917. "The Penalties of Patriotism: An Appreciation of the Life, Patriotism and Services of Francis Vigo, Pierre Gibault, George Rogers Clark, 'the Founders of the Northwest.'" *Journal of the Illinois State Historical Society* 9 (4): 401–11.

Thouin, André. 1802. "Catalogue des graines rares envoyées pendant l'an X du jardin botanique de Charleston dans la Caroline méridionale, par F. A. Michaux." *Annales du Muséum national d'historie naturelle* 1:92.

Thouin, André, and Oscar Leclerc. 1827. *Cours de culture et de naturalisation des végétaux*. Paris: Huzard.

Threlkel, Marguerite. 1927. "Mann's Lick." *Filson Club History Quarterly* 1 (4): 169–76.

Thwaites, Reuben Gold, ed. 1904. *Early Western Travels, 1748–1846*. Vol. 3. Cleveland, OH: Arthur H. Clark.

Tolzmann, Don H., ed. 1988. *The First Description of Cincinnati and the Other Ohio Settlements: The Travel Report of Johann Heckewelder (1792)*. Lanham, MD: University Press of America.

Turner, Frederick Jackson. 1898. "The Origin of Genet's Projected Attack on Louisiana and the Floridas." *American Historical Review* 3 (4): 650–71.

———, ed. 1903. *Correspondence of the French Ministers to the United States, 1791–1797*. Vol. 2 of *Annual Report of the American Historical Association for the Year 1903*. Washington, DC: Government Printing Office.

United States Bureau of the Census. 1908a. *Heads of Families at the First Census of the United States Taken in the Year 1790*. Vol. 11, *North Carolina*. Washington, DC: Government Printing Office.

———. 1908b. *Heads of Families at the First Census of the United States Taken in the Year 1790*. Vol. 12, *South Carolina*. Washington, DC: Government Printing Office.

United States Congress. 2005. *Biographical Directory of the United States Congress, 1774–2005*. Washington, DC: Government Printing Office.

University of Tennessee. 1898. "Biographical List of Trustees." *University of Tennessee Record* 5:229–68.

Uttal, Leonard J. 1984. "Type Localities of the *Flora Boreali-Americana* of André Michaux." *Rhodora* 86:1–66.

Ventenat, Étienne-Pierre. 1803. *Descriptions des plantes nouvelles et peu connues, cultivées dans le jardin de J. M. Cels*. Paris: L'imprimerie de Crapelet.

———. 1848. "Notes on the Aya-Pana." *Charleston Medical Journal and Review* 3:527–28.

Viguerie, Jean de. 2003. *Histoire et dictionnaire du temps des Lumières, 1715–1789*. Paris: Press Robert Laffont.

Vineyard, Maribeth Lang. 1997. *William Wiseman and the Davenports: Pioneers of Old Burke County, North Carolina*. Franklin, NC: Genealogy.

Virginia General Assembly House of Delegates. 1835. *Journal of the House of Delegates of the Commonwealth of Virginia*. Richmond: Samuel Shepherd.

Virginia Writers Project. 1940. *Virginia: A Guide to the Old Dominion*. American Guide Series. New York: Oxford University Press.

Vocelle, James T. 1914. *History of Camden County, Georgia*. Jacksonville, FL: Kennedy Brown-Hall.

Wadström, C. B. 1798. *Précis sur l'établissement des colonies de Sierra Leone et de Boulama*. Paris: Pougens.

Wait, Jane Wofford, John W. Wofford, and Carrie Wofford Floyd. 1928. *A History of the Wofford Family: Direct Descendants of Colonel Joseph Wofford*. Spartanburg, SC: Band and White.

Walker, Alexander McDonald, ed. 1960. *New Hanover County Court Minutes, Part 3: 1786–1793*. Bethesda, MD: printed by the author.

Walker, Charles Arthur, Jr. 1996. "The Cherokee Rose in the Southeastern United States: A Historical Perspective." PhD diss., North Carolina State University.

Wallace, Paul A. W. 1998. *Indian Paths of Pennsylvania*. Harrisburg: Pennsylvania Historical and Museum Commission.

Ward, Daniel B. 2007. "The Thomas Walter Herbarium Is Not the Herbarium of Thomas Walter." *Taxon* 56 (3): 917–26.

Waring, Alice Noble. 1962. *The Fighting Elder: Andrew Pickens, 1739–1817.* Columbia: University of South Carolina Press.

Warren, Elizabeth. 1938. "John Brown and His Influence in Kentucky Politics, 1784–1805." *Register of the Kentucky State Historical Society* 36:61–65.

———. 1962. "Senator John Brown's Role in the Kentucky Spanish Conspiracy." *Filson Club History Quarterly* 36 (2): 158–76.

Weakley, Alan. 2004. "Report from the Herbarium." *North Carolina Botanical Garden Newsletter* 32 (1): 9.

———. 2011. *Flora of the Southern and Mid-Atlantic States: Working Draft of 15 May 2011.* Chapel Hill: University of North Carolina Herbarium.

———. 2015. *Flora of the Southern and Mid-Atlantic States: Working Draft of 21 May 2015.* Chapel Hill: University of North Carolina Herbarium. http://www.herbarium.unc.edu/flora.htm.

Webber, Mable L. 1920. "Marriage and Death Notices for the *City Gazette* [Charleston]." *South Carolina Historical and Genealogical Magazine* 21 (3): 121–31.

———. 1925. "Abstracts from an Old Account Book of Georgetown District." *South Carolina Historical and Genealogical Magazine* 26:151–57.

Wells, Ann Hartwell. 1976. "Early Maps of Tennessee, 1794–1799." *Tennessee Historical Quarterly* 35 (2): 123–43.

Wells, Carol. 1990–91. *Davidson County, Tennessee County Court Minutes.* 3 vols. Bowie, MD: Heritage.

Wells, Diane. 2002. *Roadside History: A Guide to Kentucky Highway Markers.* Frankfort: Kentucky Historical Society.

West, Carroll Van. 1998. *Tennessee Encyclopedia of History and Culture.* Nashville: Tennessee Historical Society.

Wilkinson, Norman. 1972. *E. I. du Pont, Botaniste: The Beginning of a Tradition.* Charlottesville: University of Virginia Press.

Williams, Charlie. 1997. "Carolina 1796 '. . . d'un nouveau Magnolia.'" *Magnolia: Journal of the Magnolia Society* 32 (2): 15–31.

———. 1999. "André Michaux and the Discovery of *Magnola macrophylla* in North Carolina." *Castanea* 64 (1): 1–13.

———. 2004. "Explorer, Botanist, Courier, or Spy? André Michaux and the Genet Affair of 1793." In Baranski 2004, 98–106.

———. 2007. "The 'Lost *Shortia*,' a Botanical Mystery." In Sanders 2007, 19–24.

Williams, Charlie, Eliane M. Norman, and Gérard K. Aymonin. 2004. "The Type Locality of *Shortia galacifolia* Visited Once Again." In Baranski 2004, 169–73.

Williams, Roger. 2001. *Botanophilia in Eighteenth-Century France.* Dordrecht, Netherlands: Kluwer Academic.

———. 2003. *French Botany in the Enlightenment: The Ill-Fated Voyages of La Pérouse and His Rescuers.* Dordrecht, Netherlands: Kluwer Academic, 2003.

Williams, Samuel Cole. 1920. "The North Carolina-Tennessee Boundary Line Survey." *Tennessee Historical Magazine* 6 (12): 46–57.

———. 1928. *Early Travels in the Tennessee Country, 1540–1800.* Johnson City, TN: Watauga Press.

———. (1924) 1993. *History of the Lost State of Franklin.* Johnson City, TN: Overmountain Press.

Wilson, George R., and Gayle Thornburgh. 1946. *The Buffalo Trace.* Indianapolis: Indiana Historical Society.

Wood, W. Raymond. 2003. *Prologue to Lewis and Clark: Mackay and Evans Expedition.* Norman: University of Oklahoma Press.

Woods, James M. 2011. *A History of the Catholic Church in the American South, 1513–1900.* Gainesville: University Press of Florida.

Writers' Program of the Work Projects Administration in the State of West Virginia. 1941. *West Virginia: A Guide to the Mountain State.* New York: Oxford University Press.

Wulf, Andrea. 2008. *The Brother Gardeners: Botany, Empire and the Birth of an Obsession.* New York: Alfred A. Knopf.

Wyatt, Robert. 1985. "*Aesculus parviflora* in South Carolina: Phytogeographical Implications." *Bulletin of the Torrey Botanical Club* 112:194–95.

Young, Perry Deane. 2003. *Our Young Family: Descendants of Thomas and Naomi Hyatt Young, Wilson and Elizabeth Hughes Young, Moses Young, African American Youngs.* Johnson City, TN: Overmountain Press.

Zahner, Robert. 1994. *The Mountain at the End of the Trail: A History of Whiteside Mountain.* Highlands, NC: R. Zahner.

Zahner, Robert, and Steven M. Jones. 1983. "Resolving the Type Location of *Shortia galacifolia* T & G." *Castanea* 48 (3): 163–73.

Zespedes, V. M. 1787. Corresponde a Represent. on Reserbada No. 17. Descripcion de la Florida Oriental su Clima, Terreno, Productos, Rios, Barras, Bahias, Riertos, Numero, y Calidades de Gente que habitan & c. Report to José de Gálvez dated April 15, 1787. Archivo General de Indias, Seville, Spain.

Zomlefer, Wendy B., Walter S. Judd, and David E. Giannasi. 2006. "Northernmost Limit of *Rhizophora mangle* (Red Mangrove; Rhizophoraceae) in St. Johns County, Florida." *Castanea* 71 (3): 239–44.

# Index

Added information in parentheses includes other personal and place-names found in sources, other spellings André Michaux used, and the married names of women listed by their maiden names. Page numbers in italics refer to illustrations. For the plants and animals described in Michaux's journals and letters, see the appendix.

Bartrams: boards horses with, 132–33, 242, 489n2; exchanges seeds and plants with, 143, 151, 159; shares plant collections with, 262, 267–68

Bartram's Garden: *Franklinia* admired, 41, 430n34; yellow-wood given by AM or FAM, 516n53

Basin Harbor, Vermont, 198, 475n82

Bass. *See* Baas, Capt.

Batiscan, Canada, 177, 196, 468n97

Baudin, Capt. Nicholas, 21, 339–40, 342–43, 423

Baudin expedition, xxx, xxxii, 11, 18, 21–22, 339–40; AM selected as senior botanist, 340; scientists and artists desert expedition in Mauritius, 21; symposium on expedition, 339

Bauvois. *See* Palisot de Beauvois, Ambroise Marie François Joseph

Bean's Station, Tennessee, 237, 485n72

Bear's Creek, South Carolina, 137, 457n80

Beattie's Ford, North Carolina, 260, 497n107

Beaudin, Mr. (Philadelphia), 118

Beauvais. *See* St. James Beauvais

Beaver Dam Creek, Elbert County, Georgia, 111

Beaver Dam Creek, Screven County, Georgia, 65, 109, 430n38

Bedford, Georgia, 110, 444n11

Bedford, Pennsylvania, 225, 227, 481n24

Beech Fork, Kentucky, 298, 300, 515n34

Bee's Ferry and Bee's Ferry Road, Charleston, South Carolina, 427n2, 427n4

Belin, Peter (Belym), 164, 461n31

Bell, Mr. (Bel), 109, 444n9

Bellefontaine (Waterloo), Illinois, 281, 509n107

Belleville, West Virginia, 231, 482n35

Belpre, Ohio, 231, 482n35

Bénézech, Pierre, 322, 325, 333, 337, 422n55, 522n43

Bentham, James, 165

Benton, Col. Lamuel, 261; prominence and family connections, 498n119

Benton, Sen. Thomas Hart, 498n119

Beresford, Lord, 442n52

Bernard de Ste Afrique, 338, 525n35

Bernardin de Saint-Pierre, Jacques Henri, 525n35

Berthier, Canada, 196, 475n73

Bessa, Pancrae, 348

Betts, William, 296, 514n26

Big Barren River, Kentucky. *See* Barren River

Bisset, Capt. Robert (Besy), 91, 439n20

Black Creek, South Carolina, 261, 498n120

Black Mountains, North Carolina, 125–26, 138, 256, 450n12, 458nn90–91

Black Swamp, South Carolina, 261, 498n120

Blacksburg, Virginia, 486n79

Bladensburg, Maryland, 133, 489n6; winter bird studies near, 247

Bleakley, John, 208

Bledsoe, Anthony, 504n60

Bledsoe Lick and Station. *See* Castalian Springs

Block House, Virginia. *See* Anderson's Block House

Blooming Grove Creek, Tennessee, 295, 513n13

Blount, Gov. William: AM meets, 302, 304, 503n51; AM writes to, about discovery of *Cladrastis kentukea*, yellow-wood tree, 303, 517n55, 527n58; leads territory to statehood, 502n38; replies to AM's letter about yellow-wood tree, 304, 518n64

Blue Ridge Mountains, 127, 129–30, 138, 246, 269, 272, 434n76, 450n10, 451n14, 453n34, 458nn89–90, 501n29, 519n72; AM climbs Grandfather Mountain and sings "La Marseillaise," 258

Bonaparte, Lucien, 340–41

Bonaparte, Mme. (Empress Josephine), 307, 342

Bonaparte, Napoléon, 326, 340

Bonaventure, Florida, southernmost point in North America reached by AM, 440n32

Bonpland, Aimé, 340

Boott, Francis, 165, 422n55; portrait of FAM, 424n88

Bordeaux, France, 55, 77–78, 80, 144, 158, 435n91, 525n36

Bosc, Louis (Dantic), 35, 37, 117, 202, 338, 447n14, 522n46, 523n8; AM sends natural history specimens to, 213, 220, 447n49; arrives in Charleston after AM leaves for

France, 330; excerpts from AM's letters to, 213–14, 319–20, 331; origins of friendship with AM, 424n13; portrait, *214*

Botetourt Courthouse (Fincastle), Virginia, 239, 241, 486nn79–80

Bourbon County, Kentucky, 232, 483n49

Bournonville, Charles François, 242–43, 488n99

Bowling Green, Kentucky, 269, 506n76, 506n79, 514–15nn31–32

Bowling Green, Virginia, 248–49, 489n10

Brackenridge, Hugh Henry, Esq., 225–26, 228, 479n7, 481n26

Branch, Col. John (Brandt), 249, 489n15

Bright, Samuel, 495n88

Bright Settlement, North Carolina, 257, 271

Bright's Trace, North Carolina and Tennessee, 450–51n14, 495n88

Broad River, Georgia, 110–11, 445nn17–18

Broussonet, Pierre, 167, 212, 462n45

Brown, Elizabeth (Mrs. Thomas Craighead), 506n70

Brown, James, 226, 483n50; AM reports his assistance from Lexington merchants, 484n64; family connections and political career, 480n12, 483n54

Brown, John (land agent), 502n37

Brown, Sen. John, 207, 243; letter for AM to Gov. Isaac Shelby, 233; plans of Genet and Clark, awareness of, 488n105; political career and family connections, 483–84n54

Brown, Dr. Morgan, 295, 513n15

Brown Ferry, Georgia, 165

Brugmans, Professor, 332, 523n11

Bruslé, Citizen, 332

Bruton House, 108, 444n2

Buffalo Creek (Wellsburg), West Virginia, 230

Buffalo Trace, Indiana, 507–8nn91–92

Buffon, Count de, 5, 521n38; appoints André Thouin head gardener of the Jardin du Roi, 420n11

Bull's Gap, Tennessee, 274, 304, 310

Bullskin Creek, Kentucky, 233

Burke Courthouse, North Carolina, 126, 138–39, 238, 254, 259, 306, 308–9, 449n6. *See also* Morganton

Burlington, Vermont, 175, 465n41

Bushe, Mr., 171

Bussy. *See* Droict-Bussy, Citizen

Cabot, George, 207

Cahokia, Illinois, 16, 281, 309, 509nn105–6

Calhoun, John C., 432n57

Calhoun, Patrick (Squire Coom), 432n57

Calhoun, Rebecca (Mrs. Andrew Pickens), 432n58

Camden, South Carolina, 84, 104, 124, 126, 137, 246, 253–54, 269–70, 308, 310, 449n1, 456–57n78

Camden County, Georgia, 102, 442n64, 458n92; digital map of, 460n27; explorations of, 162–63. *See also* Cumberland Island; St. Marys

Campbell, Col. Arthur, 486n76

Campbell Station, Tennessee, 274, 504n56

Canada, journey, 176–97

Cane Creek, Lancaster County, South Carolina, 270–71, 457n82, 500n19

Cane Creek, Mitchell County, North Carolina, 306

Cane Creek, Oconee County, South Carolina, 74, 433n65, 435n80

canebrake: in Florida, 101; in Tennessee, 303

canoes, description of dugout, 438n8

Cape Brulé, Canada, 179, 470n118

Cape Canaveral, Florida, 84, 210, 440n32

Cape Cinque Hommes, Missouri, 292, 512n5

Cape Fear River, North Carolina, 246, 260, 455n66, 490n35, 497n113

Cape Girardeau, Missouri, 292, 512n6

Cape Hatteras, North Carolina, 77

Cape of Good Hope, origin of garden plants, 143, 152, 178

Cape Tourmente, Canada, 179, 181–82, 470n118

Carondelet, Baron de, 289, 481–82n29, 511n125

Carter, Maj. Landon, 127, *128*, 305, 450n14, 451n16

Cartier, Jacques, 468n91, 471n3

Cashiers Valley, North Carolina, 434n73, 434n76

Castalian Springs, Tennessee, 269, 275, 309, 504n60

Louis XV, 2, 419–20n10, 421n25

Louis XVI, 3, 6, 202, 208, 419–20n10, 421n25, 424n12, 477n26

Louis-Philippe, future king of the France, 506n75, 514n30

Louisière. *See* Dehault Delassus DeLuzières, Pierre Charles

Louisville, Kentucky, 16, 225, 227, 232, 234–36, 241, 269, 277–79, 291, 299, 514n29; plant lists, 277–79

Louvet, Citizen Jean Baptiste, 338, 525n34

Lower Long Cane (Hopewell) Presbyterian Church, 432n56

Lucas, John Baptiste Charles (Lucas de Pentareau), 230, 482n30

Lucayas Islands, 120, 468n63

Lumber River. *See* Drowning Creek

Lynch's Creek, South Carolina, 261

Lyon, John, 421–22n40, 457n85, 494n69

Ly-Tio Fane, Madeline, 344

MacCay. *See* McKay, William

MacGill, William (Magill), 251

MacKay, James (Mackey), 288, 511n129

MacKinsy, 299–300

MacMaster, Archibal, 47, 426n42

Madagascar, xxx, 11, 21, 59, 339, 343–44

Madison (Maddisson), 297, 514–15n32

Madison, Agatha Strother (Mrs. John), 514–15n32

Madison, Anne Lewis (Mrs. Rowland), 514–15n32

Madison, Gabriel, 514–15n32

Madison, George, 514–15n32

Madison, James, Jr., 207, 515n33

Madison, Jane Smith (Mrs. George), 514–15n32

Madison, John, 514–15n32

Madison, Rowland (Roland), 514–15n32

Madrid, Spain, 233, 512n3

Magaw, Sam, 208

Malbaie, Canada, 179, 196, 470n127

Malette, Louise Catherine (Mme. Gabriel Ravot), 430n37

Malsherbes, 35, 424n12

Manchester, South Carolina, 308–10, 519n83

Mangourit, Michel-Ange-Bernard, 208–13, 209, 218, 226, 245, 252, 265, 333, 462n8,

493n51, 524n18; energetically organizes Genet's attack on Spanish Florida, 488n102; requests governmental financial support for AM, 211; urges that AM remain in Paris to write his books, 334–35

Manigault Ferry, South Carolina, 253, 309–10, 519n84

Manigault Plantation, 251, 492n46

Mann's Lick, Kentucky, 277, 507n84

Mansker, Kaspar (Col. Mansko), 302; dispute with AM, 516n51

Mansker's Station (Mansko's Lick), 276, 516n51

Marbois, Consul. *See* Barbé-Marbois, Pierre-François de

Marchall, Widow, 110

Marie Antionette, queen of France, 17

Marietta, Ohio, 231, 482n34

Marshall, Capt., 105, 443n78

Martin, Lewis (Louis), 70, 433n65

Martin, Widow, 295

Martin, William, 260, 497n112

Mason-Dixon boundary survey, 453n38

Massie's Station, Ohio, 482n40

Matanzas Inlet, Florida, 89–90, 95–96

Matanzas River, Florida, 89

Mauduit, 140, 458n96

Mauritius (Isle de France), 21, 342–44, 526n33

Maxel, James (Maxville), 296, 514n26

May, John, 210, 270, 476–77n14, 500n19

Maysville (Limestone), Kentucky, 225, 231–32, 482n41

Mazyck (Mazie, Dr.), 251, 492n42

McAllister's Tavern, Pennsylvania, 487n92

McBee's Ferry, Tennessee, 274, 502n43, 503nn49–50

McClenegan's Ferry, South Carolina, 271, 500n22

McClung, Charles, 504n56

McDowell, Gen. Charles, 259, 497n103

McDowell, Ephraim, 486n82

McDowell, Col. James, 486n82

McFadden's Ferry (McFaddin), 506n76

McHenry, James, 419n2, 425n32

McIntosh, Gen. Lachlan, 443n69

McIntosh, Margery (Mrs. James Spaulding), 443n69

McKay, William, 261, 498n115